Modern Power Electronics and AC Drives

Bimal K. Bose
Condra Chair of Excellence in Power Electronics
The University of Tennessee, Knoxville

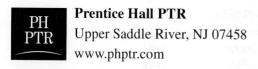

Prentice Hall PTR
Upper Saddle River, NJ 07458
www.phptr.com

ISBN 0-13-016743-6

Library of Congress Cataloging-in-Publication Data

Bose, Bimal K.
   Modern power electronics and AC drives / Bimal Bose.
     p. cm.
   Includes bibliographical references and index.
   ISBN 0-13-016743-6 (alk. paper)
    1. Electronic motors, Alternating current--Automatic control. 2. Electronic driving. 3. Power electronics. I. Title.

TK2781.B67 2001
621.46--dc21

2001032192

Production Supervisor: Wil Mara
Acquisitions Editor: Bernard Goodwin
Editorial Assistant: Michelle Vincenti
Marketing Manager: Dan DePasquale
Manufacturing Manager: Alexis Heydt
Cover Designer: Nina Scuderi
Composition: Aurelia Scharnhorst

© 2002 Prentice Hall PTR
Prentice-Hall, Inc.
Upper Saddle River, NJ 07458

All rights reserved. No part of this book may be reproduced, in any form or by any means, without permission in writing from the author and publisher.

The publisher offers discounts on this book when ordered in bulk quantities. For more information contact: Corporate Sales Department, Prentice Hall PTR, One Lake Street, Upper Saddle River, NJ 07458. Phone: 800-382-3419; FAX: 201-236-7141; E-mail: corpsales@prenhall.com.

Printed in the United States of America

10 9 8 7 6 5 4 3 2 1

ISBN 0-13-016743-6

Pearson Education LTD.
Pearson Education Australia PTY, Limited
Pearson Education Singapore, Pte. Ltd
Pearson Education North Asia Ltd
Pearson Education Canada, Ltd.
Pearson Educación de Mexico, S.A. de C.V.
Pearson Education—Japan
Pearson Education Malaysia, Pte. Ltd
Pearson Education, Upper Saddle River, New Jersey

Figures 4-20, 4-22, 4-23, 4-24, and 4-27 are reprinted with permission from *Thyristor Phase-Controlled Converters and Cycloconverters*, by B. R. Pelly; John Wiley & Sons, Inc.1971.

Figures 11-5, 11-6, 11-42, 11-43, 11-44, 11-45 and 11-46 are reprinted with permission from *Fuzzy Logic Toolbox User's Guide, Version 2*, by Math Works, Inc. 1998.

About the Author

Dr. Bimal K. Bose (*Life Fellow, IEEE*) currently holds the Condra Chair of Excellence in Power Electronics at the University of Tennessee, Knoxville since 1987. Prior to this, he was a research engineer in General Electric R & D Center in Schenectady, NY for 11 years (1976-87) and faculty member of Rensselaer Polytechnic Institute, Troy, NY for 5 years (1971-76). He has been in the power electronics area for more than 40 years and contributed widely that includes more than 150 papers, 21 U.S. Patents, 6 books (including this one) and invited presentations, tutorials and keynote addresses throughout the world. He is the recipient of a number of awards and honors that include IEEE Millennium Medal (2000), IEEE Continuing Education Award (1997), IEEE Lamme Gold Medal (1996), IEEE Industrial Electronics Society Eugene Mittelmann Award (for life-time achievement)(1994), IEEE Region 3 Outstanding Engineer Award (1994), IEEE Industry Applications Society Outstanding Achievement Award (1993), General Electric Silver Patent Medal(1986) and Publication Award (1987), and Calcutta University Mouat Gold Medal (1970).

Contents

Preface	xvii
List of Principal Symbols	xix
Chapter 1 **Power Semiconductor Devices**	1
1.1 Introduction	1
1.2 Diodes	2
1.3 Thyristors	4
1.3.1 Volt-Ampere Characteristics	5
1.3.2 Switching Characteristics	6
1.3.3 Power Loss and Thermal Impedance	6
1.3.4 Current Rating	8
1.4 Triacs	8
1.5 Gate Turn-Off Thyristors (GTOs)	10
1.5.1 Switching Characteristics	11
1.5.2 Regenerative Snubbers	14
1.6 Bipolar Power or Junction Transistors (BPTs or BJTs)	14
1.7 Power MOSFETs	17
1.7.1 V-I Characteristics	17
1.7.2 Safe Operating Area (SOA)	17
1.8 Static Induction Transistors (SITs)	19
1.9 Insulated Gate Bipolar Transistors (IGBTs)	20
1.9.1 Switching Characteristics and Thermal Impedance	22
1.10 MOS-Controlled Thyristors (MCTs)	24
1.11 Integrated Gate-Commutated Thyristors (IGCTs)	25
1.12 Large Band-Gap Materials for Devices	26
1.13 Power Integrated Circuits (PICs)	26
1.14 Summary	27

Chapter 2 AC Machines for Drives — 29

- 2.1 Introduction — 29
- 2.2 Induction Machines — 30
  - 2.2.1 Rotating Magnetic Field — 30
  - 2.2.2 Torque Production — 33
  - 2.2.3 Equivalent Circuit — 35
    - 2.2.3.1 Equivalent Circuit Analysis — 35
  - 2.2.4 Torque-Speed Curve — 39
  - 2.2.5 NEMA Classification of Machines — 42
  - 2.2.6 Variable-Voltage, Constant-Frequency Operation — 42
  - 2.2.7 Variable-Frequency Operation — 43
  - 2.2.8 Constant Volts/Hz Operation — 44
  - 2.2.9 Drive Operating Regions — 46
  - 2.2.10 Variable Stator Current Operation — 47
  - 2.2.11 The Effect of Harmonics — 49
    - 2.2.11.1 Harmonic Heating — 49
    - 2.2.11.2 Machine Parameter Variation — 53
    - 2.2.11.3 Torque Pulsation — 53
  - 2.2.12 Dynamic d-q Model — 56
    - 2.2.12.1 Axes Transformation — 57
    - 2.2.12.2 Synchronously Rotating Reference Frame—Dynamic Model (Kron Equation) — 63
    - 2.2.12.3 Stationary Frame—Dynamic Model (Stanley Equation) — 67
    - 2.2.12.4 Dynamic Model State-Space Equations — 70
- 2.3 Synchronous Machines — 74
  - 2.3.1 Wound Field Machine — 74
    - 2.3.1.1 Equivalent Circuit — 76
    - 2.3.1.2 Developed Torque — 79
    - 2.3.1.3 Salient Pole Machine Characteristics — 80
    - 2.3.1.4 Dynamic d^e-q^e Machine Model (Park Model) — 83
  - 2.3.2 Synchronous Reluctance Machine — 86
  - 2.3.3 Permanent Magnet (PM) Machine — 86
    - 2.3.3.1 Permanent Magnet Materials — 86
    - 2.3.3.2 Sinusoidal Surface Magnet Machine (SPM) — 89
    - 2.3.3.3 Sinusoidal Interior Magnet Machine (IPM) — 89
    - 2.3.3.4 Trapezoidal Surface Magnet Machine — 93
- 2.4 Variable Reluctance Machine (VRM) — 94
- 2.5 Summary — 96

Chapter 3 Diodes and Phase-Controlled Converters — 99

- 3.1 Introduction — 99
- 3.2 Diode Rectifiers — 100
  - 3.2.1 Single-Phase Bridge – *R*, *RL* Load — 100
  - 3.2.2 Effect of Source Inductance — 103
  - 3.2.3 Single-Phase Bridge – *RL*, CEMF Load — 104
  - 3.2.4 Single-Phase Bridge – *CR* Load — 105
  - 3.2.5 Distortion, Displacement, and Power Factors — 107
  - 3.2.6 Distortion Factor (DF) — 108
  - 3.2.7 Displacement Power Factor (DPF) — 108
  - 3.2.8 Power Factor (PF) — 109
  - 3.2.9 Three-Phase Full Bridge – *RL* Load — 109
  - 3.2.10 Three-Phase Bridge – *CR* Load — 112
- 3.3 Thyristor Converters — 112
  - 3.3.1 Single-Phase Bridge – *RL*, CEMF Load — 112
  - 3.3.2 Discontinuous Conduction — 118
  - 3.3.3 Three-Phase Converter – *RL*, CEMF Load — 122
  - 3.3.4 Three-Phase, Half-Wave Converter — 122
  - 3.3.5 Analysis for Line Leakage Inductance (L_c) — 124
  - 3.3.6 Three-Phase Bridge Converter — 128
  - 3.3.7 Discontinuous Conduction — 132
  - 3.3.8 Three-Phase Dual Converter — 136
  - 3.3.9 Six-Pulse, Center-Tap Converter — 136
  - 3.3.10 12-Pulse Converter — 137
  - 3.3.11 Concurrent and Sequential Control of Bridge Converters — 140
- 3.4 Converter Control — 141
  - 3.4.1 Linear Firing Angle Control — 142
  - 3.4.2 Cosine Wave Crossing Control — 142
  - 3.4.3 Phase-Locked Oscillator Principle — 145
- 3.5 EMI and Line Power Quality Problems — 148
  - 3.5.1 EMI Problems — 148
  - 3.5.2 Line Harmonic Problems — 149
- 3.6 Summary — 151

Chapter 4 Cycloconverters — 153

- 4.1 Introduction — 153
- 4.2 Phase-Controlled Cycloconverters — 154
  - 4.2.1 Operation Principles — 154
  - 4.2.2 A Three-Phase Dual Converter as a Cycloconverter — 156

	4.2.3 Cycloconverter Circuits		158
	4.2.3.1 Three-Phase, Half-Wave Cycloconverter		158
	4.2.3.2 Three-Phase Bridge Cycloconverter		161
	4.2.3.2.1 Modulation Factor		161
	4.2.4 Circulating vs. Non-Circulating Current Mode		162
	4.2.4.1 Circulating Current Mode		162
	4.2.4.2 Blocking Mode		166
	4.2.5 Load and Line Harmonics		167
	4.2.5.1 Load Voltage Harmonics		167
	4.2.5.2 Line Current Harmonics		171
	4.2.6 Line Displacement Power Factor		171
	4.2.6.1 Theoretical Derivation of Line DPF		173
	4.2.7 Control of Cycloconverter		177
	4.2.8 DPF Improvement Methods		180
	4.2.8.1 Square-Wave Operation		180
	4.2.8.2 Asymmetrical Firing Angle Control		180
	4.2.8.3 Circulating Current Control		183
4.3	Matrix Converters		185
4.4	High-Frequency Cycloconverters		186
	4.4.1 High-Frequency, Phase-Controlled Cycloconverter		187
	4.4.2 High-Frequency, Integral-Pulse Cycloconverter		187
	4.4.2.1 Sinusoidal Supply		187
	4.4.2.2 Quasi-Square-Wave Supply		188
4.5	Summary		189
4.6	References		189

Chapter 5 **Voltage-Fed Converters**   **191**

5.1	Introduction	191
5.2	Single-Phase Inverters	192
	5.2.1 Half-Bridge and Center-Tapped Inverters	192
	5.2.2 Full, or H-Bridge, Inverter	193
	5.2.2.1 Phase-Shift Voltage Control	195
5.3	Three-Phase Bridge Inverters	197
	5.3.1 Square-Wave, or Six-Step, Operation	197
	5.3.2 Motoring and Regenerative Modes	201
	5.3.3 Input Ripple	202
	5.3.4 Device Voltage and Current Ratings	203
	5.3.5 Phase-Shift Voltage Control	203
	5.3.6 Voltage and Frequency Control	205

5.4	Multi-Stepped Inverters	206
5.4.1	12-Step Inverter	207
5.4.2	18-Step Inverter by Phase-Shift Control	209
5.5	Pulse Width Modulation Techniques	210
5.5.1	PWM Principle	210
5.5.1.1	PWM Classification	210
5.5.1.1.1	Sinusoidal PWM	211
5.5.1.1.2	Selected Harmonic Elimination PWM	218
5.5.1.1.3	Minimum Ripple Current PWM	223
5.5.1.1.4	Space-Vector PWM	224
5.5.1.1.5	Sinusoidal PWM with Instantaneous Current Control	236
5.5.1.1.6	Hysteresis-Band Current Control PWM	236
5.5.1.1.7	Sigma-Delta Modulation	239
5.6	Three-Level Inverters	240
5.6.1	Control of Neutral Point Voltage	243
5.7	Hard Switching Effects	245
5.8	Resonant Inverters	247
5.9	Soft-Switched Inverters	249
5.9.1	Soft Switching Principle	249
5.9.1.1	Inverter Circuits	249
5.10	Dynamic and Regenerative Drive Braking	253
5.10.1	Dynamic Braking	253
5.10.2	Regenerative Braking	254
5.11	PWM Rectifiers	255
5.11.1	Diode Rectifier with Boost Chopper	255
5.11.1.1	Single-Phase	255
5.11.1.2	Three-Phase	257
5.11.2	PWM Converter as Line-Side Rectifier	258
5.11.2.1	Single-Phase	258
5.11.2.2	Three-Phase	259
5.12	Static VAR Compensators and Active Harmonic Filters	262
5.13	Introduction to Simulation—MATLAB/SIMULINK	264
5.14	Summary	267
5.15	References	269

Chapter 6   Current-Fed Converters   271

6.1	Introduction	271
6.2	General Operation of a Six-Step Thyristor Inverter	272

  6.2.1 Inverter Operation Modes   274
    6.2.1.1 Mode 1: Load-Commutated Rectifier ($0 < \alpha < \pi/2$)   274
    6.2.1.2 Mode 2: Load-Commutated Inverter ($\pi/2 < \alpha < \pi$)   276
    6.2.1.3 Mode 3: Force-Commutated Inverter ($\pi < \alpha < 3\pi/2$)   276
    6.2.1.4 Mode 4: Force-Commutated Rectifier ($3\pi/4 < \alpha < 2\pi$)   276
 6.3 Load-Commutated Inverters   277
  6.3.1 Single-Phase Resonant Inverter   277
    6.3.1.1 Circuit Analysis   278
  6.3.2 Three-Phase Inverter   281
    6.3.2.1 Lagging Power Factor Load   281
    6.3.2.2 Over-Excited Synchronous Machine Load   282
    6.3.2.3 Synchronous Motor Starting   284
 6.4 Force-Commutated Inverters   285
  6.4.1 Auto-Sequential Current-Fed Inverter (ASCI)   285
 6.5 Harmonic Heating and Torque Pulsation   287
 6.6 Multi-Stepped Inverters   289
 6.7 Inverters with Self-Commutated Devices   290
  6.7.1 Six-Step Inverter   290
    6.7.1.1 Load Harmonic Resonance Problem   293
  6.7.2 PWM Inverters   294
    6.7.2.1 Trapezoidal PWM   295
    6.7.2.2 Selected Harmonic Elimination PWM (SHE-PWM)   297
  6.7.3 Double-Sided PWM Converter System   299
  6.7.4 PWM Rectifier Applications   302
    6.7.4.1 Static VAR Compensator/Active Filter   302
    6.7.4.2 Superconducting Magnet Energy Storage (SMES)   303
    6.7.4.3 DC Motor Speed Control   303
 6.8 Current-Fed vs. Voltage-Fed Converters   303
 6.9 Summary   305
 6.10 References   305

Chapter 7   **Induction Motor Slip-Power Recovery Drives**   **307**

 7.1 Introduction   307
 7.2 Doubly-Fed Machine Speed Control by Rotor Rheostat   308
 7.3 Static Kramer Drive   309
  7.3.1 Phasor Diagram   313
  7.3.2 AC Equivalent Circuit   316
  7.3.3 Torque Expression   319
  7.3.4 Harmonics   321
  7.3.5 Speed Control of a Kramer Drive   322
  7.3.6 Power Factor Improvement   322

7.4	Static Scherius Drive		324
	7.4.1	Modes of Operation	326
	7.4.2	Modified Scherbius Drive for VSCF Power Generation	328
7.5	Summary		331
7.6	References		331

Chapter 8 Control and Estimation of Induction Motor Drives — 333

- 8.1 Introduction — 333
- 8.2 Induction Motor Control with Small Signal Model — 334
  - 8.2.1 Small-Signal Model — 335
- 8.3 Scalar Control — 338
  - 8.3.1 Voltage-Fed Inverter Control — 339
    - 8.3.1.1 Open Loop Volts/Hz Control — 339
    - 8.3.1.2 Energy Conservation Effect by Variable Frequency Drive — 342
    - 8.3.1.3 Speed Control with Slip Regulation — 342
    - 8.3.1.4 Speed Control with Torque and Flux Control — 345
    - 8.3.1.5 Current-Controlled Voltage-Fed Inverter Drive — 346
    - 8.3.1.6 Traction Drives with Parallel Machines — 348
  - 8.3.2 Current-Fed Inverter Control — 350
    - 8.3.2.1 Independent Current and Frequency Control — 350
    - 8.3.2.2 Speed and Flux Control in Current-Fed Inverter Drive — 351
    - 8.3.2.3 Volts/Hz Control of Current-Fed Inverter Drive — 352
  - 8.3.3 Efficiency Optimization Control by Flux Program — 352
- 8.4 Vector or Field-Oriented Control — 356
  - 8.4.1 DC Drive Analogy — 356
  - 8.4.2 Equivalent Circuit and Phasor Diagram — 358
  - 8.4.3 Principles of Vector Control — 359
  - 8.4.4 Direct or Feedback Vector Control — 360
  - 8.4.5 Flux Vector Estimation — 363
    - 8.4.5.1 Voltage Model — 363
    - 8.4.5.2 Current Model — 366
  - 8.4.6 Indirect or Feedforward Vector Control — 368
    - 8.4.6.1 Indirect Vector Control Slip Gain (K_s) Tuning — 375
  - 8.4.7 Vector Control of Line-Side PWM Rectifier — 378
  - 8.4.8 Stator Flux-Oriented Vector Control — 381
  - 8.4.9 Vector Control of Current-Fed Inverter Drive — 384
  - 8.4.10 Vector Control of Cycloconverter Drive — 385
- 8.5 Sensorless Vector Control — 388
  - 8.5.1 Speed Estimation Methods — 388
    - 8.5.1.1 Slip Calculation — 388

8.5.1.2	Direct Synthesis from State Equations	389
8.5.1.3	Model Referencing Adaptive System (MRAS)	390
8.5.1.4	Speed Adaptive Flux Observer (Luenberger Observer)	392
8.5.1.5	Extended Kalman Filter (EKF)	396
8.5.1.6	Slot Harmonics	399
8.5.1.7	Injection of Auxiliary Signal on Salient Rotor	399

- 8.5.2 Direct Vector Control without Speed Signal — 401
  - 8.5.2.1 Programmable Cascaded Low-Pass Filter (PCLPF) Stator Flux Estimation — 401
  - 8.5.2.2 Drive Machine Start-up with Current Model Equations — 404
- 8.6 Direct Torque and Flux Control (DTC) — 408
  - 8.6.1 Torque Expression with Stator and Rotor Fluxes — 408
  - 8.6.2 Control Strategy of DTC — 410
- 8.7 Adaptive Control — 413
  - 8.7.1 Self-Tuning Control — 414
    - 8.7.1.1 Load Torque Disturbance (T_L) Compensation — 415
  - 8.7.2 Model Referencing Adaptive Control (MRAC) — 416
  - 8.7.3 Sliding Mode Control — 419
    - 8.7.3.1 Control Principle — 419
    - 8.7.3.2 Sliding Trajectory Control of a Vector Drive — 424
- 8.8 Self-Commissioning of Drive — 430
- 8.9 Summary — 435
- 8.10 References — 435

Chapter 9 Control and Estimation of Synchronous Motor Drives — 439

- 9.1 Introduction — 439
- 9.2 Sinusoidal SPM Machine Drives — 440
  - 9.2.1 Open Loop Volts/Hertz Control — 440
  - 9.2.2 Self-Control Model — 444
  - 9.2.3 Absolute Position Encoder — 446
    - 9.2.3.1 Optical Encoder — 446
    - 9.2.3.2 Analog Resolver with Decoder — 448
  - 9.2.4 Vector Control — 449
    - 9.2.4.1 Field-Weakening Mode — 451
- 9.3 Synchronous Reluctance Machine Drives — 455
  - 9.3.1 Current Vector Control of SyRM Drive — 457
    - 9.3.1.1 Constant d^e-Axis Current (i_{ds}) Control — 458
    - 9.3.1.2 Fast Torque Response Control — 459
    - 9.3.1.3 Maximum Torque/Ampere Control — 463
    - 9.3.1.4 Maximum Power Factor Control — 463

9.4	Sinusoidal IPM Machine Drives		465
	9.4.1 Current Vector Control with Maximum Torque/Ampere		465
	9.4.2 Field-Weakening Control		468
	9.4.3 Vector Control with Stator Flux Orientation		471
		9.4.3.1 Feedback Signal Processing	477
		9.4.3.2 Square-Wave (SW) Mode Field-Weakening Control	479
		9.4.3.3 PWM — Square-Wave Sequencing	482
9.5	Trapezoidal SPM Machine Drives		483
	9.5.1 Drive Operation with Inverter		483
		9.5.1.1 $2\pi/3$ Angle Switch-on Mode	485
		9.5.1.2 PWM Voltage and Current Control Mode	486
	9.5.2 Torque-Speed Curve		486
	9.5.3 Machine Dynamic Model		489
	9.5.4 Drive Control		490
		9.5.4.1 Close Loop Speed Control in Feedback Mode	490
		9.5.4.2 Close Loop Current Control in Freewheeling Mode	492
	9.5.5 Torque Pulsation		493
	9.5.6 Extended Speed Operation		494
9.6	Wound-Field Synchronous Machine Drives		495
	9.6.1 Brush and Brushless dc Excitation		495
	9.6.2 Load-Commutated Inverter (LCI) Drive		496
		9.6.2.1 Control of LCI Drive with Constant γ Angle	498
		9.6.2.2 Delay angle α_d or φ' Angle Control	501
		9.6.2.3 Control with Machine Terminal Voltage Signals	504
		9.6.2.4 Phase-Locked Loop (PLL) γ Angle Control	506
	9.6.3 Scalar Control of Cycloconverter Drive		507
	9.6.4 Vector Control of Cycloconverter Drive		510
	9.6.5 Vector Control with Voltage-Fed Inverter		513
9.7	Sensorless Control		515
	9.7.1 Trapezoidal SPM Machine		515
		9.7.1.1 Terminal Voltage Sensing	515
		9.7.1.2 Stator Third Harmonic Voltage Detection	519
	9.7.2 Sinusoidal PM Machine (PMSM)		522
		9.7.2.1 Terminal Voltage and Current Sensing	522
		9.7.2.2 Inductance Variation (saliency) Effect	524
		9.7.2.3 Extended Kalman Filter (EKF)	526
9.8	Switched Reluctance Motor (SRM) Drives		529
9.9	Summary		532
9.10	References		533

Chapter 10 Expert System Principles and Applications — 535

- 10.1 Introduction — 535
- 10.2 Expert System Principles — 536
  - 10.2.1 Knowledge Base — 537
    - 10.2.1.1 Frame Structure — 539
    - 10.2.1.2 Meta-Knowledge — 540
    - 10.2.1.3 ES Language — 540
  - 10.2.2 Inference Engine — 541
  - 10.2.3 User Interface — 541
- 10.3 Expert System Shell — 543
  - 10.3.1 Shell Features — 543
  - 10.3.2 External Interface — 543
  - 10.3.3 Program Development Steps — 544
- 10.4 Design Methodology — 546
- 10.5 Applications — 546
  - 10.5.1 P-I Control Tuning of a Drive — 547
  - 10.5.2 Fault Diagnostics — 547
  - 10.5.3 Selection of Commercial ac Drive Product — 549
  - 10.5.4 Configuration Selection, Design, and Simulation of a Drive System — 549
    - 10.5.4.1 Configuration Selection — 550
    - 10.5.4.2 Motor Ratings Design — 550
    - 10.5.4.3 Converter Design — 552
    - 10.5.4.4 Control Design and Simulation Study — 554
- 10.6 Glossary — 555
- 10.7 Summary — 556
- 10.8 References — 557

Chapter 11 Fuzzy Logic Principles and Applications — 559

- 11.1 Introduction — 559
- 11.2 Fuzzy Sets — 560
  - 11.2.1 Membership Functions — 561
  - 11.2.2 Operations on Fuzzy Sets — 564
- 11.3 Fuzzy System — 566
  - 11.3.1 Implication Methods — 569
    - 11.3.1.1 Mamdani Type — 569
    - 11.3.1.2 Lusing Larson Type — 570
    - 11.3.1.3 Sugeno Type — 571
  - 11.3.2 Defuzzification Methods — 573
    - 11.3.2.1 Center of Area (COA) Method — 573
    - 11.3.2.2 Height Method — 575

11.3.2.3	Mean of Maxima (MOM) Method	575
11.3.2.4	Sugeno Method	576

11.4 Fuzzy Control   576
    11.4.1 Why Fuzzy Control?   576
    11.4.2 Historical Perspective   576
    11.4.3 Control Principle   577
    11.4.4 Control Implementation   581
11.5 General Design Methodology   581
11.6 Applications   582
    11.6.1 Induction Motor Speed Control   582
    11.6.2 Flux Programming Efficiency Improvement of Induction Motor Drive   585
        11.6.2.1 Pulsating Torque Compensation   589
    11.6.3 Wind Generation System   591
        11.6.3.1 Wind Turbine Characteristics   592
        11.6.3.2 System Description   592
        11.6.3.3 Fuzzy Control   593
    11.6.4 Slip Gain Tuning of Indirect Vector Control   597
        11.6.4.1 Derivation of Q^* and v_{ds}^* 598
    11.6.5 Stator Resistance R_s Estimation   602
    11.6.6 Estimation of Distorted Waves   606
        11.6.6.1 Mamdani Method   608
        11.6.6.2 Sugeno Method   609
11.7 Fuzzy Logic Toolbox   609
    11.7.1 FIS Editor   611
    11.7.2 Membership Function Editor   611
    11.7.3 Rule Editor   612
    11.7.4 Rule Viewer   612
    11.7.5 Surface Viewer   613
    11.7.6 Demo Program for Synchronous Current Control   613
11.8 Glossary   619
11.9 Summary   622
11.10 References   623

Chapter 12   Neural Network Principles and Applications   625

12.1 Introduction   625
12.2 The Structure of a Neuron   626
    12.2.1 The Concept of a Biological Neuron   626
    12.2.2 Artificial Neuron   627
        12.2.2.1 Activation Functions of a Neuron   628

12.3	Artificial Neural Network	629
12.3.1	Example Application: Y = Asin X	632
12.3.2	Training of Feedforward Neural Network	632
12.3.2.1	Learning Methods	634
12.3.2.2	Alphabet Character Recognition by an ANN	634
12.3.3	Back Propagation Training	637
12.3.4	Back propagation Algorithm for Three-Layer Network	637
12.3.4.1	Weight Calculation for Output Layer Neurons	637
12.3.4.2	Weight Calculation for Hidden Layer Neurons	641
12.3.5	On-Line Training	643
12.4	Other Networks	644
12.4.1	Radial Basis Function Network	644
12.4.2	Kohonen's Self-Organizing Feature Map Network	645
12.4.3	Recurrent Neural Network for Dynamic System	646
12.4.3.1	Training an RNN by the EKF Algorithm	647
12.5	Neural Network in Identification and Control	650
12.5.1	Time-Delayed Neural Network	650
12.5.2	Dynamic System Models	650
12.5.3	ANN Identification of Dynamic Models	652
12.5.4	Inverse Dynamics Model	654
12.5.5	Neural Network-Based Control	655
12.6	General Design Methodology	657
12.7	Applications	658
12.7.1	PWM Controller	658
12.7.1.1	Selected Harmonic Elimination (SHE) PWM	658
12.7.1.2	Instantaneous Current Control PWM	659
12.7.1.3	Space Vector PWM	660
12.7.2	Vector-Controlled Drive Feedback Signal Estimation	667
12.7.3	Estimation of Distorted Waves	670
12.7.4	Model Identification and Adaptive Drive Control	671
12.7.5	Speed Estimation by RNN	675
12.7.6	Adaptive Flux Estimation by RNN	676
12.8	Neuro-Fuzzy Systems	678
12.8.1	ANNBased Fuzzy Inference System (ANFIS)	678
12.9	Demo Program With Neural Network Toolbox	682
12.9.1	Introduction to Neural Network Toolbox	682
12.9.2	Demo Program	683
12.10	Glossary	684
12.11	Summary	689
12.12	References	690

Index   **691**

Preface

It is with pride, excitement and a lot of expectations, I am presenting this book to the professional community of the world. As you know, power electronics and motor drives constitute a complex and interdisciplinary subject which have gone through spectacular evolution in the last three decades. Recently, artificial intelligence (AI) techniques are extending the frontier of this technology. It is without any doubt that the power electronics will play a dominant role in the 21st century in industrial, commercial, residential, aerospace, utility and military applications with the emphasis for energy saving and solving environmental pollution problems.

I have been in the power electronics area for more than forty years (since the technology was born) through my career pursuits in academia and industry, and have followed the technology evolution very aggressively. In the past, I contributed a number of books (authored and edited) in power electronics area of which Power Electronics and AC Drives (Prentice Hall-1986) is most important. It was taken as an advanced text in many universities in the world. This new book can be considered as significant updating and expansion of the previous book where I have tried to embed practically whatever knowledge I have in power electronics and ac drives. It contains the subject from A-to-Z, i.e., power semiconductor devices, electrical machines, different classes of converters, induction and synchronous motor drives with control and estimation, and AI techniques (expert system, fuzzy logic and neural networks). In essence, I have tried to incorporate practically all the aspects of state-of-the-art technology of power electronics and motor drives in the book. The content of the book is essentially based on my lecture notes of one senior course and three graduate courses, which I have developed and taught in the University of Tennessee during the last fourteen years. It will be my deep satisfaction if I can see that the book is being considered as a text in more universities than the previous one. The universities, which are already following my previous book, can now safely accept this new book.

The content of the book can be summarized as follows: Chapter 1 contains description of different types of power semiconductor devices including the recent IGCT, where IGBT device has been emphasized. Chapter 2 describes induction and synchronous machine theories in somewhat detail from the viewpoint of variable frequency drive applications, which include dynamic d-q machine models. Complex space vectors have been introduced but avoided in much of the text because, in the author's opinion, they tend to frighten most of the students. Switched reluctance machine has been included for completeness. Chapters 3 and 4 discuss the classical phase-controlled thyristor converters and cycloconverters, respectively. For completeness, high frequency link converters are included in Chapter 4. Chapter 5 covers voltage-fed converters and PWM techniques where space vector PWM has been emphasized. More recent topics, such as soft-switching, power factor compensation, multi-level converters, static VAR compensators and active filters are included. Chapter 6 deals with current-fed converters that include PWM converters. Chapter 6 describes slip power recovery drives with wound-rotor induction motors, and mainly consist of Kramer and Scherbius drives. Chapter 8 covers control and estimation of cage type induction motor drives which includes discussion on speed sensorless control and drive self-commissioning. Induction motor drive is a dominant theme in the book. Chapter 9 describes control and estimation of synchronous machine drives that includes sensorless control and a brief description of switched reluctance motor drive. Chapter 10 gives a brief description of expert system and its applications. In the author's opinion, ES has a lot of potentiality but has been practically ignored by the power electronics community. Chapter 11 deals with fuzzy logic and its applications, and finally, Chapter 12 gives description of neural network and its applications. In the author's opinion, the ANN technology will have a large impact on power electronics area in future. A set of questions has been formulated for different chapters which will be forwarded to readers on request [bbose@utk.edu].

This book could not be possible without the help of some of my professional colleagues and students. First, I am deeply grateful to Burak Ozpineci, my graduate student, for his enormous help in the manuscript preparation of the book. Next, my gratitude goes to the student Joao Pinto who helped me in revising the manuscript. The two demo programs in Chapters 11 and 12 were developed by him for the book. I would like to express my thanks to Dr. In-Dong Kim of Pukyong National University, Korea (who was formerly visiting professor in my laboratory) for supplying the Corel Flow software that helped us to draw most of the art work. I am very grateful to Prof. Paresh Sen of Queen's University, Canada for his constant encouragement. Also, I thank Prof. Marian Kazmierkowsky of Warsaw University of Technology, Poland; Prof. Marcelo Simoes of Colorado School of Mines, and Dr. Ned Mohan of University of Minnesota for their help.

Finally, I thank Wil Mara, Prentice Hall PTR's production editor, for doing this enormous job so efficiently. Also on the production end, I thank Aurelia Scharnhorst for her superb page composition, and Corinne Ovadia for her skillful creation of hundreds of new drawings.

I am deeply grateful to the brilliant scientists and engineers on whose scholarly contributions this book is based. Finally, I am very grateful to my wife Arati for her immense patience and sacrifice while preparing this book during the last three years.

Bimal K. Bose
University of Tennessee

List of Principal Symbols

Symbols are generally defined locally. The list of principal symbols used in the text is given below.

d^e - q^e	Synchronously rotating reference frame (or rotating frame) direct and quadrature axes
d^s - q^s	Stationary reference frame direct and quadrature axes (also known as α - β axes)
f	Frequency (Hz)
I_d	Dc current (Ampere)
I_f	Machine field current
I_L	Rms load current
I_m	Rms magnetizing current
I_P	Rms active current
I_Q	Rms reactive current
I_r	Rms rotor current (referred to stator)
I_s	Rms stator current
$i_{dr}^{\,s}$	d^s – axis rotor current
$i_{ds}^{\,s}$	d^s – axis stator current
i_{qr}	q^e – axis rotor current
i_{qs}	q^e – axis stator current
J	Moment of inertia (Kg-m^2)
X_r	Rotor reactance (Ohm)
X_s	Synchronous reactance

X_{ds}	d^e – axis synchronous reactance
X_{lr}	Rotor leakage reactance
X_{ls}	Stator leakage reactance
X_{qs}	q^e – axis synchronous reactance
α	Firing angle
β	Advance angle
γ	Turn-off angle
δ	Torque or power angle of synchronous machine
θ	Thermal impedance (Ohm) (also torque angle or angle)
θ_e	Angle of synchronously rotating frame ($\omega_e t$)
θ_r	Rotor angle
θ_{sl}	Slip angle
μ	Overlap angle
τ	Time constant (s)
L_c	Commutating inductance (Henry)
L_d	Dc link filter inductance
L_m	Magnetizing inductance
L_r	Rotor inductance
L_s	Stator inductance
L_{lr}	Rotor leakage inductance
L_{ls}	Stator leakage inductance
L_{dm}	d^e – axis magnetizing inductance
L_{qm}	q^e – axis magnetizing inductance
N	Turns ratio (primary to secondary)
N_e	Synchronous speed (rpm)
N_r	Rotor speed
P	Number of poles (also active power)
P_g	Airgap power (watts)
P_m	Mechanical output power
Q	Reactive power
R_r	Rotor resistance (Ohm)
R_s	Stator resistance
S	Slip (per unit)
T	Time period (s) (also temperature) (°C)
T_e	Developed torque (Nm)

T_L	Load torque
t_{off}	Turn-off time (s)
V_d	Dc voltage
V_I	Inverter dc voltage
V_f	Induced emf
V_m	Peak phase voltage (volt)
V_g	Rms airgap voltage
V_R	Rectifier dc voltage
v_s	Instantaneous supply voltage
v_d	Inst. Dc voltage
v_f	Inst. Field voltage
v_{dr}^s	d^s – axis rotor voltage
v_{ds}^s	d^s – axis stator voltage
v_{qr}	q^e – axis rotor voltage
v_{qs}	q^e – axis stator voltage
ϕ	Displacement power factor angle
ψ_a	Armature reaction flux linkage (Weber-turns)
ψ_f	Field flux linkage
ψ_m	Airgap flux linkage
ψ_r	Rotor flux linkage
ψ_s	Stator flux linkage
ψ_{dr}^s	d^s – axis rotor flux linkage
ψ_{ds}^s	d^s – axis stator flux linkage
ψ_{qr}	q^e – axis rotor flux linkage
ψ_{qs}	q^e – axis stator flux linkage
ω_e (or ω)	Stator or line frequency (r/s)
ω_m	Rotor mechanical speed
ω_r	Rotor electrical speed
ω_{sl}	Slip frequency
$\hat{X}$	Peak value of a sinusoidal phasor or sinusoidal space vector magnitude (also estimated parameter) (X is any arbitrary variable)
	Space vector (X is arbitrary variable)

CHAPTER 1

Power Semiconductor Devices

1.1 INTRODUCTION

Power semiconductor devices constitute the heart of modern power electronic apparatus. They are used in power electronic converters in the form of a matrix of on-off switches, and help to convert power from ac-to-dc (rectifier), dc-to-dc (chopper), dc-to-ac (inverter), and ac-to-ac at the same (ac controller) or different frequencies (cycloconverter). The switching mode power conversion gives high efficiency, but the disadvantage is that due to the nonlinearity of switches, harmonics are generated at both the supply and load sides. The switches are not ideal, and they have conduction and turn-on and turn-off switching losses. Converters are widely used in applications such as heating and lighting controls, ac and dc power supplies, electrochemical processes, dc and ac motor drives, static VAR generation, active harmonic filtering, etc. Although the cost of power semiconductor devices in power electronic equipment may hardly exceed 20–30 percent, the total equipment cost and performance may be highly influenced by the characteristics of the devices. An engineer designing equipment must understand the devices and their characteristics thoroughly in order to design efficient, reliable, and cost-effective systems with optimum performance. It is interesting to note that the modern technology evolution in power electronics has generally followed the evolution of power semiconductor devices. The advancement of microelectronics has greatly contributed to the knowledge of power device materials, processing, fabrication, packaging, modeling, and simulation.

Today's power semiconductor devices are almost exclusively based on silicon material and can be classified as follows:

- Diode
- Thyristor or silicon-controlled rectifier (SCR)
- Triac

- Gate turn-off thyristor (GTO)
- Bipolar junction transistor (BJT or BPT)
- Power MOSFET
- Static induction transistor (SIT)
- Insulated gate bipolar transistor (IGBT)
- MOS-controlled thyristor (MCT)
- Integrated gate-commutated thyristor (IGCT)

In this chapter, we will briefly study the operational principles and characteristics of these devices.

1.2 DIODES

Power diodes provide uncontrolled rectification of power and are used in applications such as electroplating, anodizing, battery charging, welding, power supplies (dc and ac), and variable-frequency drives. They are also used in feedback and the freewheeling functions of converters and snubbers. A typical power diode has P-I-N structure, that is, it is a P-N junction with a near-intrinsic semiconductor layer (I-layer) in the middle to sustain reverse voltage.

Figure 1.1 shows the diode symbol and its volt-ampere characteristics. In the forward-biased condition, the diode can be represented by a junction offset drop and a series-equivalent resistance that gives a positive slope in the V-I characteristics. The typical forward conduction drop is 1.0 V. This drop will cause conduction loss, and the device must be cooled by the appropriate heat sink to limit the junction temperature. In the reverse-biased condition, a small

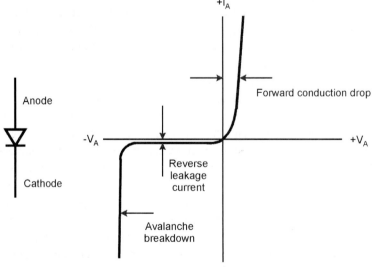

Figure 1.1 Diode symbol and volt-ampere characteristics

Diodes

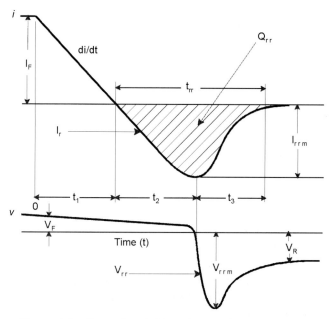

Figure 1.2 Turn-off switching characteristics of a diode

leakage current flows due to minority carriers, which gradually increase with voltage. If the reverse voltage exceeds a threshold value, called the breakdown voltage, the device goes through avalanche breakdown, which is when reverse current becomes large and the diode is destroyed by heating due to large power dissipation in the junction.

The turn-off voltage and current characteristics as functions of time, which are indicated in Figure 1.2, are particularly important for a diode. In the forward high-conduction region, the conduction drop (V_F) is small, as mentioned before. At this condition, the P and N regions near the junction and the I-layer remain saturated with minority carriers.

If the device is open-circuited, the carriers die by a recombination process, which takes a reasonably long time. Normally, a reverse dc voltage (V_R) is applied to turn off the device, as indicated in Figure 1.2. At time $t = 0$, the reverse voltage is applied when the current goes down linearly because of series-circuit inductance. During time t_2, the current is negative and the minority carriers sweep out across the junction, but the excess carrier concentration keeps the junction saturated, and therefore, the same negative current slope is maintained. The conduction drop decreases during t_1 and t_2 due to the reduction of Ohmic (equivalent resistance) drop. At the end of t_2, the device sustains voltage, and steady-state voltage appears at the end of t_3. During t_3, the reverse current falls quickly partly due to sweeping out and partly by recombination. The fast decay of negative current creates an inductive drop that adds with the reverse voltage V_R as shown. The reverse recovery time $t_{rr} = t_2 + t_3$ and the corresponding recovery charge Q_{rr} (shown by the hatched area) that are affected by the recombination process are important parameters of a diode. The snappiness by which the recovery current falls to zero determines the voltage boost V_{rr}. This voltage may be destructive and can be softened by a resistance-capacitance snubber, which will be discussed later. The recovery current causes additional loss (switching loss) in the diode, which can be determined graphically from Figure 1.2.

Power diodes can be classified as follows:

- Standard or slow-recovery diode
- Fast-recovery diode
- Schottky diode

Slow- and fast-recovery diodes have P-I-N geometry, as mentioned above. In a fast-recovery diode, as the name indicates, the recovery time t_{rr} and the recovery charge Q_{rr} (shown by the hatched area) are reduced by the minority carrier lifetime control that enhances the recombination process. However, the adverse effect is a higher conduction drop. For example, the POWEREX fast-recovery diode type CS340602, which has a dc current rating ($I_F(dc)$) of 20 A and a blocking voltage rating (V_{rrm}) of 600 V, has the following ratings: $V_{FM} = 1.5$ V, $I_{rrm} = 5.0$ mA, $t_{rr} = 0.8$ µs, and $Q_{rr} = 15$ µC. The standard slow-recovery diodes are used for line frequency (50/60 Hz) power rectification. They have a lower conduction drop, but a higher t_{rr}. These diodes are available with ratings of several kilovolts and several kiloamperes. A Schottky diode is basically a majority carrier diode and is formed by a metal-semiconductor junction. As a result, the diode has a lower conduction drop (typically 0.5 V) and faster switching time, but the limitations are a lower blocking voltage (typically up to 200 V) and higher leakage current. For example, the International Rectifier Schottky diode type 6TQ045 has ratings of $V_{rrm} = 45$ V, $I_{F(AV)} = 6$ A, $V_F = 0.51$ V, and reverse leakage current $I_{rm} = 0.8$ mA (at 25 °C). These diodes are used in high-frequency circuits.

The electrical and thermal characteristics of diodes are somewhat similar to thyristors, which will be discussed next. Specific circuit applications of different types of diodes will be discussed in later chapters.

1.3 THYRISTORS

Thyristors, or silicon-controlled rectifiers (SCRs) have been the traditional workhorses for bulk power conversion and control in industry. The modern era of solid-state power electronics started due to the introduction of this device in the late 1950s. Chapters 3, 4, and 6 will discuss thyristor converters and their applications. The term "thyristor" came from its gas tube equivalent, thyratron. Often, it is a family name that includes SCR, triac, GTO, MCT, and IGCT. Thyristors can be classified as standard, or slow phase-control-type and fast-switching, voltage-fed inverter-type. The inverter-type has recently become obsolete and will not be discussed further.

Thyristors

1.3.1 Volt-Ampere Characteristics

Figure 1.3 shows the thyristor symbol and its volt-ampere characteristics. Basically, it is a three-junction P-N-P-N device, where P-N-P and N-P-N component transistors are connected in regenerative feedback mode. The device blocks voltage in both the forward and reverse directions (symmetric blocking). When the anode is positive, the device can be triggered into conduction by a short positive gate current pulse; but once the device is conducting, the gate loses its control to turn off the device. A thyristor can also turn on by excessive anode voltage, its rate of rise (dv/dt), by a rise in junction temperature (T_J), or by light shining on the junctions.

The volt-ampere characteristics of the device indicate that at gate current $I_G = 0$, if forward voltage is applied on the device, there will be a leakage current due to blocking of the middle junction. If the voltage exceeds a critical limit (breakover voltage), the device switches into conduction. With increasing magnitude of I_G, the forward breakover voltage is reduced, and eventually at I_{G3}, the device behaves like a diode with the entire forward blocking region removed. The device will turn on successfully if a minimum current, called a latching current, is maintained. During conduction, if the gate current is zero and the anode current falls below a critical limit, called the holding current, the device reverts to the forward blocking state. With reverse voltage, the end P-N junctions of the device become reverse-biased and the V-I curve becomes essentially similar to that of a diode rectifier. Modern thyristors are available with very large voltage (several KV) and current (several KA) ratings.

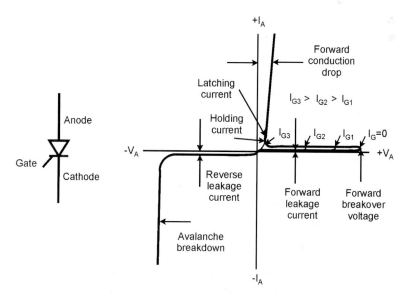

Figure 1.3 Thyristor symbol and volt-ampere characteristics

1.3.2 Switching Characteristics

Initially, when forward voltage is applied across a device, the off-state, or static dv/dt, must be limited so that it does not switch on spuriously. The dv/dt creates displacement current in the depletion layer capacitance of the middle junction, which induces emitter current in the component transistors and causes switching action. When the device turns on, the anode current di/dt can be excessive, which can destroy the device by heavy current concentration. During conduction, the inner P-N regions remain heavily saturated with minority carriers and the middle junction remains forward-biased. To recover the forward voltage blocking capability, a reverse voltage is applied across the device to sweep out the minority carriers and the phenomena are similar to that of a diode (see Figure 1.2). However, when the recovery current goes to zero, the middle junction still remains forward-biased. This junction eventually blocks with an additional delay when the minority carriers die by the recombination process. The forward voltage can then be applied successfully, but the reapplied dv/dt will be somewhat less than the static dv/dt because of the presence of minority carriers. For example, POWEREX SCR/diode module CM4208A2 (800 V, 25 A) has limiting di/dt = 100 A/µs and off-state dv/dt = 500 V/µs parameters. A suitably-designed snubber circuit (discussed later) can limit di/dt and dv/dt within acceptable limits. In a converter circuit, a thyristor can be turned off (or commutated) by a segment of reverse ac line or load voltage (defined as line or load commutation, respectively), or by an inductance-capacitance circuit-induced transient reverse voltage (defined as forced commutation).

1.3.3 Power Loss and Thermal Impedance

A thyristor has dominant conduction loss like a diode, but its switching loss (to be discussed later) is very small. The device specification sheet normally gives information on power dissipation for various duty cycles of sinusoidal and rectangular current waves. Figure 1.4 shows the power dissipation characteristics for a rectangular current wave. The reverse blocking loss and gate circuit loss are also included in the figure. These curves are valid up to 400 Hz supply frequency. The heat due to power loss in the vicinity of a junction flows to the case and then to the ambient through the externally mounted heat sink, causing a rise in the junction temperature T_J. The maximum T_J of a device is to be limited because of its adverse effect on device performance. For steady power dissipation P, T_J can be calculated as

$$T_J - T_A = P(\theta_{JC} + \theta_{CS} + \theta_{SA}) \tag{1.1}$$

where T_A is the ambient temperature, and θ_{JC}, θ_{CS}, and θ_{SA} represent the thermal resistance from junction to case, case to sink, and sink to ambient, respectively. The resistance θ_{SA} is determined by the cooling system design, and the methods of cooling may include heat sink with natural convection cooling, forced air cooling, or forced liquid cooling. From Equation (1.1), it is evident that for a limited T_{Jmax} (usually 125 °C), the dissipation P can be increased by reducing θ_{SA}. This means that a more efficient cooling system will increase power dissipation, that is, the

Thyristors

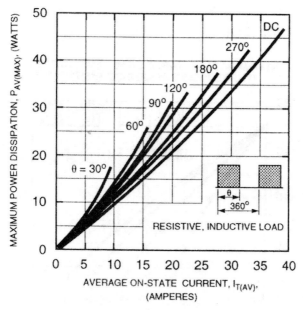

Figure 1.4 Average on-state power dissipation of thyristor for rectangular current wave (POWEREX CM4208A2)

power-handling capability of a device. An infinite heat sink is defined when $\theta_{SA} = 0$, that is, the case temperature $T_C = T_A$.

In practical operation, the power dissipation P is cyclic, and the thermal capacitance or storage effect delays the junction temperature rise, thus permitting higher loading of the device. The transient thermal equivalent circuit can be represented by a parallel RC circuit, where P is equivalent to the current source and the resulting voltage across the circuit represents the temperature T_J. Figure 1.5(a) shows the T_J curve for the dissipation of a single power pulse. Considering the complementary nature of heating and cooling curves, the following equations can be written:

$$T_J(t_1) = T_A + P\theta(t_1) \tag{1.2}$$

$$T_J(t_2) = T_A + P[\theta(t_2) - \theta(t_2 - t_1)] \tag{1.3}$$

where $\theta(t_1)$ is the transient thermal impedance at time t_1. The device specification sheet normally gives thermal impedance between junction and case. The additional effect due to heat sink can be added if desired. Figure 1.5(b) shows typical junction temperature build-up for three repeated pulses. The corresponding T_J expressions by the superposition principle can be given as

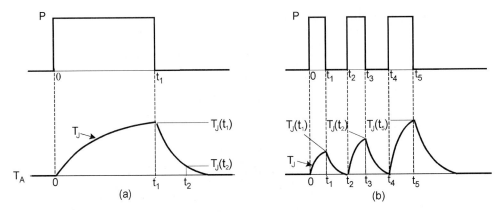

Figure 1.5 Junction temperature rise with pulsed power dissipation
(a) Single pulse, (b) Multiple pulses

$$T_J(t_1) = T_A + P\theta(t_1) \tag{1.4}$$

$$T_J(t_3) = T_A + P[\theta(t_3) - \theta(t_3 - t_1) + \theta(t_3 - t_2)] \tag{1.5}$$

$$T_J(t_5) = T_A + P[\theta(t_5) - \theta(t_5 - t_1) + \theta(t_5 - t_2) - \theta(t_5 - t_3) + \theta(t_5 - t_4)] \tag{1.6}$$

Figure 1.6 illustrates the transient thermal impedance curve ($\theta_{JC}(t)$) of a thyristor (type CM4208A2) as a function of time. The device has the rated thermal resistances of θ_{JC} = 0.8 °C/W and θ_{CS} = 0.2 °C/W. Note that the device cooling and thermal impedance concepts discussed here are also valid for all power semiconductor devices.

1.3.4 Current Rating

Based on the criteria of limiting T_J as discussed above, Figure 1.7 shows the average current rating $I_{T(AV)}$ vs. permissible case temperature T_C for various duty cycles of rectangular current wave. For example, if T_C is limited to 110°C, the thyristor can carry 12 A average current for = 120°. If a better heat sink limits T_C to 100 °C, the current can be increased to 18 A. Figure 1.7 can be used with Figure 1.4 to design the heat sink thermal resistance.

1.4 TRIACS

A triac has a complex multiple-junction structure, but functionally, it is an integration of a pair of phase-controlled thyristors connected in inverse-parallel on the same chip. Figure 1.8(a) shows the triac symbol and (b) shows its volt-ampere characteristics. The three-terminal device can be triggered into conduction in both positive and negative half-cycles of supply voltage by applying gate trigger pulses. In I+ mode, the terminal T_2 is positive and the device is switched on by positive gate current pulse. In III- mode, the terminal T_1 is positive and it is switched on

Triacs

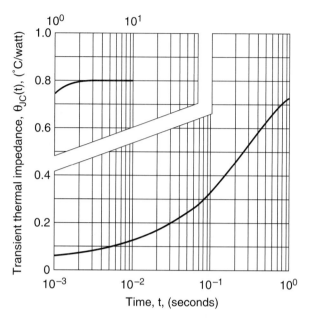

Figure 1.6 Transient thermal impedance curve of thyristor (CM4208A2)

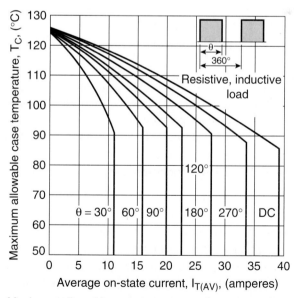

Figure 1.7 Maximum allowable case temperature for rectangular current wave (CM4208A2)

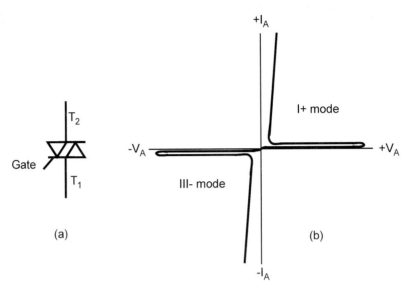

Figure 1.8 Triac symbol and volt-ampere characteristics

by negative gate current pulse. A triac is more economical than a pair of thyristors in anti-parallel and its control is simpler, but its integrated construction has some disadvantages. The gate current sensitivity of a triac is poorer and the turn-off time is longer due to the minority carrier storage effect. For the same reason, the reapplied dv/dt rating is lower, thus making it difficult to use with inductive load. A well-designed RC snubber is essential for a triac circuit. Triacs are used in light dimming, heating control, appliance-type motor drives, and solid-state relays with typically 50/60 Hz supply frequency.

Figure 1.9 shows a popular incandescent lamp dimmer circuit using a triac and the corresponding waveforms. The gate of the triac gets the drive pulse from an RC circuit through a diac, which is a symmetric voltage-blocking device. The capacitor voltage v_c lags the line voltage wave. When v_c exceeds the threshold voltage $\pm V_s$ of the diac, a pulse of current in either polarity triggers the triac at angle α_f, giving full-wave ac phase-controlled output to the load. The firing angle can be varied in the range α_1 to α_2 to control light intensity by varying the resistance R_1.

1.5 GATE TURN-OFF THYRISTORS (GTOS)

A gate turn-off thyristor (GTO), as the name indicates, is basically a thyristor-type device that can be turned on by a small positive gate current pulse, but in addition, has the capability of being turned off by a negative gate current pulse. The turn-off capability of a GTO is due to the diversion of P-N-P collector current by the gate, thus breaking the P-N-P / N-P-N regenerative feedback effect. GTOs are available with asymmetric and symmetric voltage-blocking capabilities, which are used in voltage-fed and current-fed converters, respectively. The turn-off current gain of a GTO, defined as the ratio of anode current prior to turn-off to the negative gate current required for turn-off, is very low, typically 4 or 5. This means that a 6000 A GTO requires as

Gate Turn-Off Thyristors (GTOs)

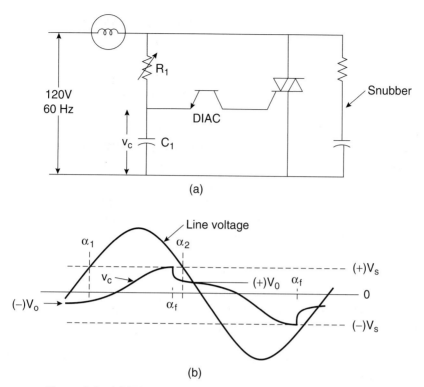

Figure 1.9 (a) Triac light dimmer circuit, (b) Control waveforms

high as −1500 A gate current pulse. However, the duration of the pulsed gate current and the corresponding energy associated with it is small and can easily be supplied by low-voltage power MOSFETs. GTOs are used in motor drives, static VAR compensators (SVCs), and ac/dc power supplies with high power ratings. When large-power GTOs became available, they ousted the force-commutated, voltage-fed thyristor inverters.

1.5.1 Switching Characteristics

The switching characteristics of GTOs are somewhat different from those of thyristors and therefore require some explanation. Figure 1.10 shows a GTO chopper (dc-to-dc converter) circuit with a polarized snubber. The snubber consists of a turn-on component (L_L) called a series snubber and a turn-off component (R_s, C_s, and D_s), called a shunt snubber. This type of converter is typically used for a subway dc motor propulsion drive.

Figure 1.11 shows the turn-on and turn–off characteristics of the circuit with the snubber. The turn-on characteristics of a GTO are essentially similar to those of a thyristor. Initially, before turn-on, the capacitor C_s is charged to supply voltage V_d and the load current is flowing through the freewheeling diode. At turn-on, the series snubber limits di/dt through the device, and the supply voltage V_d is applied across the load. At the same time, C_s discharges through R_s

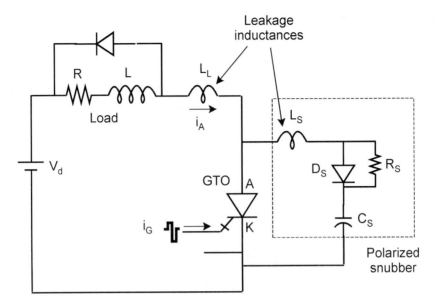

Figure 1.10 GTO chopper with polarized snubber

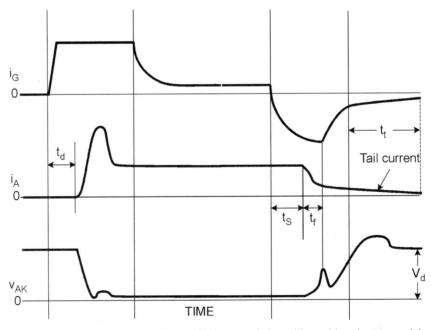

Figure 0.1 GTO turn-on and turn-off characteristics with snubber (not to scale)

Gate Turn-Off Thyristors (GTOs)

and the GTO in series (neglecting the L_s effect), dumping most of its energy into the resistor R_s. The power dissipation P_s in the resistor is approximately given as

$$P_s = 0.5 C_s V_d^2 f \qquad (1.7)$$

where f = chopper operating frequency. Obviously, the turn-on switching loss of the device is reduced because of delayed build-up of the device current. When the GTO is turned off by negative gate current pulse, the controllable anode current i_A begins to fall after a short time delay, defined as storage time (t_s). The fall time t_f is somewhat abrupt, and is typically less than 1.0 μs. As the forward voltage begins to develop, anode current tends to bypass through the shunt capacitor, limiting dv/dt across the device. The leakage inductance L_s in the snubber creates a spike voltage, as shown. A large voltage spike is extremely harmful because current concentration may create localized heating, causing what is known as second breakdown failure. This emphasizes the need to minimize the shunt snubber leakage inductance. After the spike voltage, the anode voltage overshoots due to underdamped resonance before settling to normal forward blocking voltage V_d. GTO has a long tail current, as shown, mainly due to sweeping out of the minority carriers. This tail current at large anode voltage causes large turn-off switching loss unless voltage build-up is slowed with large C_s.

However, large C_s increases snubber dissipation, as given by Equation (1.7). The trade-off between snubber loss and turn-off switching loss, and the corresponding total loss curves with increasing snubber capacitance are given approximately by Figure 1.12. The curves indicate that

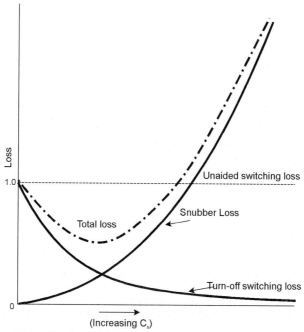

Figure 1.11 Trade-off between snubber loss and turn-off switching loss with increasing capacitor

the device-switching loss is diverted to snubber loss, and the total loss may be higher than the intrinsic switching loss of the device. Since GTO power losses are somewhat higher during switching, the converter switching frequency is low and is typically restricted within 1.0 kHz.

1.5.2 Regenerative Snubbers

The snubber loss may be substantial in a high-power and/or high-frequency converter. This, in turn, may reduce the converter's efficiency and put a burden on the cooling system. To combat this problem, various regenerative or energy recovery schemes have been proposed where the stored energy in the snubber capacitor is pumped back to the source or load. Figure 1.13(a) shows a passive energy recovery scheme. When the GTO is turned off, the snubber capacitor C_s charges to the full supply voltage, as usual. At the subsequent turn-on of the GTO, the stored energy is transferred to capacitor C resonantly through the inductance L and diode D. When the GTO turns off again, the energy in C is absorbed in the load and C_s charges again to voltage V_d. Figure 1.13(b) shows a regenerative snubber that uses an auxiliary chopper. At GTO turn-off, the snubber capacitor C_s charges to supply voltage. At subsequent turn-on of the device, the energy is resonantly transferred to capacitor C, as before. The energy in C is then pumped to the source through a dc-to-dc boost converter.

The discussion on dissipative and regenerative snubbers given in this section is also valid for other devices. The idea of a regenerative snubber appears very attractive, but its application should be carefully weighed against the extra cost, loss, complexity, and equipment reliability. High-power GTO converters normally use regenerative snubbers. Otherwise, RC snubbers are commonly used. Snubberless converters, which will be discussed later, are also possible.

1.6 BIPOLAR POWER OR JUNCTION TRANSISTORS (BPTS OR BJTS)

A bipolar junction transistor (BPT or BJT), unlike a thyristor-like device, is a two-junction, self-controlled device where the collector current is under the control of the base drive current. Bipolar junction transistors have recently been ousted by IGBTs (insulated gate bipolar transistors) in the higher end and by power MOSFETs in the lower end. The dc current gain (h_{FE}) of a power transistor is low and varies widely with collector current and temperature. The gain is increased to a high value in the Darlington connection, as shown in Figure 1.14. However, the disadvantages are higher leakage current, higher conduction drop, and reduced switching frequency. The shunt resistances and diode in the base-emitter circuit help to reduce collector leakage current and establish base bias voltages. A transistor can block voltage in the forward direction only (asymmetric blocking). The feedback diode, as shown, is an essential element for chopper and voltage-fed converter applications. Double or triple Darlington transistors are available in module form with matched parallel devices for higher power rating.

Power transistors have an important property known as the second breakdown effect. This is in contrast to the avalanche breakdown effect of a junction, which is also known as first breakdown effect. When the collector current is switched on by the base drive, it tends to crowd on the base-emitter junction periphery, thus constricting the collector current in a narrow area of the

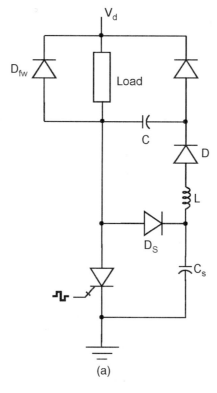

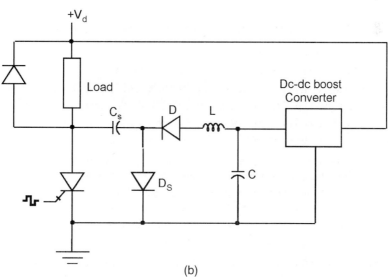

Figure 1.12 Regenerative snubbers (a) Passive, (b) Active

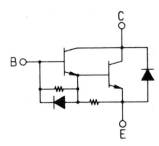

Figure 1.13 Darlington transistor symbol

reverse-biased collector junction. This tends to create a hot spot and the junction fails by thermal runaway, which is known as second breakdown. The rise in junction temperature at the hot spot accentuates the current concentration owing to the negative temperature coefficient of the drop, and this regeneration effect causes collapse of the collector voltage, thus destroying the device. A similar problem arises when an inductive load is turned off. As the base-emitter junction becomes reverse-biased, the collector current tends to concentrate in a narrow area of the collector junction.

Manufacturers provide specifications in the form of safe operating areas (SOAs) during turn-on (FBSOA) and turn-off (RBSOA), as shown in Figure 1.15. Obviously, a well-designed polarized RC snubber is indispensable in a transistor converter.

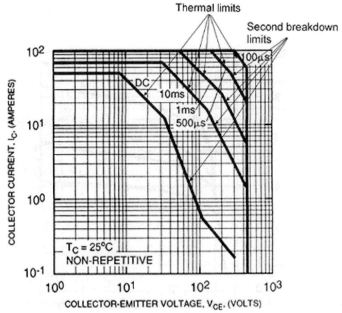

Figure 1.14 Forward-bias safe operating area (SOA) of a transistor (POWEREX KS524505) (600 V, 50 A)

1.7 POWER MOSFETS

1.7.1 V-I Characteristics

Unlike the devices discussed so far, a power MOSFET (metal-oxide semiconductor field-effect transistor) is a unipolar, majority carrier, "zero junction," voltage-controlled device. Figure 1.16(a) shows the symbol of an N-type MOSFET and (b) shows its volt-ampere characteristics. If the gate voltage is positive and beyond a threshold value, an N-type conducting channel will be induced that will permit current flow by majority carrier (electrons) between the drain and the source. Although the gate impedance is extremely high at steady state, the effective gate-source capacitance will demand a pulse current during turn-on and turn-off. The device has asymmetric voltage-blocking capability, and has an integral body diode, as shown, which can carry full current in the reverse direction. The diode is characterized by slow recovery and is often bypassed by an external fast-recovery diode in high-frequency applications.

The V-I characteristics of the device have two distinct regions, a constant resistance ($R_{DS(on)}$) region and a constant current region. The $R_{DS(on)}$ of a MOSFET is an important parameter which determines the conduction drop of the device. For a high voltage MOSFET, the longer conduction channel makes this drop large ($R_{DS(on)} \alpha V^{2.5}$). It is interesting to note that modern trench gate technology tends to lower the conduction resistance [10]. The positive temperature coefficient of this resistance makes parallel operation of MOSFET easy. In fact, large MOSFETS are fabricated by parallel connection of many devices.

While the conduction loss of a MOSFET is large for higher voltage devices, its turn-on and turn-off switching times are extremely small, causing low switching loss. The device does not have the minority carrier storage delay problem associated with a bipolar device, and its switching times are determined essentially by the ability of the drive to charge and discharge a tiny input capacitance $C_{ISS} = C_{GS} + C_{GD}$ (defined as Miller capacitance) with C_{DS} shorted, where C_{GS} = gate-to-source capacitance, C_{GD} = gate-to-drain capacitance, and C_{DS} = drain-to-source capacitance. Although a MOSFET can be controlled statically by a voltage source, it is normal practice to drive it by a current source dynamically followed by a voltage source to minimize switching delays. MOSFETs are extremely popular in low-voltage, low-power, and high-frequency (hundreds of kHz) switching applications. Application examples include switching mode power supplies (SMPS), brushless dc motors (BLDMs), stepper motor drives, and solid-state dc relays.

1.7.2 Safe Operating Area (SOA)

An important property of a MOSFET is that it does not have the second breakdown problem of a BJT. If localized and potentially destructive heating occurs within the device, the positive temperature coefficient effect of resistance forces local current concentration to be uniformly distributed across the total area. The maximum SOA of a MOSFET is shown in Figure 1.17. The device in the figure has a maximum dc current rating of 8 A at saturated condition, which can be

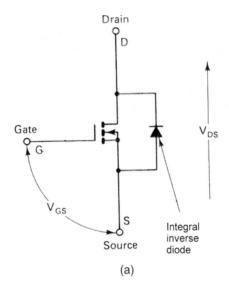

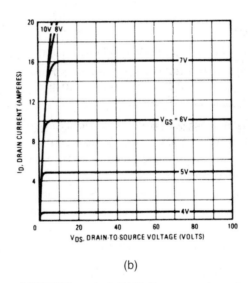

Figure 1.15 (a) Power MOSFET symbol, (b) Volt-ampere characteristics (Harris 2N6757); (150 V, 8 A)

increased to an absolute maximum value of 12 A on a single-pulse basis. The maximum drain-to-source voltage V_{DS} is limited to 150 V without causing avalanche breakdown.

The SOAs of a MOSFET are determined solely by the junction temperature (T_J) rise, which can be calculated from thermal impedance information. As shown in Figure 1.17, the dc current can be carried in the full voltage range by 75 watts of power dissipation. This corresponds to junction-to-case thermal resistance $\theta_{JC} = 1.7$ °C/W for $T_{Jmax} = 150$ °C and infinite

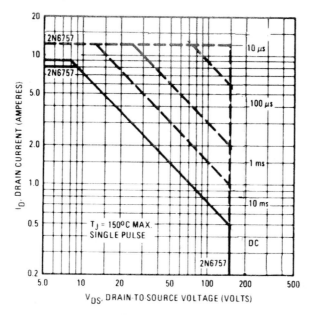

Figure 1.16 MOSFET safe operating area (SOA)

heat sink with ambient temperature $T_A = T_C = 25\,°C$ (see Equation 1.4). If the case temperature increases due to finite heat sink thermal resistance, the power dissipation is to be linearly derated, as shown in Figure 1.18. Note that there is no secondary slope on the SOA curves, indicating the absence of second breakdown as commonly observed on BJT SOA curves (Figure 1.15). Figure 1.17 also shows SOA curves based on single current pulse-operation. For example, at $V_{DS} = 100$ V, a dc current of 0.75 A (75 W) can be increased to 9.0 A (900 W) with 100 μs pulse. With the transient thermal impedance information of a device, safe dissipation limits for multi-pulse operation on a duty-cycle basis within the constraint of T_{Jmax} can be easily calculated. With square SOA and the absence of second breakdown effect, MOSFET converters with minimal stray inductance can be designed without snubbers if desired.

1.8 STATIC INDUCTION TRANSISTORS (SITS)

A static induction transistor (SIT) is a high-voltage, high-power, high-frequency device that can be considered essentially the solid-state version of a triode vacuum tube, which has been known for a long time. It is a short N-channel majority carrier device where the P-type gate electrodes are buried within the drain and source N-type layers. A drawback of the device is that it is normally on, but if gate voltage is negative, the reverse-biased P-N junction will inhibit drain current flow. Functionally, it is almost identical to a junction-FET, except that its lower channel resistance causes lower conduction drop. The reliability, noise, and radiation hardness of an SIT are claimed to be superior to a MOSFET. Although the conduction drop of the device is lower than that of an equivalent series-parallel combination of MOSFETs, the excessively large con-

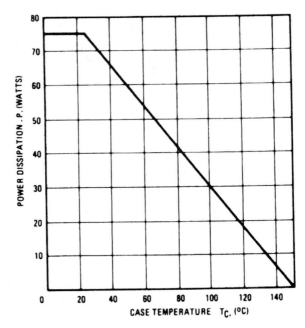

Figure 1.17 MOSFET power vs. temperature derating curves

duction drop makes it unsuitable in most power electronics applications unless justified by a need for FET-like switching frequency. These devices have been used in AM/FM transmitters, high-frequency induction heating, high-voltage, low-current power supplies, ultrasonic generators, and linear power amplifiers.

1.9 INSULATED GATE BIPOLAR TRANSISTORS (IGBTS)

The introduction of insulated gate biploar transistors (IGBTs) in the mid-1980s was an important milestone in the history of power semiconductor devices. They are extremely popular devices in power electronics up to medium power (a few kWs to a few MWs) range and are applied extensively in dc/ac drives and power supply systems. They ousted BJTs in the upper range, as mentioned before, and are currently ousting GTOs in the lower power range. An IGBT is basically a hybrid MOS-gated turn-on/off bipolar transistor that combines the advantages of both a MOSFET and BJT. Figure 1.19(a) shows the basic structure of an IGBT and (b) shows the device symbol.

Its architecture is essentially similar to that of a MOSFET, except an additional P^+ layer has been added at the collector over the N^+ drain layer of the MOSFET. The device has the high-input impedance of a MOSFET, but BJT-like conduction characteristics. If the gate is positive with respect to the emitter, an N-channel is induced in the P region. This forward-biases the base-emitter junction of the P-N-P transistor, turning it on and causing conductivity modulation of the N^- region, which gives a significant reduction of conduction drop over that of a MOSFET.

Insulated Gate Bipolar Transistors (IGBTs)

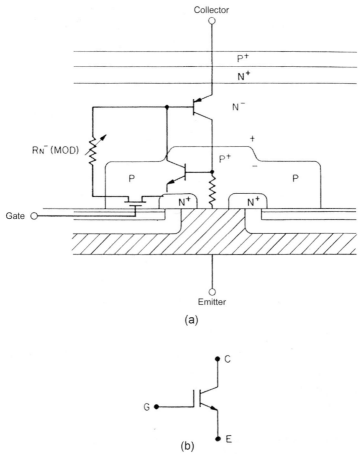

Figure 1.18 (a) IGBT structure with equivalent circuit, (b) Device symbol

At the on-condition, the driver MOSFET in the equivalent circuit of the IGBT carries most of the total terminal current. The thyristor-like latching action caused by the parasitic N-P-N transistor is prevented by sufficiently reducing the resistivity of the P⁺ layer and diverting most of the current through the MOSFET. The device is turned off by reducing the gate voltage to zero or negative, which shuts off the conducting channel in the P region. The device has higher current density than that of a BJT or MOSFET. Its input capacitance (C_{iss}) is significantly less than that of a MOSFET. Also, the ratio of gate-collector capacitance to gate-emitter capacitance is lower, giving an improved Miller feedback effect.

Figure 1.20 shows the volt-ampere characteristics of an IGBT near the saturation region, indicating its BJT-like characteristics. A modern IGBT uses trench-gate technology to reduce the conduction drop further. The device does not show any second breakdown characteristics of a BJT and its square SOA is limited thermally like a MOSFET. Therefore, an IGBT converter can be designed with or without a snubber.

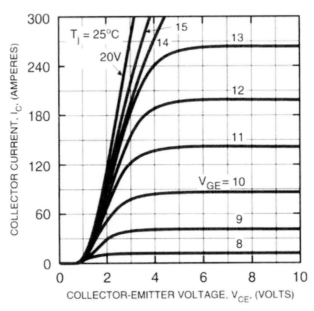

Figure 1.19 Volt-ampere characteristics of an IGBT (POWEREX IPM CM150TU-12H); (600 V, 150 A)

1.9.1 Switching Characteristics and Thermal Impedance

Figure 1.21(b) shows the typical switching voltage and current waves for the "hard-switched," snubberless, half-bridge converter shown in (a) of the same figure. The profile of conduction and switching losses is indicated in the lower part of (b). Initially, the IGBT Q_1 is off and the inductive load current is taken by the free wheeling diode D_2. When Q_1 is turned on, the load current is initially taken by the device at full voltage (with a small leakage inductance drop), as shown. In fact, the diode D_2 recovery current is added to it before the diode sustains reverse voltage and Q_1 voltage falls to zero. Similarly, at turn-off, the device voltage builds up at full current and then the diode D_2 takes over the line current as shown. The fall time t_f is very short and is dictated by turn-off of the MOSFET section of the IGBT. The device shows a tail time t_t, which is due to minority carrier storage in the N$^-$ region. The loss curve indicates that the average switching loss becomes high at high switching frequency. Obviously, a turn-on/turn-off snubber can reduce the device's switching losses, as discussed before.

Figure 1.22 shows the transient thermal impedance characteristics of an IGBT in normalized form, which indicates that $\theta_{JC}(t)$ is always a fraction of the corresponding thermal resistance $\theta_{JC}(dc)$. For this particular device (POWEREX CM150TU-12H), $\theta_{JC}(dc) = 0.21$ °C/W. This figure helps to design heat sink for certain duty cycle power pulses, as indicated in Figure 1.5. The power rating (currently 3500 V, 1200 A) and electrical characteristics of the device are

Insulated Gate Bipolar Transistors (IGBTs)

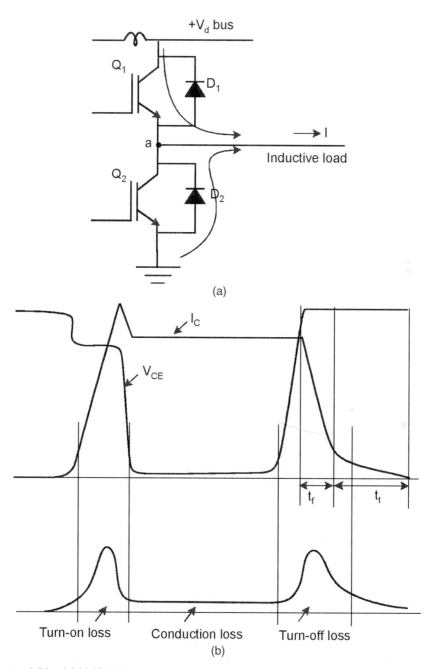

Figure 1.20 (a) Half-bridge converter configuration, (b) Typical switching characteristics of an IGBT

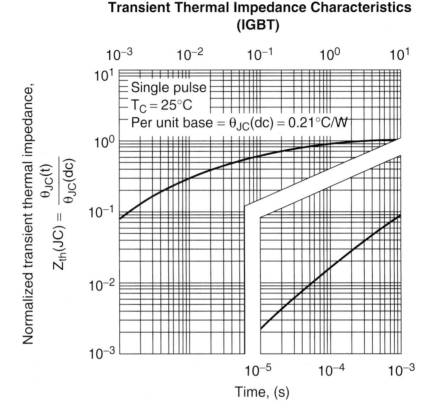

Figure 1.21 Transient thermal impedance characteristics of an IGBT (CM150TU-12H)

continuously improving. IGBT intelligent power modules (IPMs) are available with built-in gate drivers, controls, and protection for up to a several hundred kW power rating.

1.10 MOS-CONTROLLED THYRISTORS (MCTS)

An MOS-controlled thyristor (MCT), as the name indicates, is a thyristor-like, trigger-into-conduction hybrid device that can be turned on or off by a short voltage pulse on the MOS gate. The device has a microcell construction, where thousands of microdevices are connected in parallel on the same chip. The cell structure is somewhat complex. Figure 1.23 shows the equivalent circuit and symbol of the device. It is turned on by a negative voltage pulse at the gate with respect to the anode and is turned off by a positive voltage pulse. The MCT has a thyristor-like P-N-P-N structure, where the P-N-P and N-P-N transistor components are connected in regenerative feedback, as shown in the figure. However, unlike a thyristor, it has unipolar (or asymmetric) voltage-blocking capability. If the gate of an MCT is negative with respect to the anode, a P-channel is induced in the P-FET, which causes forward-biasing of the N-P-N transistor. This

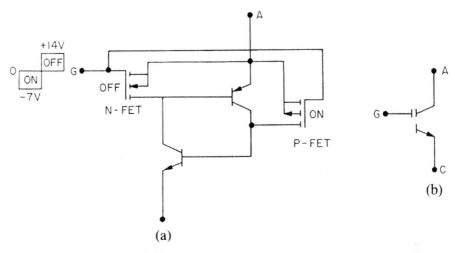

Figure 1.22 MCT symbol and equivalent circuit

also forward-biases the P-N-P transistor and the device goes into saturation by positive feedback effect. At conduction, the drop is around one volt (like a thyristor). If the gate is positive with respect to the anode, the N-FET will saturate and short-circuit the emitter-base junction of the P-N-P transistor. This will break the positive feedback loop for thyristor operation and the device will turn off. The turn-off occurs purely by recombination effect and therefore the tail time of the MCT is somewhat large. The device has a limited SOA, and therefore a snubber circuit is mandatory in an MCT converter. Recently, the device has been promoted for "soft-switched" converter applications (to be discussed in Chapter 5), where the SOA is not utilized. In spite of complex geometry, the current density of an MCT is high compared to a power MOSFET, BJT, and IGBT, and therefore it needs a smaller die area.

The MCT was commercially introduced in 1992, and currently, medium-power devices are available commercially. The future acceptance of the device remains uncertain at this point.

1.11 INTEGRATED GATE-COMMUTATED THYRISTORS (IGCTS)

The integrated gate-commutated thyristor (IGCT) is the newest member of the power semiconductor family at this time, and was introduced by ABB in 1997 [12]. Basically, it is a high-voltage, high-power, hard-driven, asymmetric-blocking GTO with unity turn-off current gain. This means that a 4500 V IGCT with a controllable anode current of 3000 A requires turn-off negative gate current of 3000 A. Such a gate current pulse of very short duration and very large di/dt has small energy content and can be supplied by multiple MOSFETS in parallel with ultra-low leakage inductance in the drive circuit. The gate drive circuit is built-in on the device module. The device is fabricated with a monolithically integrated anti-parallel diode. The conduction drop, turn-on di/dt, gate driver loss, minority carrier storage time, and turn-off dv/dt of the device are claimed to be superior to GTO. Faster switching of the device permits snubberless operation and higher-than-GTO switching frequency. Multiple IGCTs can be connected in series or in parallel for higher power applications. The device has been applied in power system intertie installations (100 MVA) and medium-power (up to 5 MW) industrial drives.

1.12 LARGE BAND-GAP MATERIALS FOR DEVICES

So far, all the power semiconductor devices discussed exclusively use silicon as the basic raw material, and this will possibly continue, at least in the near future. However, new types of large band-gap materials, such as silicon carbide and semiconducting diamond, are showing high promise for the future generation of power devices. Silicon carbide is particularly more promising because its technology is more "mature" than for diamond. The material has high carrier mobility, faster minority carrier lifetime, and high electrical and thermal conductivities. These properties permit high-voltage and high-power capabilities, fast switching (i.e., high switching frequency), low conduction drop, good radiation hardness, and high junction temperature. All key devices, such as diodes, power MOSFETs, thyristors, GTOs, etc. are possible with this new material.

Silicon carbide power MOSFETs with T_{Jmax} up to 350 °C appear particularly interesting as replacements for high-power silicon IGBTs in the future. High-voltage silicon carbide Schottky diodes with close to one-volt drop and negligible leakage and recovery current are available at this time.

1.13 POWER INTEGRATED CIRCUITS (PICS)

A discussion on power semiconductor devices is incomplete without some mention of power integrated circuits (PICs). In a PIC, the control and power electronics are generally integrated monolithically on the same chip. Loosely, a PIC is defined as "smart power." The motivations behind a PIC are reductions in size and cost, and improvement in reliability. The main problems in PIC fabrication are isolation between high-voltage and low-voltage devices and thermal management. A PIC is often differentiated from a high-voltage integrated circuit (HVIC), where the voltage is high but the current is small, that is, the loss is low. Low-voltage NMOS, CMOS, and bipolar devices can be conveniently integrated with MOS-gated power devices. Recently, a large family of PICs that includes power MOSFETs or IGBT smart switches, half-bridge inverter drivers, H-bridge inverters, two-phase step motor drivers, one-quadrant choppers for dc motor drives, three-phase brushless dc motor drivers, etc. has become available. Figure 1.24 shows a monolithic PIC (within the dotted rectangle) for driving a brushless dc motor. The simplified block diagram of the 40 V, 2 A PIC consists of a six-switch power stage, Hall sensor decoding logic, current-regulated PWM (pulse width modulated) control of the lower switches, and thermal/undervoltage protection features.

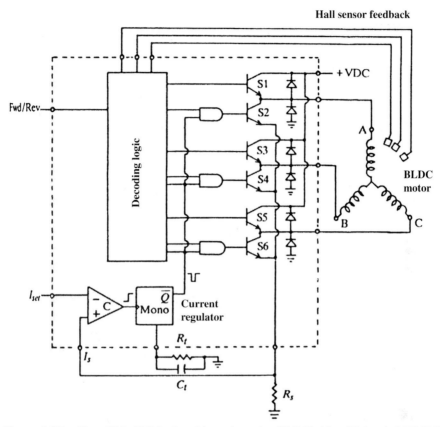

Figure 1.23 Monolithic PIC for brushless dc motor (BLDC) drive (Unitrode UC3620)

1.14 SUMMARY

This chapter has given a brief, but comprehensive review of the different types of power semiconductor devices that are used in power electronic systems. The devices covered include the diode, thyristor, triac, GTO, BJT, power MOSFET, IGBT, MCT, and IGCT. Power semiconductor devices constitute a vast and complex subject, and the technology is going through continuous evolution. Within the scope of this chapter, we only gave brief descriptions of the devices and their characteristics. It is needless to say that a power electronics engineer responsible for designing an apparatus should be thoroughly familiar with the different devices to achieve optimum cost and the performance goals of the system. Traditionally, thyristor-type devices have been very popular in power electronic systems and many applications can be found in this area. Therefore, these types of devices and their characteristics were described at the beginning of the chapter. MOS-gated devices, particularly power MOSFETs and IGBTs, have been applied extensively in recent years, and therefore these devices were covered in more detail. Although the BJT is now practically obsolete, and SIT and MCT devices are seldom used, these were briefly described for completeness of the subject. The IGCT is a recent member in the device family with good potential, and we are yet to see its growth of applications in competition with IGBTs and GTOs. Finally, large band-gap materials and PICs were discussed. It is interesting to see that the advent of new power semiconductor devices, growth of their power ratings, and improvement of their characteristics are continually driving the power electronics and motor drives technologies forward. Suffice it to say that if the device evolution would have stopped at the SCR level, the power electronics technology would have stalled hopelessly in the primitive stage. In the next several chapters, we will gradually develop applications of power semiconductor devices.

References

1. N. Mohan, T. M. Undeland, and W. P. Robins, *Power Electronics*, John Wiley, 1995.
2. B. W. Williams, *Power Electronics*, John Wiley, 1987.
3. A. M. Trzynadlowski, *Modern Power Electronics*, John Wiley, 1998.
4. B. K. Bose, "Evaluation of modern power semiconductor devices and future trends of converters", *IEEE Trans. on Ind. Appl.*, vol. 28, pp. 403–412, March/April 1992.
5. B. K. Bose, "Power electronics – a technology review", *Proceedings of the IEEE*, vol. 80, pp. 1303–1334, August 1992.
6. B. J. Baliga, "Power semiconductor devices for variable frequency drives", *Proc. of the IEEE*, vol. 82, pp. 1112–1121, August 1994.
7. *POWEREX Power Semiconductor Data Book*, vol. 3, August 1988.
8. *HARRIS Semiconductor MOSFETs*, 1994.
9. *POWEREX IGBT Intelligent Power Modules, Appl. and Data Book*, 1998.
10. B. J. Baliga, "Trends in power semiconductor devices", *IEEE Trans. on Elec. Devices*, vol. 43, pp. 1717–1731, October 1996.
11. V. A. K. Temple, "Power semiconductor devices: faster, smarter and more efficient", *Int'l. Power Elec. Conf. Rec.*, Yokohama, pp. 6–12, 1995.
12. P. K. Steimer et al., "IGCT – a new emerging technology for high power, low cost inverters", *IEEE IAS Annu. Meet. Conf. Rec.*, pp. 1592–1599, 1997.

CHAPTER 2

AC Machines for Drives

2.1 INTRODUCTION

The electrical machine that converts electrical energy into mechanical energy, and vice versa, is the workhorse in a drive system. Drive systems are widely used in applications such as pumps, fans, paper and textile mills, elevators, electric vehicle and subway transportation, home appliances, wind generation systems, servos and robotics, computer peripherals, steel and cement mills, ship propulsion, etc. A machine is a complex structure electrically, mechanically, and thermally. Although machines were introduced more than one hundred years ago, the research and development (R&D) in this area appears to be never-ending. However, the evolution of machines has been slow compared to that of power semiconductor devices and power electronic converters. An engineer designing a high-performance drive system must have intimate knowledge about machine performance, the dynamic model, and parameter variations. Industrial drive applications are generally classified into constant-speed and variable-speed drives. Traditionally, ac machines with a constant frequency sinusoidal power supply have been used in constant-speed applications, whereas dc machines were preferred for variable-speed drives. Dc machines have the disadvantages of higher cost, higher rotor inertia, and maintenance problems with commutators and brushes. Commutators and brushes, in addition, limit the machine speed and peak current, cause EMI problems, and do not permit a machine to operate in dirty and explosive environments. However, dc machine drive converters and controls are simple, and the machine torque response is very fast. Ac machines do not have the disadvantages of dc machines as mentioned above. In the last two or three decades, we have seen extensive research and development efforts for variable-frequency, variable-speed ac machine drive technology. Although currently, the majority of variable-speed drive applications use dc machines, they are progressively being replaced by ac drives. In most cases, new applications use ac drives.

Ac machines can generally be classified as follows:

- Induction machines
    Cage or wound rotor (doubly-fed)
    Rotating or linear
- Synchronous machines
    Rotating or linear
    Reluctance
    Wound field or permanent magnet
    Radial or axial gap (disk)
    Surface magnet or interior (buried) magnet
    Sinusoidal or trapezoidal
- Variable reluctance machines
    Switched reluctance
    Stepper

Ac machines for drives constitute a vast and complex subject. In this chapter, we will study the basic static and dynamic performance characteristics of induction and synchronous motors with particular relevance to variable-speed applications. The rotating radial-type machines that are most commonly used will be emphasized. Further details of machine properties will be covered in the later chapters.

2.2 INDUCTION MACHINES

Among all types of ac machines, the induction machine, particularly the cage type, is most commonly used in industry. These machines are very economical, rugged, and reliable, and are available in the ranges of fractional horse power (FHP) to multi-megawatt capacity. Low-power FHP machines are available in single-phase, but poly-phase (three-phase) machines are used most often in variable-speed drives.

Figure 2.1 shows an idealized three-phase, two-pole induction motor where each phase winding in the stator and rotor is represented by a concentrated coil. The three-phase windings, either in wye or delta form, are distributed sinusoidally and embedded in slots. In a wound-rotor machine, the rotor winding is similar to that of the stator, but in a cage machine, the rotor has a squirrel cage-like structure with shorted end rings. Basically, the machine can be looked upon as a three-phase transformer with a rotating and short-circuited secondary. Both stator and rotor cores are made with laminated ferromagnetic steel sheets. The air gap in the machine is practically uniform (non-salient pole).

2.2.1 Rotating Magnetic Field

One of the most fundamental principles of induction machines is the creation of a rotating and sinusoidally distributed magnetic field in the air gap. Neglecting the effect of slots and space harmonics due to nonideal winding distribution, it can be shown that a sinusoidal three-phase

Induction Machines

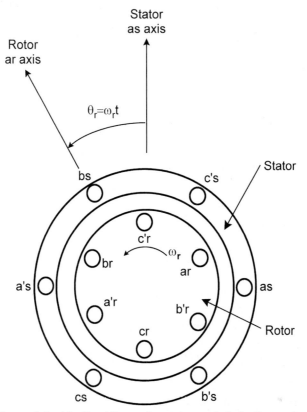

Figure 2.1 Idealized three-phase, two-pole induction motor

balanced power supply in the three-phase stator winding creates a synchronously rotating magnetic field. The derivation can be made either by a graphical or analytical method.

Consider that three-phase sinusoidal currents are impressed in the three-phase stator windings, which are given as

$$i_a = I_m \cos \omega_e t \tag{2.1}$$

$$i_b = I_m \cos\left(\omega_e t - \frac{2\pi}{3}\right) \tag{2.2}$$

$$i_c = I_m \cos\left(\omega_e t + \frac{2\pi}{3}\right) \tag{2.3}$$

Each phase winding will independently produce a sinusoidally distributed mmf (magnetomotive force) wave, which pulsates about the respective axes. Figure 2.2 shows the mmf waves at time $t = 0$ when $i_a = I_m$, $i_b = -I_m/2$, and $I_c = -I_m/2$.

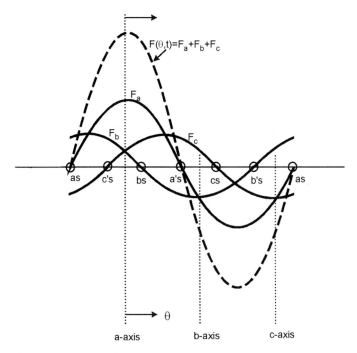

Figure 2.2 Mmf distribution in three-phase windings of stator

At spatial angle θ, the instantaneous mmf expressions can be given as

$$F_a(\theta) = Ni_a \cos\theta \tag{2.4}$$

$$F_b(\theta) = Ni_b \cos\left(\theta - \frac{2\pi}{3}\right) \tag{2.5}$$

$$F_c(\theta) = Ni_c \cos\left(\theta + \frac{2\pi}{3}\right) \tag{2.6}$$

where N = number of turns in a phase winding. Note that the mmf waves are phase-shifted in space by a $2\pi/3$ angle. The resultant mmf at angle θ is given as

$$\begin{aligned} F(\theta) &= F_a(\theta) + F_b(\theta) + F_c(\theta) \\ &= Ni_a \cos\theta + Ni_b \cos\left(\theta - \frac{2\pi}{3}\right) + Ni_c \cos\left(\theta + \frac{2\pi}{3}\right) \end{aligned} \tag{2.7}$$

Induction Machines

Substituting Equations (2.1) through (2.3) in (2.7) gives

$$F(\theta,t) = NI_m \left[\cos\omega_e t \cos\theta + \cos\left(\omega_e t - \frac{2\pi}{3}\right)\cos\left(\theta - \frac{2\pi}{3}\right) \right. \\ \left. + \cos\left(\omega_e t + \frac{2\pi}{3}\right)\cos\left(\theta + \frac{2\pi}{3}\right) \right]$$ (2.8)

Simplifying Equation (2.8), the $F(\theta, t)$ expression can be written as

$$F(\theta,t) = \frac{3}{2} NI_m \cos(\omega_e t - \theta)$$ (2.9)

Equation (2.9) indicates that a sinusoidally distributed mmf wave of peak value $3/2\ NI_m$ is rotating in the air gap at synchronous speed ω_e. In a two-pole machine, $F(\theta, t)$ makes one revolution per cycle of current variation. This means that for a P-pole machine, the rotational speed can be given as

$$N_e = \frac{120 f_e}{P}$$ (2.10)

where N_e = synchronous speed in rpm and $f_e = \omega_e/2\pi$ is the stator frequency in Hz.

2.2.2 Torque Production

If the rotor is initially stationary, its conductors will be subjected to a sweeping magnetic field, inducing current in the short-circuited rotor at the same frequency. The interaction of air gap flux and rotor mmf produces torque, as explained by the waveforms of Figure 2.3.

At synchronous speed of the machine, the rotor cannot have any induction, and therefore, torque cannot be produced. At any other speed Nr, the speed differential $N_e - N_r$, called slip speed, induces rotor current and torque is developed. The rotor moves in the same direction as that of the rotating magnetic field to reduce the induced current (Lenz's law). In fact, the rotor-induced magnetic poles, slipping with respect to the rotor, lock with the stator poles. The per unit slip S is defined as

$$S = \frac{N_e - N_r}{N_e} = \frac{\omega_e - \omega_r}{\omega_e} = \frac{\omega_{sl}}{\omega_e}$$ (2.11)

where ω_e = stator supply frequency (r/s), ω_r = rotor electrical speed (r/s), and ω_{sl} = slip frequency (r/s). The rotor mechanical speed is $\omega_m = (2/P)\ \omega_r$ (r/s), where P = number of poles of the machine. Evidently, the rotor voltage is induced at slip frequency, which correspondingly produces slip frequency current in the rotor. In Figure 2.3(a), the sinusoidal air gap flux density wave moving at speed ω_e induces voltage in the rotor conductors, as shown by the vertical lines. The resulting rotor current wave lags the voltage wave by the rotor power factor angle θ_r. The stepped rotor mmf wave (indicating the space harmonics) can be constructed from the current

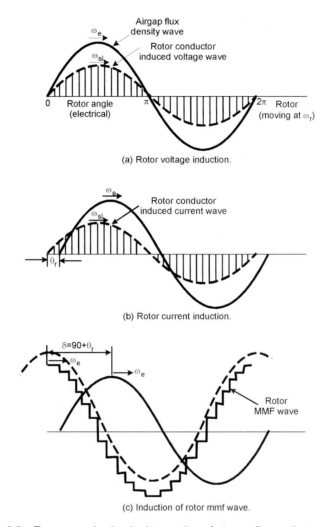

Figure 2.3 Torque production by interaction of air gap flux and rotor mmf waves

wave, which can be approximated by the dashed curve shown in the figure. Since the rotor is moving at speed ω_r and its current wave is moving at speed ω_{sl} relative to the rotor, the rotor mmf wave moves at the same speed as that of the air gap flux wave. The torque expression can be derived [1] as

$$T_e = \pi \left(\frac{P}{2}\right) lr B_p F_p \sin\delta \qquad (2.12)$$

where P = number of poles, l = axial length of the machine, r = machine radius, B_p = peak value of air gap flux density, F_p = peak value of rotor mmf, and $\delta = \pi/2 + \theta_r$ is defined as the torque angle. Other forms of torque expression will be given later.

2.2.3 Equivalent Circuit

A simple per phase equivalent circuit model of an induction motor is a very important tool for analysis and performance prediction at steady-state condition. Figure 2.4 shows the development of a per phase transformer-like equivalent circuit. The synchronously rotating air gap flux wave generates a counter emf (CEMF) V_m, which is then converted to slip voltage $V_r' = nSV_m$ in rotor phase, where n = rotor-to-stator turns ratio and S = per unit slip. The stator terminal voltage V_s differs from voltage V_m by the drops in stator resistance R_s and stator leakage inductance L_{ls}. The excitation current I_0 consists of two components: a core loss component $I_c = V_m/R_m$ and a magnetizing component $I_m = V_m/\omega_e L_m$, where R_m = equivalent resistance for core loss and L_m = magnetizing inductance. The rotor-induced voltage V_r' causes rotor current I_r' at slip frequency ω_{sl}, which is limited by the rotor resistance R_r' and the leakage reactance $\omega_{sl} L_{lr}'$. The stator current I_s consists of excitation component I_0 and the rotor-reflected current I_r. Figure 2.4(b) shows the equivalent circuit with respect to the stator, where I_r is given as

$$I_r = nI_r' = \frac{n^2 SV_m}{R_r' + j\omega_{sl} L_{lr}'} = \frac{V_m}{\left(\dfrac{R_r}{S}\right) + j\omega_e L_{lr}} \quad (2.13)$$

and parameters $R_r (= R_r'/n^2)$ and $L_{lr} (= L_{lr}'/n^2)$ are referred to the stator. At standstill, $S = 1$, and therefore, Figure 2.4(b) corresponds to the short-circuited transformer-equivalent circuit. At synchronous speed, $S = 0$, current $I_r = 0$ and the machine takes excitation current I_0 only. At any subsynchronous speed, $0 < S < 1.0$, and with a small value of S, the rotor current I_r is principally influenced by the R_r/S ($R_r/S \gg \omega_e L_{lr}$) parameter.

The phasor diagram for the equivalent circuit in Figure 2.4(b) is shown in Figure 2.5, where all the variables are in rms values.

The torque expression can be given in the form

$$T_e = \frac{3}{2}\left(\frac{P}{2}\right)\hat{\psi}_m \hat{I}_r \sin\delta \quad (2.14)$$

where $\hat{\psi}_m$ = peak value of air gap flux linkage/pole and $\hat{I}_r$ = peak value of rotor current.

2.2.3.1 Equivalent Circuit Analysis

The various power expressions can be written from the equivalent circuit of Figure 2.4(b) as follows:

Input power: $\quad P_{in} = 3V_s I_s \cos\phi \quad (2.15)$

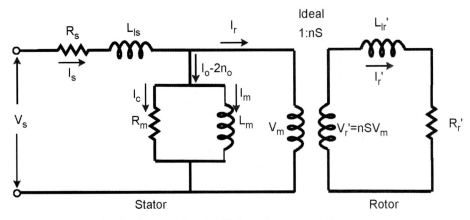

(a) Equivalent circuit with transformer coupling.

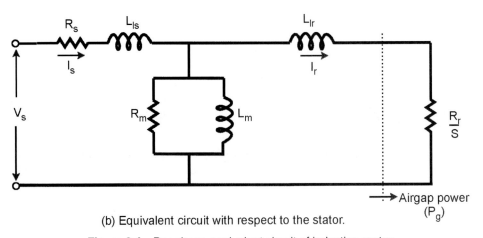

(b) Equivalent circuit with respect to the stator.

Figure 2.4 Per phase equivalent circuit of induction motor

Stator copper loss: $\qquad P_{ls} = 3I_s^2 R_s \qquad$ (2.16)

Core loss: $\qquad P_{lc} = \dfrac{3V_m^2}{R_m} \qquad$ (2.17)

Power across air gap: $\qquad P_g = 3I_r^2 \dfrac{R_r}{S} \qquad$ (2.18)

Induction Machines

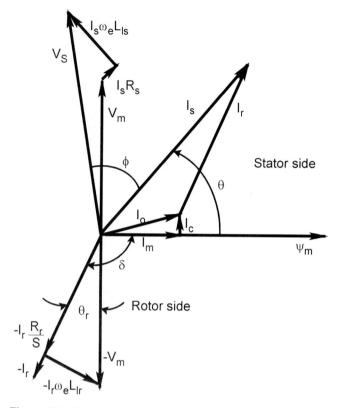

Figure 2.5 Phasor diagram for equivalent circuit of Figure 2.4(b)

Rotor copper loss:
$$P_{lr} = 3I_r^2 R_r \tag{2.19}$$

Output power:
$$P_o = P_g - P_{lr} = 3I_r^2 R_r \frac{1-S}{S} \tag{2.20}$$

Shaft power:
$$P_{sh} = P_o - P_{FW} \tag{2.21}$$

where $\cos \varphi$ = input power factor and P_{FW} = friction and windage loss of the machine. Since the output power is the product of developed torque T_e and speed ω_m, T_e can be expressed as

$$T_e = \frac{P_o}{\omega_m} = \frac{3}{\omega_m} I_r^2 R_r \frac{1-S}{S} = 3\left(\frac{P}{2}\right) I_r^2 \frac{R_r}{S\omega_e} \tag{2.22}$$

where $\omega_m = (2/P)\omega_r = (2/P)(1-S)\omega_e$ is the rotor mechanical speed (r/s). Substituting Equation (2.18) in (2.22) yields

$$T_e = \left(\frac{P}{2}\right)\frac{P_g}{\omega_e} \tag{2.23}$$

which indicates that torque can be calculated from the air gap power by knowing the stator frequency. The power P_g is often defined as torque in synchronous watts. Again, neglecting the core loss, we can write

$$P_g = 3V_m I_s \sin\theta \tag{2.24}$$

where

$$V_m = \omega_e \psi_m \tag{2.25}$$

$$\psi_m = L_m I_m \tag{2.26}$$

and

$$I_s \sin\theta = I_r \sin\delta \tag{2.27}$$

Substituting Equations (2.24) through (2.27) in (2.23), we can write torque expressions in the following forms:

$$T_e = 3\left(\frac{P}{2}\right)\psi_m I_r \sin\delta \tag{2.28}$$

$$= \frac{3}{2}\left(\frac{P}{2}\right)\hat{\psi}_m \hat{I}_r \sin\delta \tag{2.29}$$

$$= 3\left(\frac{P}{2}\right)L_m I_m I_a \tag{2.30}$$

where $\hat{\psi}_m$ and $\hat{I}_r$ are the peak values given by $\sqrt{2}\psi_m$ and $\sqrt{2}I_r$, respectively, and $I_a = I_r \sin\delta$. Equation (2.29) verifies the same as Equation (2.14). The torque expression (2.30) is analogous to that of a dc machine, where I_m = magnetizing or flux component of stator current, I_a = armature or torque component of stator current, and $3(P/2)L_m$ = torque constant. Note that I_m and I_a are orthogonal, or mutually decoupled.

The equivalent circuit of Figure 2.4(b) can be simplified to that shown in Figure 2.6, where the core loss resistor R_m has been dropped and the magnetizing inductance L_m has been shifted at the input. This approximation is easily justified for an integral horsepower machine

Induction Machines

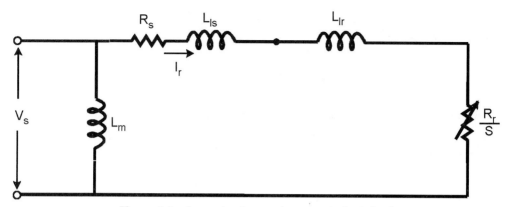

Figure 2.6 Approximate per phase equivalent circuit

where $|(R_s + j\omega_e L_{ls})| \ll \omega_e L_m$. The performance prediction by the simplified circuit typically varies within 5 percent from that of the actual machine.

In Figure 2.6, the magnitude of current I_r can be solved as

$$I_r = \frac{V_s}{\sqrt{(R_s + R_r/S)^2 + \omega_e^2(L_{ls} + L_{lr})^2}} \tag{2.31}$$

Substituting Equation (2.31) in (2.22) yields

$$T_e = 3\left(\frac{P}{2}\right)\frac{R_r}{S\omega_e} \cdot \frac{V_s^2}{(R_s + R_r/S)^2 + \omega_e^2(L_{ls} + L_{lr})^2} \tag{2.32}$$

Equation (2.32) is a function of slip S for constant frequency and supply voltage.

2.2.4 Torque-Speed Curve

The torque T_e can be calculated as a function of slip S from Equation (2.32). Figure 2.7 shows the torque-speed ($\omega_r/\omega_e = 1 - S$) curve, where the value of the slip is extended beyond the region $0 < S < 1.0$. The zones can be defined as plugging ($1.0 < S < 2.0$), motoring ($0 < S < 1.0$), and regenerating ($S < 0$). In the normal motoring region, $T_e = 0$ at $S = 0$, and as S increases (i.e., speed decreases), T_e increases in a quasi-linear curve until breakdown, or maximum torque T_{em} is reached. In this region, the stator drop is small and air gap flux remains approximately constant. Beyond the breakdown torque, T_e decreases with the increase of S. The machine starting torque T_{es} at $S = 1$ can be written from Equation (2.32) as

$$T_{es} = 3\left(\frac{P}{2}\right)\frac{R_r}{\omega_e} \cdot \frac{V_s^2}{(R_s + R_r)^2 + \omega_e^2(L_{ls} + L_{lr})^2} \tag{2.33}$$

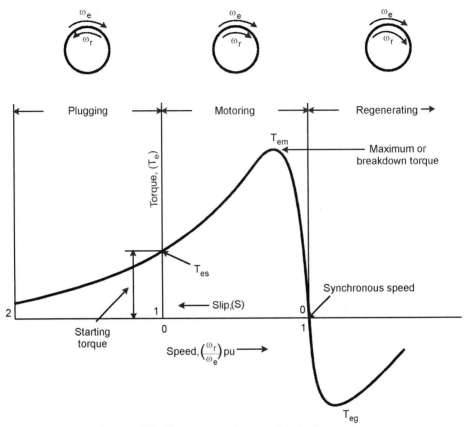

Figure 2.7 Torque-speed curve of induction motor

In the plugging region, the rotor rotates in the opposite direction to that of the air gap flux so that $S > 1$. This condition may arise if the stator supply phase sequence is reversed when the rotor is moving, or because of an overhauling type of load which drives the rotor in the opposite direction. Since the torque is positive but the speed is negative, the plugging torque appears as braking torque. However, the energy due to the plugging torque is dissipated within the machine, causing excessive machine heating. In the regenerating region, as the name indicates, the machine acts as a generator. The rotor moves at supersynchronous speed in the same direction as that of the air gap flux so that the slip becomes negative, creating negative, or regeneration torque. The negative slip corresponds to negative equivalent resistance R_r/S in Figure 2.6. The positive resistance R_r/S consumes energy during motoring, but the negative R_r/S generates energy and supplies it back to the source. With a variable-frequency power supply, the machine stator frequency can be controlled to be lower than the rotor speed ($\omega_e < \omega_r$) to obtain a regenerative braking effect. An induction motor can, of course, continually operate as a generator (induction generator) if its shaft is rotated at supersynchronous speed by a prime mover, such as a wind turbine.

Induction Machines

If Equation (2.32) is differentiated with respect to S and equated to zero, then

$$S_m = \pm \frac{R_r}{\sqrt{R_s^2 + \omega_e^2 (L_{ls} + L_{lr})^2}} \quad (2.34)$$

where S_m is the slip corresponding to breakdown torque T_{em}. Substituting the $+S_m$ expression in (2.32), the motoring breakdown torque is

$$T_{em} = \frac{3}{4} \frac{P}{\omega_e} \frac{V_s^2}{\sqrt{R_s^2 + \omega_e^2 (L_{ls} + L_{lr})^2} + R_s} \quad (2.35)$$

and by substituting the $-S_m$, the regenerative breakdown torque is

$$T_{eg} = -\frac{3}{4} \frac{P}{\omega_e} \frac{V_s^2}{\sqrt{R_s^2 + \omega_e^2 (L_{ls} + L_{lr})^2} - R_s} \quad (2.36)$$

As expected, $|T_{em}| = |T_{eg}|$ if stator resistance R_s is neglected. A further simplification of the equivalent circuit of Figure 2.6 can be made by neglecting the stator parameters R_s and L_{ls}. This assumption is not unreasonable for an integral horsepower machine, particularly if the speed is typically above 10 percent. Then, the Equation (2.32) can be simplified as

$$T_e = 3\left(\frac{P}{2}\right)\left(\frac{V_s}{\omega_e}\right)^2 \frac{\omega_{sl} R_r}{R_r^2 + \omega_{sl}^2 L_{lr}^2} \quad (2.37)$$

Equation (2.37) can be shown to be the same as (2.28) by substituting the following relations:

$$I_r = \frac{V_s}{\sqrt{(R_r/S)^2 + \omega_e^2 L_{lr}^2}} \quad (2.38)$$

$$\cos\theta_r = -\sin\delta = \frac{R_r/S}{\sqrt{(R_r/S)^2 + \omega_e^2 L_{lr}^2}} \quad (2.39)$$

From Figure 2.6 (neglecting R_s and L_{ls}) and recognizing that the air gap flux can be given by

$$\psi_m = V_s / \omega_e \quad (2.40)$$

in a low-slip region, (2.37) can be approximated as

$$T_e = 3\left(\frac{P}{2}\right)\frac{1}{R_r} \psi_m^2 \omega_{sl} \quad (2.41)$$

where $R_r^2 \gg \omega_{sl}^2 L_{lr}^2$. Equation (2.41) is very important. It indicates that at constant flux ψ_m, the torque T_e is proportional to ω_{sl}, or at constant ω_{sl}, T_e is proportional to ψ_m^2.

2.2.5 NEMA Classification of Machines

The National Electrical Manufacturers Association (NEMA) of the U.S. has classified cage-type inducton machines into different categories to meet the diversified application needs of the industry. These are characterized by torque-speed curves, as shown in Figure 2.8. The most significant machine parameter in this classification is the effective rotor resistance. Class A machines are characterized by somewhat low starting torque, high starting current, and low operating slip. A Class A machine has low rotor resistance, and therefore, operating efficiency is high at low slip. Class B machines are most commonly used for constant-speed industrial drives. The starting torque, starting current, and breakdown torque of a Class B machine are somewhat lower than those of a Class A machine, but a Class B machine has somewhat higher slip characteristics. The machine is designed with higher rotor leakage inductance. Both Class C and Class D machines are characterized by higher starting torque and lower starting current due to higher rotor resistance. Most recently, high-efficiency Class E-type machines have been introduced.

2.2.6 Variable-Voltage, Constant-Frequency Operation

A very simple and economical method of controlling speed in a cage-type induction motor is to vary the stator voltage at constant supply frequency. The three-phase stator voltage at line frequency can be controlled by the firing angle control of anti-parallel thyristors connected to

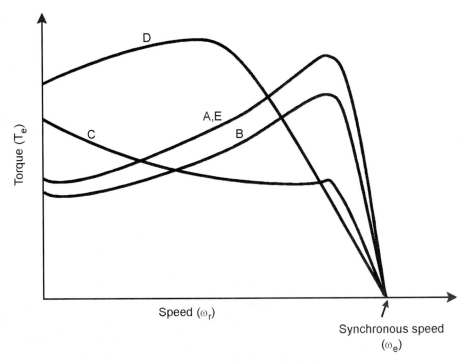

Figure 2.8 NEMA classification of induction motors

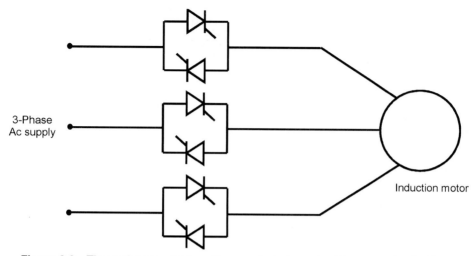

Figure 2.9 Three-phase variable voltage control at constant frequency by thyristors

each phase, as shown in Figure 2.9. This type of circuit has been used extensively as a solid-state "soft starter" for constant-speed induction motors, where the stator voltage is applied gradually to limit the stator current.

Figure 2.10 shows the torque-speed curves with variable stator voltage which have been plotted from Equation (2.32). A load-torque curve for a pump or fan-type drive ($T_L = k\omega_r^2$) is also shown in the figure, where the points of intersection define stable points for variable-speed operation. The motors of NEMA Class D category are generally suitable for this type of speed control, which gives high rotor copper loss and correspondingly low machine efficiency. With a low-slip machine, such as Class B, or with a constant torque-type load, the range of speed control will evidently be diminished. On the other hand, if the machine is designed to have $S_m > 1$, a constant- or variable-load torque-type load can be controlled in the full range of speed. Since the air gap flux is reduced at lower supply voltage, the stator current tends to be excessive at low speeds, giving high copper loss. In addition, distorted phase current in the machine and line and a poor line power factor are also disadvantages. Often, single-phase, low-power, appliance-type motor drives where the efficiency is not an important consideration use this type of drive for speed control.

2.2.7 Variable-Frequency Operation

If the stator frequency of a machine is increased beyond the rated value, but the voltage is kept constant, the torque-speed curves derived from Equation (2.32), can be plotted as shown in Figure 2.11. The air gap flux and rotor current decrease as the frequency increases, and corre-

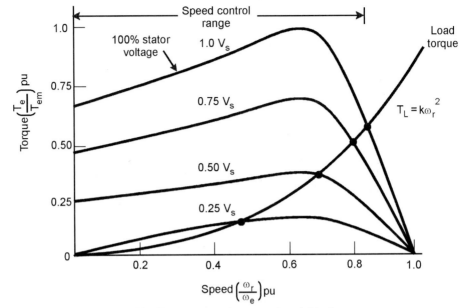

Figure 2.10 Torque-speed curves at variable frequency

spondingly, the developed torque also decreases. The breakdown torque as a function of slip (at constant frequency) can be derived by differentiating Equation (2.37) as

$$T_{em} = 3\left(\frac{P}{2}\right)\left(\frac{V_s}{\omega_e}\right)^2 \frac{\omega_{slm} R_r}{R_r^2 + \omega_{slm}^2 L_{lr}^2} \qquad (2.42)$$

where $\omega_{slm} = R_r/L_{lr}$ is the slip frequency at maximum torque. The equation shows that $T_{em}\omega_e^2 = $ constant (i.e., the machine behaves like a dc series motor in variable-frequency operation).

2.2.8 Constant Volts/Hz Operation

If an attempt is made to reduce the supply frequency at the rated supply voltage, the air gap flux ψ_m will tend to saturate, causing excessive stator current and distortion of flux wave. Therefore, the region below the base or rated frequency (ω_b) should be accompanied by the proportional reduction of stator voltage so as to maintain the air gap flux constant. Figure 2.12 shows the plot of torque-speed curves at volts/Hz = constant. Note that the breakdown torque T_{em} given by Equation (2.42) remains approximately valid, except in the low-frequency region where the air gap flux is reduced by the stator impedance drop ($V_m < V_s$). Therefore, in this region, the stator drop must be compensated by an additional boost voltage so as to restore the T_{em} value, as shown in Figure 2.12.

If the air gap flux of the machine is kept constant (like a dc shunt motor) in the constant torque region, as indicated in Figure 2.12, it can be shown that the torque sensitivity per ampere

Induction Machines

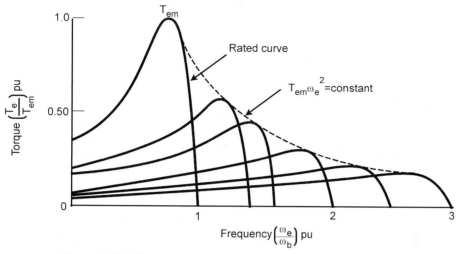

Figure 2.11 Torque-speed curves with variable stator voltage

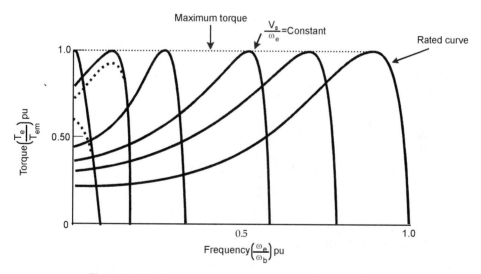

Figure 2.12 Torque-speed curves at constant volts/Hz

of stator current is high, permitting fast transient response of the drive with stator current control. In variable-frequency, variable-voltage operation of a drive system, the machine usually has low slip characteristics (i.e., low rotor resistance), giving high efficiency. In spite of the low inherent starting torque for base frequency operation, the machine can always be started at maximum torque, as indicated in Figure 2.12. The absence of a high in-rush starting current in a direct-start drive reduces stress and therefore improves the effective life of the machine. By far, the majority of variable-speed ac drives operate with a variable-frequency, variable-voltage power supply. This will be discussed further in Chapters 8 and 9.

2.2.9 Drive Operating Regions

The different operating regions of torque-speed curves for a variable-speed drive system with a variable-frequency, variable-voltage supply are shown in Figure 2.13, and the corresponding voltage-frequency relation is shown in Figure 2.14. Figure 2.14 also shows torque, stator current, and slip as functions of frequency. The inverter maximum, but short-time or transient torque capability, is limited by the peak inverter current and is somewhat lower than the machine torque capability (shown by the dotted profile in Figure 2.13). The margin permits machine breakdown torque variation by a variation of machine parameters.

The drive operating point can be anywhere within the inverter torque envelope. Since the inverter is more expensive than the machine, this margin is not too uneconomical. The steady-state torque envelope, further limited by the power semiconductor junction temperature T_J, is indicated in Figure 2.13.

Typically, the inverter peak or short-time limit torque is 50 percent higher than steady-state torque for 60-seconds duration (NEMA standard). For a longer time duration of the transient, this peak torque must be reduced. Obviously, with improved cooling system design, the steady-state torque envelope can be increased.

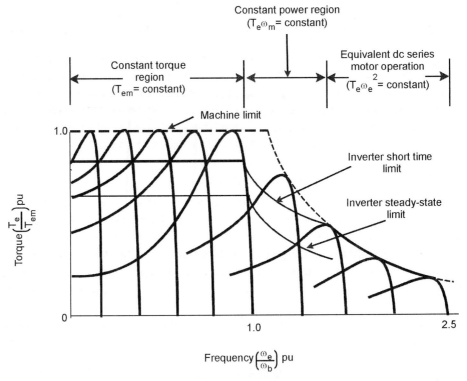

Figure 2.13 Torque-speed curves at variable voltage and variable frequency up to field-weakening region

Induction Machines

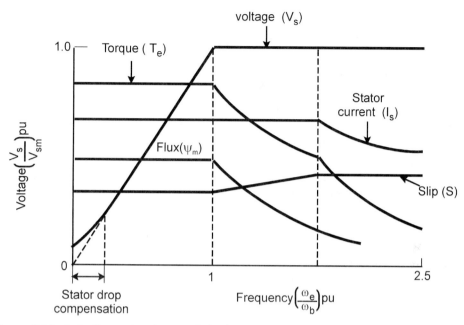

Figure 2.14 Induction motor characteristics in constant torque and field-weakening regions

At the right edge of the constant torque region, the stator voltage reaches its full or rated value, and then the machine enters into the constant power region at higher frequency. In this region, the air gap flux decreases, but the stator current is maintained constant by increasing the slip. This is equivalent to the field-weakening mode of operation of a dc separately excited motor. Note that stable operation at any operating point within the envelope can be obtained by controlling the voltage and frequency and will be discussed in Chapter 8.

2.2.10 Variable Stator Current Operation

Instead of controlling the stator voltage, the stator current can be controlled directly to control the developed torque. Figure 2.15 shows the simplified per phase equivalent circuit with a variable-stator current source. Since the Thevenin impedance is large for a current source, the stator circuit impedance has been neglected. With current control, the developed torque depends on the relative distribution of magnetizing current and rotor current, which are both affected by the inverse ratio of parallel circuit impedances, which in turn are dependent on frequency and slip.

Neglecting the rotor leakage reactance ($\omega_e L_{lr} \ll R_r/S$) and core loss, the distribution of currents can be given as

$$I_m = \frac{R_r/S}{\sqrt{(R_r/S)^2 + \omega_e^2 L_m^2}} I_s \qquad (2.43)$$

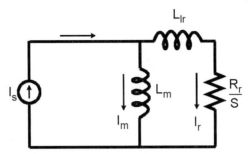

Figure 2.15 Simplified equivalent circuit for variable stator current operation

$$I_r = I_a = \frac{\omega_e L_m}{\sqrt{(R_r/S)^2 + \omega_e^2 L_m^2}} I_s \qquad (2.44)$$

where I_a = active or in-phase component of stator current. Substituting Equations (2.43) and (2.44) in (2.30), the torque expression is

$$T_e = K' I_s^2 \frac{S\omega_e}{R_r^2 + S^2 \omega_e^2 L_m^2} \qquad (2.45)$$

where

$$K' = 3\left(\frac{P}{2}\right) R_r L_m^2$$

Equation (2.45) gives torque as a function of stator current, frequency, and slip. Typical motor torque-speed curves at different stator currents but fixed frequencies are shown in Figure 2.16. If, for example, the machine is operated at rated current (I_s = 1.0 pu), the starting torque will be very low compared to that of a voltage-fed machine at V_s = 1.0 pu. The reason is that the air gap flux will be very low due to the rotor's short-circuiting effect. As the speed increases (i.e., the slip decreases), the stator voltage increases, and as a result, the torque increases with higher air gap flux. If saturation of the machine is neglected, the torque increases to a high value, as shown by the dashed line, and then decreases to zero with steep slope at synchronous speed. In a practical machine, however, the saturation will limit the developed torque, as shown by the solid line. A torque curve with rated stator voltage is superimposed in Figure 2.16, where the part with the negative slope can be considered to have stable operation with the rated air gap flux. This curve intersects the I_s = 1.0 pu torque curve at point A. Theoretically, the machine can be operated either at A or B for the same torque demand. Because of lower slip at point B, the rotor current will be lower and the gap flux will be somewhat higher, causing partial saturation. This will result in higher core loss and harmonic torque pulsation, which will be discussed later. The stator copper loss is the same at A and B, but the rotor copper loss is somewhat higher at A. Overall, operation at A is more desirable. However, since A is in the unstable region of the

Induction Machines

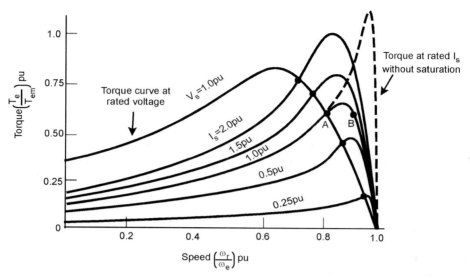

Figure 2.16 Torque-speed curves with variable stator current

torque curve, effective close-loop stable operation of the machine is mandatory. The developed torque can be varied by varying the stator current and slip so that the air gap flux remains constant (i.e., the locus shown by the dots is on the negative slope of the equivalent voltage-generated torque curve). Note that for constant flux, slip increases with higher stator current. For variable-frequency, variable-current operation, similar families of torque-speed curves can be described on the torque-frequency plane, as shown in Figure 2.17.

2.2.11 The Effect of Harmonics

Although machines like to have a power supply with sinusoidal voltage or current waves, the practical converter-fed power supply using a matrix of switches is hardly sinusoidal. The waveforms are generally either pulse width modulated (PWM) or square and will be discussed in Chapter 5. Such waveforms can be analyzed by Fourier series and are shown to have a fundamental or useful component and undesirable harmonic components. The harmonics generally have two undesirable effects:

- Harmonic heating
- Torque pulsation

2.2.11.1 Harmonic Heating

For simplicity, here we will consider only the balanced three-phase, square-wave voltage supply (see Figure 5.9). Analysis with square-wave current supply will be similar. However, the analysis with the PWM wave is somewhat involved and should preferably be done with the help of a computer program.

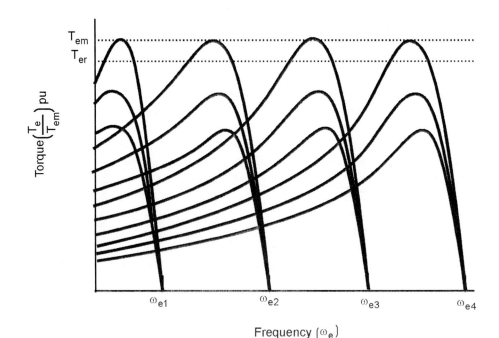

Figure 2.17 Torque-speed curves with variable stator current and variable frequency

For a balanced, three-phase, square-wave voltage supply, only the odd harmonics should be considered. Again, the third and its multiple (triplen) harmonics are in the same phase (cophasal—often defined as zero sequence component), and therefore cannot cause any triplen harmonic current in a wye- or delta-connected machine winding with isolated neutral. Considering only the non-triplen harmonics, the Fourier series of the phase voltages can be given as

$$v_{as} = V_{1m} \sin\omega_e t + V_{5m} \sin 5\omega_e t + V_{7m} \sin 7\omega_e t + \cdots \qquad (2.46)$$

$$v_{bs} = V_{1m} \sin\left(\omega_e t - \frac{2\pi}{3}\right) + V_{5m} \sin 5\left(\omega_e t - \frac{2\pi}{3}\right) + V_{7m} \sin 7\left(\omega_e t - \frac{2\pi}{3}\right) + \cdots \qquad (2.47)$$

$$v_{cs} = V_{1m} \sin\left(\omega_e t + \frac{2\pi}{3}\right) + V_{5m} \sin 5\left(\omega_e t + \frac{2\pi}{3}\right) + V_{7m} \sin 7\left(\omega_e t + \frac{2\pi}{3}\right) + \cdots \qquad (2.48)$$

where V_{1m}, V_{5m}, etc. are the peak values. Equations (2.47) and (2.48) can be simplified as

$$v_{bs} = V_{1m} \sin\left(\omega_e t - \frac{2\pi}{3}\right) + V_{5m} \sin\left(5\omega_e t + \frac{2\pi}{3}\right) + V_{7m} \sin\left(7\omega_e t - \frac{2\pi}{3}\right) + \cdots \qquad (2.49)$$

$$v_{cs} = V_{1m} \sin\left(\omega_e t + \frac{2\pi}{3}\right) + V_{5m} \sin\left(5\omega_e t - \frac{2\pi}{3}\right) + V_{7m} \sin\left(7\omega_e t + \frac{2\pi}{3}\right) + \cdots \qquad (2.50)$$

For each harmonic component, the machine can be approximately represented by a constant-parameter, linear-equivalent circuit and the resultant ripple current can be solved by the superposition principle. The per phase equivalent circuit of Figure 2.6 can be converted to a harmonic-equivalent circuit as shown in Figure 2.18 where the core loss resistor R_m has been omitted. In fact, there will be some skin effect increase of rotor resistance with harmonics which will also be ignored. In this figure, n is the order of the harmonic and S_n is the slip at the nth harmonic. Equations (2.46), (2.49), and (2.50) indicate that the 5th harmonic voltage component has negative phase sequence, whereas the 7th harmonic has positive phase sequence like the fundamental component. Obviously, the 5th harmonic will create air gap flux which will rotate in the backward direction, whereas the fundamental and the 7th harmonic will create flux that will rotate in the forward direction. The effect of higher harmonics can be derived in a similar manner. Since rotor speed is related to fundamental frequency only, the rotor appears practically stationary with respect to a fast-moving harmonic field, that is, $S_n \simeq 1.0$.

Mathematically, the slip at the nth harmonic field can be given as

$$S_n = \frac{n\omega_e \mp \omega_r}{n\omega_e} \qquad (2.51)$$

where negative and positive signs relate to forward- and backward-rotating fields, respectively. Substituting $\omega_r/\omega_e = (1 - S)$ in Equation (2.51) and simplifying yields

$$S_n = \frac{(n-1) + S_1}{n} \qquad (2.52)$$

$$S_n = \frac{(n+1) - S_1}{n} \qquad (2.53)$$

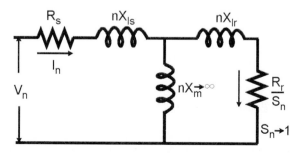

Figure 2.18 Per phase equivalent circuit for harmonics calculation

corresponding to forward- and backward-rotating fields, respectively. In these equations, S_1 is the fundamental frequency slip. For example, if S_1 varies from 0 to 1, S_5 and S_7 will vary in ranges from 1.2 to 1.0 and 0.857 to 1.0, respectively. Note that the harmonic currents are not influenced by the fundamental frequency operating condition (i.e., they are independent of the torque and speed of the machine). Assuming that nX_m approaches infinity and $(nX_{ls} + nX_{lr}) \gg (R_s + R_r)$,

$$I_n \simeq \frac{V_n}{n(X_{ls} + X_{lr})} \tag{2.54}$$

where V_n and I_n are the nth harmonic rms voltage and current, respectively, the corresponding expression of rms harmonic ripple current I_h is

$$I_h = \sqrt{I_5^2 + I_7^2 + I_{11}^2 + I_{13}^2 + \cdots} \\ = \sqrt{\sum_{n=5,7,\ldots} I_n^2} \tag{2.55}$$

where I_5, I_7, etc. are the rms harmonic components of the current. The total stator and rotor copper losses due to fundamental and ripple currents can be given as

$$P_{ls} = 3(I_{sl}^2 + I_h^2)R_s \tag{2.56}$$

$$P_{lr} = 3(I_{rl}^2 + I_h^2)R_r \tag{2.57}$$

where I_{sl} and I_{rl} are the fundamental stator and rotor currents, respectively.

The core loss in the machine created by time-varying flux in the machine laminations consists of hysteresis and eddy current losses. This loss is contributed by the fundamental as well as harmonic components of the supply voltage. The per phase stator core loss expression at fundamental frequency can be given as

$$P_{cs} \simeq K_h \omega_e \psi_m^2 + K_e \omega_e^2 \psi_m^2 \tag{2.58}$$

where ψ_m = fundamental air gap flux linkage, K_h = hysteresis loss coefficient, and K_e = eddy current loss coefficient. The parameter K_h depends on the magnetic properties of iron and machine size, and K_e depends on machine geometry, size, lamination thickness, and resistivity of iron. Under normal operating conditions, the fundamental rotor frequency $S\omega_e$ is a small fraction of the stator frequency and the corresponding rotor core loss P_{cr} is given by

$$P_{cr} = K_h S\omega_e \psi_m^2 + K_e (S\omega_e)^2 \psi_m^2 \tag{2.59}$$

where $S\omega_e$ has been substituted for ω_e. Obviously, the rotor core loss is small compared to that of the stator. Substituting Equation (2.40) in (2.59), the total core loss is given as

$$P_c = P_{cs} + P_{cr} = \left[K_h\left(\frac{1+S}{\omega_e}\right) + K_e(1+S^2)\right]V_m^2 = \frac{V_m^2}{R_m} \tag{2.60}$$

Induction Machines

therefore, the equivalent core loss resistance can be written as

$$R_m = \frac{1}{K_h\left(\frac{1+S}{\omega_e}\right) + K_e(1+S^2)} \tag{2.61}$$

Assuming that the core losses due to harmonic fluxes are governed by the same principles that determine the fundamental core losses, the coefficients K_h and K_e remain the same at harmonic frequency. Since the harmonic slip $S_n \cong 1$, the equivalent core loss resistance R_{mn} at harmonic frequency $n\omega_e$ can be given from (2.61) as

$$R_{mn} = \frac{1}{\left(\frac{K_h}{n\omega_e} + K_e\right)} \tag{2.62}$$

The superposition principle can be applied to find the total core loss. In a variable-frequency drive, harmonic core loss is normally smaller than harmonic copper loss.

In addition to copper and core losses, there are stray losses in the machine. By definition, stray losses are those losses that occur in excess of copper loss, no-load core loss, and friction and windage loss. These losses are essentially due to hysteresis and eddy current losses induced by stator and rotor leakage fluxes. Approximate expression of stray loss can be derived in a similar manner as core loss modeling.

2.2.11.2 Machine Parameter Variation

It is needless to say that the machine parameters in the equivalent circuit hardly remain constant during operating condition. Both stator and rotor resistances increase linearly with temperature, depending on the temperature coefficient of the resistance of the material. Besides, there is skin effect due to harmonics. The skin effect causes current crowding on the conductor surface, which causes an increase of resistance, but a decrease of leakage inductance. The skin effect is negligible on the stator winding, but its effect is dominant on the rotor bars. The magnetizing inductance is subjected to saturation with higher magnetizing current. Both the stator and rotor leakage inductances can also have some amount of saturation at higher currents.

2.2.11.3 Torque Pulsation

Pulsating torque is produced by air gap flux at one frequency interacting with rotor mmf at a different frequency. However, like frequency air gap flux and rotor mmf waves produce unidirectional torque components. The general torque expression as a function of air gap flux, rotor current, and the phase angle between the air gap flux and rotor current is given in Equation (2.28a). Considering only the fundamental frequency, or any harmonic frequency alone, the angle δ

remains fixed and therefore only unidirectional torque is produced. A harmonic component of air gap flux induces rotor current at the same frequency, and therefore torque is developed in the same direction as that of the rotating flux. For example, the 7th harmonic frequency torque will add with the fundamental frequency torque, but the 5th harmonic frequency torque will oppose it.

The torque pulsation will occur when angle δ varies with time. This occurs, as mentioned before, when ψ_m of one frequency interacts with I_r of another frequency, modulating δ at a rate which is the difference between speeds of the corresponding rotating phasors. The pulsating torques can be calculated by superimposing the flux and rotor current phasors of various frequencies on a single diagram, as shown in Figure 2.19(a). In the figure, the effects of only the fundamental, 5th, and 7th harmonic voltages are considered, and the flux phasors are assumed to be cophasal at an instant $t = 0$. Each harmonic voltage will produce the corresponding flux and rotor current components as shown in the figure. The equivalent circuit resistances for the 5th and 7th harmonics are neglected [$(R_s + R_r) \ll n(X_{ls} + X_{lr})$], and therefore, harmonic currents will lag the respective flux components by angle π (see Figure 2.5). The fundamental and 7th harmonic phasors rotate in the counter-clockwise direction at speeds ω_e and $7\omega_e$, respectively, whereas the 5th harmonic phasors rotate in the clockwise direction at speed $5\omega_e$.

Figure 2.19(b) has been constructed from Figure 2.19(a) by giving the whole diagram a clockwise rotation at ω_e to make the fundamental phasors stationary. From the diagram, it can be seen that the 6th harmonic torque is contributed by the interaction of fundamental flux with the 5th and 7th harmonic currents, and of the fundamental current with the 5th and 7th harmonic fluxes. Mathematically, the 6th harmonic torque expression can be written as

$$T_{e6} = K[\psi_{1m}I_{7r}\sin(\pi - 6\omega_e t) + \psi_{7m}I_{1r}\sin(\delta + 6\omega_e t) \\ + \psi_{1m}I_{5r}\sin(\pi + 6\omega_e t) + \psi_{5m}I_{1r}\sin(\delta - 6\omega_e t)] \quad (2.63) \\ \simeq K[\psi_{1m}(I_{7r} - I_{5r})\sin 6\omega_e t + I_{1r}(\psi_{7m} + \psi_{5m})\cos 6\omega_e t]$$

$$\simeq K\psi_{1m}(I_{7r} - I_{5r})\sin 6\omega_e t \quad (2.64)$$

whereas the fundamental component of torque is given as

$$T_{e1} = K\psi_{rm}I_{1r}\sin\delta \quad (2.28a)$$

where K = torque constant and δ is approximately taken as $\pi/2$ in Equation (2.64). Since the harmonic flux components ψ_{7m} and ψ_{5m} are very small, the contribution of the second term can be neglected.

The pulsating torque will tend to cause jitter in the machine speed, but the effect of high-frequency components will be smoothed (filtering effect) due to rotor inertia.

Induction Machines

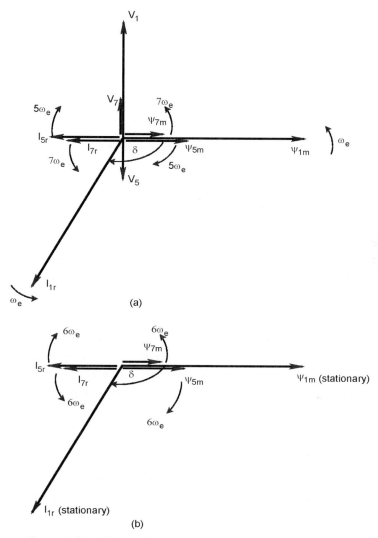

Figure 2.19 Phasor diagram explaining 6th harmonic torque

Assuming pure inertia load, the speed jitter at the 6th harmonic torque can be given as

$$J \frac{d\omega_m}{dt} = T_m \sin 6\omega_e t \qquad (2.65)$$

or

$$\omega_m = \frac{T_m}{J 6\omega_e} \cos 6\omega_e t \qquad (2.66)$$

where J = rotor moment of inertia, ω_m = rotor speed, and T_m = peak value of the 6th harmonic torque from Equation (2.64). Equation (2.66) indicates that at higher inertia and high harmonic frequency, the speed pulsation will be highly attenuated. On the other hand, at low frequency, mechanical resonance may be induced, causing severe shaft vibration, fatigue, wearing of gear teeth, and instability of the feedback control system.

2.2.12 Dynamic d-q Model

So far, we have considered the per phase equivalent circuit of the machine, which is only valid in steady-state condition. In an adjustable-speed drive, the machine normally constitutes an element within a feedback loop, and therefore its transient behavior has to be taken into consideration. Besides, high-performance drive control, such as vector- or field-oriented control, is based on the dynamic *d-q* model of the machine. Therefore, to understand vector control (to be discussed in Chapter 8) principles, a good understanding of the *d-q* model is mandatory.

The dynamic performance of an ac machine is somewhat complex because the three-phase rotor windings move with respect to the three-phase stator windings as shown in Figure 2.20(a).

Basically, it can be looked on as a transformer with a moving secondary, where the coupling coefficients between the stator and rotor phases change continuously with the change of rotor position θ_r. The machine model can be described by differential equations with time-varying mutual inductances, but such a model tends to be very complex. Note that a three-phase machine can be represented by an equivalent two-phase machine as shown in Figure 2.20(b), where d^s - q^s correspond to stator direct and quadrature axes, and $d^r - q^r$ correspond to rotor direct and quadrature axes. Although it is somewhat simple, the problem of time-varying parameters still remains. R. H. Park, in the 1920s, proposed a new theory of electric machine analysis to solve this problem. He formulated a change of variables which, in effect, replaced the variables (voltages, currents, and flux linkages) associated with the stator windings of a synchronous machine with variables associated with fictitious windings rotating with the rotor at synchronous speed. Essentially, he transformed, or referred, the stator variables to a synchronously rotating reference frame fixed in the rotor. With such a transformation (called Park's transformation), he showed that all the time-varying inductances that occur due to an electric circuit in relative motion and electric circuits with varying magnetic reluctances can be eliminated. Later, in the 1930s, H. C. Stanley showed that time-varying inductances in the voltage equations of an induction machine due to electric circuits in relative motion can be eliminated by transforming the rotor variables to variables associated with fictitious stationary windings. In this case, the rotor variables are transformed to a stationary reference frame fixed on the stator. Later, G. Kron proposed a transformation of both stator and rotor variables to a synchronously rotating reference frame that moves with the rotating magnetic field. This model is extremely important, and will be discussed later in detail. D. S. Brereton proposed a transformation of stator variables to a rotating reference frame that is fixed on the rotor. In fact, it was shown later by Krause and Thomas that time-varying inductances can be eliminated by referring the stator and rotor variables to a common reference frame which may rotate at any speed (arbitrary reference frame). Without

Induction Machines

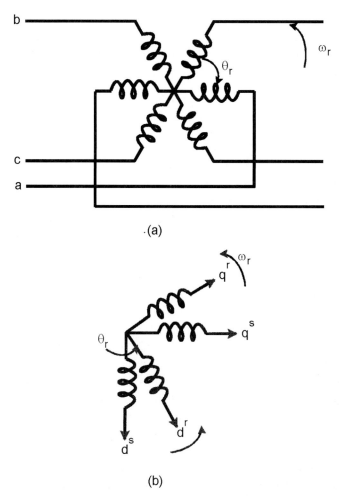

Figure 2.20 (a) Coupling effect in three-phase stator and rotor windings of motor, (b) Equivalent two-phase machine

going deep into the rigor of machine analysis, we will try to develop a dynamic machine model in synchronously rotating and stationary reference frames.

2.2.12.1 Axes Transformation

Consider a symmetrical three-phase induction machine with stationary *as-bs-cs* axes at $2\pi/3$-angle apart, as shown in Figure 2.21. Our goal is to transform the three-phase stationary reference frame (*as-bs-cs*) variables into two-phase stationary reference frame (d^s-q^s) variables and then transform these to synchronously rotating reference frame (d^e-q^e), and vice versa.

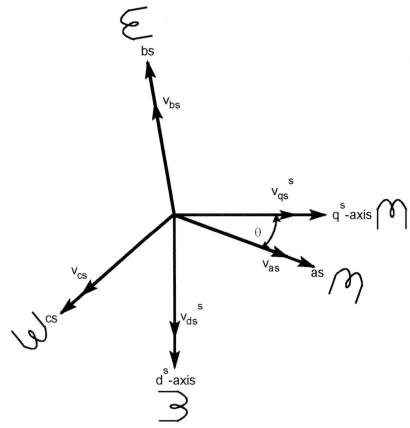

Figure 2.21 Stationary frame *a-b-c* to *ds–qs* axes transformation

Assume that the d^s-q^s axes are oriented at θ angle, as shown in Figure 2.21. The voltages $v_{ds}{}^s$ and $v_{qs}{}^s$ can be resolved into *as-bs-cs* components and can be represented in the matrix form as

$$\begin{bmatrix} v_{as} \\ v_{bs} \\ v_{cs} \end{bmatrix} = \begin{bmatrix} \cos\theta & \sin\theta & 1 \\ \cos(\theta-120°) & \sin(\theta-120°) & 1 \\ \cos(\theta+120°) & \sin(\theta+120°) & 1 \end{bmatrix} \begin{bmatrix} v_{qs}{}^s \\ v_{ds}{}^s \\ v_{os}{}^s \end{bmatrix} \qquad (2.67)$$

Induction Machines

The corresponding inverse relation is

$$\begin{bmatrix} v_{qs}^s \\ v_{ds}^s \\ v_{os}^s \end{bmatrix} = \frac{2}{3} \begin{bmatrix} \cos\theta & \cos(\theta - 120°) & \cos(\theta + 120°) \\ \sin\theta & \sin(\theta - 120°) & \sin(\theta + 120°) \\ 0.5 & 0.5 & 0.5 \end{bmatrix} \begin{bmatrix} v_{as} \\ v_{bs} \\ v_{cs} \end{bmatrix} \quad (2.68)$$

where v_{os}^s is added as the zero sequence component, which may or may not be present. We have considered voltage as the variable. The current and flux linkages can be transformed by similar equations.

It is convenient to set $\theta = 0$, so that the q^s-axis is aligned with the *as*-axis. Ignoring the zero sequence component, the transformation relations can be simplified as

$$v_{as} = v_{qs}^s \quad (2.69)$$

$$v_{bs} = -\frac{1}{2} v_{qs}^s - \frac{\sqrt{3}}{2} v_{ds}^s \quad (2.70)$$

$$v_{cs} = -\frac{1}{2} v_{qs}^s + \frac{\sqrt{3}}{2} v_{ds}^s \quad (2.71)$$

and inversely

$$v_{qs}^s = \frac{2}{3} v_{as} - \frac{1}{3} v_{bs} - \frac{1}{3} v_{cs} = v_{as} \quad (2.72)$$

$$v_{ds}^s = -\frac{1}{\sqrt{3}} v_{bs} + \frac{1}{\sqrt{3}} v_{cs} \quad (2.73)$$

Figure 2.22 shows the synchronously rotating d^e-q^e axes, which rotate at synchronous speed ω_e with respect to the $d^s - q^s$ axes and the angle $\theta_e = \omega_e t$. The two-phase $d^s - q^s$ windings are transformed into the hypothetical windings mounted on the $d^e - q^e$ axes. The voltages on the $d^s - q^s$ axes can be converted (or resolved) into the $d^e - q^e$ frame as follows:

$$v_{qs} = v_{qs}^s \cos\theta_e - v_{ds}^s \sin\theta_e \quad (2.74)$$

$$v_{ds} = v_{qs}^s \sin\theta_e + v_{ds}^s \cos\theta_e \quad (2.75)$$

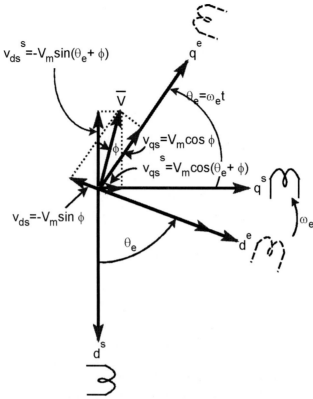

Figure 2.22 Stationary frame $d^s - q^s$ to synchronously rotating frame $d^e - q^e$ transformation

For convenience, the superscript *e* has been dropped from now on from the synchronously rotating frame parameters. Again, resolving the rotating frame parameters into a stationary frame, the relations are

$$v_{qs}^{\,s} = v_{qs} \cos\theta_e + v_{ds} \sin\theta_e \tag{2.76}$$

$$v_{ds}^{\,s} = -v_{qs} \sin\theta_e + v_{ds} \cos\theta_e \tag{2.77}$$

As an example, assume that the three-phase stator voltages are sinusoidal and balanced, and are given by

$$v_{as} = V_m \cos(\omega_e t + \phi) \tag{2.78}$$

Induction Machines

$$v_{bs} = V_m \cos\left(\omega_e t - \frac{2\pi}{3} + \phi\right) \tag{2.79}$$

$$v_{cs} = V_m \cos\left(\omega_e t + \frac{2\pi}{3} + \phi\right) \tag{2.80}$$

Substituting Equations (2.78)–(2.80) in (2.72)–(2.73) yields

$$v_{qs}^{\,s} = V_m \cos(\omega_e t + \phi) \tag{2.81}$$

$$v_{ds}^{\,s} = -V_m \sin(\omega_e t + \phi) \tag{2.82}$$

Again, substituting Equations (2.74)– (2.75) in (2.81)–(2.82), we get

$$v_{qs} = V_m \cos\phi \tag{2.83}$$

$$v_{ds} = -V_m \sin\phi \tag{2.84}$$

Equations (2.81)–(2.82) show that $v_{qs}^{\,s}$ and $v_{ds}^{\,s}$ are balanced, two-phase voltages of equal peak values and the latter is at $\pi/2$ angle phase lead with respect to the other component. Equations (2.83)–(2.84) verify that sinusoidal variables in a stationary frame appear as dc quantities in a synchronously rotating reference frame. This is an important derivation. Note that the stator variables are not necessarily balanced sinusoidal waves. In fact, they can be any arbitrary time functions.

The variables in a reference frame can be combined and represented by a complex space vector (or phasor). For example, from Equations (2.81) – (2.82),

$$\begin{aligned}
\overline{V} = v_{qds}^{\,s} &= v_{qs}^{\,s} - j v_{ds}^{\,s} \\
&= V_m \left[\cos(\omega_e t + \phi) + j \sin(\omega_e t + \phi)\right] \\
&= \hat{V}_m e^{j\phi} e^{j\omega_e t} \\
&= \sqrt{2} V_s e^{j(\theta_e + \phi)}
\end{aligned} \tag{2.85}$$

which indicates that the vector $\overline{V}$ rotates counter-clockwise at speed ω_e from the initial ($t = 0$) angle of ϕ to the q^e-axis. Equation (2.85) also indicates that for a sinusoidal variable, the vector magnitude is the peak value ($\hat{V}_m$), which is $\sqrt{2}$ times the rms phasor magnitude (V_s). The $q^e - d^e$ components can also be combined into a vector form:

$$v_{qds}^e = v_{qs}^e - jv_{ds}^e = (v_{qs}^s \cos\theta_e - v_{ds}^s \sin\theta_e) - j(v_{qs}^s \sin\theta_e + v_{ds}^s \cos\theta_e)$$
$$= (v_{qs}^s - jv_{ds}^s)e^{-j\theta_e} = \bar{V}e^{-j\theta_e} \tag{2.86}$$

or inversely

$$\bar{V} = v_{qs}^s - jv_{ds}^s = (v_{qs} - jv_{ds})e^{+j\theta_e} \tag{2.87}$$

Note that the vector magnitudes in stationary and rotating frames are equal, that is,

$$|\bar{V}| = \hat{V}_m = \sqrt{v_{qs}^{s^2} + v_{ds}^{s^2}} = \sqrt{v_{qs}^2 + v_{ds}^2} \tag{2.88}$$

The factor $e^{j\theta_e}$ may be interpreted as a vector rotational operator (defined as a vector rotator (VR) or unit vector) that converts rotating frame variables into stationary frame variables. Cos θ_e and sin θ_e are the cartesian components of the unit vector. In Equation (2.86), $e^{-j\theta_e}$ is defined as the inverse vector rotator (VR^{-1}) that converts $d^s - q^s$ variables into $d^e - q^e$ variables. The vector $\bar{V}$ and its components projected on rotating and stationary axes are shown in Figure 2.22. The as-bs-cs variables can also be expressed in vector form. Substituting Equations (2.72)–(2.73) into (2.85)

$$\bar{V} = v_{qs}^s - jv_{ds}^s$$
$$= \left(\frac{2}{3}v_{as} - \frac{1}{3}v_{bs} - \frac{1}{3}v_{cs}\right) - j\left(-\frac{\sqrt{3}}{2}v_{bs} + \frac{\sqrt{3}}{2}v_{cs}\right)$$
$$= \frac{2}{3}\left[v_{as} + \left(-\frac{1}{2} + j\frac{\sqrt{3}}{2}\right)v_{bs} + \left(-\frac{1}{2} - j\frac{\sqrt{3}}{2}\right)v_{cs}\right] \tag{2.89}$$
$$= \frac{2}{3}\left[v_{as} + av_{bs} + a^2v_{cs}\right]$$

where $a = e^{j2\pi/3}$ and $a^2 = e^{-j2\pi/3}$. The parameters a and a^2 can be interpreted as unit vectors aligned to the respective bs and cs axes of the machine, and the reference axis corresponds to the v_{as}-axis. Similar transformations can be made for rotor circuit variables also.

Induction Machines

2.2.12.2 Synchronously Rotating Reference Frame—Dynamic Model (Kron Equation)

For the two-phase machine shown in Figure 2.20(b), we need to represent both $d^s - q^s$ and $d^r - q^r$ circuits and their variables in a synchronously rotating $d^e - q^e$ frame. We can write the following stator circuit equations:

$$v_{qs}^s = R_s i_{qs}^s + \frac{d}{dt}\psi_{qs}^s \tag{2.90}$$

$$v_{ds}^s = R_s i_{ds}^s + \frac{d}{dt}\psi_{ds}^s \tag{2.91}$$

where ψ_{qs}^s and ψ_{ds}^s are q-axis and d-axis stator flux linkages, respectively. When these equations are converted to d^e-q^e frame, the following equations can be written [4]:

$$v_{qs} = R_s i_{qs} + \frac{d}{dt}\psi_{qs} + \omega_e \psi_{ds} \tag{2.92}$$

$$v_{ds} = R_s i_{ds} + \frac{d}{dt}\psi_{ds} - \omega_e \psi_{qs} \tag{2.93}$$

where all the variables are in rotating form. The last term in Equations (2.92) and (2.93) can be defined as speed emf due to rotation of the axes, that is, when $\omega_e = 0$, the equations revert to stationary form. Note that the flux linkage in the d^e and q^e axes induce emf in the q^e and d^e axes, respectively, with $\pi/2$ lead angle.

If the rotor is not moving, that is, $\omega_r = 0$, the rotor equations for a doubly-fed wound-rotor machine will be similar to Equations (2.92)–(2.93):

$$v_{qr} = R_r i_{qr} + \frac{d}{dt}\psi_{qr} + \omega_e \psi_{dr} \tag{2.94}$$

$$v_{dr} = R_r i_{dr} + \frac{d}{dt}\psi_{dr} - \omega_e \psi_{qr} \tag{2.95}$$

where all the variables and parameters are referred to the stator. Since the rotor actually moves at speed ω_r, the d-q axes fixed on the rotor move at a speed $\omega_e - \omega_r$ relative to the synchronously rotating frame. Therefore, in $d^e - q^e$ frame, the rotor equations should be modified as

$$v_{qr} = R_r i_{qr} + \frac{d}{dt}\psi_{qr} + (\omega_e - \omega_r)\psi_{dr} \tag{2.96}$$

$$v_{dr} = R_r i_{dr} + \frac{d}{dt}\psi_{dr} - (\omega_e - \omega_r)\psi_{qr} \tag{2.97}$$

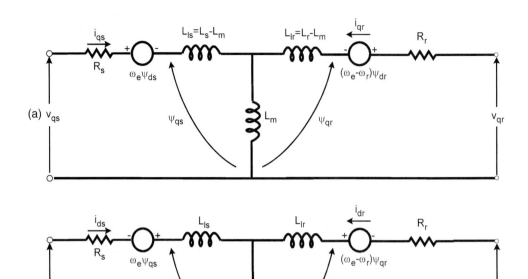

Figure 2.23 Dynamic $d^e - q^e$ equivalent circuits of machine (a) q^e- axis circuit, (b) d^e – axis circuit

Figure 2.23 shows the $d^e - q^e$ dynamic model equivalent circuits that satisfy Equations (2.92)–(2.93) and (2.96)–(2.97). A special advantage of the $d^e - q^e$ dynamic model of the machine is that all the sinusoidal variables in stationary frame appear as dc quantities in synchronous frame, as discussed before.

The flux linkage expressions in terms of the currents can be written from Figure 2.23 as follows:

$$\psi_{qs} = L_{ls}i_{qs} + L_m(i_{qs} + i_{qr}) \tag{2.98}$$

$$\psi_{qr} = L_{lr}i_{qr} + L_m(i_{qs} + i_{qr}) \tag{2.99}$$

$$\psi_{qm} = L_m(i_{qs} + i_{qr}) \tag{2.100}$$

$$\psi_{ds} = L_{ls}i_{ds} + L_m(i_{ds} + i_{dr}) \tag{2.101}$$

$$\psi_{dr} = L_{lr}i_{dr} + L_m(i_{ds} + i_{dr}) \tag{2.102}$$

$$\psi_{dm} = L_m(i_{ds} + i_{dr}) \tag{2.103}$$

Induction Machines

Combining the above expressions with Equations (2.92), (2.93), (2.96), and (2.97), the electrical transient model in terms of voltages and currents can be given in matrix form as

$$\begin{bmatrix} v_{qs} \\ v_{ds} \\ v_{qr} \\ v_{qr} \end{bmatrix} = \begin{bmatrix} R_s + SL_s & \omega_e L_s & SL_m & \omega_e L_m \\ -\omega_e L_s & R_s + SL_s & -\omega_e L_m & SL_m \\ SL_m & (\omega_e - \omega_r)L_m & R_r + SL_r & (\omega_e - \omega_r)L_r \\ -(\omega_e - \omega_r)L_m & SL_m & -(\omega_e - \omega_r)L_r & R_r + SL_r \end{bmatrix} \begin{bmatrix} i_{qs} \\ i_{ds} \\ i_{qr} \\ i_{dr} \end{bmatrix} \quad (2.104)$$

where S is the Laplace operator. For a singly-fed machine, such as a cage motor, $v_{qr} = v_{dr} = 0$.

If the speed ω_r is considered constant (infinite inertia load), the electrical dynamics of the machine are given by a fourth-order linear system. Then, knowing the inputs v_{qs}, v_{ds}, and ω_e, the currents i_{qs}, i_{ds}, i_{qr}, and i_{dr} can be solved from Equation (2.104). If the machine is fed by current source, i_{qs}, i_{ds}, and ω_e are independent. Then, the dependent variables v_{qs}, v_{ds}, i_{qr}, and i_{dr} can be solved from Equation (2.104).

The speed ω_r in Equation (2.104) cannot normally be treated as a constant. It can be related to the torques as

$$T_e = T_L + J\frac{d\omega_m}{dt} = T_L + \frac{2}{P}J\frac{d\omega_r}{dt} \quad (2.105)$$

where T_L = load torque, J = rotor inertia, and ω_m = mechanical speed.

Often, for compact representation, the machine model and equivalent circuits are expressed in complex form. Multiplying Equation (2.93) by $-j$ and adding with Equation (2.92) gives

$$v_{qs} - jv_{ds} = R_s(i_{qs} - ji_{ds}) + \frac{d}{dt}(\psi_{qs} - j\psi_{ds}) + j\omega_e(\psi_{qs} - j\psi_{ds}) \quad (2.106)$$

or

$$v_{qds} = R_s i_{qds} + \frac{d}{dt}\psi_{qds} + j\omega_e \psi_{qds} \quad (2.107)$$

where v_{qds}, i_{qds}, etc. are complex vectors (the superscript e has been omitted). Similarly, the rotor Equations (2.96) – (2.97) can be combined to represent

$$v_{qdr} = R_r i_{qdr} + \frac{d}{dt}\psi_{qdr} + j(\omega_e - \omega_r)\psi_{qdr} \quad (2.108)$$

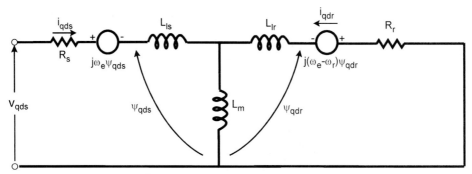

Figure 2.24 Complex synchronous frame dqs equivalent circuit

Figure 2.24 shows the complex equivalent circuit in rotating frame where $v_{qdr} = 0$. Note that the steady-state equations can always be derived by substituting the time derivative components to zero. Therefore from Equations (2.107)–(2.108), the steady-state equations can be derived as

$$V_s = R_s I_s + j\omega_e \psi_s \qquad (2.109)$$

$$0 = \frac{R_r}{S} I_r + j\omega_e \psi_r \qquad (2.110)$$

where the complex vectors have been substituted by the corresponding rms phasors. These equations satisfy the steady-state equivalent circuit shown in Figure 2.4 if the parameter R_m is neglected.

The development of torque by the interaction of air gap flux and rotor mmf was discussed earlier in this chapter. Here it will be expressed in more general form, relating the d-q components of variables. From Equation (2.29), the torque can be generally expressed in the vector form as

$$T_e = \frac{3}{2}\left(\frac{P}{2}\right)\bar{\psi}_m \times \bar{I}_r \qquad (2.111)$$

Resolving the variables into d^e-q^e components, as shown in Figure 2.25,

$$T_e = \frac{3}{2}\left(\frac{P}{2}\right)(\psi_{dm} i_{qr} - \psi_{qm} i_{dr}) \qquad (2.112)$$

Several other torque expressions can be derived easily as follows:

$$T_e = \frac{3}{2}\left(\frac{P}{2}\right)(\psi_{dm} i_{qs} - \psi_{qm} i_{ds}) \qquad (2.113)$$

Induction Machines

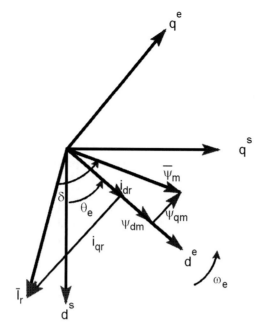

Figure 2.25 Flux and current vectors in $d^e - q^e$ frame

$$= \frac{3}{2}\left(\frac{P}{2}\right)(\psi_{ds}i_{qs} - \psi_{qs}i_{ds}) \tag{2.114}$$

$$= \frac{3}{2}\left(\frac{P}{2}\right)L_m(i_{qs}i_{dr} - i_{ds}i_{qr}) \tag{2.115}$$

$$= \frac{3}{2}\left(\frac{P}{2}\right)(\psi_{dr}i_{qr} - \psi_{qr}i_{dr}) \tag{2.116}$$

Equations (2.104), (2.105), and (2.115) give the complete model of the electro-mechanical dynamics of an induction machine in synchronous frame. The composite system is of the fifth order and nonlinearity of the model is evident. Figure 2.26 shows the block diagram of the machine model along with input voltage and output current transformations.

2.2.12.3 Stationary Frame—Dynamic Model (Stanley Equation)

The dynamic machine model in stationary frame can be derived simply by substituting $\omega_e = 0$ in Equation (2.104) or in (2.92), (2.93), (2.96), and (2.97). The corresponding stationary frame equations are given as

$$v_{qs}^s = R_s i_{qs}^s + \frac{d}{dt}\psi_{qs}^s \tag{2.90}$$

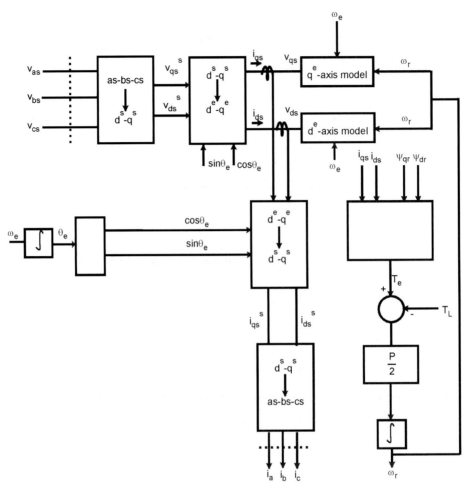

Figure 2.26 Synchronously rotating frame machine model with input voltage and output current transformations

$$v_{ds}^{\,s} = R_s i_{ds}^{\,s} + \frac{d}{dt}\psi_{ds}^{\,s} \tag{2.91}$$

$$0 = R_r i_{qr}^{\,s} + \frac{d}{dt}\psi_{qr}^{\,s} - \omega_r \psi_{dr}^{\,s} \tag{2.117}$$

$$0 = R_r i_{dr}^{\,s} + \frac{d}{dt}\psi_{dr}^{\,s} + \omega_r \psi_{qr}^{\,s} \tag{2.118}$$

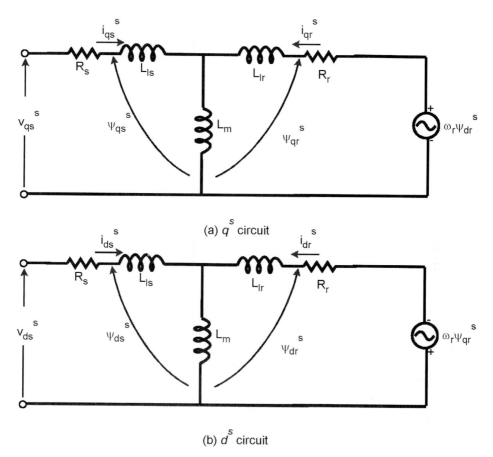

Figure 2.27 d^s-q^s equivalent circuits

where $v_{qr} = v_{dr} = 0$. Figure 2.27 shows the corresponding equivalent circuits. As mentioned before, in the stationary frame, the variables appear as sine waves in steady state with sinusoidal inputs.

The torque Equations (2.112)–(2.116) can also be written with the corresponding variables in stationary frame as

$$T_e = \frac{3}{2}\left(\frac{P}{2}\right)(\psi_{dm}{}^s i_{qr}{}^s - \psi_{qm}{}^s i_{dr}{}^s) \tag{2.119}$$

$$= \frac{3}{2}\left(\frac{P}{2}\right)(\psi_{dm}{}^s i_{qs}{}^s - \psi_{qm}{}^s i_{ds}{}^s) \tag{2.120}$$

$$= \frac{3}{2}\left(\frac{P}{2}\right)(\psi_{ds}{}^s i_{qs}{}^s - \psi_{qs}{}^s i_{ds}{}^s) \tag{2.121}$$

$$= \frac{3}{2}\left(\frac{P}{2}\right) L_m (i_{qs}^{\,s} i_{dr}^{\,s} - i_{ds}^{\,s} i_{qr}^{\,s}) \tag{2.122}$$

$$= \frac{3}{2}\left(\frac{P}{2}\right) (\psi_{dr}^{\,s} i_{qr}^{\,s} - \psi_{qr}^{\,s} i_{dr}^{\,s}) \tag{2.123}$$

Equations (2.90)–(2.91) and (2.117)–(2.118) can easily be combined to derive the complex model as

$$v_{qds}^{\,s} = R_s i_{qds}^{\,s} + \frac{d}{dt}\psi_{qds}^{\,s} \tag{2.124}$$

$$0 = R_r i_{qds}^{\,s} + \frac{d}{dt}\psi_{qdr}^{\,s} - j\omega_r \psi_{qdr}^{\,s} \tag{2.125}$$

where $v_{qds}^{\,s} = v_{qs}^{\,s} - j v_{ds}^{\,s}$, $\psi_{qds}^{\,s} = \psi_{qs}^{\,s} - j\psi_{ds}^{\,s}$, $i_{qds}^{\,s} = i_{qs}^{\,s} - ji_{ds}^{\,s}$, $\psi_{qdr}^{\,s} = \psi_{qr}^{\,s} - j\psi_{dr}^{\,s}$, etc. The complex equivalent circuit in stationary frame is shown in Figure 2.28(a). Often, a per phase equivalent circuit with CEMF ($\omega_r \bar{\psi}_r$) and sinusoidal variables is described in the form of Figure 2.28(b) omitting the parameter L_m.

2.2.12.4 Dynamic Model State-Space Equations

The dynamic machine model in state-space form is important for transient analysis, particularly for computer simulation study. Although the rotating frame model is generally preferred, the stationary frame model can also be used. The electrical variables in the model can be chosen as fluxes, currents, or a mixture of both. In this section, we will derive state-space equations of the machine in rotating frame with flux linkages as the main variables. A hybrid model in terms of stationary frame stator currents and rotor fluxes will be discussed in Chapter 8.

Let's define the flux linkage variables as follows:

$$F_{qs} = \omega_b \psi_{qs} \tag{2.126}$$

$$F_{qr} = \omega_b \psi_{qr} \tag{2.127}$$

$$F_{ds} = \omega_b \psi_{ds} \tag{2.128}$$

$$F_{dr} = \omega_b \psi_{dr} \tag{2.129}$$

where ω_b = base frequency of the machine.

Substituting the above relations in Equations (2.92)–(2.93) and (2.96)–(2.97), we can write

$$v_{qs} = R_s i_{qs} + \frac{1}{\omega_b}\frac{dF_{qs}}{dt} + \frac{\omega_e}{\omega_b} F_{ds} \tag{2.130}$$

Induction Machines

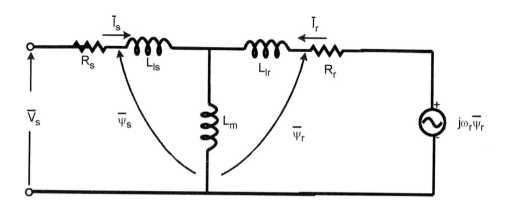

(a) Stationary frame complex equivalent circuit

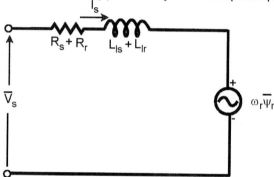

(b) Simplified per phase equivalent circuit

Figure 2.28 Complex stationary frame with dqs equivalent circuits

$$v_{ds} = R_s i_{ds} + \frac{1}{\omega_b}\frac{dF_{ds}}{dt} - \frac{\omega_e}{\omega_b}F_{qs} \qquad (2.131)$$

$$0 = R_r i_{qr} + \frac{1}{\omega_b}\frac{dF_{qr}}{dt} + \frac{(\omega_e - \omega_r)}{\omega_b}F_{dr} \qquad (2.132)$$

$$0 = R_r i_{dr} + \frac{1}{\omega_b}\frac{dF_{dr}}{dt} - \frac{(\omega_e - \omega_r)}{\omega_b}F_{qr} \qquad (2.133)$$

where it is assumed that $v_{qr} = v_{dr} = 0$.

Multiplying Equations (2.98)–(2.103) by ω_b on both sides, the flux linkage expressions can be written as

$$F_{qs} = \omega_b \psi_{qs} = X_{ls} i_{qs} + X_m (i_{qs} + i_{qr}) \tag{2.134}$$

$$F_{qr} = \omega_b \psi_{qr} = X_{lr} i_{qr} + X_m (i_{qs} + i_{qr}) \tag{2.135}$$

$$F_{qm} = \omega_b \psi_{qm} = X_m (i_{qs} + i_{qr}) \tag{2.136}$$

$$F_{ds} = \omega_b \psi_{ds} = X_{ls} i_{ds} + X_m (i_{ds} + i_{dr}) \tag{2.137}$$

$$F_{dr} = \omega_b \psi_{dr} = X_{lr} i_{dr} + X_m (i_{ds} + i_{dr}) \tag{2.138}$$

$$F_{dm} = \omega_b \psi_{dm} = X_m (i_{ds} + i_{dr}) \tag{2.139}$$

where $X_{ls} = \omega_b L_{ls}$, $X_{lr} = \omega_b L_{lr}$, and $X_m = \omega_b L_m$, or

$$F_{qs} = X_{ls} i_{qs} + F_{qm} \tag{2.140}$$

$$F_{qr} = X_{lr} i_{qr} + F_{qm} \tag{2.141}$$

$$F_{ds} = X_{ls} i_{ds} + F_{dm} \tag{2.142}$$

$$F_{dr} = X_{lr} i_{dr} + F_{dm} \tag{2.143}$$

From Equations (2.140)–(2.143), the currents can be expressed in terms of the flux linkages as

$$i_{qs} = \frac{F_{qs} - F_{qm}}{X_{ls}} \tag{2.144}$$

$$i_{qr} = \frac{F_{qr} - F_{qm}}{X_{lr}} \tag{2.145}$$

$$i_{ds} = \frac{F_{ds} - F_{dm}}{X_{ls}} \tag{2.146}$$

$$i_{dr} = \frac{F_{dr} - F_{dm}}{X_{lr}} \tag{2.147}$$

Induction Machines

Substituting Equations (2.144)–(2.145) in (2.140)–(2.141), respectively, the F_{qm} expression is given as

$$F_{qm} = X_m \left[\frac{(F_{qs} - F_{qm})}{X_{ls}} + \frac{(F_{qr} - F_{qm})}{X_{lr}} \right] \quad (2.148)$$

or

$$F_{qm} = \frac{X_{m1}}{X_{ls}} F_{qs} + \frac{X_{m1}}{X_{lr}} F_{qr} \quad (2.149)$$

where

$$X_{m1} = \frac{1}{\left(\dfrac{1}{X_m} + \dfrac{1}{X_{ls}} + \dfrac{1}{X_{lr}} \right)} \quad (2.150)$$

Similar derivation can be made for F_{dm} as follows:

$$F_{dm} = \frac{X_{m1}}{X_{ls}} F_{ds} + \frac{X_{m1}}{X_{lr}} F_{dr} \quad (2.151)$$

Substituting the current Equations (2.144)–(2.147) into the voltage Equations (2.130)–(2.133),

$$v_{qs} = \frac{R_s}{X_{ls}} (F_{qs} - F_{qm}) + \frac{1}{\omega_b} \frac{dF_{qs}}{dt} + \frac{\omega_e}{\omega_b} F_{ds} \quad (2.152)$$

$$v_{ds} = \frac{R_s}{X_{ls}} (F_{ds} - F_{dm}) + \frac{1}{\omega_b} \frac{dF_{ds}}{dt} - \frac{\omega_e}{\omega_b} F_{qs} \quad (2.153)$$

$$0 = \frac{R_r}{X_{lr}} (F_{qr} - F_{qm}) + \frac{1}{\omega_b} \frac{dF_{qr}}{dt} + \frac{(\omega_e - \omega_r)}{\omega_b} F_{dr} \quad (2.154)$$

$$0 = \frac{R_r}{X_{lr}} (F_{dr} - F_{dm}) + \frac{1}{\omega_b} \frac{dF_{dr}}{dt} - \frac{(\omega_e - \omega_r)}{\omega_b} F_{qr} \quad (2.155)$$

which can be expressed in state-space form as

$$\frac{dF_{qs}}{dt} = \omega_b \left[v_{qs} - \frac{\omega_e}{\omega_b} F_{ds} - \frac{R_s}{X_{ls}} (F_{qs} - F_{qm}) \right] \quad (2.156)$$

$$\frac{dF_{ds}}{dt} = \omega_b \left[v_{ds} + \frac{\omega_e}{\omega_b} F_{qs} - \frac{R_s}{X_{ls}} (F_{ds} - F_{dm}) \right] \quad (2.157)$$

$$\frac{dF_{qr}}{dt} = -\omega_b \left[\frac{(\omega_e - \omega_r)}{\omega_b} F_{dr} + \frac{R_r}{X_{lr}} (F_{qr} - F_{qm}) \right] \quad (2.158)$$

$$\frac{dF_{dr}}{dt} = -\omega_b \left[-\frac{(\omega_e - \omega_r)}{\omega_b} F_{qr} + \frac{R_r}{X_{lr}} (F_{dr} - F_{dm}) \right] \quad (2.159)$$

Finally, from Equation (2.114)

$$T_e = \frac{3}{2} \left(\frac{P}{2} \right) \frac{1}{\omega_b} (F_{ds} i_{qs} - F_{qs} i_{ds}) \quad (2.160)$$

Equations (2.156)–(2.160), along with Equation (2.105), describe the complete model in state-space form where F_{qs}, F_{ds}, F_{qr}, and F_{dr} are the state variables. Simulation of the machine will be discussed in Chapter 5.

2.3 SYNCHRONOUS MACHINES

A synchronous machine, as the name indicates, must rotate at synchronous speed; that is, the speed is uniquely related to the supply frequency, as indicated in Equation (2.10). It is a serious competitor to the induction machine in variable-speed drive applications. Both machine types are similar in many respects, and much of the discussion in the previous section holds true for the synchronous machine. Therefore, only the salient features of the synchronous machine will be reviewed here.

2.3.1 Wound Field Machine

Figure 2.29 shows an idealized three-phase, two-pole wound field synchronous machine. The stator winding of the machine is identical to that of the induction machine, but the rotor has a winding that carries dc current and produces flux in the air gap that helps the stator-induced rotating magnetic field to drag the rotor along with it. The dc field current is supplied to the rotor from a static rectifier through slip rings and brushes, or by brushless excitation, which will be discussed later. Since the rotor always moves at synchronous speed (i.e., the slip is zero), the synchronously rotating $d^e - q^e$ axes are fixed with the rotor, where the d^e axis corresponds to the north pole, as shown. There is no stator-induced induction in the rotor, and therefore, the rotor mmf is supplied exclusively by the field winding. This permits the machine to run at an arbitrary power factor at the stator terminal, that is, leading, lagging, or unity. On the other hand, in an induction machine, the stator supples the rotor excitation that makes the machine power factor always lagging.

The mechanism of torque production is somewhat similar to that of an induction machine. The machine shown is characterized as a salient pole because of the nonuniform air gap around the rotor, which contributes to asymmetrical magnetic reluctance in the d and q axes. This is in contrast to a machine with a cylindrical rotor structure having a uniform air gap (such as an

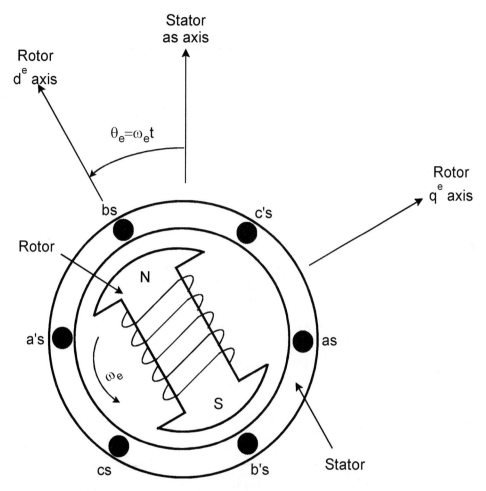

Figure 2.29 Idealized three-phase, two-pole synchronous machine (salient pole)

induction motor), defined as a nonsalient pole machine. For example, low-speed synchronous generators in hydro-electric power stations use salient pole machines, whereas high-speed generators in steam-power stations use nonsalient pole machines. In addition to field winding, the rotor usually contains an amortisseur, or damper winding, which is like short-circuited squirrel-cage bars in an induction motor. The machine is more expensive, but efficiency is somewhat higher. Wound field machines are normally used for high-power (multi-megawatt) drives. A typical comparison between a high-power induction motor and a synchronous motor for a rolling mill application [10] is given in Table 2.1.

Table 2.1 Comparison of Induction and Wound Field Synchronous Machines (6MW size)

	IM	SM
Displacement factor (at rated operation)	0.89	1.0
Efficiency (incl. excitation)	93.9%	95.5%
Max. output at 200% speed	240%	240%
Moment of inertia	134%	100%
Total weight of motor	101%	100%
Torque response time	<10ms	<10ms
Time constant for flux change	3.0s	0.35s
Required converter kVA	354%	258%
Excitation rectifier kVA (at peak load)	-	10%

2.3.1.1 Equivalent Circuit

A simple per phase steady-state equivalent circuit for a nonsalient pole, synchronous machine can be derived from the same physical considerations as those for an induction motor, and it is shown in Figure 2.30. Figure 2.30(a) shows the transformer-like coupled equivalent circuit linking the stator and the moving rotor winding. The rotor is supplied by a field current I_f due to the supply voltage V_f. The rotor section can be substituted in terms of the stator by a current source I_f' at frequency ω_e, as shown in Figure 2.30(b), where n is the ratio relating the rms magnitude of I_f' to the magnitude of the dc field current I_f. At steady-state operation, the power transferred to the rotor winding is zero, and all the power across the air gap is converted to mechanical power. Neglecting the core-loss resistor R_m, Figure 2.30 (b) can be drawn in the form of Figure 2.30(c) using the Thevenin theorem, where $V_f = \omega_e L_m n I_f = \omega_e \psi_f$ is defined as excitation or speed emf due to flux linkage ψ_f induced by field current I_f. The sum of leakage reactance $\omega_e L_{ls}$ and magnetizing reactance $\omega_e L_m$ is known as the synchronous reactance ($X_s = \omega_e L_s = \omega_e (L_{ls} + L_m)$), and the total impedance $Z_s = R_s + jX_s$ is known as synchronous impedance.

As mentioned before, a synchronous machine can operate at any desired power factor: leading, lagging, or unity. The power factor can be controlled by the magnitude of the field excitation. At a given frequency ω_e, the air gap voltage is $V_m = \omega_e \psi_m$, where $\psi_m = L_m I_m$ is the air gap flux linkage. This tends to balance the fixed supply voltage V_s, and correspondingly, the magnetizing current I_m tends to be constant. The current I_m is contributed by field component I_f

Synchronous Machines

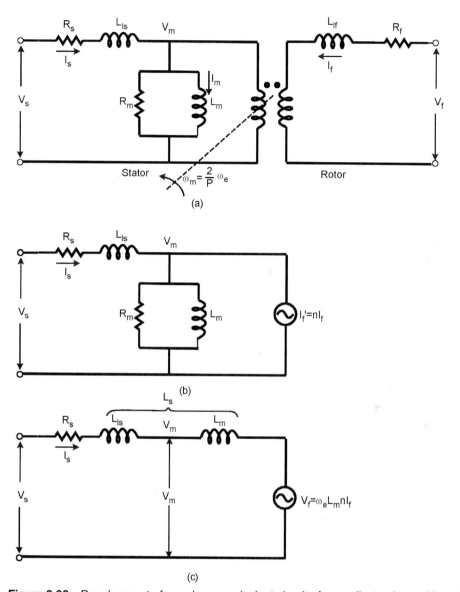

Figure 2.30 Development of per phase equivalent circuit of nonsalient pole machine

and the reactive current component of stator current I_s. If the machine is overexcited, the lagging reactive current is supplied to the output (i.e., the terminal power factor is leading). On the other hand, if the machine is underexcited, it takes lagging current from the line to supplement the excitation.

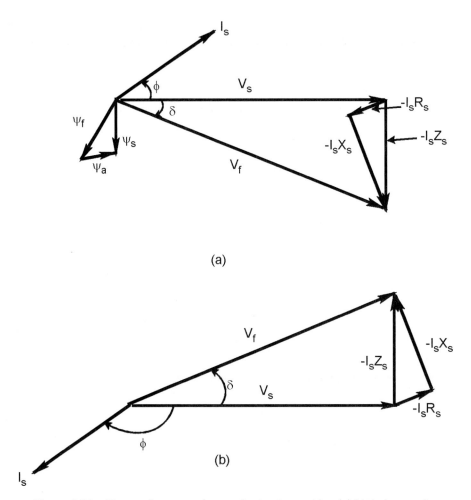

Figure 2.31 Phasor diagrams of nonsalient pole machine (a) Motoring mode, (b) Generating mode

Figure 2.31 shows the phasor diagrams for the equivalent circuit of Figure 2.30(c) under both motoring and generating conditions. The motoring mode is shown with a leading power factor.

The resistance drop is small and is often neglected. Neglecting R_s, the flux linkage phasor diagram is added where

$$\psi_s = \left|\frac{V_s}{\omega_e}\right| \angle -\frac{\pi}{2} \tag{2.161}$$

$$\psi_a = I_s L_s \tag{2.162}$$

Synchronous Machines

the angle δ between V_s and V_f is known as the power or torque angle of a synchronous machine, and it is negative in the motoring mode (with V_s as reference phasor), but is positive in the generating mode. The phasor diagram in the generating mode is shown for a lagging power factor.

2.3.1.2 Developed Torque

Neglecting R_s in Figure 2.31, we can write the I_s expression as

$$I_s = \frac{V_s \angle 0 - V_f \angle -\delta}{X_s \angle -\frac{\pi}{2}}$$

$$= \frac{V_s \angle -\frac{\pi}{2}}{X_s} - \frac{V_f \angle -\left(\delta + \frac{\pi}{2}\right)}{X_s} \tag{2.163}$$

or

$$I_s \cos\phi = \frac{V_s}{X_s}\cos\left(-\frac{\pi}{2}\right) - \frac{V_f}{X_s}\cos\left(-\delta - \frac{\pi}{2}\right)$$

$$= -\frac{V_f}{X_s}\cos\left(\delta + \frac{\pi}{2}\right) \tag{2.164}$$

The power input to the machine is

$$P_i = 3 V_s I_s \cos\phi \tag{2.165}$$

Substituting Equation (2.164) in (2.165) yields

$$P_i = 3\frac{V_s V_f}{X_s}\sin\delta \tag{2.166}$$

If machine losses are ignored, the power P_i is also delivered to the shaft.

$$P_s = P_i = \frac{2}{P}\omega_e T_e \tag{2.167}$$

Combining Equations (2.166) and (2.167) gives

$$T_e = 3\left(\frac{P}{2}\right)\frac{V_s}{\omega_e}\frac{V_f}{X_s}\sin\delta \tag{2.168}$$

$$= 3\left(\frac{P}{2}\right)\frac{\psi_s \psi_f}{L_s}\sin\delta \tag{2.169}$$

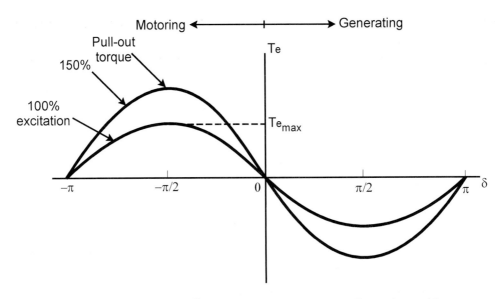

Figure 2.32 Torque-δ angle characteristics of nonsalient pole machine

Equation (2.169) gives the developed torque as a function of torque angle δ, which is plotted in Figure 2.32 for both motoring and generating modes. The torque is zero at $\delta = 0$ and becomes maximum at $\delta = \pm \pi/2$. Stability considerations dictate that the machine should be operated with δ angle within $\pm \pi/2$. At a fixed frequency and supply voltage, the torque curve is proportional to the field excitation current as shown in the figure, where the saturation effect is neglected. Or, for a fixed torque angle and field excitation, the torque remains unchanged if the supply voltage-to-frequency ratio (i.e., ψ_s) remains constant.

Equations (2.165) and (2.167) can be combined to write the torque expression as

$$T_e = 3\left(\frac{P}{2}\right)\psi_s I_s \cos\phi$$
$$= 3\left(\frac{P}{2}\right)\psi_s I_T \quad (2.170)$$

where ψ_s = stator flux linkage and $I_T = I_s \cos\varphi$ is the in-phase or torque component of stator current.

2.3.1.3 Salient Pole Machine Characteristics

So far we have discussed the characteristics of a nonsalient pole cylindrical rotor machine. The characteristics of a salient pole machine differ from those of a nonsalient pole machine because of the nonuniform air gap reluctances in the d^e and q^e axes. The resulting asymmetry in the direct and quadrature axes magnetizing reactances causes the corresponding synchronous reactances to be unsymmetrical (i.e., $X_{ds} \neq X_{qs}$). Figure 2.33 shows phasor diagrams of a salient

Synchronous Machines

pole machine for the motoring and generating modes, and also includes flux linkages. Again, for simplicity, the stator resistance has been dropped. The excitation or speed emf V_f is shown aligned with the q^e axes, whereas ψ_f is aligned with the d^e axes. The phase voltage V_s and phase current I_s are resolved into corresponding d^e and q^e components, and a voltage phasor diagram is drawn with the corresponding reactive drops. In the phasor diagram, the armature reaction flux ψ_a aids the field flux to result in the stator flux ψ_s as shown. The motoring mode phasor diagram, which is drawn for lagging power factor, $\psi_s > \psi_f$, whereas in the generating mode, $\psi_s < \psi_f$, because it is operating at leading power factor. Note that d^e-q^e axes phasor diagrams can also be drawn for a nonsalient pole machine where $X_{ds} = X_{qs}$.

From the phasor diagram, Figure 2.33(a), we can write

$$I_s \cos\phi = I_{qs} \cos\delta - I_{ds} \sin\delta \tag{2.171}$$

The figure can also be a vector diagram if all the rms phasors are multiplied by the factor $\sqrt{2}$, as mentioned before.

Substituting Equation (2.171) in (2.165), the input power P_i can be given as

$$P_i = 3V_s \left(I_{qs} \cos\delta - I_{ds} \sin\delta \right) \tag{2.172}$$

Again, from the phasor diagram we can write

$$I_{ds} = \frac{V_s \cos\delta - V_f}{X_{ds}} \tag{2.173}$$

$$I_{qs} = \frac{V_s \sin\delta}{X_{qs}} \tag{2.174}$$

Substituting Equations (2.173)–(2.174) in (2.172) yields

$$P_i = 3\frac{V_s V_f}{X_{ds}} \sin\delta + 3V_s^2 \frac{(X_{ds} - X_{qs})}{2X_{ds}X_{qs}} \sin 2\delta \tag{2.175}$$

or

$$T_e = 3\left(\frac{P}{2}\right)\frac{1}{\omega_e}\left(\frac{V_s V_f}{X_{ds}} \sin\delta + V_s^2 \frac{(X_{ds} - X_{qs})}{2X_{ds}X_{qs}} \sin 2\delta \right) \tag{2.176}$$

$$T_e = 3\left(\frac{P}{2}\right)\left(\frac{\psi_s \psi_f}{L_{ds}} \sin\delta + \psi_s^2 \frac{(L_{ds} - L_{qs})}{2L_{ds}L_{qs}} \sin 2\delta \right) \tag{2.177}$$

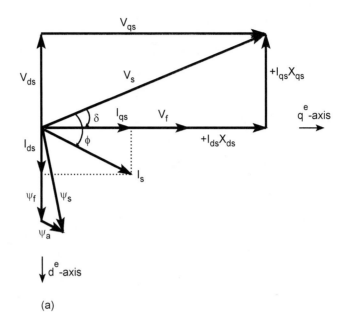

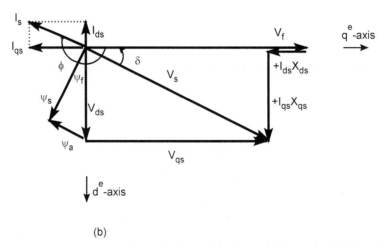

Figure 2.33 Phasor diagram of salient pole machine (a) Motoring mode, (b) Generating mode

Equation (2.177) gives the developed torque with torque angle δ for a salient pole machine. The first component of the equation is contributed by the field ψ_f and is identical to Equation (2.169) except L_s is replaced by L_{ds}. The second component is defined as reluctance torque, which arises due to rotor saliency (i.e., $X_{ds} \neq X_{qs}$), where the rotor tends to align with the position of minimum reluctance and is not influenced by the field excitation. For a standard salient pole machine ($L_{ds} > L_{qs}$), this is an additive torque component.

Synchronous Machines

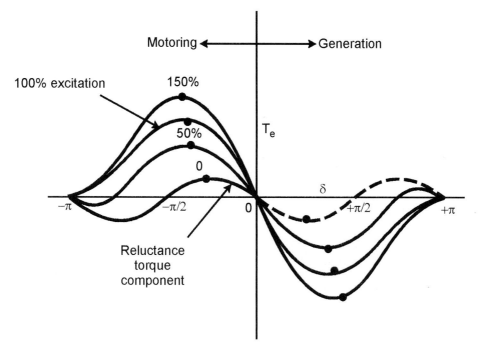

Figure 2.34 Torque-δ angle characteristics of salient pole machine

The torque (T_e)-δ angle curves for different field excitations are plotted in Figure 2.34 for both motoring and generating modes. The steady-state stability limit corresponds to the maximum points and is indicated by the dots. The reluctance torque component is the lowest curve, where the stability limit is reached at $\delta = \pm\pi/4$. It is evident from Equation (2.177) that if V_s/ω_e is maintained constant (i.e., the supply voltage is changed proportional to frequency), for a fixed excitation and torque angle, the developed torque remains constant.

2.3.1.4 Dynamic d^e-q^e Machine Model (Park Model)

A nonsalient pole synchronous machine without damper winding can be represented by an approximate per phase transient equivalent circuit as shown in Figure 2.35. The voltage V_m is the air gap voltage, which is also defined as the voltage behind the subtransient impedance $R_s + \omega_e L_{ls}$. The air gap flux linkage is somewhat sluggish to change during transient, and therefore, impedance beyond the air gap voltage does not appear in the circuit.

A more comprehensive, dynamic performance of a salient pole synchronous machine can be studied by synchronously rotating d^e-q^e frame model developed by Park [6]. The model can be derived following the same procedure discussed for an induction machine. The damper winding is equivalent to the rotor cage winding of an induction motor. Including the effects of the $d^e - q^e$ components of damper winding and the field excitation circuit, which is active in the

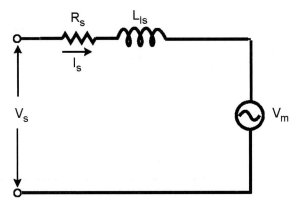

Figure 2.35 Per phase approximate transient equivalent circuit

d^e axis only, the equivalent circuits are shown in Figure 2.36, where the field circuit parameters V_{fr}, I_{fr}, R_{fr}, and L_{lfr} are referred to the stator circuit and all other symbols are given in standard notation. The electrical model of the machine in matrix form can be given as

$$\begin{bmatrix} v_{qs} \\ v_{ds} \\ 0 \\ 0 \\ v_{fr} \end{bmatrix} = \begin{bmatrix} R_s + SL_{qs} & \omega_e L_{ds} & SL_{qm} & \omega_e L_{dm} & \omega_e L_{dm} \\ -\omega_e L_{qs} & R_s + SL_{ds} & -\omega_e L_{qm} & SL_{dm} & SL_{dm} \\ SL_{qm} & 0 & R_{qr} + SL_{qr} & 0 & 0 \\ 0 & SL_{dm} & 0 & R_{dr} + SL_{dr} & SL_{dm} \\ 0 & SL_{dm} & 0 & SL_{dm} & R_{fr} + S(L_{lfr} + L_{dm}) \end{bmatrix} \begin{bmatrix} i_{qs} \\ i_{ds} \\ i_{qr} \\ i_{dr} \\ I_{fr} \end{bmatrix} \quad (2.178)$$

and

$$T_e = \frac{3}{2}\left(\frac{P}{2}\right)(\psi_{ds} i_{qs} - \psi_{qs} i_{ds}) \quad (2.179)$$

$$T_e = T_L + \frac{2}{P} J \frac{d\omega_e}{dt} \quad (2.180)$$

The above equations indicate that the electro-mechanical dynamics of the machine are described by a sixth-order nonlinear system. All the transformation equations from d^s-q^s frame to d^e-q^e frame, and vice versa, remain the same as in an induction motor. The equations can easily be described in state-space form for simulation. Note that the induction machine model becomes identical with the model of a synchronous machine if: (1) the motor runs at synchronous speed; (2) there is no saliency; and (3) excitation of the synchronous machine is ignored.

Synchronous Machines

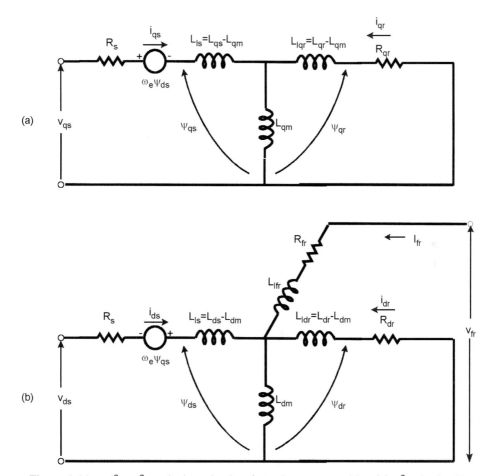

Figure 2.36 $d^e - q^e$ equivalent circuits of synchronous machine (a) q^e-axis circuit, (b) d^e-axis circuit

The steady-state model of the machine can be derived by equating all the time derivatives or S-related terms to zero. The following are the steady-state equations:

$$V_{qs} = R_s I_{qs} + \omega_e \left(\psi_f + L_{ds} I_{ds} \right)$$
$$= R_s I_{qs} + V_f + X_{ds} I_{ds} \tag{2.181}$$

$$V_{ds} = R_s I_{ds} - X_{qs} I_{qs} \tag{2.182}$$

where $\psi_f = L_{dm} I_{fr}$, and all the variables are shown as rms phasors. If stator resistance is neglected, these equations describe the phasor diagram shown in Figure 2.33(a). Since the steady-state rms phasor diagram is a special case of the d^e-q^e model, it is now clear why Figure 2.33 was drawn on d^e-q^e axes.

2.3.2 Synchronous Reluctance Machine

The idealized structure of a reluctance motor is the same as that of the salient pole synchronous machine shown in Figure 2.29, except that the rotor does not have any field winding. The stator has a three-phase symmetrical winding, which creates sinusoidal rotating magnetic field in the air gap, and reluctance torque is developed because the induced magnetic field in the rotor has a tendency to cause the rotor to align with the stator field at a minimum reluctance position. The developed torque of the reluctance machine included in Equation (2.177) can be given as

$$T_e = 3\left(\frac{P}{2}\right)\left(\psi_s^2 \frac{(L_{ds} - L_{qs})}{2L_{ds}L_{qs}} \sin 2\delta\right) \qquad (2.183)$$

The plotting of Equation (2.183) in Figure 2.34 indicates that the stability limit is reached at $\delta = \pm \pi/4$. The rotor of the modern reluctance machine is designed with iron laminations in the axial direction separated by nonmagnetic material, as shown in Figure 2.37, to increase the reluctance to flux in the q^e-axis. Compared to the induction motor, it is slightly heavier and has a lower power factor. With proper design, the performance of the reluctance motor may approach that of an induction machine. With a high saliency ratio (L_{ds}/L_{qs}), a power factor of 0.8 can be reached. The efficiency of a reluctance machine may be higher than an induction motor because there is no rotor copper loss. Because of inherent simplicity, robustness of construction, and low cost, reluctance machines have been popularly used in many low-power applications, such as fiber-spinning mills, where a number of motors operate synchronously with a common power supply. The interest in reluctance motor drives is growing.

2.3.3 Permanent Magnet (PM) Machine

In a permanent magnet synchronous machine, the dc field winding of the rotor is replaced by a permanent magnet. The advantages are elimination of field copper loss, higher power density, lower rotor inertia, and more robust construction of the rotor. The demerits are loss of flexibility of field flux control and possible demagnetization effect. The machine has higher efficiency than an induction motor, but generally its cost is higher, which makes the life cycle cost of the drive somewhat lower. Permanent magnet machines, particularly at low-power range, are widely used in industry. Recently, the interest in their application is growing, particularly up to 100 kW.

2.3.3.1 Permanent Magnet Materials

The property of a permanent magnet and the selection of the proper materials are very important in the design of a permanent magnet (PM) synchronous machine. Figure 2.38 shows

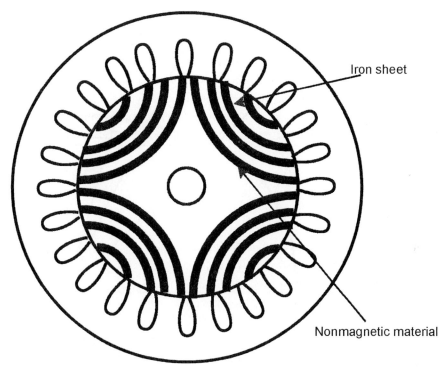

Figure 2.37 Cross-section of synchronous reluctance motor

the demagnetization segment of the B-H curve where the permanent magnet is usually designed to operate. The maximum flux density B_r corresponding to point A' will be available initially if the magnet is short-circuited with steel keepers (no air gap). When the magnet is installed in the machine, the air gap will have some demagnetization effect and the operating point B' will correspond to the no-load line shown in the figure. The slope of the no-load line (with respect to the H-axis) will be smaller with higher air gap. With current flowing in the stator winding, the magnetic axis (d^e) armature reaction effect can have a further demagnetization effect, which will further reduce the air gap flux density. A load line corresponding to worst-case demagnetization, which may be due to a starting, transient, or machine fault condition, is also shown in the figure. Once the operating point reaches D and the demagnetization effect is removed, the magnet will recover along the recoil line, which has approximately the same slope as the original B-H curve near $H = 0$. In a subsequent operation, the stable operating point will be determined by the intersection of the load line and the recoil line. The magnet is therefore permanently demagnetized at no-load operation, corresponding to the vertical distance between A' and A. The worst-case demagnetization point is therefore vitally important for machine performance and should be closely controlled. Alternatively, if the material of the permanent magnet is selected to have a straight-line demagnetization curve, the recoil line will coincide with the demagnetization line

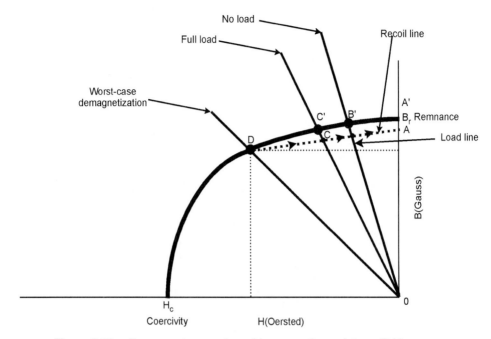

Figure 2.38 Permanent magnet machine operating points on B-H curve

irrespective of the worst-case demagnetization point (i.e., permanent demagnetization will be negligible).

Figure 2.39 shows the characteristics of several possible PM materials. Alnico has high service temperature, good thermal stability, and high flux density, but the disadvantage is low coercive force coupled with squarish B-H characteristics, which makes the permanent demagnetization high so that it is practically unsuitable for a PM machine. Barium and strontium ferrites are widely used as permanent magnets. Ferrite has the advantages of low cost and plentiful supply of raw material. They are also easy to produce, and their process is suited for high volume, as well as moderately high service temperature (400°C). The magnet has a practically linear demagnetization curve, but its remnance (B_r) is low. Therefore, the volume and weight of the machine tends to be high. The Cobalt-Samarium (CoSm) magnet is made of iron, nickel, cobalt, and rare-Earth Samarium. It has the advantages of high remnance, high energy density defined by (BH_{max}), and linear demagnetization characteristics. The service temperature can be as high as 300 °C, and the temperature stability (% change in B per °C) is very good (–0.03%). But, the material is very expensive because of an inadequate supply of Samarium. The Neodymium-iron-boron (Nd-Fe-B) magnet has the highest energy density, highest remnance, and very good coercivity (H_c). The disadvantages are low service temperature (150 °C) and susceptibility to oxidation unless protected by a coating. Besides, the temperature stability (-0.13%) is inferior to that of a CoSm magnet. The material is expensive compared to ferrite, but because of higher energy

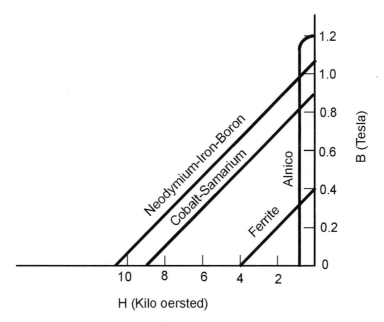

Figure 2.39 Permanent magnet characteristics

density, the machine weight is reduced. The application of Nd-Fe-B magnets is growing in PM machines.

2.3.3.2 Sinusoidal Surface Magnet Machine (SPM)

In this machine, as shown in Figure 2.40, the stator has a three-phase sinusoidal winding as before, which creates a synchronously rotating air gap flux. The PMs are glued on the rotor surface using epoxy adhesive. The rotor has an iron core, which may be solid or made of punched laminations for simplicity of manufacture. Line-start 60 Hz PM machines may have a squirrel-cage winding to start as an induction motor. For variable-speed operation, PM machines may or may not have a cage or damper winding, which has an additional loss due to harmonics. If the machine is rotated by a prime mover, the stator windings generate balanced three-phase sinusoidal voltages. Since the relative permeability of a PM is very close to one ($\mu_r > 1$), and magnets are mounted on the rotor surface, the effective air gap of the machine is large and the machine is a nonsalient pole ($L_{dm} = L_{qm}$). This contributes to a low armature reaction effect due to low magnetizing inductance.

2.3.3.3 Sinusoidal Interior Magnet Machine (IPM)

Unlike an SPM, in an interior or buried magnet synchronous machine (IPM), the magnets are mounted inside the rotor. Although a number of geometries are possible, a typical configuration is shown in Figure 2.41. The stator has the usual three-phase sinusoidal winding. The difference in the geometry gives the following characteristics to the IPM machine: (1) the machine is

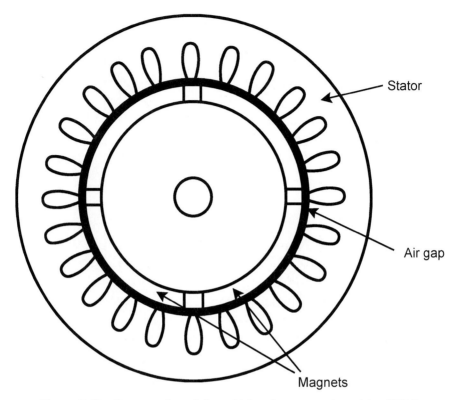

Figure 2.40 Cross-section of sinusoidal surface magnet machine (SPM)

more robust, permitting a much higher speed of operation, (2) the effective air gap in the d^e-axis is larger than that in the q^e-axis, which makes the machine a salient pole with $L_{dm} < L_{qm}$ (unlike a standard wound field synchronous machine), and (3) with the effective air gap being low, the armature reaction effect becomes dominant.

The steady-state analysis of a sinusoidal PM machine with an equivalent circuit and phasor diagram remains the same as a wound field machine except that the equivalent field current I_f should be considered constant, that is, the flux linkage $\psi_f = L_m I_f' =$ constant. The synchronously rotating frame transient equivalent circuits, shown in Figure 2.36, also hold true here, except the machine may not have any damper winding. Figure 2.42 shows the equivalent circuits where the finite core loss is represented by the dotted damper windings. Ignoring the core loss, the circuit equations can be written as

$$v_{qs} = R_s i_{qs} + \omega_e \psi_{ds}' + \omega_e \hat{\psi}_f + \frac{d}{dt}\psi_{qs} \tag{2.184}$$

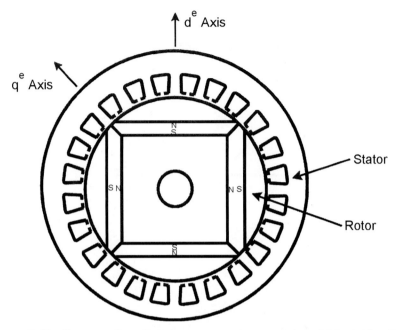

Figure 2.41 Cross-section of interior permanent magnet sinusoidal machine (IPM)

$$v_{ds} = R_s i_{ds} - \omega_e \psi_{qs} + \frac{d}{dt}\psi_{ds} \quad (2.185)$$

where

$$\hat{\psi}_f = L_{dm} I'_f \quad (2.186)$$

$$\psi'_{ds} = i_{ds}(L_{ls} + L_{dm}) = i_{ds} L_{ds} \quad (2.187)$$

$$\psi_{ds} = \hat{\psi}_f + \psi'_{ds} \quad (2.188)$$

$$\psi_{qs} = i_{qs}(L_{ls} + L_{qm}) = i_{qs} L_{qs} \quad (2.189)$$

and the torque equation is

$$T_e = \frac{3}{2}\left(\frac{P}{2}\right)\left(\psi_{ds} i_{qs} - \psi_{qs} i_{ds}\right) \quad (2.114)$$

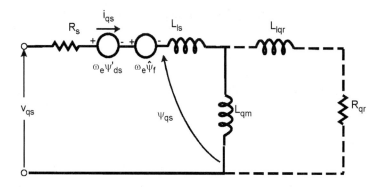

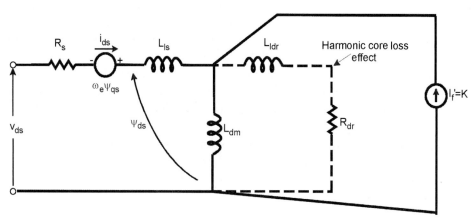

Figure 2.42 Synchronously rotating frame ($d^e - q^e$) equivalent circuits of IPM machine

Substituting Equations (2.186)–(2.189) in (2.184), (2.185), and (2.114) and simplifying, we can write

$$\frac{di_{qs}}{dt} = \frac{\omega_b}{X_{qs}}\left[v_{qs} - R_s i_{qs} - \frac{\omega_e}{\omega_b}X_{ds}i_{ds} - \frac{\omega_e}{\omega_b}V_f\right] \quad (2.190)$$

$$\frac{di_{ds}}{dt} = \frac{\omega_b}{X_{ds}}\left[v_{ds} - R_s i_{ds} - \frac{\omega_e}{\omega_b}X_{qs}i_{qs}\right] \quad (2.191)$$

$$T_e = \frac{3P}{4\omega_b}\left[\left(F'_{ds} + V_f\right)i_{qs} - F_{qs}i_{ds}\right] \quad (2.192)$$

where $V_f = \omega_b \hat{\psi}_f$, $X_{qs} = \omega_b L_{qs}$, $X_{ds} = \omega_b L_{ds}$, $F_{ds}' = \omega_b \psi_{ds}'$, $F_{qs} = \omega_b \psi_{qs}$, and ω_b = base frequency. These equations, which are valid for IPM as well as SPM (except $L_{dm} = L_{qm}$), can be used for computer simulation study.

Again, for steady-state operation of the machine, the time derivative components of Equations (2.184) and (2.185) are zero, that is, these can be written in the form of Equations (2.181) and (2.182), respectively, which correspond to Figure 2.33, except the resistance drops.

2.3.3.4 Trapezoidal Surface Magnet Machine

A trapezoidal SPM machine is a nonsalient pole, surface-mounted PM machine similar to a sinusoidal SPM machine except its three-phase stator winding (normally wye-connected) has concentrated full-pitch distribution instead of sinusoidal winding distribution. Figure 2.43 shows the cross-section of the machine; its three stator phases are shown at the right side. The two-pole machine is shown with a gap to reduce the flux fringing effect, and the stator is shown with four slots per pole per phase. As the machine rotates, most of the time flux linkage in a phase winding varies linearly, except when the magnet gap passes through the phase axis. If the machine is rotated by a prime mover, the stator phase voltages will have symmetrical trapezoidal wave shape as shown in Figure 2.44. An electronic inverter is required in the front end to establish a six-step current wave at the center of each half-cycle to develop torque. Since converter use is mandatory, it is often defined as an electronic motor. With the help of an inverter and an absolute-position sensor mounted on the shaft, both sinusoidal and trapezoidal SPM machines can be controlled to have "brushless dc motor" (BLDM) performance. However, a trapezoidal machine gives closer to dc machine-like performance. The machine is simple, inexpensive, and has somewhat higher power density than the sinusoidal machine. Further discussion of the machine and the drive will be given in Chapter 9. Low-power (up to a few kW) drives using this machine are commonly used in servo and appliance drives where the commutators and brushes of a dc motor are not desirable.

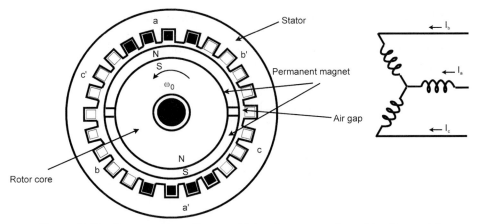

Figure 2.43 Cross-section of trapezoidal surface magnet machine (two-pole)

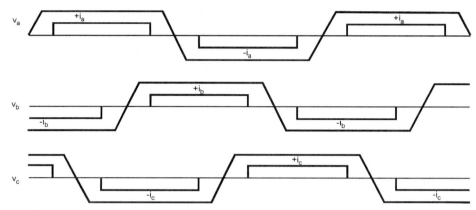

Figure 2.44 Stator phase voltage and current waves in trapezoidal PM machine

2.4 VARIABLE RELUCTANCE MACHINE (VRM)

A variable or double reluctance machine (VRM), as the name indicates, has double saliency, meaning it has saliency in the stator as well as in the rotor. As mentioned before, the VRM has two classifications: switched reluctance machine (SRM) and stepper motor. The stepper motor is basically a digital motor, i.e., it moves by a fixed step or angle with a digital pulse. Small stepper motors are widely used for computer peripheral-type applications. However, since the machine is not suitable for variable-speed applications, there will not be any further discussion of it.

There has been interest in switched reluctance motor drives in the literature, and recently, great effort has been made to commercialize them in competition with induction motors. Figure 2.45(a) shows the cross-section of a four-phase machine with four stator-pole pairs and three rotor-pole pairs (8/6 motor). The machine rotor does not have any winding or PM. The stator poles have concentrated winding (instead of sinusoidal winding), and each stator-pole pair winding, as shown in the figure, is excited by a converter phase. For example, the stator-pole pair A-A' is energized when the rotor pole-pair a-a' approaches it to produce the torque by magnetic pull, but is de-energized when pole alignment occurs. All four machine phases are excited sequentially and synchronously with the help of a rotor position encoder to get unidirectional torque. The inductance profile of a stator-pole pair with respect to rotor angular position and the corresponding phase-current waves is shown in Figure 2.45(b). In the forward direction, as shown, the motoring torque is developed by establishing the stator current pulse where the inductance profile has positive slope; whereas for regenerative braking, the current pulse is at the negative slope. For the 60° cycle period shown in the figure, a particular phase is excited every 60°, and four consecutive phases are excited at 15° intervals. The magnitude of torque can be given as

$$T_e = \frac{1}{2}mi^2 \qquad (2.193)$$

Variable Reluctance Machine (VRM)

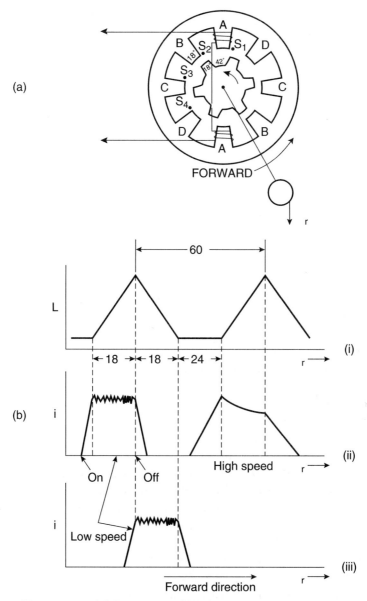

Figure 2.45 (a) Construction of switched reluctance machine (SRM), (b) Inductance profile and phase current waves

where m = inductance slope and i = instantaneous current. The current i can be maintained constantly by control during the inductance slope. At high speeds, the rotor-induced CEMF is high, which makes the current wave as indicated. Note that the backward slope at the end of the current pulse contributes to negative torque.

The favorable attributes of this electronic motor are simplicity and robustness of construction; potentially, it is somewhat cheaper than other classes of machines. However, the torque generation is pulsating in nature and there are serious acoustic noise problems.

2.5 SUMMARY

In this chapter, we attempted to give a comprehensive review of the different types of ac machines, including induction machines, wound field synchronous machines (WFSM), surface magnet synchronous machines (SPM), interior magnet synchronous machines (IPM), surface magnet trapezoidal synchronous machines, and switched reluctance machines (SRM). Cage-type induction machines, which are commonly used, were emphasized over wound rotor machines.

In the beginning, steady-state performance of induction motors was discussed, which was then followed by the dynamic d-q model in both synchronously rotating reference frame and stationary reference frame. Then, state-space equations in terms of flux linkages were derived mainly for simulation. For easy understanding, the author represented the dynamic equations and corresponding equivalent circuits in cartesian form with d and q coordinates instead of complex space vector representation. In this context, the difference between space vectors and rms phasors for sinusoidal variables was clarified.

For synchronous machines, steady-state analyses with an rms phasor diagram and then a d^e-q^e dynamic model were discussed in detail. Since PM machines are widely used, a dynamic model with state-space equations was derived. Finally, switched reluctance machine principles were discussed briefly.

An intimate understanding of machine performance, including parameter variation characteristics, is necessary to design modern high-performance drives. Often, for a particular application, more than one type of machine can be used. The final selection will depend on a cost/performance trade-off, where the various factors to consider are initial cost, size and weight, efficiency, dynamic response, power factor, rotor inertia, reliability, need of position or speed encoder, etc. Further discussion of machines will be continued in Chapters 8 and 9.

References

1. A. E. Fitzgerald, C. Kingslay, and S. D. Umans, *Electric Machinery*, McGraw-Hill, New York, 1983.
2. G. R. Slemon, "Electrical machines for variable frequency drives", *Proceedings of the IEEE*, vol. 82, pp. 1123–1139, Aug. 1994.
3. P. C. Krause, *Analysis of Electric Machinery,* McGraw-Hill, New York, 1986.
4. C. M. Ong, *Dynamic Simulation of Electric Machinery*, Prentice Hall, New Jersey, 1998.
5. B. K. Bose, "Power electronics and motion control – technology status and recent trends", *IEEE Trans. on Ind. Appl.*, vol. 29, pp. 902–909, Sept./Oct. 1993.
6. R. H. Park, "Two-reaction theory of synchronous machines – generalized method of analysis- Part 1", *AIEE Trans.*, vol. 48, pp. 716–727, July 1929.
7. H. C. Stanley, "An analysis of induction motor", *AIEE Trans.*, vol. 57 (Supplement), pp. 751–755, 1938.
8. G. Kron, *Equivalent Circuits of Electric Machinery*, John Wiley, New York, 1951.

9. S. D. T. Robertson and K. M. Hebber, "Torque pulsations in induction motors with inverter drives", *IEEE Trans. Ind. Appl.*, vol. 7, pp. 318–323, Mar./Apr. 1971.
10. R. Hagmann, "AC cycloconverter drives for cold and hot rolling mill applications", *IEEE IAS Annu. Meet. Conf. Rec.*, pp. 1134–1140, 1991.

CHAPTER 3

Diodes and Phase-Controlled Converters

3.1 INTRODUCTION

Diodes and phase-controlled converters constitute the largest segment of power electronics that interface to the electric utility system today. The history of these converters extends nearly one hundred years, and they are often defined as classical power electronics. Before the advent of solid-state diodes and thyristors, which are invariably used presently, gas-filled glass-bulb devices such as mercury-arc rectifiers, phanotrons, thyratrons, and ignitrons were dominant in the early part of this century. Then, during World War II, saturable-core magnetic amplifiers were introduced. This class of converters mainly converts 50/60 Hz ac to dc (rectification), but a select group can also function for dc to ac conversion (inversion). The efficiency of the converters is very high, typically in the vicinity of 98%, because device conduction loss is low and switching loss is practically negligible. However, the disadvantage is that they generate harmonics in the utility system creating a power quality problem for other consumers. Besides, thyristor converters constitute a low lagging power factor load on the utility system. The application of these converters may include the following:

- Electrochemical processes such as electroplating, anodizing, metal refining, and chemical gas production (hydrogen, oxygen, chlorine, etc.)
- Adjustable-speed dc and ac motor drives
- High-voltage dc (HVDC) systems
- Dc and ac general-purpose power supplies, including UPS (uninterruptable power supply) systems
- Dc-to-ac power conversion from solar cells, fuel cells, etc. with interface to the utility system

In this chapter, we will describe the key configurations of diodes and phase-controlled thyristor converters. Although the dc motor drive is not a theme of this book, the dc motor speed control with thyristor converter will be included here for completeness. The background in this chapter will be an important ingredient for phase-controlled cycloconverters, which will be discussed in Chapter 4, and current-fed converters, which will be described in Chapter 6. While explaining the principles of different converters, we will consider the devices as ideal, meaning zero conduction drop, no reverse recovery current, and instantaneous turn-on and turn-off switching.

3.2 DIODE RECTIFIERS

Diode rectifiers are the simplest and possibly the most important power electronics circuits. They are rectifiers because power can flow only from the ac side to the dc side. In this section, we will discuss only the most important circuit configurations, that is, the single-phase diode bridge and three-phase diode bridge. Again, the commonly used loads such as resistance, resistance-inductance, and capacitance-resistance will be considered. Note that thyristor converters with zero firing angle also behave as diode rectifiers. Therefore, understanding the more complex diode circuits by extrapolating the performance of thyristor converters is left as an exercise to the reader.

3.2.1   Single-Phase Bridge – *R*, *RL* Load

The single-phase diode bridge rectifier is one of the simplest power electronic circuits, and it is shown in Figure 3.1 with resistance-inductance load. The performance of this circuit is similar to the rectifier with a center-tapped transformer shown in Figure 3.2. The latter circuit uses only two diodes; in addition, a transformer that can provide voltage level change and electrical isolation from the primary is used. In Figure 3.1, the ac supply is represented by a sinusoidal voltage source (v_s) in series with a Thevinin leakage inductance L_c (the resistance is neglected), which will be ignored in the present analysis. In the positive polarity of supply voltage, as shown, the diodes D_1 and D_2 will conduct current through the load, whereas in the negative polarity, diodes D_3 and D_4 will conduct. When a pair of diodes is conducting, the voltage v_s will appear as reverse voltage to the other devices.

Figure 3.3(a) shows the waveforms for the *R* load where the load voltage (v_d) is single-phase, full-wave rectified, which contains a dc component, and the load current (i_d) is proportional to this voltage. Each of the diodes on the positive side (D_1D_3) and negative side (D_2D_4) of the bridge conduct for a half-cycle so that the continuity of current flow is maintained. The line current (i_s) is sinusoidal (no distortion), and it is in phase with the line voltage, as shown in the

Diode Rectifiers

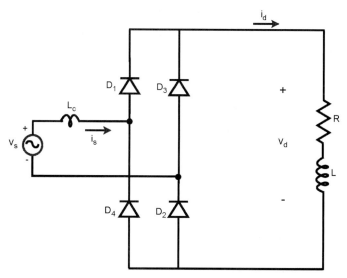

Figure 3.1 Single-phase diode bridge rectifier with R or RL load

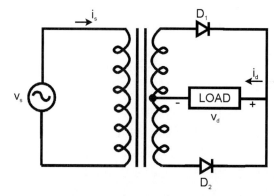

Figure 3.2 Single-phase rectifier with center-tapped transformer

figure. The load voltage and current waves are very rich in harmonics, and the order of harmonics are even (since $f(\omega t) = f(-\omega t)$). The Fourier analysis of v_d wave gives

$$v_d(\omega t) = a_0 + \sum_{n=2,4,\ldots}^{\infty} (a_n \cos n\omega t + b_n \sin n\omega t) \tag{3.1}$$

$$a_n = \frac{1}{\pi} \int_0^{2\pi} v_d(\omega t) \cos n\omega t \, d\omega t \tag{3.2}$$

$$b_n = \frac{1}{\pi} \int_0^{2\pi} v_d(\omega t) \sin n\omega t \, d\omega t \tag{3.3}$$

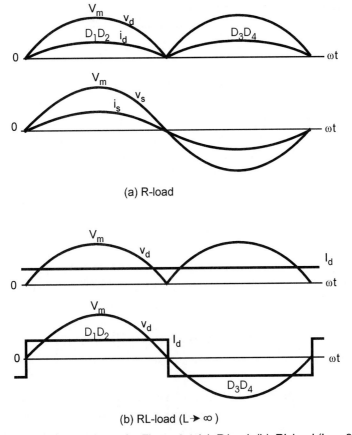

Figure 3.3 Waveforms for Figure 3.1 (a) R-load, (b) RL-load (L → 90)

$$a_0 = V_d = \frac{1}{\pi}\int_0^\pi V_m \sin\omega t\, d\omega t = \frac{2}{\pi}V_m = 0.9V_s \qquad (3.4)$$

$$I_d = \frac{1}{\pi}\int_0^\pi i_d\, d\omega t = \frac{V_d}{R} = 0.9\frac{V_s}{R} \qquad (3.5)$$

where V_d = average or dc output voltage, V_m = peak supply voltage, V_s = rms supply voltage, and I_d = average or dc output current. The b_n or sine components will disappear if cosine symmetry is assumed. Therefore, substituting $v_d(\omega t) = V_m \sin\omega t$ in Equation (3.2) gives

$$\begin{aligned} a_n &= \frac{2}{\pi}\int_0^\pi V_m \sin\omega t \cos n\omega t\, d\omega t \\ &= \frac{4V_m}{\pi}\sum_{n=2,4,\ldots}^\infty \frac{-1}{(n-1)(n+1)} \end{aligned} \qquad (3.6)$$

Diode Rectifiers

The complete Fourier series expression is

$$v_d(\omega t) = \frac{2}{\pi} V_m - \frac{4V_m}{3\pi}\cos 2\omega t - \frac{4V_m}{15\pi}\cos 4\omega t - \frac{4V_m}{35\pi}\cos 6\omega t ... \quad (3.7)$$

which indicates magnitudes of the even harmonic components.

Figure 3.3(b) shows the waveforms with RL load, where L is assumed to be infinity for simplicity. The output voltage waveform will remain the same, but the load current will be smooth dc as shown because all the harmonic voltages will be absorbed by the load inductance. With the load current constrained to be pure dc, the source current will be a square wave of amplitude I_d. The current will contain only odd harmonics (since $f(\omega t) = -f(-\omega t)$) and is given by the Fourier series

$$i_s(\omega t) = \frac{4I_d}{\pi}\left[\sin \omega t + \frac{1}{3}\sin 3\omega t + \frac{1}{5}\sin 5\omega t + ...\right] \quad (3.8)$$

where the fundamental (I_{s1}) and total rms (I_s) values are given by $4I_d/\sqrt{2}\pi$ and I_d, respectively. The ripple or distortion factor (also defined as total harmonic distortion (THD)) for this wave can be calculated from Equation (3.8) as 0.482 or 48.2 percent.

$$RF = \sqrt{\frac{I_s^2 - I_{s1}^2}{I_{s1}^2}} = \sqrt{\left(\frac{I_s^2}{I_{s1}^2}\right) - 1} \quad (3.9)$$

This is a considerable departure from the ideal sine current waveform desired in the line. Obviously, if the load inductance is limited, a ripple current will flow in the load and this will reflect in the line current. The waveforms given above help to design the rectifier easily.

3.2.2 Effect of Source Inductance

So far, we have neglected the leakage inductance L_c of the line. This parameter affects the current transfer (or commutation) from the outgoing device to the incoming device and distorts the square line current wave, as shown in Figure 3.4. Again, it is assumed that the load is highly inductive so that the current I_d is maintained constant. Consider, for example, the current commutation from D_1D_2 to the D_3D_4 pair. In the presence of L_c, this current transfer will be slowed down. Initially, $i_s = +I_d$ and it is flowing through L_c. As the supply voltage becomes negative, D_3D_4 will begin to conduct. This will short-circuit the bridge and the voltage v_s will be impressed across L_c, which will start decreasing the line current i_s. The current in D_3D_4 will gradually increase and the current in D_1D_2 will gradually decrease (the total I_d remaining constant) until the current transfer is complete. The total commutation angle μ, as shown, will depend on the load current, inductance value, and supply voltage magnitude. Note that the dc output voltage V_d will be somewhat smaller because of the loss of a voltage segment during commutation. A detailed analysis for commutation overlap will be given later.

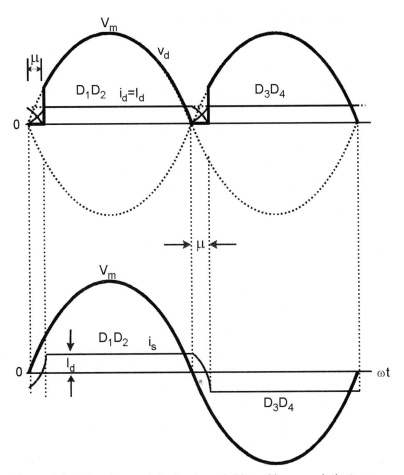

Figure 3.4 Waveforms of single-phase bridge with source inductance

3.2.3 Single-Phase Bridge – *RL*, CEMF Load

Figure 3.1 may contain a CEMF load in addition to resistance and inductance. The CEMF may be due to a battery or dc motor load. With CEMF load and assuming that the current is always flowing in the load (continuous conduction), the dc (or average) load voltage equation can be given as

$$V_d = I_d R + V_c \tag{3.10}$$

where V_c = CEMF. In such applications, the load inductance is finite and therefore will contribute some ripple current. With large CEMF, the load current may be intermittent or discontinuous, which will be discussed later.

3.2.4 Single-Phase Bridge – CR Load

A very important configuration is the diode bridge rectifier with capacitance-resistance (CR) load shown in Figure 3.5. The capacitor C_F filters the rectified voltage to make the dc output voltage smooth. The equivalent resistance R represents the load. The dc voltage may be used for a voltage-fed inverter or dc-to-dc converter. The waveforms of the circuit are shown in Figure 3.6. The capacitor voltage v_d behaves as a CEMF of the rectifier. The capacitor will charge with pulse current every half-cycle, when $v_d < v_s$ near the peak voltage V_m, as shown, and the current becomes limited by the line inductance L_c. The capacitor then discharges exponentially with the load time constant $C_F R$. The load current is always proportional to the capacitor voltage. The discontinuous load current pulses cause pulsating line current as shown in the figure. Note that the initial charging of the capacitor should be done through a series resistor, which can be shorted with a bypass switch at steady state. Otherwise, the diodes will be overloaded with large in-rush current. The ripple in the capacitor voltage will be reduced with higher capacitance value, causing narrow current pulses, but the pulses will widen with higher line inductance. Obviously, at no load (R = infinity), the capacitor will remain charged to peak value V_m of the supply.

An approximate analysis of converter operation can be made if the ripple in the capacitor voltage v_d is neglected, that is, the capacitor size is very large. With large energy storage, the

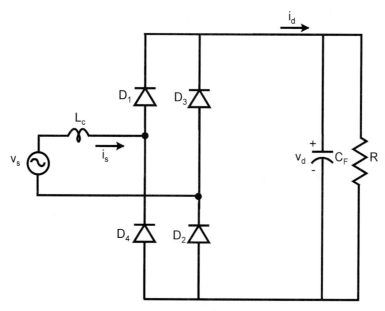

Figure 3.5 Single-phase diode bridge rectifier with CR load

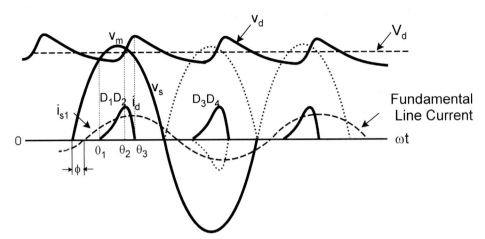

Figure 3.6 Waveforms for Figure 3.5

load current will always be smooth, although the capacitor charging current is pulsating. The capacitor will start charging at angle θ_1 when

$$V_m \sin\theta_1 = V_d \quad (3.11)$$

where $v_d = V_d$.

During conduction, the equation for charging current i_d is given as

$$L_c \frac{di_d}{dt} = V_m \sin\omega t - V_d \quad (3.12)$$

which can be expressed as

$$i_d = \frac{1}{\omega L_c} \int_{\theta_1}^{\theta} (V_m \sin\omega t - V_d)\, d\omega t \quad (3.13)$$

The current will reach the peak value at angle θ_2 when $V_d = v_s$ again.

For the discontinuous conduction shown, $i_d = 0$ at $\omega t = \theta_3$. Therefore,

$$0 = \frac{1}{\omega L_c} \int_{\theta_1}^{\theta_3} (V_m \sin\omega t - V_d)\, d\omega t \quad (3.14)$$

Equation (3.14) gives the relation between θ_1 and θ_3 for a given V_d. The average current I_d can be given by

$$I_d = \frac{1}{\pi} \int_{\theta_1}^{\theta_3} i_d\, d\omega t \quad (3.15)$$

Equations (3.11), (3.13), (3.14), and (3.15) can be combined to derive a relation between the current I_d and the voltage V_d. A plot of this relation in per unit values is shown in Figure 3.7 where $I_{base} = V_s/\omega L_c$ (short-circuit current) and $V_{base} = 2/\pi\, V_m$ (see Equation (3.4)). Note that with higher load current I_d, the level of V_d will be adjusted so that the average capacitor charging current is the same as I_d.

At no load, the capacitor will charge to peak voltage V_m, ($V_d\,(pu) = 1.57$), so that the current falls to zero. The detailed analysis of a practical converter and the corresponding waveforms is quite involved and requires computer simulation.

3.2.5 Distortion, Displacement, and Power Factors

So far, we have seen that the line current wave flowing in the utility line is usually distorted. This is natural because the converter is essentially a nonlinear load. The harmonics in the distorted current load the power line equipment. Besides, the distorted current flowing through

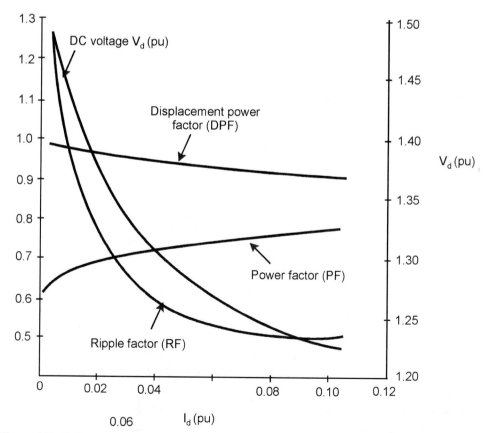

Figure 3.7 DC voltage, RF, DPF, and PF characteristics for Figure 3.5 (CF → ∞) (not in scale)

3.2.6 Distortion Factor (DF)

The degree of line current distortion can be determined by the distortion factor (DF). It is defined as

$$DF = \frac{Rms\ value\ of\ fundamental\ current}{Rms\ value\ of\ total\ current}$$

$$= \frac{I_{s1}}{\sqrt{I_{s1}^2 + \sum_{n=1,2,3,\ldots}^{\infty} I_{sn}^2}} \tag{3.16}$$

For square-wave current, as shown in Figure 3.8,

$$DF = \frac{\frac{4}{\pi}\left(\frac{1}{\sqrt{2}}\right)I_d}{I_d} = \frac{2\sqrt{2}}{\pi} = 0.9 \tag{3.17}$$

3.2.7 Displacement Power Factor (DPF)

It will be shown later that fundamental line current in a phase-controlled converter always lags the fundamental voltage by a displacement angle ϕ, as indicated in Figure 3.8. The displacement power factor (DPF) can be defined as

$$DPF = \frac{Average\ power}{Fundamental\ rms\ voltage \times Fundamental\ rms\ current} \tag{3.18}$$

$$= \frac{P_1}{V_s I_{s1}} = \frac{V_s I_{s1} \cos\phi}{V_s I_{s1}} = \cos\phi$$

where V_s = fundamental rms voltage and I_{s1} = fundamental rms current.

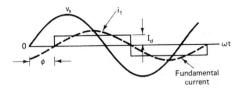

Figure 3.8 Line voltage and current waves

Diode Rectifiers

3.2.8 Power Factor (PF)

The line power factor (PF) can be defined as

$$PF = \frac{Average\ power}{Supply\ rms\ voltage \times Supply\ rms\ current}$$

$$= \frac{P_1}{V_s\sqrt{I_{s1}^2 + \sum_{n=1,2,3,\ldots}^{\infty} I_n^2}} \quad (3.19)$$

Substituting Equations (3.16) and (3.18) in (3.19) gives

$$PF = \frac{V_s I_{s1} \cos\phi}{V_s\sqrt{I_{s1}^2 + \sum_{n=1,2,3,\ldots}^{\infty} I_n^2}} = \cos\phi \frac{I_{s1}}{\sqrt{I_{s1}^2 + \sum_{n=1,\ldots}^{\infty} I_n^2}} \quad (3.20)$$

$$= DPF \times DF$$

Since the circuit of Figure 3.5 is very important for household appliances and low-power (typically up to 10kW) industrial applications, the analysis of pulsating line current i_s is very important. With computer simulation, the Fourier series of the current wave can be analyzed and its RF (see equation (3.9)), DPF, and PF can be determined as functions of load current I_d and included in Figure 3.7. Note that the DPF is near unity, as indicated in Figure 3.6, but severe harmonic distortion of the current wave makes high RF and low PF, which tend to improve with higher loading.

3.2.9 Three-Phase Full Bridge – RL Load

For higher power applications and where three-phase power supply is available, a three-phase bridge rectifier, as shown in Figure 3.9, should be used. The converter is shown with a delta-wye transformer at the input and resistance-inductance load. Line inductance L_c is neglected for simplicity. As mentioned before, the use of a transformer is optional; one should be used when the voltage level changes and isolation from the supply line is desirable. The circuit is redrawn in (b) of the same figure, where the upper + converter consists of devices $D_1 D_3 D_5$ and the lower converter consists of $D_2 D_4 D_6$. Figure 3.10 shows the waveforms for the bridge rectifier assuming highly inductive load ($L \rightarrow$ infinity). As shown in (a), the diode in the + converter, which has the highest voltage, and the diode in the – converter, which has the lowest voltage, will conduct at any instant. Each diode will conduct for $2\pi/3$ angle, and the load current will be smooth dc, as shown in the figure. The line currents at the output and input of the transformer are also constructed graphically. The output dc voltage wave v_d, which is between the

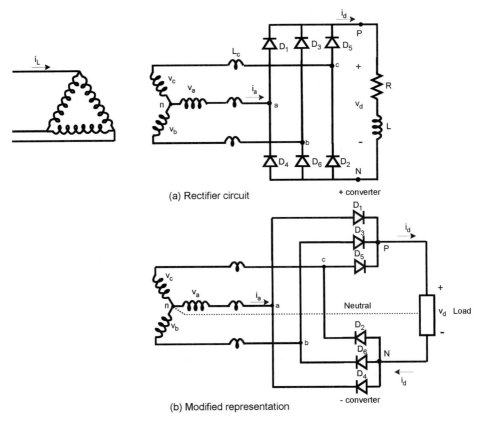

Figure 3.9 Three-phase diode bridge rectifier with R or RL load

positive and negative envelopes of (a), is shown in (b). The wave is symmetrical and repeats every $\pi/3$ angle. The average dc voltage V_d can be calculated as

$$V_d = \frac{3}{\pi} \int_{-\pi/6}^{+\pi/6} \sqrt{2} V_L \cos\omega t \, d\omega t = 1.35 V_L \tag{3.21}$$

where $V_L = \sqrt{3} V_s$ is the rms line voltage at the rectifier input. It can be easily seen that each diode experiences a peak reverse voltage of $\sqrt{3} V_m$. The input current i_a in (e) has the characteristic six-step wave shape, which can be expressed by the Fourier series as

$$i_s = \frac{2\sqrt{3}}{\pi} I_d \left[\sin\omega t - \frac{1}{5}\sin 5\omega t - \frac{1}{7}\sin 7\omega t + \frac{1}{11}\sin 11\omega t + \ldots \right] \tag{3.22}$$

This indicates that the fundamental rms current is $(\sqrt{6}/\pi) I_d$, and harmonics of the order $(6n \pm 1)$ [n = integer] only are present. The fundamental input current will be in phase with the

Diode Rectifiers

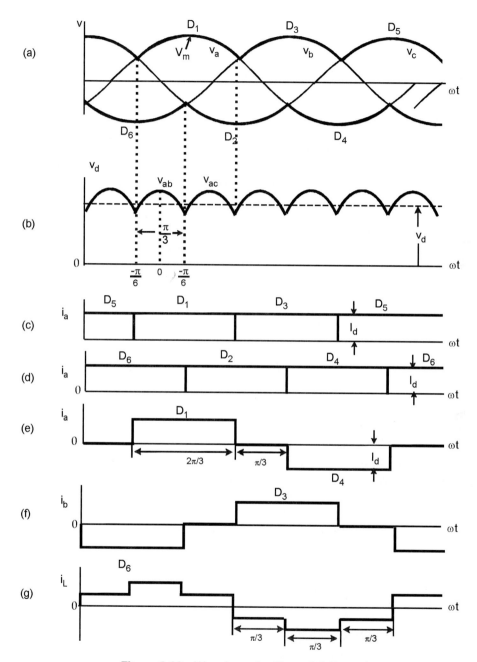

Figure 3.10 Waveforms for Figure 3.9 ($L \to \infty$)

phase voltage, giving DPF = 1. Note that the three-phase bridge has the advantage of less harmonic content in load voltage and line current.

3.2.10 Three-Phase Bridge – *CR* Load

A three-phase bridge rectifier with capacitance-resistance load is shown in Figure 3.11. It is basically an extension of Figure 3.5 for higher power load. Figure 3.12 shows the waveforms assuming the filter capacitor C_F is very large. The envelope of the peak-to-peak line voltage is the same as that of Figure 3.10(b). For example, when line voltage v_{ab} exceeds the dc voltage V_d, diodes D_1 and D_6 will conduct to contribute a pulse of load current flowing in phase a and phase b. Then again, when $v_{ac} > V_d$, another load current pulse will flow through D_1 and D_2. It can be shown easily that each line contributes two identical current pulses in each half-cycle, as indicated in Figure 3.12(b). With higher load current I_d, the level of V_d will decrease. Again, as in Figure 3.6, the RF of the line current will be high, contributing low PF, but DPF will be near unity. The RF, DPF, and PF can be determined by Fourier analysis of the line current wave. Figure 3.13 shows the plot of these parameters where $V_{base} = 1.35\ V_L$ and I_{base} = short-circuit line current given by

$$I_{base} = \frac{V_s}{\omega L_c} \tag{3.23}$$

Comparing with Figure 3.7, it can be seen that in a three-phase bridge, the dc voltage drop is much smaller, the RF is much lower, and the PF is somewhat improved.

3.3 THYRISTOR CONVERTERS

In a diode rectifier, as discussed above, power flows only from input to output and it cannot be controlled. Of course, the fluctuation of power will occur naturally with the fluctuation of supply voltage and load. In many applications, the power flow requires control. Even for a load that requires constant voltage or constant current, control is required to compensate supply and load fluctuation. A thyristor converter can control the power flow by the phase control principle, which delays the firing angle of the gate. The converter has the additional capability that power can be converted from the dc to the ac side, in other words, it is an inverter function. There are many possible configurations of thyristor converter. In fact, every diode rectifier has an equivalent thyristor configuration. In this section, only the topologies that are commonly used will be studied. Much of the discussion for the diode rectifier is also valid for the thyristor converter.

3.3.1 Single-Phase Bridge – *RL*, CEMF Load

Figure 3.14 shows the single-phase thyristor bridge configuration with resistance, inductance and CEMF load. The line leakage inductance L_c has been neglected for simplicity. Figure 3.15 shows the waveforms for the rectifier mode of operation. The load inductance is assumed to be very large so that the load current I_d is ripple-free. In the positive half-cycle, the thyristors Q_1Q_2

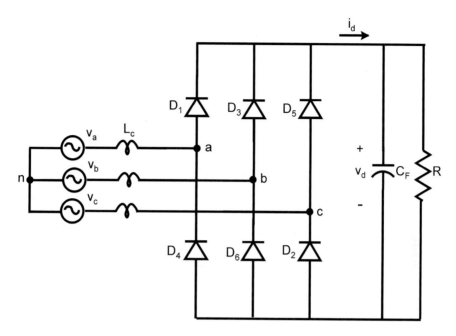

Figure 3.11 Three-phase diode bridge rectifier with CR load

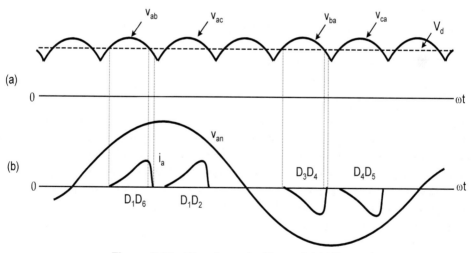

Figure 3.12 Waveforms for Figure 3.11 ($C_F \to \infty$)

are forward-biased, and when these two devices are triggered into conduction at firing angle α, the load current will flow through these devices. Since the load is inductive, the thyristor current will continue to flow beyond π angle when the voltage v_s reverses its polarity. Thyristors Q_3Q_4 are fired symmetrically at α angle in the next half-cycle. This places a reverse voltage across Q_1Q_2, which turns off, or commutates, and the load current is taken by Q_3Q_4. This method of commutation is

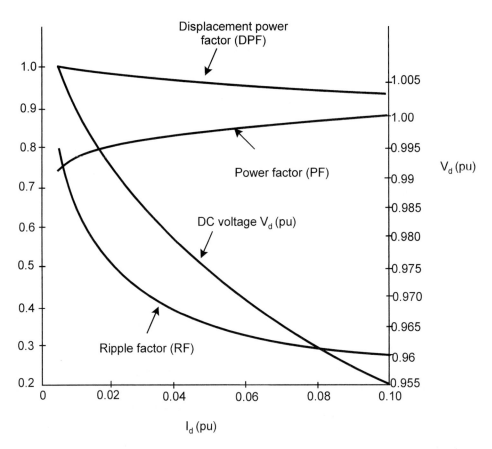

Figure 3.13 DC voltage, RF, DPF, and PF for Figure 3.11 ($C_F \to \infty$) (not in scale)

known as, natural, or line commutation. Each thyristor pair conducts for a half-cycle and the load voltage wave v_d is shown in (a). The thyristors can be fired by single gate current pulse, but generally, pulse-train firing during the entire half-cycle is preferred so that the device does not turn off inadvertently if the anode current is fluctuating and falls below the holding current. Note that during every half-cycle, the instantaneous power flows to the output for part of the cycle when v_d is positive, but for another part of the cycle, the power flow is negative when part of the inductive energy stored in the load is returned back to the source. However, the average power flow is positive, giving the rectifier mode of operation. The average dc output voltage V_d can be calculated as

$$V_d = \frac{1}{\pi} \int_{\alpha}^{\alpha+\pi} v_d \, d\omega t = \frac{1}{\pi} \int_{\alpha}^{\alpha+\pi} \sqrt{2} V_s \sin \omega t \, d\omega t \qquad (3.24)$$
$$= V_{d0} \cos \alpha$$

where $V_{do} = 2V_m/\pi$, which is the same as Equation (3.4). The voltage V_d can be controlled by controlling the firing angle α. At $\alpha = 0$, $V_d = V_{d0}$ and the circuit operates like a diode-bridge rec-

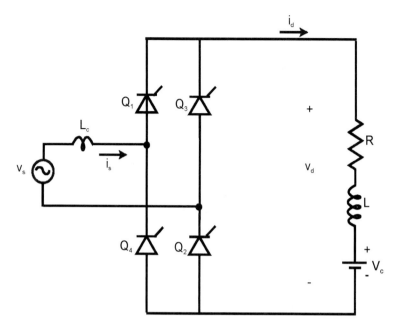

Figure 3.14 Single-phase thyristor bridge with *RL* and CEMF load

tifier. The α angle can be retarded to the maximum value of $\pi/2$ for the rectifier mode of operation when $V_d = 0$. This condition is ideal because it requires infinity-load inductance and zero-load resistance to maintain continuous conduction. At this condition, the load inductance L absorbs energy during the first quarter-cycle, but is then returned back to the source in the next quarter-cycle. The line current wave always remains a square-wave, but its phase is shifted by α in the lagging direction. Note that with finite load inductance, the load current may be continuous, but it will contain harmonics.

The firing angle α can be retarded beyond the $\pi/2$ angle, and continuous conduction can be maintained if the load contains a CEMF of negative polarity. In this mode, the converter operates as an inverter, and average power from the dc source is supplied to the ac line. Figure 3.16 shows the inversion mode of operation with $\alpha = 3\pi/4$ angle. Here, in the initial part of the cycle, the instantaneous power flow is positive with $+v_d$, but in the major part of the cycle, the power flow is negative with $-v_d$. Ideally, the α angle can be extended up to angle π, but in practice, it should be limited to a few degrees below π. The reason for this limited angle is that reverse voltage must be impressed for a definite minimum time across the outgoing thyristors for successful line commutation. The commutating thyristor voltage wave for Q_1Q_2 is shown in (b) of the same figure. A converter that can operate both as a rectifier and as an inverter is defined as a two-quadrant converter. This is in contrast to a diode rectifier, which functions as one-quadrant converter. A two-quadrant converter, for example, is desirable for dc motor speed control, where α

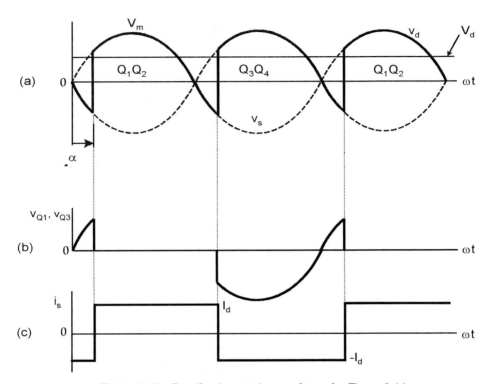

Figure 3.15 Rectification mode waveforms for Figure 3.14

can be varied between 0 and $\pi/2$ in the motoring mode, but regenerative braking mode is possible by reversing the CEMF and simultaneously retarding the firing angle beyond $\pi/2$.

Figure 3.17(a) shows the plot of V_d (pu) with α angle, where $0 \leq \alpha \leq \pi/2$ range indicates rectifier mode operation and $\pi/2 \leq \alpha \leq \pi$ range indicates inverter mode operation. Two-quadrant operation is indicated in (b) of the same figure. Note that two single-phase bridges can be connected in anti-parallel across the load to construct a dual-bridge, which can give four-quadrant mode of operation. This will be discussed later for three-phase bridge operation.

If the load inductance is beyond a critical value such that continuous conduction is maintained, the load equation can be written as

$$v_d = Ri_d + L\frac{di_d}{dt} + V_c \tag{3.25}$$

Thyristor Converters

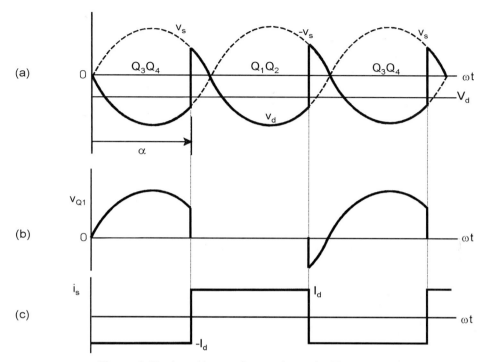

Figure 3.16 Inversion mode waveforms for Figure 3.14

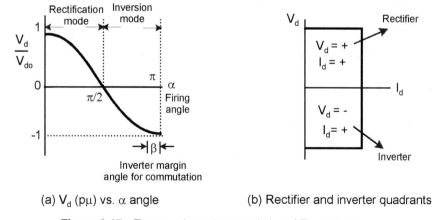

(a) V_d (pu) vs. α angle　　　　(b) Rectifier and inverter quadrants

Figure 3.17 Two-quadrant characteristics of Figure 3.14

The corresponding average voltage V_d equation is

$$V_d = \frac{1}{\pi} \int v_d d\omega t = I_d R + V_c \tag{3.26}$$

or

$$I_d = \frac{V_d - V_c}{R} \tag{3.27}$$

because inductance cannot sustain any average voltage. The average power delivered to the load is given by

$$P_d = \frac{1}{\pi} \int_\alpha^{\alpha+\pi} v_d i_d \, d\omega t = I_0^2 R + V_c I_d \tag{3.28}$$

where I_0 = rms load current. With $L \to \infty$, $I_0 = I_d$.

The load voltage v_d and line current i_s contain harmonics which can be analyzed by Fourier series. The unsymmetrical nature of v_d wave indicates that it contains even harmonics and its fundamental frequency is twice that of the ac source. The harmonics are absorbed by the load inductance. The superposition principle may be applied to solve harmonic components of current. The line current, as shown, is a pure square-wave, which is given by the following Fourier series:

$$i_s(\omega t) = \frac{4I_d}{\pi} \left[\sin \omega t + \frac{1}{3} \sin 3\omega t + \frac{1}{5} \sin 5\omega t + \ldots \right] \tag{3.8}$$

A special characteristic of the phase-controlled, line-commutated converter is that it takes lagging reactive current from the line, irrespective of rectification or inversion mode. Since the fundamental line current is in phase with the square line current wave shown in Figures 3.15(c) and 3.16(c), the displacement power factor angle is the same as the firing angle, meaning DPF = cos ϕ = cos α.

3.3.2 Discontinuous Conduction

The thyristor converter discussed so far is assumed to operate in continuous conduction, where the load voltage wave is fabricated by the segments of the supply voltage wave. In practice, a converter may also operate in discontinuous conduction mode, which is more likely with a CEMF load, such as in dc motor speed control. A two-quadrant converter may have discontinuous conduction both in rectification and inversion, which will be studied later for the three-phase bridge converter. Here, only the rectification mode will be studied.

Figure 3.18 shows the waveforms for a single-phase bridge rectifier under discontinuous conduction with resistance-inductance-CEMF load. The thyristor pair $Q_1 Q_2$ in Figure 3.14 can be fired to conduct when the supply voltage v_s is positive and exceeds the CEMF V_c. The load

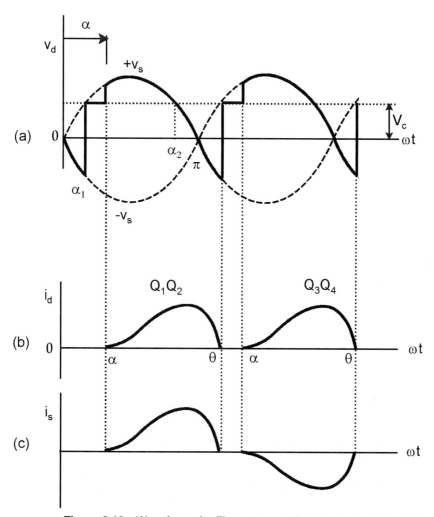

Figure 3.18 Waveforms for Figure 3.14 at discontinuous conduction

current i_d will grow up to angle α_2, but will continue up to angle θ due to the load inductance effect. Then, the output voltage v_d will be equal to V_c, as shown, until the Q_3Q_4 pair is fired symmetrically at α angle in the next half-cycle. Obviously, the range of α can be given as $\alpha_1 \leq \alpha \leq \alpha_2$, where $\alpha_1 = \sin^{-1}(V_c/\sqrt{2}V_s)$ and $\alpha_2 = \pi - \alpha_1$. The conduction may change from discontinuous to continuous if α or V_c is low, or load inductance L is high. The voltage equation during conduction can be given as

$$L\frac{di_d}{dt} + Ri_d = \sqrt{2}V_s \sin\omega t - V_c \quad (3.29)$$

The current i_d can be solved as

$$i_d = Ae^{-\frac{R}{L}t} + \frac{\sqrt{2}V_s}{|Z|}\sin(\omega t - \phi) - \frac{V_c}{R} \tag{3.30}$$

where

$$|Z| = \sqrt{R^2 + \omega^2 L^2}$$

$$\phi = \tan^{-1}\frac{\omega L}{R}$$

Noting that $i_d = 0$ at $\omega t = \alpha$, Equation (3.30) can be written as

$$i_d = \frac{\sqrt{2}V_s}{R}\left\{\frac{R}{|Z|}\sin(\omega t - \phi) - m + \left[m - \frac{R}{|Z|}\sin(\alpha - \phi)\right]e^{\frac{-R}{\omega L}(\omega t - \alpha)}\right\} \tag{3.31}$$

where $m = V_c/\sqrt{2}V_s$ is the CEMF coefficient. Equation (3.31) is valid in the range $\alpha \leq \omega t \leq \theta$ for discontinuous conduction. From the current waveform, $i_d = 0$ again at $\omega t = \theta$. Substituting this condition in Equation (3.31) yields

$$e^{\frac{-R}{\omega L}(\theta - \alpha)} = \frac{\cos\phi \sin(\theta - \phi) - m}{\cos\phi \sin(\alpha - \phi) - m} \tag{3.32}$$

This is a transcendental equation relating parameters α, m, θ, and $\omega L/R$, which can also be solved by a computer program.

Figure 3.19 gives a plot showing the relationships among the parameters. The curves are bounded on the lower side by the α-limit curves, where $\alpha_1 = \sin^{-1}(V_c/\sqrt{2}V_s)$ and $\alpha_2 = \pi - \alpha_1$, and on the upper left side by the continuous conduction boundary $\alpha = \theta - \pi$. The load dc and rms currents can be calculated from Equation (3.31) with the help of Figure 3.19. However, since the inductance L cannot sustain any average voltage, the dc load current I_d can be calculated as

$$\begin{aligned}I_d &= \frac{1}{\pi}\int_\alpha^\theta \frac{\sqrt{2}V_s \sin\omega t - V_c}{R} d\omega t \\ &= \frac{\sqrt{2}V_s}{\pi R}[\cos\alpha - \cos\theta - m(\theta - \phi)]\end{aligned} \tag{3.33}$$

The discontinuous conduction case can be extrapolated to reach continuous conduction. Writing Equation (3.30) for both $\omega t = \alpha$ and $\omega t = \pi + \alpha$ and equating yields

Thyristor Converters

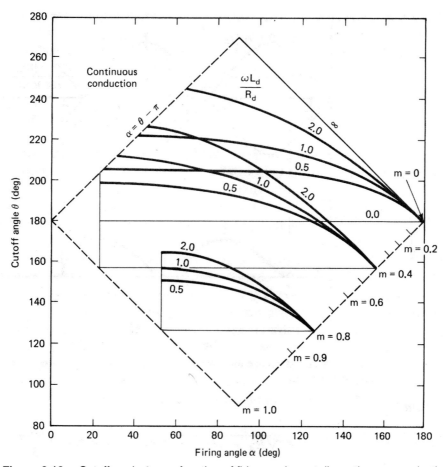

Figure 3.19 Cutoff angle θ as a function of firing angle α at discontinuous conduction

$$A = \frac{\sqrt{2}V_s}{|Z|} \cdot \frac{\sin(\alpha - \phi)}{\left(e^{\frac{-R}{\omega L}\pi} - 1\right) e^{\frac{R}{\omega L}\alpha}} \tag{3.34}$$

and therefore i_d expression for continuous conduction is given as

$$i_d = \frac{\sqrt{2}V_s}{R}\left[\cos\phi\sin(\omega t - \phi) - m - \frac{2\cos\phi\sin(\alpha - \phi)}{\left(1 - e^{\frac{-R}{\omega L}\pi}\right)} \cdot e^{\frac{-R}{\omega L}(\omega t - \alpha)}\right] \tag{3.35}$$

Therefore the dc current I_d can be derived from Equation (3.35) as

$$I_d = \frac{1}{\pi}\int_\alpha^{\alpha+\pi} i_d \, d\omega t = \frac{\sqrt{2}V_s}{\pi R}(2\cos\alpha - m\pi) \qquad (3.36)$$

which can be shown as identical with Equation (3.27).

3.3.3 Three-Phase Converter – *RL*, CEMF Load

For higher load power (typically above 10 kW), a single-phase converter should be replaced by a three-phase converter. A polyphase converter, in general, not only balances the power flow in a three-phase utility system, but provides considerable improvement in load voltage and line current harmonics, as will be shown later. As a result, the harmonic filtering requirement becomes a nominal problem.

3.3.4 Three-Phase, Half-Wave Converter

Among all the configurations of polyphase converters, the three-phase bridge converter as shown in Figure 3.20 is most commonly used. The applications include speed controls of dc motors, current-fed induction and synchronous motor drives, general-purpose dc power supplies, UPS systems, etc. It is also an element in phase-controlled cycloconverters, which are used for large ac motor drives. Cycloconverters will be discussed in Chapter 4. In many respects, the operation of a three-phase bridge is similar to that of a single-phase bridge, as explained before. The converter consists of six thyristors arranged in the form of three legs, the center points of which are connected to a three-phase ac power supply. The transformer connection is optional, but useful for shifting the voltage level or isolation from the primary. A three-phase bridge converter can be constructed by series-cascading a three-phase, half-wave + converter and a three-

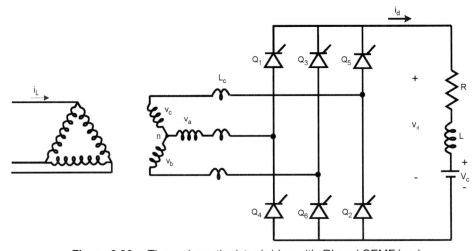

Figure 3.20 Three-phase thyristor bridge with *RL* and CEMF load

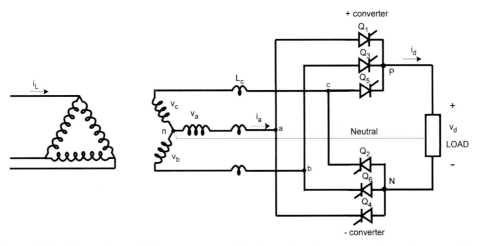

Figure 3.21 Three-phase bridge as a cascaded connection of a + converter and a − converter

phase, half-wave − converter with the neutral isolated, as shown in Figure 3.21. The operation of the two converter components are identical except the + converter operates in the upper voltage envelope whereas the − converter operates in the lower voltage envelope. We will study the + converter only in detail.

Consider the + converter only in Figure 3.21 with the neutral connected and assume continuous conduction with perfect load filtering ($L \to \infty$). Figure 3.22 shows the waveforms of the converter. The three thyristors $Q_1 Q_3 Q_5$ conduct symmetrically, each for $2\pi/3$ angle through the load, and provide a common return to the transformer neutral point n. A thyristor can be fired to conduct when its anode voltage is positive with respect to the cathode (i.e., with respect to the load voltage), and conduction will continue until the next thyristor is fired after $2\pi/3$ angle. The commutation from an outgoing to an incoming thyristor occurs naturally (natural or line commutation) by a segment of negative line voltage. The firing angle α is defined from the crossover point of phase voltages ($\pi/6$ angle), which is the earliest point when a thyristor can conduct. At $\alpha = 0$, the thyristors can be considered to be operating as diodes. The average dc load voltage V_d' can be calculated as

$$V_d' = \frac{3}{2\pi} \int_{\pi/6+\alpha}^{(\pi/6+\alpha)+2\pi/3} \sqrt{2} V_s \sin \omega t \, d\omega t = V_{d0}' \cos \alpha \qquad (3.37)$$

where $V_{d0}' = 0.675 V_L$ and V_L is the rms line voltage. The voltage V_d' can be varied by controlling the firing angle α. Up to $\alpha = \pi/6$, the instantaneous dc voltage v_d' is always positive, and therefore conduction will be continuous even with resistive load if there is no CEMF. For $\alpha > \pi/6$, the instantaneous v_d' wave will be negative for part of the cycle, and discontinuous conduction may occur if the load is not sufficiently inductive. At $\alpha = \pi/2$, positive and negative segments of v_d'

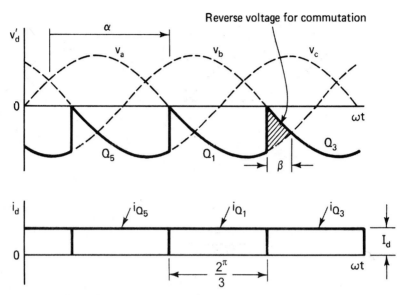

Figure 3.22 Waveforms for three-phase, half-wave + converter in inversion mode ($\alpha = 5\pi/6$)

wave will balance, giving $V_d' = 0$. The angle α can be controlled beyond $\pi/2$, giving a sustained inverter mode of operation, as discussed before, provided an energy source exists in the load.

Figure 3.23 shows the waveforms for inverter operation at $\alpha = 5\pi/6$. The angle β, known as the retard limit angle, is important for line commutation because the reverse voltage shown by the shaded area is impressed across the outgoing thyristor. Typically, $\beta_{min} = 10°$ to $15°$. With the load current always positive, the converter has two-quadrant characteristics, as shown in Figure 3.17.

An examination of the output v_d' wave indicates that it contains triplen harmonic frequencies (i.e., third, sixth, ninth, etc.). The increase in pulse number from two to three also increases the dc voltage V_d' at the output. In a three-phase converter, each device conducts for one-third of the cycle and therefore carries the average current $I_d/3$ and rms current $I_d/\sqrt{3}$. A unidirectional current pulse for $2\pi/3$ in the transformer secondary winding is harmful because it may cause dc saturation in the core. The problem may be avoided by providing a zigzag connection in the secondary. Although the half-wave converter alone is not used in practice, its analysis is important because the circuit is a basic functional element in all polyphase converters and cycloconverters.

3.3.5 Analysis for Line Leakage Inductance (L_c)

A finite leakage inductance L_c in the line will cause gradual current transfer from the outgoing thyristor to the incoming thyristor, as shown in Figure 3.24. During the commutation overlap angle μ as indicated, the line-to-line voltage is shorted and the supply volt-seconds area is absorbed by the two leakage inductances L_c in series until the current transfer is completed. Dur-

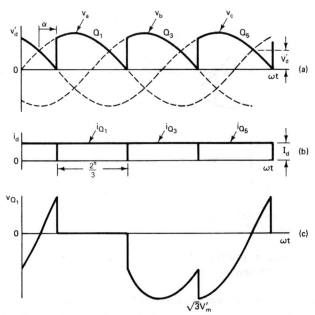

Figure 3.23 Waveforms of three-phase, half-wave + converter in rectification mode ($\alpha = \pi/6$)

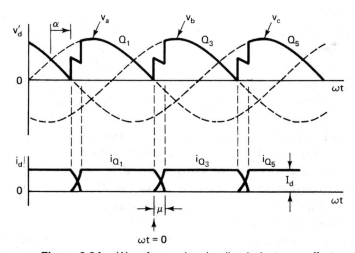

Figure 3.24 Waveforms showing line inductance effect

ing this period, the load voltage dwells at an intermediate level between the two phase voltages. Considering commutation from Q_1 to Q_3 in Figure 3.21, we can write the following equations:

$$v_a = L_c \frac{di_{Q1}}{dt} + v'_d \tag{3.38}$$

$$v_b = L_c \frac{di_{Q3}}{dt} + v'_d \tag{3.39}$$

During commutation, the load current I_d can be assumed to be constant, that is,

$$i_{Q1} + i_{Q3} = I_d \tag{3.40}$$

This gives

$$\frac{di_{Q_1}}{dt} + \frac{di_{Q_3}}{dt} = 0 \tag{3.41}$$

Combining Equations (3.38), (3.39), and (3.41) yields

$$v'_d = \frac{1}{2}(v_a + v_b) \tag{3.42}$$

Thus, the instantaneous load voltage is the mean of the two phase voltages. From Figure 3.24, it is evident that every $2\pi/3$ interval, some volt-second area is lost during commutation and, as a result, the effective dc voltage is reduced. Combining Equations (3.38) and (3.42) yields

$$\frac{di_{Q1}}{dt} = -\frac{1}{2L_c}(v_b - v_a) \tag{3.43}$$

or

$$i_{Q1} = -\frac{1}{2L_c}\int(v_b - v_a)\,dt \tag{3.44}$$

Substituting the line voltage $v_{ba} = v_b - v_a = \sqrt{3}\sqrt{2}\,V_s \sin(\omega t + \alpha)$ in Equation (3.44) and solving gives

$$i_{Q1} = \frac{\sqrt{6}V_s}{2\omega L_c}\cos(\omega t + \alpha) + A \tag{3.45}$$

Assuming $\omega t = 0$ at the beginning of commutation, where $i_{Q1} = I_d$, the constant A can be evaluated as

$$A = I_d - \frac{\sqrt{6}V_s}{2\omega L_c}\cos\alpha \tag{3.46}$$

Substituting this in (3.45) yields

$$i_{Q1} = I_d - \frac{\sqrt{6}V_s}{2\omega L_c}[\cos\alpha - \cos(\omega t + \alpha)] \tag{3.47}$$

Substituting Equation (3.47) in (3.40), we have

$$i_{Q3} = \frac{\sqrt{6}V_s}{2\omega L_c}\left[\cos\alpha - \cos(\omega t + \alpha)\right] \tag{3.48}$$

Again, substituting $i_{Q3} = I_d$ at $\omega t = \mu$ in Equation (3.48) gives

$$\cos\alpha - \cos(\mu + \alpha) = \frac{2\omega L_c I_d}{\sqrt{6}V_s} \tag{3.49}$$

Therefore, the commutation angle μ can be expressed as

$$\mu = \cos^{-1}\left(\cos\alpha - \frac{2\omega L_c I_d}{\sqrt{6}V_s}\right) - \alpha \tag{3.50}$$

Equation (3.50) shows that the overlap angle will increase if L_c or I_d increases, or the voltage v_s at the instant of commutation decreases. The mean dc voltage loss due to the commutation notch can be given as

$$\begin{aligned}V_x &= \frac{3}{2\pi}\int_0^\mu \frac{1}{2}(v_b - v_a)\,d\omega t \\ &= \frac{3}{4\pi}\int_0^\mu \sqrt{6}V_s \sin(\omega t + \alpha)\,d\omega t \\ &= -\frac{3\sqrt{6}V_s}{4\pi}\left[\cos(\mu + \alpha) - \cos\alpha\right]\end{aligned} \tag{3.51}$$

Substituting Equation (3.49) in (3.51) gives

$$V_x = L_c I_d \frac{3\omega}{2\pi} = 3L_c I_d f \tag{3.52}$$

where f = supply frequency in Hz. Therefore, with loading, the dc voltage V_{dl} can be given from Equations (3.37) and (3.52) as

$$V_{dl} = V_d' - V_x = 0.674 V_L \cos\alpha - 3L_c I_d f \tag{3.53}$$

Equation (3.53) indicates that the load voltage is reduced linearly with dc current, and the equivalent Thevenin resistance of the converter is given as $R_{Th} = 3L_c f$.

Figure 3.25 shows the inverter operation of a three-phase, half-wave positive converter, where the commutation notch makes the dc voltage more negative, that is,

$$V_{dl} = -\left|V_d'\right| - \left|V_x\right| \tag{3.54}$$

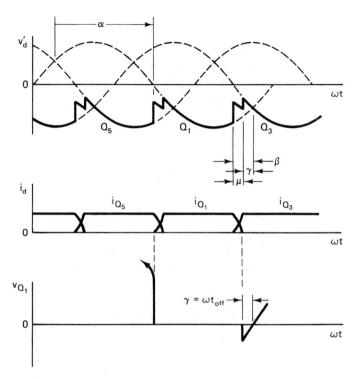

Figure 3.25 Waveforms showing inverter mode of operation with line inductance effect

The overlap angle has more significance in inverter operation since it determines how far angle α can be increased (i.e., the minimum β angle for safe commutation). In the figure,

$$\beta = \mu + \gamma \tag{3.55}$$

where γ = turn-off angle, as indicated. Substituting $\alpha = \pi - (\mu + \gamma)$ in Equation (3.49) gives

$$\cos\gamma - \cos(\mu + \gamma) = \frac{2\omega L_c I_d}{\sqrt{6}V_s} \tag{3.56}$$

The thyristors require minimum turn-off time t_{off} for successful commutation, which correspondingly determines the angle $\gamma = \omega t_{off}$. For fluctuating load, adequate β angle margin should be provided to ensure successful commutation.

3.3.6 Three-Phase Bridge Converter

Three-phase bridge converter operation can be analyzed by superpositioning the waveforms of both a positive half-wave converter and a negative half-wave converter and isolating the load neutral connection.

Figure 3.26 shows waveforms of the bridge for firing angle $\alpha = \pi/6$, where the effect of leakage inductance has been ignored. Note that the current waves are the same as in Figure 3.10. The – converter, consisting of thyristors $Q_4 Q_6 Q_2$, operates on the negative voltage envelope and is fired symmetrically at $2\pi/3$ angle intervals like the + converter, except that it is phase-shifted by $\pi/3$, as shown. The load voltage v_d is enclosed within a line-to-line voltage envelope and has the six-pulse waveshape shown in (b). At any instant, a thyristor from the + converter and another from the – converter must conduct to complete the load circuit. The average dc voltage V_d is twice that of a half-wave converter, which can be given from Equation (3.37) as

$$V_d = 2 \times 0.675 V_L \cos\alpha = 1.35 V_L \cos\alpha \tag{3.57}$$

Since the full bridge operates on line-to-line voltage, the line voltage waves have been constructed and the negative half cycles have been flipped to the positive side for convenience. The α angle can be symmetrically controlled for both the component converters in the sequence Q_1-Q_2-Q_3-Q_4-Q_5-Q_6 to regulate the dc voltage V_d. The Fourier analysis of a v_d wave indicates that it contains harmonics of the order $6n$, where $n = 1, 2, 3$, and so on. Evidently, the waveform with an increasing pulse number is easier to filter and a nominal value of load inductance is required for a smooth i_d wave. The phase currents i_a and i_b can be constructed by the superposition of thyristor currents and have the characteristic six-step waveform, which contains harmonics of the order $6n \pm 1$ (i.e., fifth, seventh, eleventh, thirteenth, etc.), as shown by Equation (3.22). If no transformer is used, i_a and i_b will constitute the line current waves. With a delta-wye transformer of unity turns ratio, the input line current i_L can be constructed by superpositioning the i_a and i_b waves, as shown. The transformer does not have a dc saturation problem because of mmf balancing, and the order of the input harmonic currents is the same as that of the i_a and i_b waves. If the firing angle is retarded beyond $\pi/2$ ($\pi/2 \leq \alpha \leq \pi$), the converter will operate in the inverter mode, as explained in Figure 3.27. Again, the full bridge will have two-quadrant characteristics as shown in Figure 3.17.

The current wave i_a in Figure 3.26 and its fundamental component have been plotted in the correct phase position with the supply phase voltage wave v_a in Figure 3.28. Figure 3.29 shows the corresponding active and reactive current relations of the fundamental current in both the rectification and inversion modes. It is assumed that the dc current I_d remains constant, the load is highly inductive, and it has the correct polarity and magnitude of emf to balance the dc output voltage. Since the input displacement angle ϕ is equal to the firing angle α, the rms values of the active and reactive current components can be given as

$$I_P = R \cos\alpha \tag{3.58}$$

$$I_Q = R \sin\alpha \tag{3.59}$$

where $R = (2\sqrt{3}/\sqrt{2}\pi)$ and I_d is the rms value of the fundamental current. For the firing angle in the range $\pi/2 < \alpha < \pi$ (i.e., in the inverting mode), the active current I_P becomes negative, but the reactive current I_Q always remains lagging.

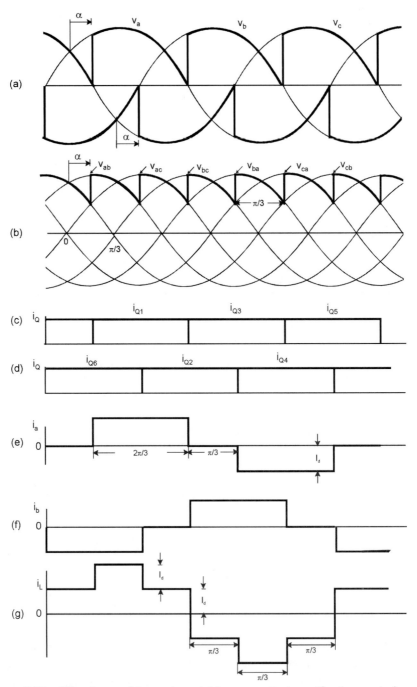

Figure 3.26 Waveforms of three-phase bridge converter in rectification mode ($\alpha = \pi/6$)

Thyristor Converters

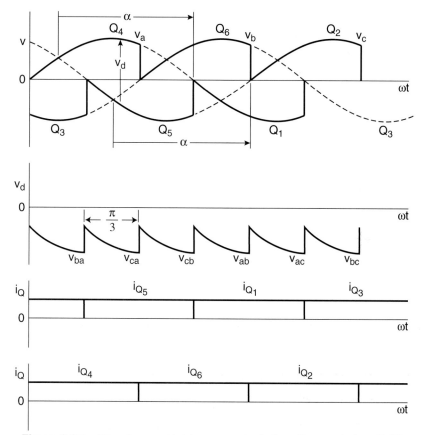

Figure 3.27 Waveforms of bridge converter in inverting mode ($\alpha = 5\pi/6$)

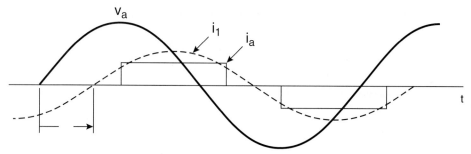

Figure 3.28 Phase relation of input phase voltage and line current

So far, the voltage and current waves in a bridge converter have been considered ideal, neglecting the line leakage inductance effect. Figure 3.30 shows the typical waveforms with overlap angle μ. The positive and negative converters operate independently, and therefore, the volt-second area loss per commutation remains the same as that of a half-wave converter. Since

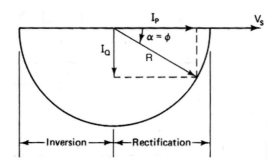

Figure 3.29 Input line active and reactive current characteristics of bridge converter (I_d = constant)

the number of commutations are twice per cycle, the dc voltage loss will also be doubled; that is, from Equations (3.53) and (3.57)

$$V_{dl} = V_d - 2V_x = 1.35 V_L \cos\alpha - 6 L_c I_d f \qquad (3.60)$$

Again, as discussed before, the overlap angle is particularly important in the inverting mode, where the safe minimum γ angle must be maintained under the worst-case load condition.

3.3.7 Discontinuous Conduction

A three-phase bridge converter can fall into discontinuous conduction in both the rectification and inversion modes with CEMF load, as indicated in Figure 3.31. This mode is particularly important in two-quadrant speed control of a dc motor drive, where the CEMF varies directly with speed. The operation is somewhat similar to single-phase bridge operation, which was explained previously. Consider the waveforms in rectification mode shown in Figure 3.31(a). For discontinuous conduction, the thyristor pair can be fired when the respective line voltage is higher than the CEMF. A pulse of load current will flow every $\pi/3$ angle, and at any instant, one device from the + converter and another device from the – converter will conduct. Each thyristor will carry two consecutive current pulses during the $2\pi/3$ interval, and the resulting line current wave will be similar to Figure 3.12, except with a phase lag. Consider the load voltage v_d segment in the interval $\alpha + \pi/3$ to $\alpha + 2\pi/3$. The load current equation in this interval can be given as

$$L\frac{di_d}{dt} + R i_d = V_m \sin\omega t - V_c \qquad \left(\alpha + \frac{\pi}{3}\right) < \alpha < \theta \qquad (3.61)$$

where $V_m = \sqrt{2} V_L$ is the peak line voltage and V_c = CEMF.

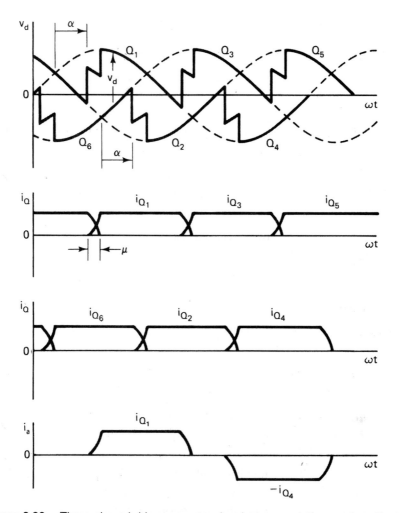

Figure 3.30 Three-phase bridge converter showing commutation overlap effect

Since $i_d = 0$ at angle $\alpha + \pi/3$, substituting this condition in Equation (3.61), the i_d expression can be solved as

$$i_d = \frac{V_m}{|Z|}\left[\sin(\omega t - \phi) - \sin\left(\alpha + \frac{\pi}{3} - \phi\right)\exp\left\{\left((\alpha + \frac{\pi}{3}) - \omega t\right)\cot\phi\right\}\right] \\ - \frac{V_c}{R}\left[1 - \exp\left\{\left(\alpha + \frac{\pi}{3} - \omega t\right)\cot\phi\right\}\right] \quad (3.62)$$

where

$$|Z| = \sqrt{R^2 + \omega^2 L^2} \quad (3.63)$$

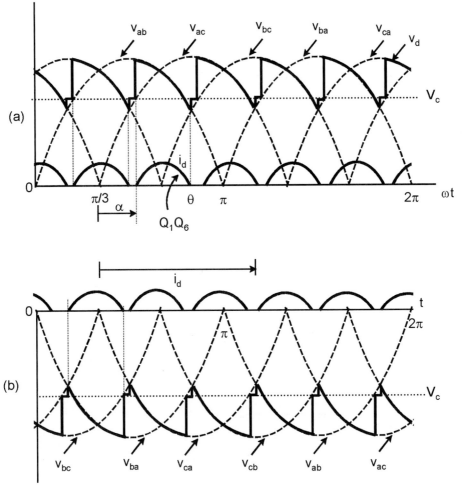

Figure 3.31 Waveforms for Figure 3.20 in discontinuous conduction mode (a) Rectification mode, (b) Inversion mode

$$\phi = \tan^{-1}\frac{\omega L}{R} \qquad (3.64)$$

Again, $i_d = 0$ at $\omega t = \theta$, as shown in the figure. Substituting this in Equation (3.62) gives

$$\frac{V_m}{|Z|}\left[\sin(\theta-\phi)-\sin\left(\alpha+\frac{\pi}{3}-\phi\right)\exp\left\{\left(\alpha+\frac{\pi}{3}-\theta\right)\cot\phi\right\}\right] \\ = \frac{V_c}{R}\left[1-\exp\left\{\left(\alpha+\frac{\pi}{3}-\theta\right)\cot\phi\right\}\right] \qquad (3.65)$$

or

$$\sin(\theta-\phi)-\sin\left(\alpha+\frac{\pi}{3}-\phi\right)\exp\left\{\left(\alpha+\frac{\pi}{3}-\theta\right)\cot\phi\right\}$$
$$=\frac{|Z|}{R}m\left[1-\exp\left\{\left(\alpha+\frac{\pi}{3}-\theta\right)\cot\phi\right\}\right] \quad (3.66)$$

where $m = V_c/V_m$ is the CEMF coefficient. This transcendental equation relates the parameters α, m, θ, and $\omega L/R$ similar to Equation (3.32). The average dc voltage expression can be derived as

$$V_d = \frac{3}{\pi}\left[\int_{\alpha+\pi/3}^{\theta} V_m \sin\omega t\, d\omega t + \int_{\theta}^{\alpha+2\pi/3} V_c\, d\omega t\right]$$
$$= \frac{3}{\pi}\left[V_m\left\{\cos\left(\alpha+\frac{\pi}{3}\right)-\cos\theta\right\}+V_c\left(\alpha+\frac{2\pi}{3}-\theta\right)\right] \quad (3.67)$$

But,

$$V_d = \frac{1}{\pi}\int v_d\, d\omega t = I_d R + V_c \quad (3.26)$$

Combining Equations (3.67) and (3.26), the CEMF expression can be given as

$$V_c = \frac{V_m\left[\cos\left(\alpha+\frac{\pi}{3}\right)-\cos\theta\right]}{\left(\theta-\alpha-\frac{\pi}{3}\right)}-\frac{\pi R I_d}{3\left(\theta-\alpha-\frac{\pi}{3}\right)} \quad (3.68)$$

or, the current I_d expression can be given as

$$I_d = \frac{1}{\pi R}\left[3V_m\left\{\cos\left(\alpha+\frac{\pi}{3}\right)-\cos\theta\right\}-3V_c\left(\theta-\alpha-\frac{\pi}{3}\right)\right] \quad (3.69)$$

Equations (3.67), (3.68), and (3.69) can be combined to derive a relation between $V_d(pu) = V_d/V_m$ and $I_d(pu) = I_d/(3V_m/\omega L)$ as a function of firing angle α and $\omega L/R$ parameters. Figure 3.32 shows the plot for $\omega L/R = 5.0$ in both rectification and inversion modes, which also combines the continuous conduction regions [4]. In continuous conduction,

$$V_d = 1.35 V_L \cos\alpha \quad (3.57)$$

where $V_L = \sqrt{2}V_m$.

The $V_d(pu)$ lines are horizontal and are functions of angle α only. The boundary between continuous and discontinuous regions is shown by a dotted curve. If the resistance drop $I_d R$ is neglected, we can write the following relations:

$$V_d(pu) = \frac{V_d}{V_m} = \frac{K}{V_m}\omega_m = K'\omega_m \tag{3.70}$$

and

$$I_d(pu) = \frac{I_d}{3V_m/\omega L} = K''T_e \tag{3.71}$$

where ω_m = angular speed, T_e = developed torque of the motor, and K = CEMF or torque constant. The finite resistance drop (increasing with I_d) will cause increasing droop of the speed curves with torque. Equations (3.70) and (3.71) indicate that Figure 3.32 also gives profiles of speed vs. torque as functions of firing angle α.

3.3.8 Three-Phase Dual Converter

A dual converter basically consists of two converters that are connected in anti-parallel. Figure 3.33(a) shows the three-phase half-wave configuration, and (b) shows the corresponding bridge configuration. Here, the + converter supplies the positive load current, whereas the negative load current is taken up by the – converter. Since each converter component can operate in two quadrants (see Figure 3.17), a dual converter has four-quadrant characteristics, as shown in Figure 3.34. Dual converters (both single-phase and three-phase versions) are popularly used in "Thyristor Leonard" speed control of dc motors, where reversible and regenerative operations of the drive can be obtained. The circuit is used for phase-controlled cycloconverters, which will be discussed in Chapter 4. Dual converters can be operated in blocking mode or in circulating current mode through an inter-group reactor (IGR). In blocking mode, only one component converter is permitted to conduct while inhibiting the other to prevent a line-to-line short-circuit. However, in circulating current mode, current may be allowed to circulate continuously between the converter components. This will be further discussed later.

3.3.9 Six-Pulse, Center-Tap Converter

Two three-phase, half-wave converters with a phase shift of π can be connected in parallel through an IGR to constitute a six-pulse, center-tap converter as shown in Figure 3.35. The circuit is commonly used in low-voltage, high-current, electrochemical-type applications. The dc load voltage v_d is the same as that of a half-bridge converter, and each component converter contributes 50% of the load current I_d. Each component operates independently, with each thyristor conducting for $2\pi/3$, and the IGR absorbs the instantaneous potential difference between the common cathodes G and H. As a result, the load voltage v_d is a six-stepped wave, as in a bridge converter. Note that if the circuit is used without IGR, a conventional six-phase, half-wave con-

Thyristor Converters

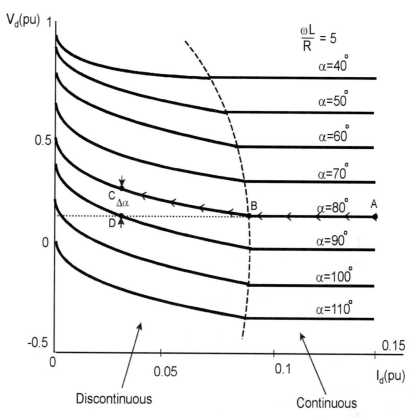

Figure 3.32 Load characteristics of thyristor bridge (Figure 3.20) under continuous and discontinuous conduction

verter operation will result, where each thyristor and transformer secondary winding will conduct symmetrically for $\pi/3$ per cycle. Such a circuit operation, although satisfactory, is not desirable due to poor utilization of the transformer and thyristors. One demerit of the center-tap circuit is that the load current must not fall below a minimum limit, which is the peak magnetizing current of the IGR. If this happens, the circuit reverts to six-pulse wye converter operation and the load voltage rises by 15% at $\alpha = 0$.

3.3.10 12-Pulse Converter

If the converter current or voltage rating is high so that a single thyristor device is not adequate, multiple devices may be connected in parallel or in series, respectively. The parallel connection of devices is particularly difficult because of the matching problem in static and dynamic conditions. Instead, parallel or series operation of converters with phase-shifting transformers is advantageous because of harmonic reduction on the load and line sides, although the additional cost and losses of the transformer are involved. An example of phase-shifted parallel operation

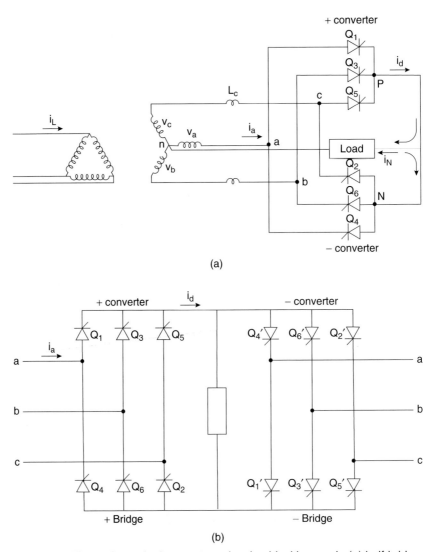

Figure 3.33 Three-phase dual converters showing blocking mode (a) half-bridges, (b) full-bridges

was given in Figure 3.35. An example of phase-shifted series operation of bridges for higher voltage operation is given in Figure 3.36. A single bridge gives 6-pulse operation, but a series connection of two bridges with transformer secondaries at $\pi/6$ phase shift gives 12-pulse operation. The load voltage wave v_d and line current wave i_L are shown in Figure 3.37. It can be shown that the order of load voltage harmonics is $12n$ (i.e., 12th, 24th, 36th, etc.) and the corresponding order of line current harmonics are $12n \pm 1$ (i.e., 11th, 13th, 23rd, 25th, etc.) where n is an integer. For higher current operation, the two bridges can be connected in parallel through an

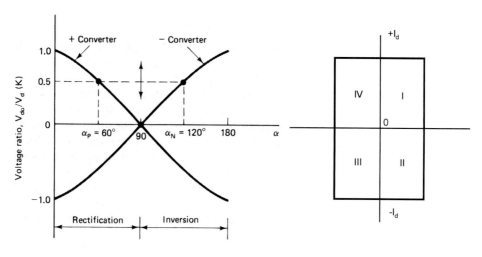

Figure 3.34 Output voltage characteristics of dual-bridge converter indicating four-quadrant operation

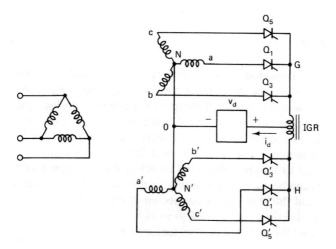

Figure 3.35 Six-pulse center-tap converter with inter-group reactor (IGR)

IGR to give 12-pulse operation. The principle can be extended to 24-pulse and 48-pulse converters. For high-power converter applications, such as high-voltage dc (HVDC) and large dc power supplies for motor drives, operation with an increasing pulse number is very desirable. Note that every thyristor converter topology becomes an equivalent diode converter topology with firing angle $\alpha = 0$.

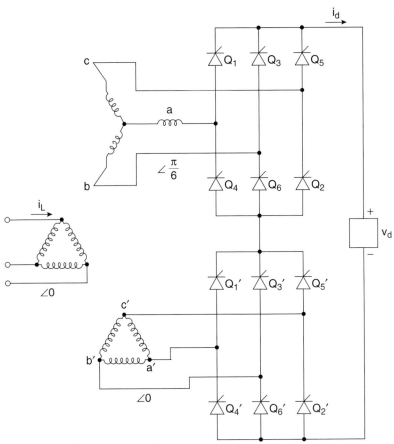

Figure 3.36 12-pulse series connected bridge converter

3.3.11 Concurrent and Sequential Control of Bridge Converters

The double-bridge converter shown in Figure 3.36 is normally controlled symmetrically or concurrently for two-quadrant operation, that is, the consecurive thyristors are fired at $\pi/6$ interval (for 12-pulse) in the entire range of α between 0 and π. With such a concurrent firing angle control, the line displacement factor deteriorates as the α angle approaches $\pi/2$ as shown in Figure 3.29. The DPF can be considerably improved if one converter is phase-controlled while the other remains at full advance ($\alpha = 0$) or full retard ($\alpha = \pi$) for rectifier or inverter operation, respectively. This means that in the rectification mode, one bridge operates as a diode rectifier while the other bridge is phase-controlled. For example, in rectifier mode, α_1 (upper bridge) = $2\pi/3$ and α_2 (lower bridge) = 0. These give $V_d = (1.35 - 0.675)V_L = 0.675\ V_L$. Figure 3.38 shows the characteristics in sequential control in comparison to concurrent control in both the rectification and inversion modes. Note that $V_d = 0$ when $\alpha_1 = \pi$ and $\alpha_2 = 0$, giving unity DPF in the

Converter Control

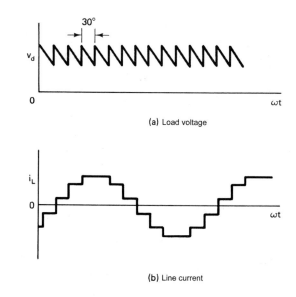

(a) Load voltage

(b) Line current

Figure 3.37 Waveforms of 12-pulse converter

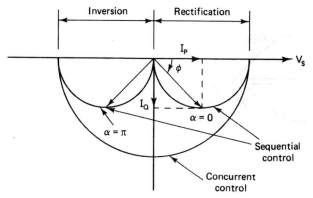

Figure 3.38 Characteristics of series bridge converter with concurrent and sequential control

line. However, this advantage gives a penalty in the ripple of line current and load voltage; a 12-pulse converter operates in 6-pulse mode.

3.4 CONVERTER CONTROL

The function of a controller is to control the firing angle α of a converter symmetrically in response to a demand of output dc voltage or current. Any other control loop can be applied as desired. The controller usually incorporates the following functions:

- Line synchronization
- Control of firing angle α
- Advance limit of angle α
- Retard limit of angle α

Line synchronization permits the establishment of a symmetrical firing angle control to all thyristors with respect to the fixed angular position of line ac voltage wave. The firing angle controller alters angle α angle in response to a variable input control voltage. The advance and retard limit controls restrict the angle α within safe limits. Theoretically, the advance limit angle can be established as early as $\alpha = 0$, but the retard limit angle must provide a sufficient margin so that a minimum turn-off angle γ is maintained to avoid commutation failure.

3.4.1 Linear Firing Angle Control

Figure 3.39 explains the simple linear firing angle control method for a single-phase bridge converter. The line supply voltage v_{ab} is stepped down through a transformer and converted to a square wave through a zero-crossing detector. A sawtooth wave of twice the supply frequency can be generated as shown in (c) such that it remains synchronized with the square wave. The sawtooth wave starts with an initial voltage A at zero angle, linearly decreases to zero at angle π, and restarts again. The control voltage V_c is compared with the sawtooth wave and the firing angle is generated at the crossover point by the following linear relation:

$$\alpha = -\frac{\pi}{A}V_c + \pi \qquad (3.72)$$

The long firing pulse train in the interval $\pi - \alpha$ is created for the respective thyristor through a steering circuit. As the control voltage V_c is increased, α advances, giving a higher dc voltage at the output until α approaches zero at $V_c = A$. On the other extreme, α approaches π at $V_c = 0$ in the absence of a retard limit control. The relation between control voltage V_c and output V_d can be derived by substituting equation (3.72) in $V_d = V_{d0} \cos \alpha$ as

$$V_d = V_{d0} \cos\left(\pi - \frac{\pi}{A}V_c\right) \qquad (3.73)$$

Because of nonlinear transfer characteristics, linear firing angle control is hardly used.

3.4.2 Cosine Wave Crossing Control

A popularly used control method where linearity of transfer characteristics is achieved is known as the cosine wave-crossing method. Figure 3.40 illustrates this method for a single-phase bridge converter. The sine wave supply voltage v_{ab} is phase-advanced by $\pi/2$ to generate a cosine wave; it is phase-inverted every second half-cycle to construct the "cosine wave" shown

Converter Control

in (b) of the figure. The firing angle α is generated by the crossover point of control voltage V_c and cosine wave at every half-cycle as

$$\cos\alpha = \frac{V_c}{V_p} \tag{3.74}$$

where V_p = peak value of the cosine wave. Substituting (3.74) in $V_d = V_{d0} \cos\alpha$ gives

$$V_d = \frac{V_{d0}}{V_p} V_c = K V_c \tag{3.75}$$

indicating a linear relation between output and input with a gain factor K. In this case, the converter looks like a switching-mode linear amplifier. Note that if the cosine wave magnitude varies with the fluctuation of supply voltage, the gain K remains unaltered. It should be noted,

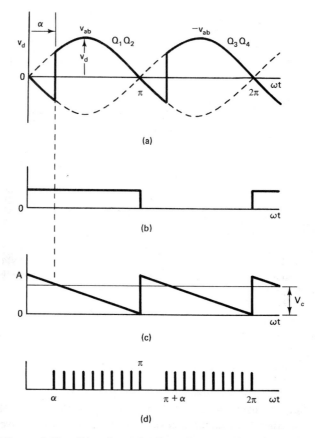

Figure 3.39 Waveforms for linear firing angle control

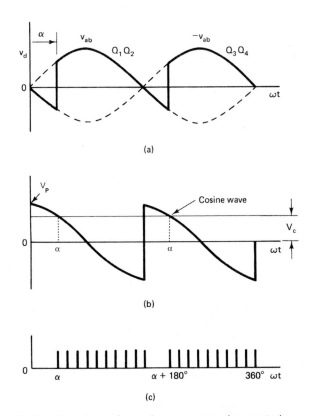

Figure 3.40 Waveforms for cosine wave-crossing control

however, that Equation (3.75) is valid only for continuous conduction. With discontinuous conduction, as discussed before, the gain becomes nonlinear and depends on angle α and load parameters.

Figure 3.41 shows the cosine wave-crossing control method for a three-phase bridge converter and Figs. 3.42 and 3.43 explain its operation. The derivation of firing logic signals for thyristor Q_1 is shown only, but the principle can be easily extended to other thyristors. The line voltage v_{ac} is the reference wave in which the angle 0 to π corresponds to the firing angle range of Q_1. The phase voltage $-v_b$ leads v_{ac} by $\pi/2$ and constitutes the cosine reference wave for thyristor Q_1. The phase and line voltages are stepped down through transformers and connected to the comparators, as shown. The comparator 1, which compares the control voltage V_c with phase voltage $-v_b$, transitions to logic 1 at firing angle α. The output of comparators 1 and 2 are logically ANDed to trigger flip-flop 9 at the leading edge, which in turn couples a pulse train (not shown) to the gate of Q_1. The flip-flop is reset at the firing of Q_3, thus limiting the gate pulse

Converter Control

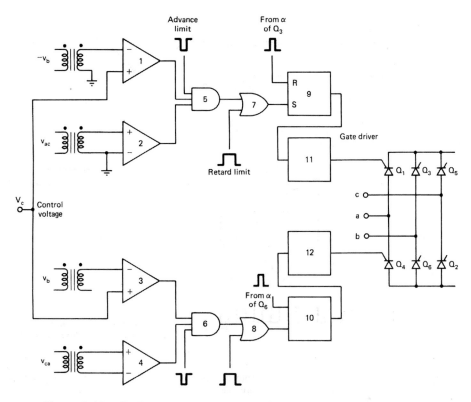

Figure 3.41 Cosine wave-crossing control of three-phase bridge converter

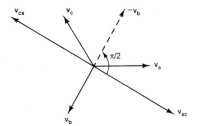

Figure 3.42 Phasor diagram for cosine wave-crossing control

duration to $2\pi/3$. The firing angle of Q_1 can be advanced or retarded by increasing or decreasing, respectively, the magnitude of V_c. The advance-limit notch, as indicated, is coupled to AND gate 5, and the retard-limit pulse is coupled to OR gate 7, as shown in Figure 3.41.

3.4.3 Phase-Locked Oscillator Principle

The cosine wave-crossing method described in the previous section derives cosine reference waves directly from power supply voltages. The harmonics generated by the converter flow

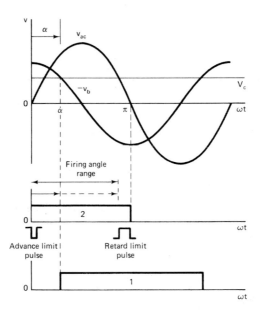

Figure 3.43 Waveforms explaining control circuit operation of thyristor Q_1

through the line source impedance and distort the line voltage. Similar distortion and transients can be introduced in the system by the power supply itself or by converters operating in parallel. A large filter to eliminate the distortion may not be satisfactory because of phase shift, which is also sensitive to supply frequency variation. A method to eliminate this problem is to digitally synthesize the cosine wave using the phase-locked loop (PLL) technique. Figure 3.44 explains the principle of digital synthesis on a biased cosine wave for a single-phase bridge converter. The frequency synthesizer generates a 30.72 kHz clock from a 60 Hz line frequency. The output frequency $f_0 = Nf$ can be programmed by selecting the frequency-divider ratio N ($=512$) so that f_0 tracks the supply frequency f within a definite locking range. The PLL is essentially a digital feedback system where the reference frequency f^* and feedback frequency f are compared in the phase frequency detector (PFD) and an analog error signal proportional to the phase difference is generated at the input of a loop filter. The amplified error signal drives a voltage-controlled oscillator (VCO) to generate the desired output frequency. If the output wave tends to fall back in phase (or frequency), the error voltage builds up to correct the VCO output such that the input and feedback waves lock together with a small phase error.

In Figure 3.45, the 30.72 kHz output clock, which has frequency and phase synchronization with the ac phase voltage, generates a biased cosine wave through a counter and ROM (read-only memory) lookup table. The counter is synchronized to the 60Hz wave and retrieves the ROM output repetitively every half-cycle. The control voltage can be compared with the ROM output either digitally or by an analog method by converting the ROM output through a digital-to-analog (D/A) converter. This technique can be easily extended to control a three-phase bridge converter.

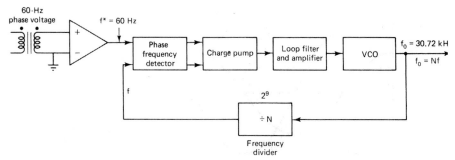

Figure 3.44 Phase-locked loop (PLL) frequency synthesizer

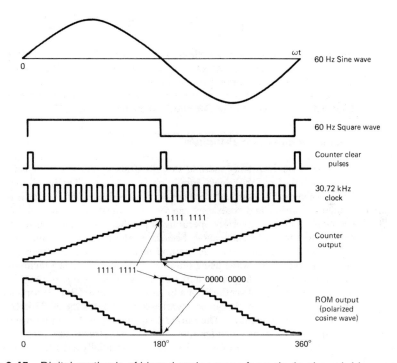

Figure 3.45 Digital synthesis of biased cosine wave for a single-phase bridge converter

The phase-locked oscillator principle can be used directly to control the firing pulses of a converter. At steady-state condition of the converter, the firing pulses for successive thyristors are applied at evenly spaced intervals of time. Thus, the time interval between any two consecutive firing signals is equal to the period of the input voltage wave, divided by the pulse number of the converter. An analog phase-locked oscillator method to generate such a pulse train for a three-phase bridge converter is shown in Figure 3.46. The method removes the dependence of the controller from the line voltage waves, and therefore it is particularly attractive for a soft ac

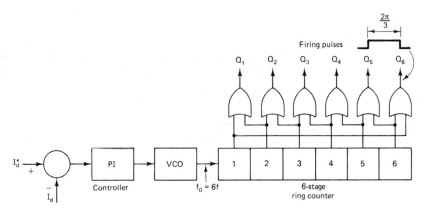

Figure 3.46 Phase-locked oscillator control of three-phase bridge converter

line supply. The converter current control is illustrated by a feedback method where the error signal drives a voltage-controlled oscillator (VCO) through a proportional-integral (P-I) controller. The VCO output drives a six-stage ring counter and the thyristor firing logic signals are derived by OR-coupling the consecutive stages as shown. At steady state, the loop error is zero and the P-I controller locks the output voltage such that the VCO operates at exactly six times the supply frequency, but the phase angle of the firing pulses can change. As the load changes or the command current I_d^* is changed, the VCO frequency tends to drift, but a change in angle α compensates I_d so that the system stabilizes at a new α angle. The response of the phase-locked oscillator is sluggish, and therefore, commutation failure may occur if the transient is fast.

3.5 EMI AND LINE POWER QUALITY PROBLEMS

Diode and thyristor converters, as discussed previously, essentially constitute a nonlinear load on electric utility systems, and this type of load has grown enormously in recent years. The line waveforms generated by these converters are far from sinusoidal, and as a result, serious EMI (electromagnetic interference) and harmonic problems are created.

3.5.1 EMI Problems

EMI problems arise due to the sudden changes in voltage (dv/dt) or current (di/dt) levels in a waveform. For example, in a diode rectifier, the line current can be pulses of short duration, and the diode recovery current pulse can generate transient voltage spikes in the line inductance. Similarly, a thyristor can generate recovery current transient, line voltage spike during commutation overlap, and high di/dt in line current wave during fast commutation. In fact, any fast-switching power semiconductor device creates similar high dv/dt and di/dt in the waveforms. A conductor carrying a high dv/dt wave acts like an antenna, and the radiated high-frequency wave may couple to a sensitive signal circuit and appear as noise (radiated EMI). Or, a parasitic coupling capacitor may carry this noise signal through the ground wire (conducted EMI). Similarly, a high *di/dt* current wave may create conducted EMI by coupling through a parasitic mutual inductance. The EMI problems create communication line interference and malfunctions to sensitive signal electronic circuits. Proper shielding, noise filtering, careful equipment layout, and grounding can solve EMI problems. Various EMC (electromagnetic compatibility) standards [5, 6] have been defined to regulate EMI problems.

3.5.2 Line Harmonic Problems

The harmonic currents generated by the converters flow through the utility system and cause various power quality problems. The distorted current flowing through line source inductance distorts the distribution bus voltage. The non-sinusoidal bus voltage may create a problem on sensitive loads operating on the same bus. Additionally, harmonic currents create additional loading and losses in line equipment, such as generator, transmission, and distribution lines, transformers, and circuit breakers. A simple example of line loading is a household appliance supplied by a front-end diode-rectifier capacitor filter where a peaky current wave may severely limit the outlet power capacity. Of course, in a phase-controlled converter, the additional VAR due to low DPF may also load the line equipment. The harmonics also give error in meter reading, protective relay malfunction, and can cause spurious line resonance with distributed inductance and capacitance parameters.

As the nonlinear power electronic loads are growing on the utility systems, various IEEE and IEC standards have been developed to protect power quality. For example, the IEEE-519 Standards [7] (amended in 1992) limit a line's individual harmonic and total current distortion (THD) at a point depending on the short circuit current ratio (I_{sc}/I_L), as shown in Table 3.1. A bus with lower short-circuit current means higher Thevenin impedance, and therefore, more distortion of the bus voltage for a specified load current I_L. This means that for a specified bus voltage distortion, the permissible harmonic current loading should be less. For example, with $I_{SC}/I_L < 20$, the Standard permits THD = 5% of the fundamental and individual odd harmonics, as indicated in the table. Note that the harmonic Standards are valid at the point of common coupling (PCC), as shown in Figure 3.47. This means that, for example, if a factory consumes linear and nonlinear loads, the Standards will be valid only at the PCC point, permitting the advantage of nonlinear load compensation by the linear load. The IEC 1000 Standards (developed in the mid-1990s) are much more rigorous and are applied to harmonic emission by individual equipment. The harmonic problems on the line can be mitigated by passive (nonresonant and resonant) and active filters, or by active wave shaping in the line converter itself. Similarly, lagging VAR compensation is also possible by the line converter itself or by separate passive and active line VAR conditioners. Active filtering and VAR compensation will be discussed in Chapter 5.

Table 3.1 Current Distortion Limits for General Distribution System at PCC (2.4 to 69 kV)

	Maximum Harmonic Current Distortion in % of Fundamental					
	Harmonic Order (ODD Harmonics)					
I_{SC}/I_L	<11	11<h<17	17<h<23	23<h<35	35<h	THD
<20*	4.0	4.0	1.5	0.6	0.3	5.0
20-50	7.0	3.5	2.5	1.0	0.5	8.0
50-100	10.0	4.5	4.0	1.5	0.7	12.0
100-1000	12.0	5.5	5.0	2.0	1.0	15.0
>1000	15.0	7.0	6.0	2.5	1.4	20.0

*I_{SC}, I_L = Short-circuit and load current, respectively, at PCC; h = Order of current harmonics.
The table is for the worst-case load at time more than one hour. Even harmonics, if present, are limited to 25% of the odd harmonic limits above. All power generation equipment is limited to these values of current distortion, irrespective of actual I_{SC}/I_L.

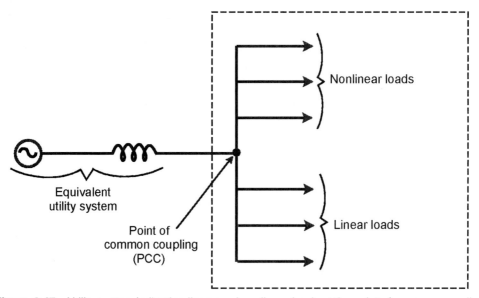

Figure 3.47 Utility system indicating linear and nonlinear loads at the point of common coupling

3.6 SUMMARY

This chapter has given a broad review of different types of diode rectifiers and thyristor phase-controlled converters which are extensively used in industry. Among diode rectifiers, emphasis has been given to single-phase and three-phase bridge configurations with inductive and capacitive loads. Similar circuit configurations and loads have also been emphasized for phase-controlled converters. Thyristor converters can operate as rectifiers as well as inverters. At zero firing angle, a thyristor converter operates as a diode rectifier. Note that this class of converter switches at zero current (defined as soft switching), and therefore, switching loss of the devices is negligible. The efficiency of the converters is high because of soft switching, accompanied with low conduction loss. Since the circuits are nonlinear, emphasis has been put on graphical analysis with the help of waveforms rather than rigorous mathematical analysis. Diode and thyristor converters constitute by far the major power electronics load on utility system and create harmonic distortion and lagging VAR loading problems. Stringent power quality standards have been recently trying to restrict this type of load directly on utility systems.

References

1. N. Mohan, T. M. Undeland, and W. P. Robbins, *Power Electronics*, John Wiley, New York, 1995.
2. B. R. Pelly, *Thyristor Phase-Controlled Converters and Cycloconverters*, John Wiley, New York, 1971.
3. G. K. Dubey, *Power Semiconductor Controlled Drives*, Prentice Hall, New Jersey, 1989.
4. T. Ohmae, T. Matsuda, T. Suzuki, N. Azusawa, K. Kamiyama, and T. Konishi, "A microprocessor-controlled fast-response speed regulator with dual current loop for dcm drives", *IEEE Trans. on Ind. Appl.*, vol. 16, pp. 388–394, May/June 1980.
5. G. Skibinski, J. Pankau, R. Sladky, and J. Campbell, "Generation, control and regulation of EMI from ac drives", *IEEE IAS Annu. Meet. Conf. Rec.*, pp. 1571–1583, 1997.
6. L. Rossetto, P. Tenti, and A. Zuccato, "Electromagnetic compatibility of industrial equipment", *IEEE FEPPCON III Conf. Rec.*, S. Africa, pp. B-14 to B-26, July 11–14, 1998.
7. C. K. Duffey and R. P. Stratford, "Update of harmonic Standard IEEE-519: IEEE recommended practices and requiurements for harmonic control in electric power systems", *IEEE IA Trans.*, vol. 25, pp. 1025–1034, Nov./Dec. 1989.

CHAPTER 4

Cycloconverters

4.1 INTRODUCTION

A cycloconverter is a frequency changer that converts ac power at one input frequency to output power at a different frequency with a one-stage conversion process. The phase-controlled thyristor converters, described in Chapter 3, can be easily extended for cycloconverter operation. Self-controlled ac switches, usually based on IGBT, can also be used in high-frequency link cycloconverters. In large-power industrial applications, thyristor phase-controlled cycloconverters are widely used. It is interesting to note that in the 1930s, phase-controlled cycloconverters with grid-controlled mercury-arc rectifiers were used in Germany to converter three-phase 50 Hz power to single-phase 16 2/3 Hz power for railway traction using universal motors. Historically, around the same time, the first known variable-frequency, adjustable-speed ac drive using a thyratron gas tube cycloconverter was used in the U.S. to control a 400 hp synchronous motor for a thermal power station auxiliary drive. Today, high-power, multi-megawatt, thyristor-based cycloconverters are very popular for driving induction and wound field synchronous motors. In fact, the general applications of cycloconverters may include the following:

- Cement and ball mill drives
- Rolling mill drives
- Slip-power recovery Scherbius drives
- Variable-speed, constant-frequency (VSCF) power generation for aircraft 400 Hz power supplies

In this chapter, we will study phase-controlled cycloconverter operation principles, converter configurations, harmonics and power factor considerations, and control principles. Matrix converters and high-frequency link cycloconverters will be briefly reviewed for completeness.

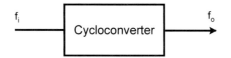

Figure 4.1 Block diagram of cycloconverter

4.2 PHASE-CONTROLLED CYCLOCONVERTERS

4.2.1 Operation Principles

Figure 4.1 shows the block diagram of a cycloconverter. In an industrial cycloconverter driving an ac motor, the input 50/60 Hz power is converted to variable-frequency, variable-voltage ac at the output to control the motor speed. The output frequency may vary from zero (rectifier operation) to an upper limit, which is always lower than the input frequency (step-down cycloconverter), and the power flow can be in either direction for four-quadrant motor speed control. In a VSCF system, the input power is usually generated by a synchronous machine that is coupled to a variable-speed turbine. The generated voltage can be regulated if the synchronous machine is wound field, but the output frequency is always proportional to the turbine speed. The function of the cycloconverter is to regulate the output frequency to be constant (typically 60 or 400 Hz). Figure 4.2 shows alternate-frequency conversion schemes. Figure 4.2(a) is a commonly used scheme where the input ac is rectified to dc and then inverted to variable-frequency ac through an inverter. In Figure 4.2(b), the input ac is converted to high-frequency ac through a step-up cycloconverter, and then converted to variable-frequency ac by a step-down cycloconverter. If the input power is dc, then the step-up cycloconverter is replaced by a high-frequency inverter.

The basic principle of cycloconversion can be explained with the help of the single-phase-to-single-phase converter circuit shown in Figure 4.3. A positive center-tap thyristor converter (see Figure 3.2) is connected in anti-parallel with a negative converter of a similar type so that

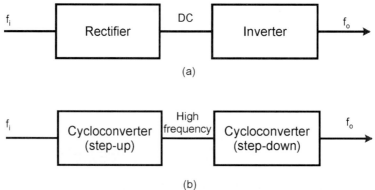

Figure 4.2 Alternate schemes of frequency conversion (a) DC link, (b) High-frequency link

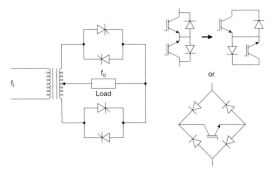

Figure 4.3 Single-phase-to-single-phase cycloconverter circuit showing ac switch configurations

the voltage and current of either polarity can be controlled in the load. The waveforms are shown in Figure 4.4, assuming ideal resistive load. In Figure 4.4(a), an integral half-cycle output wave is fabricated, which has a fundamental frequency $f_0 = (1/n)f_i$, where n ($n = 3$ in this case) is the number of input half-cycles per half-cycle of the output. The thyristor firing angle can be modulated to control the fundamental component output voltage, as shown in part (b). Instead of step-down frequency conversion, step-up frequency conversion is also possible, as indicated in (c) of the same figure. In this case, the devices are switched alternately between the positive and negative envelopes at a high frequency to generate carrier-frequency, modulated output. Here, the anti-parallel thyristor pair is replaced by a high-frequency ac switch, as indicated at the right of

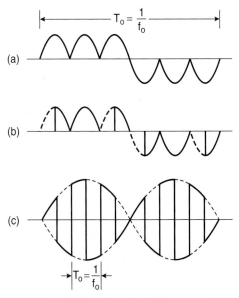

Figure 4.4 Fabricated waveforms for Figure 4.3 (a) Integral half-cycle control, (b) Waveform with firing angle modulation, (c) Waveform with step-up cycloconversion

Figure 4.3. The ac switch is basically an anti-parallel connection of IGBTs with a series diode and connected at center point, as shown, so that voltage can be blocked in either polarity, but current can flow in either direction. It can also be a diode bridge with a single IGBT as shown in the figure.

4.2.2 A Three-Phase Dual Converter as a Cycloconverter

The three-phase dual-converter, described in Chapter 3, can be used for three-phase-to-single-phase cycloconversion. The three-phase, half-wave circuit (see Figure 3.33(a)) and dual-bridge circuit (see Figure 3.33(b)) are repeated here in Figures 4.5 and 4.6, respectively. The four-quadrant operation modes of a dual converter are summarized in Figure 4.7, assuming only continuous conduction mode. With bipolar and controllable output voltage and current capability, it can be easily seen that a dual converter can be operated as a three-phase-to-single-phase cycloconverter. A dual converter can be represented by the Thevenin-equivalent circuit shown in Figure 4.8, where the harmonics and the Thevenin impedance of each converter component are neglected. The diodes, as indicated, permit the converter current in the direction shown. The two

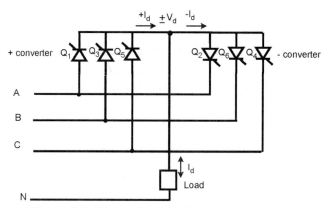

Figure 4.5   Three-phase, half-wave dual converter

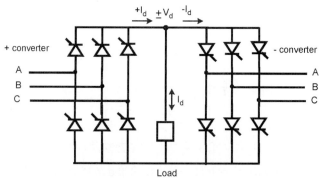

Figure 4.6   Three-phase, dual-bridge converter

Phase-Controlled Cycloconverters

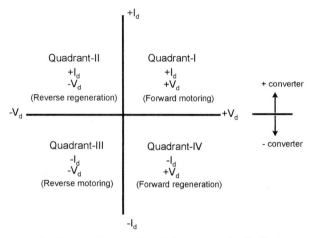

Figure 4.7 Four-quadrant operation modes of dual converter (indicates dc motor operation modes)

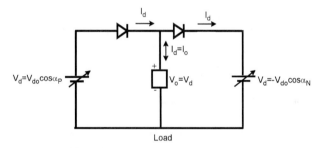

Figure 4.8 Thevenin-equivalent circuit of dual converter

converter voltages V_d are controlled to be equal so that the output voltage $V_0 = V_d$ at all conditions and the load current I_d is free to flow in either direction. This gives

$$V_o = V_d = V_{do} \cos \alpha_P \tag{4.1}$$

$$= -V_{do} \cos \alpha_N \tag{4.2}$$

where V_{d0} = dc output voltage of each converter at zero firing angle, and α_P and α_N are the respective firing angles. For a three-phase, half-wave converter, $V_{d0} = 0.675 V_L$ (see Equation (3.37)), whereas for the bridge, $V_{d0} = 1.35 V_L$ (see Equation (3.57)). The voltage-tracking control of the two converters is explained in Figure 4.9, where the converter transfer characteristics are indicated as functions of the respective firing angle. The horizontal dotted line representing the output voltage can be varied in either polarity and modulated in a sinusoidal manner by mod-

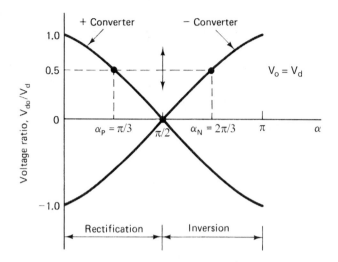

Figure 4.9 Voltage-tracking control in dual converter showing firing angle relations

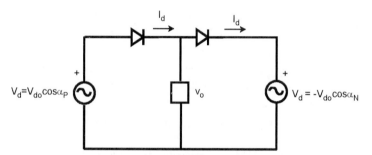

Figure 4.10 Cycloconverter equivalent circuit

ulation of the firing angles to generate a sinusoidal output voltage v_0. For tracking conditions of the two converters, Equations (4.1) and (4.2) indicate

$$\alpha_P + \alpha_N = \pi \tag{4.3}$$

For the particular output condition shown in Figure 4.9, $V_d/V_{do} = 0.5$, $\alpha_P = \pi/3$, and $\alpha_N = 2\pi/3$. Figure 4.10 shows the cycloconverter-equivalent circuit, where variable dc sources are replaced by sinusoidal sources.

4.2.3 Cycloconverter Circuits

4.2.3.1 Three-Phase, Half-Wave Cycloconverter

A practical and commonly used cycloconverter uses the three-phase, half-wave configuration shown in Figure 4.11, which is also known as a 18-thyristor, three-pulse cycloconverter. The

Phase-Controlled Cycloconverters

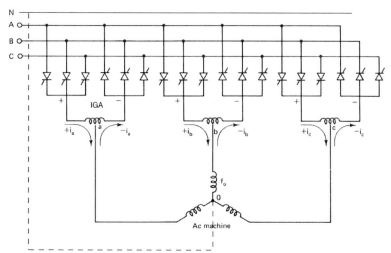

Figure 4.11 Three-phase-to-three-phase cycloconverter with 18 thyristors

circuit consists of three identical half-wave, anti-parallel phase groups and is shown with a wye-connected ac machine load. The load neutral is usually disconnected, as indicated. If the neutral is connected, the phase groups operate independently and create a large neutral current, which loads the machine. Each phase group functions as a dual converter, but the firing angle of each group is modulated sinusoidally with $2\pi/3$ phase angle shift so as to fabricate three-phase balanced voltage at the machine terminal. An inter-group reactor (IGR) is connected to each phase group to restrict circulating current, which will be explained later. Figure 4.12 explains the synthesis of an output phase voltage wave by sinusoidal modulation of the firing angle α. The output frequency and depth of modulation can be varied to generate a variable-frequency, variable-voltage power supply for the motor. The fabricated output voltage wave contains complex harmonics, which may be adequately filtered by the machine's effective leakage inductance.

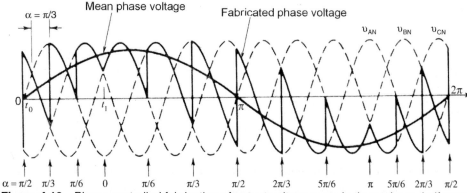

Figure 4.12 Phase-controlled fabrication of output voltage wave in three-phase, half-wave cycloconverter

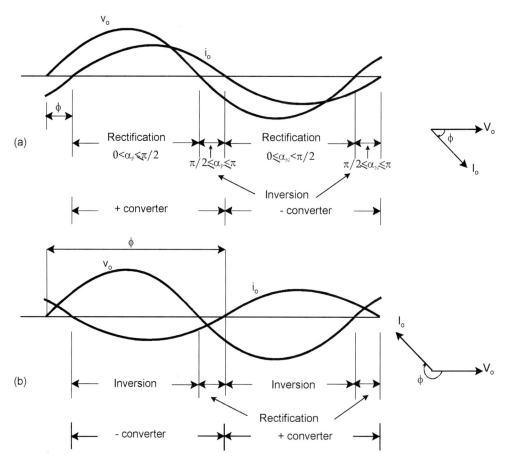

Figure 4.13 Voltage and current waves for motoring and regenerative modes (a) Motoring mode, (b) Regenerative mode

Any arbitrary DPF load can be supplied by a cycloconverter. Figure 4.13(a) shows the phase voltage and current waves in active or motoring mode, where the current wave lags the voltage wave by DPF angle φ. The positive half-cycle of current flows through the positive converter, whereas the negative converter takes the negative half-cycle of current. A component converter operates in the rectification mode if the voltage and current are of same polarity, but in the inversion mode these are of opposite polarity. Both the component converters in a phase group can be controlled simultaneously to fabricate the mean output voltage. This will permit the bidirectional phase current to flow freely through either converter. There will, of course, be an instantaneous potential difference (due to harmonics) between the outputs of the two converters, which will be discussed later. Figure 4.13(b) shows the voltage and current waves in regenerative mode of the drive, where the polarity of the current wave is reversed.

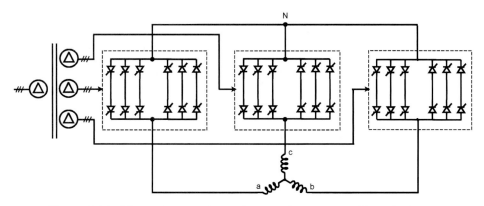

Figure 4.14 Three-phase-to-three-phase cycloconverter with 36 thyristors

4.2.3.2 Three-Phase Bridge Cycloconverter

Among the many other possible cycloconverter circuit configurations, the three-phase bridge (six-pulse) configuration, or 36-thyristor circuit, as shown in Figure 4.14, is very popular in multi-megawatt drive applications. Each phase group of the cycloconverter consists of a dual-bridge converter, which is connected to a transformer-isolated, three-phase supply. The wye-connected machine is supplied from the lower output points of the bridges, as shown, whereas the upper points are shorted to make the load neutral point. No transformer is required for isolation if the machine windings are isolated. The cycloconverter is shown without any IGR. The waveform fabrication of a bridge cycloconverter is similar to that of an 18-thyristor circuit and is left as an exercise to the reader.

Modulation Factor The output phase voltage v_0 of a cycloconverter, can be written as (see Figure 4.12)

$$v_o = \sqrt{2} V_o \sin \omega_o t \qquad (4.4)$$

where V_0 = rms voltage and $\omega_0 = 2\pi f_0$ is the output angular frequency. Since $\alpha_P + \alpha_N = \pi$, we can write

$$v_o = V_{do} \cos \alpha_P \qquad (4.5)$$

$$= -V_{do} \cos \alpha_N \qquad (4.6)$$

$$= m_f V_{do} \sin \omega_o t \qquad (4.7)$$

where m_f = modulation factor, and $V_{d0} = 0.675 V_L$ for an 18-thyristor circuit and $1.35 V_L$ for a 36-thyristor circuit. From Equations (4.4) and (4.7), the modulation factor is given as

$$m_f = \frac{\sqrt{2}V_o}{V_{do}} \tag{4.8}$$

which means m_f is the ratio of peak output phase voltage and maximum possible peak phase voltage. The parameter m_f varies in the range 0 to 1. From Equations (4.5) and (4.7), we can write

$$\cos\alpha_P = m_f \sin\omega_o t \tag{4.9}$$

that is,

$$\alpha_P = \cos^{-1}\left[m_f \sin\omega_o t\right] \tag{4.10}$$

and

$$\alpha_N = \pi - \alpha_P \tag{4.11}$$

The equations indicate that for zero output voltage, $m_f = 0$; that is, $\alpha_P = \alpha_N = \pi/2$ at all conditions. On the other hand, for maximum phase voltage, $m_f = 1$, which gives $\alpha_P = 0$ and $\alpha_N = \pi$ at positive peak value and $\alpha_P = \pi$ and $\alpha_N = 0$ at negative peak value. Figure 4.15 gives the output voltage waves at $m_f = 0.5$ and 1.0 indicating the firing angles

4.2.4 Circulating vs. Non-Circulating Current Mode

4.2.4.1 Circulating Current Mode

Like a dual converter, the phase group of a cycloconverter can be operated in either circulating current mode or non-circulating current, or blocking, mode. In circulating current mode, a current always circulates between the positive converter and the negative converter. Although the fundamental output voltage waves of the component converters are always equal, the harmonics will cause instantaneous potential difference, which will cause a short-circuit unless an IGR is

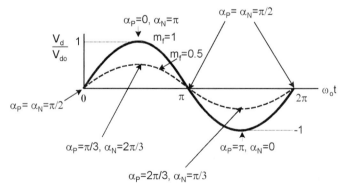

Figure 4.15 Output phase voltage wave showing modulation of firing angles

Phase-Controlled Cycloconverters

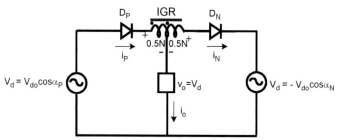

Figure 4.16 Circulating current mode equivalent circuit with IGR

connected (see Figure 4.11). Figure 4.16 shows the equivalent circuit of a phase group with an IGR. It can be shown that with an IGR, a self-induced circulating current is developed between the positive converter and the negative converter. This is explained with the help of waveforms in Figure 4.17. Assume that the load circuit is highly inductive and the load current i_0 is constrained to be sinusoidal by the sinusoidal applied voltage v_0. Assume again that at time $t = 0$, the positive load current is switched on, as shown in the figure. The positive load current is taken by the positive converter only ($i_P = i_0$). The increasingly rising positive load current during the angular interval 0 and $\pi/2$ will create a positive voltage drop $v_L = di_0/dt$ in the primary winding segment of the IGR. With the IGR polarity markings shown, the induced voltage polarity in the secondary segment will be negative and reverse-bias the diode D_N. This will inhibit any current flow in the negative converter. However, at $\pi/2$ angle, $v_L = 0$ when i_0 reaches the peak value I_m. From this point onward, v_L will tend to reverse in polarity, which will induce current in the negative converter, clamping the voltage across the IGR to zero. From now on, the IGR voltage will remain clamped to zero and the mmf $0.5NI_m$ (N = number of turns of IGR) will remain trapped in it. As a result, there will be a self-induced circulating current between the positive and negative converters, as shown in the figure. Since the total mmf (or flux linkage) in the IGR remains constant ($0.5NI_m$) at any instant (conservation of mmf or flux linkage), we can write the mmf balancing equation as

$$0.5Ni_P + 0.5Ni_N = 0.5NI_m \tag{4.12}$$

or

$$i_P + i_N = I_m \tag{4.13}$$

But,

$$i_P - i_N = i_o = I_m \sin\omega_o t \tag{4.14}$$

From Equations (4.13) and (4.14), we can solve i_P and I_N as

$$i_P = 0.5I_m + 0.5I_m \sin\omega_o t \tag{4.15}$$

$$i_N = 0.5I_m - 0.5I_m \sin\omega_o t \tag{4.16}$$

The waveforms of i_P and i_N, shown in Figure 4.17, indicate that the difference between the converter current and the load component of current constitutes the self-induced circulating current. In a practical circuit, of course, a ripple component of current will add to each current component. Note that the waveforms are valid for steady-state operation. Any load-induced transient will eventually decay to give steady-state condition with a new level of trapped mmf in the IGR. Figure 4.18 explains the voltage and current waves in a phase group of a 36-thyristor cycloconverter. Figure 4.18(a) shows the fabrication of a raw-output phase-voltage wave of a positive converter from the supply line voltages by modulation of the firing angles. The corresponding wave for the negative converter is shown in (b). The mean voltage wave that appears at the output of the IGR is shown in part (c). This voltage wave is somewhat smoother than the component voltage waves. The instantaneous potential difference between the positive and negative converter voltages appears across the IGR, which is shown in (d). Figure 4.18(e), (f), and (g) show the currents in a positive converter, negative converter, and load, respectively.

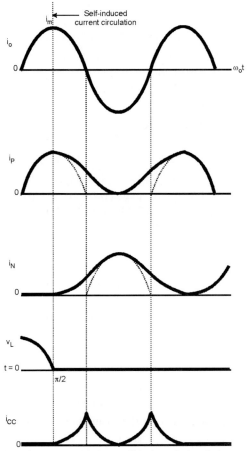

Figure 4.17 Waveforms explaining self-induced circulating current with IGR

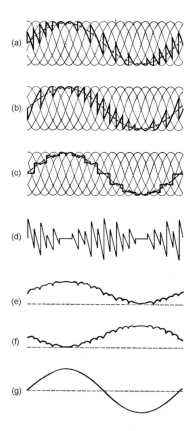

Figure 4.18 Waveforms in three-phase bridge cycloconverter in circulating current mode ($m_f = 1$) (a) Fabrication of output phase voltage (v_0) of positive converter from input line voltages, (b) Fabrication of v_0 of negative converter, (c) Mean output voltage at load terminal, (d) Voltage across IGR, (e) Positive converter current (i_P), (f) Negative converter current (i_N), (g) Output load current (i_0) [1]

The circulating current mode operation of a cycloconverter may provide some advantages and disadvantages (comparing to blocking mode), which can be summarized as follows

Advantages:
- The output phase voltage wave (v_0) is somewhat smoother, contributing less harmonic content to the load.
- The output frequency range is higher.
- The load displacement power factor condition does not affect harmonics in output voltage.
- There is less subharmonic problem in the load.
- Harmonics injected to input line are somewhat less.
- Intentional control of circulating current provides a method of controlling line displacement power factor (will be explained later).
- The control is simple.

Disadvantages:
- Bulky IGR increases cost. It also increases losses.
- The circulating current creates additional loading to the thyristors, which increases losses.
- The over-design increases cost.

Some of these points will be discussed later.

4.2.4.2 Blocking Mode

Although the circulating current mode operation of a cycloconverter provides a number of advantages, as discussed above, the cost and efficiency penalties do not make it very popular, except in special cases. In blocking or non-circulating current mode, no IGR is used and only one converter (positive or negative) component is permitted to conduct at any instant. However, the firing angles are controlled so that the output voltages track all the time, in other words, $\alpha_P + \alpha_N = \pi$. Figure 4.19 explains converter component selection with load current zero crossing detection. The basic principle is as follows: Since the positive load current is taken by the positive converter only, this converter is enabled when the load current is positive. A similar principle holds true for the selection of the negative converter. With sinusoidal output voltage, the load current will tend to be sinusoidal and the selection of a converter based on the polarity of load current is not difficult. Assume that in the beginning, the load current is positive, and therefore, the positive converter is enabled by a current polarity-sensing flip-flop. As the current i_0 decreases and goes below a threshold amplitude, the positive converter is disabled. Both converters remain inhibited for time t_g, and then the negative converter is enabled, as shown. The lock-out time gap t_g prevents a short-circuit by giving enough time for the thyristors of the outgoing converter to turn off before enabling the incoming converter. Evidently, such a control will introduce some cross-over distortion of load current.

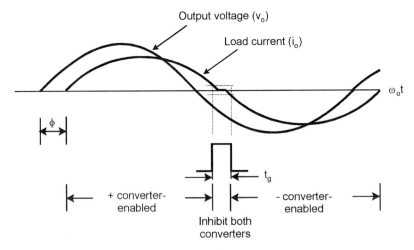

Figure 4.19 Converter component selection in blocking mode with \current zero crossing detection

Phase-Controlled Cycloconverters

The simple converter selection based on current zero crossing, as shown in the figure, has a disadvantage. Near current zero crossing, the converter may fall into discontinuous conduction mode, particularly due to back emf of the motor. With discontinuous conduction, an outgoing converter will be turned off prematurely, introducing additional distortion in the load current. More elegant converter selection principles are described in the literature [3].

In blocking mode, the fabricated voltage waves of the component converters are impressed directly across the load. This causes more severe load and source harmonic problems, in addition to the dead-band effect on the load current. The output frequency range is somewhat smaller in blocking mode. The cycloconverter control is also more complex because of the converter bank selection problem. However, there are the advantages of lower cost and improved efficiency compared to circulating current mode operation.

4.2.5 Load and Line Harmonics

As explained before, the fabricated output voltages of a cycloconverter contain complex harmonics. These harmonics are to be filtered before applying them across the load. A machine is normally connected directly at the cycloconverter output. The machine leakage inductance acts as a low-pass filter, and the phase current waves tend to be nearly sinusoidal. However, the ripple component of current will cause additional machine heating and torque pulsation, as discussed in Chapter 2. Again, a cycloconverter is basically a matrix of switches without any energy storage capability (neglecting IGR effect). Assuming the devices are loss-less, there will be a balance of instantaneous output power (p_0) with instantaneous input power (p_i). Therefore, the voltage harmonics at the output will have the corresponding reflection of harmonics at the input line currents, even with the assumption of balanced and sinusoidal supply voltages and load currents. Basically, the nonlinear switching mode operation of a converter introduces distortion at the input and output, as explained in Chapter 3. In general, the load and line harmonics will be influenced by the following:

- Circulating current or blocking mode of operation
- Pulse number (p)
- Modulation factor (m_f) of output voltage
- Output–to–input frequency ratio (f_0/f_i)
- Load DPF
- Continuous or discontinuous conduction
- Commutation overlap effect
- Feedback control and its bandwidth

4.2.5.1 Load Voltage Harmonics

Since the harmonic patterns are very complex, it is difficult to make an analysis graphically or mathematically. The most convenient method to study harmonics is to make a computer simulation study and analyze the waveforms by FFT. A simplified analytical study made in [1] will be cited here. Consider a general p-pulse, phase-controlled thyristor converter. It was explained in Chapter 3 that for ideal conditions of continuous conduction and high load inductance, the harmonic families of the load voltage and line current are given by pn and $pn \pm 1$, respectively, where p = pulse number and n = an integer. In a cycloconverter, the firing angles, instead of remaining constant, are sinusoidally modulated. Therefore, we expect the load and line harmonics to have carrier frequencies related to line frequency and sideband frequencies related to output frequency.

Figure 4.20 shows the harmonic family (f_H/f_i) of an output voltage wave in blocking mode with the output frequency ratio f_0/f_i. Each harmonic group starts as a single number in rectifier mode and then branches out with sidebands (ideally of an infinite number) as f_0/f_i increases. The harmonic lines reflect on the f_0/f_i axis with equal slope. The output fundamental frequency (f_0) line is indicated by the dotted slope. Any harmonic below this slope is a subharmonic $(f_H < f_0)$. The harmonic family is given by the expression $pnf_i \pm nf_0$, where $pn \pm n$ = odd integer. In rectifier mode operation, that is, $f_0 = 0$, the harmonic family for an 18-thyristor cycloconverter is $3f_i$,

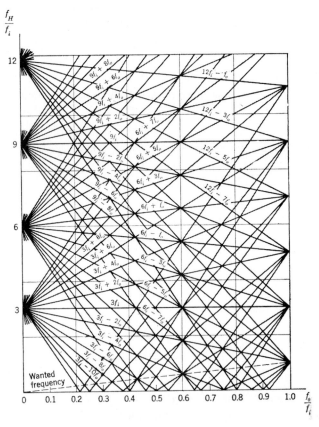

Figure 4.20 Harmonic family of output voltage in blocking mode [1]

Phase-Controlled Cycloconverters

Table 4.1 Listing of Output Voltage Harmonics in 18-thyristor and 36-thyristor Cycloconverters in Blocking Mode

18-thyristor	$3f_i \pm 0$	$6f_i \pm f_o$	$9f_i \pm 0$	
	$3f_i \pm 2f_o$	$6f_i \pm 3f_o$	$9f_i \pm 2f_o$	
	$3f_i \pm 4f_o$	$6f_i \pm 5f_o$	$9f_i \pm 4f_o$	...
	$3f_i \pm 6f_o$	$6f_i \pm 7f_o$	$9f_i \pm 6f_o$	
	...	...	...	
36-thyristor	$6f_i \pm f_o$	$12f_i \pm f_o$	$18f_i \pm f_o$	
	$6f_i \pm 3f_o$	$12f_i \pm 3f_o$	$18f_i \pm 3f_o$	
	$6f_i \pm 5f_o$	$12f_i \pm 5f_o$	$18f_i \pm 5f_o$	...
	...	...	...	

$6f_i$, $9f_i$, $12f_i$, etc., and for a 36-thyristor cycloconverter, it is $6f_i$, $12f_i$, etc. At any finite f_o/f_i, the harmonic family can be found on the vertical line.

Table 4.1 gives the harmonic families for both 18-thyristor and 36-thyristor cycloconverters. The amplitude of the harmonics (not shown) is convergent with higher sidebands and higher carrier frequencies. Consider, for example, the output frequency $f_o = 17$ Hz, which corresponds to $f_o/f_i = 0.283$ with input frequency $f_i = 60$ Hz. The actual harmonic frequencies can be found by substituting these values of f_o and f_i in Table 4.1. Locating the point $f_o/f_i = 0.283$ on Figure 4.20 indicates the presence of a subharmonic component $|3f_i - 10f_o| = 10$ Hz. This subharmonic component with typical magnitude is shown in Figure 4.21(a). The subharmonic voltage wave,

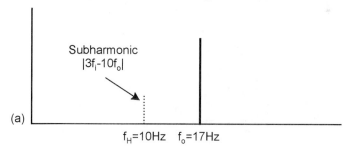

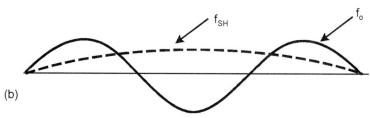

Figure 4.21 (a) Subharmonic voltage component, (b) waveform

shown in (b), with a volt-sec. area comparable to that of the fundamental wave, can cause a serious problem by saturation of the magnetics. Figure 4.20 indicates that the subharmonic problem becomes worse with higher fundamental frequency. Higher harmonic content and the subharmonic problem usually restrict the output frequency ratio range f_0/f_i from zero to 1/3, that is, dc to 20 Hz with 60 Hz line frequency.

Figure 4.22 shows the output harmonic family in circulating current mode. As discussed before, the harmonic problem in this condition is less severe, which is evident by the harmonic lines being less crowded. Each harmonic group has limited sidebands and can be summarized as follows:

- 18-thyristor circuit: $3f_i \pm 4f_0$, $6f_i \pm 7f_0$, $9f_i \pm f_0$, etc.
- 36-thyristor circuit: $6f_i \pm 7f_0$, $12f_i \pm 13f_0$, $18f_i \pm 19f_0$, etc.

The harmonic components of the output voltage can be easily listed from Figure 4.22. Note that the subharmonic problem is much less severe in this case. In fact, there is no subharmonic at the output for frequency ratios of zero to 0.6. Low or zero subharmonics and reduced harmonic content permit a circulating current cycloconverter to operate at much higher output frequency range, typically from zero to $2/3f_i$, that is, dc to 40 Hz with supply frequency of 60 Hz.

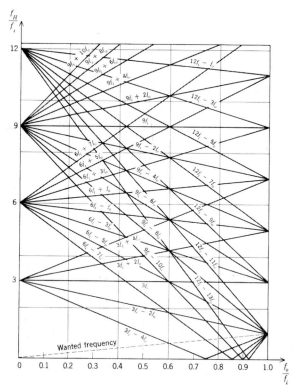

Figure 4.22 Harmonic family of output voltage in circulating current mode [1]

Phase-Controlled Cycloconverters

4.2.5.2 Line Current Harmonics

Like the graphical fabrication of the output voltage wave piece by piece from the sinusoidal line voltages, the input line current wave can also be fabricated graphically piece by piece from the sinusoidal load current wave through the closed switches. Figure 4.23 shows the fabrication of output voltage and line current waves for a 36-thyristor, blocking-mode cycloconverter. Figure 4.23(a), (c), and (e) show the fabrication of the output phase voltages from the input line voltages by modulation of the firing angles. Figure 4.23(b), (d), and (f) show the synthesis of three-phase line currents for the respective output phase group operation. This means that for three-phase-to-single-phase cycloconverter operation, Figure 4.23(b), for example, gives the line current waves for the loading of phase a output only. Note that the component line current waves are symmetrical, but they have even harmonic components. This is because the single-phase loading creates second harmonic load power that is reflected as even harmonics at the input. For three-phase-to-three-phase operation, the component line currents are added to give the total line currents shown in Figure 4.23(g). Note that although the load DPF = 1, the fundamental input currents lag the respective input phase voltages.

Figure 4.24 shows the family of line harmonics that is given by the relation $(np \pm 1)f_i \pm mf_0$, where $(np \pm 1) \pm m$ = odd integer. A listing of the harmonics is given in Table 4.2 for clarity.

Table 4.2 Listing of Line Harmonics in Circulating Current Mode

18-thyristor	36-thyristor
f_i	f_i
$f_i \pm 6f_o$	$f_i \pm 6f_o$
$f_i \pm 12f_o$	$f_i \pm 12f_o$
⋮	⋮
$2f_i \pm 3f$	$5f_i$
$2f_i \pm 6f$	$5f_i \pm 6f_o$
⋮	⋮
$4f_i \pm 3f$	$7f_i$
$4f_i \pm 6f$	$7f_i \pm 6f_o$
⋮	⋮

Note that the fundamental component only can carry the average power supplied to the load, whereas the harmonics components generate only pulsating power, the average of which is zero. Figure 4.24 indicates the presence of a small subharmonic current component, which can be generally ignored.

4.2.6 Line Displacement Power Factor

It was explained in Chapter 3 that a phase-controlled, line-commutated converter always draws lagging reactive current from the line. A cycloconverter is basically a phase-controlled,

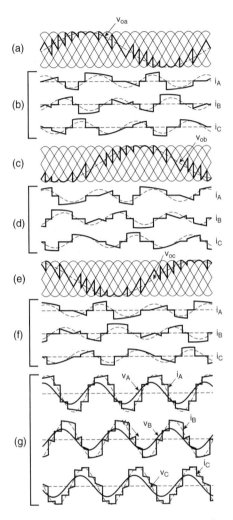

Figure 4.23 Fabrication of load voltage and line current waves for 36-thyristor cycloconverter in blocking mode ($f_0/f_i = 1/3$, $m_f = 1$, $\varphi = 0°$) (a), (c), (e) Fabrication of output phase voltages, (b) Line currents due to phase group *a* only, (d) Line currents due to phase group *b* only, (f) Line currents due to phase group *c* only, (g) Line currents for the operation of all phase groups [1]

line-commutated converter, and therefore, it is natural that it would also draw lagging reactive line current. In fact, the modulating firing angles have an averaging effect on the line DPF. This poor DPF, along with the complex harmonics in the line and load, is a big disadvantage of the cycloconverter. It can be shown that even with unity load power factor and output modulation factor $m_f = 1$, the cycloconverter draws lagging line current. The line DPF deteriorates as the load DPF decreases or the modulation factor decreases. In fact, the line DPF is lagging even when the load DPF is leading.

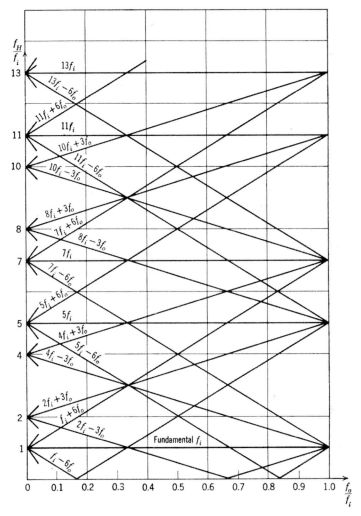

Figure 4.24 Harmonic family of input current of cycloconverter [1]

4.2.6.1 Theoretical Derivation of Line DPF

It is not difficult to derive a quantitative expression for the input DPF of a cycloconverter [4]. Consider a phase group of an 18-thyristor cycloconverter, as shown in Figure 4.25(a), and assume that the positive converter is conducting only at continuous conduction with a high inductance load. Consider again that the cycloconverter is operating at very low output frequency. Figure 4.25(b) shows a small segment of the output voltage (v_0) and current (i_0) waves, plus the expanded view of converter operation, where v_i = input phase voltage and α_P = firing angle of positive converter. Within the small output angular interval as shown, the converter can be considered to be operating as a rectifier. The line current i consists of i_0 load current pulses of

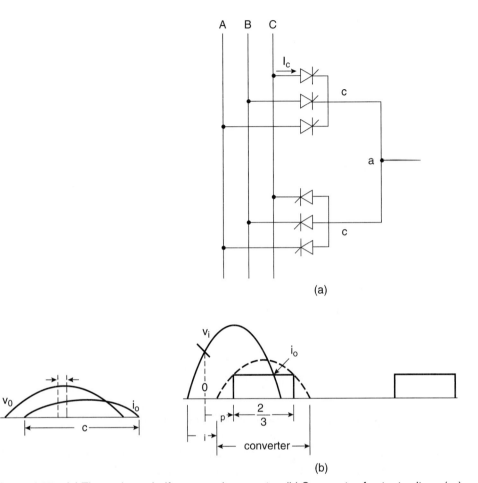

Figure 4.25 (a) Three-phase, half-wave cycloconverter, (b) Segments of output voltage (v_o) and output current (i_o) waves

$2\pi/3$ width contributed in every cycle of the input voltage, as indicated in the figure. The other input phases of the positive converter will operate in a similar manner. The Fourier series of the line current can be given as

$$i = \frac{i_o}{3} + \frac{\sqrt{3}}{\pi} i_o [\sin(\omega t - \alpha_P) - \frac{1}{2}\cos 2(\omega t - \alpha_P) - \frac{1}{4}\cos(4\omega t - \alpha_P) \\ - \frac{1}{5}\sin 5(\omega t - \alpha_P) + \cdots] \quad (4.17)$$

where ω = supply frequency. The current wave has a dc component and a fundamental component with a lagging phase angle α_P ($\alpha_P = \phi_i$, as shown) that is inherent in a phase-controlled converter. Since the supply's active and reactive power components are contributed by the funda-

Phase-Controlled Cycloconverters

mental current only, the instantaneous active and reactive power expressions for the whole positive converter (as a rectifier) can be given as

$$P'_i = 3V_s \left(\frac{\sqrt{3}i_o}{\sqrt{2}\pi}\right)\cos\alpha_P \qquad (4.18)$$

$$Q'_i = 3V_s \left(\frac{\sqrt{3}i_o}{\sqrt{2}\pi}\right)\sin\alpha_P \qquad (4.19)$$

where V_s = rms phase voltage of line. The equations can be written in the following form:

$$\begin{aligned}P'_i &= (1.17V_s \cos\alpha_P)i_o \\ &= (0.675V_L \cos\alpha_P)i_o \\ &= V_{do}\cos\alpha_P i_o \\ &= v_o i_o\end{aligned} \qquad (4.20)$$

$$\begin{aligned}Q'_i &= (1.17V_s \sin\alpha_P)i_o \\ &= V_{do}\sin\alpha_P i_o\end{aligned} \qquad (4.21)$$

where V_L = rms line voltage and v_0 = instantaneous fundamental output voltage. If the firing angle α_P remains constant, the converter operates as rectifier, that is, $V_{do}\cos\alpha_P = V_d$ and $i_0 = I_d$. At this condition, P'_i and Q'_i expressions give the steady-state input active and reactive powers, respectively. In a cycloconverter, α_P and i_0 vary sinusoidally, and therefore, P'_i and Q'_i are correspondingly modulated. It is necessary to average these parameters to determine loading on the source.

Figure 4.26 shows the output phase voltage and current waves indicating the region for positive converter operation. The waveform $V_{d0}\sin\alpha_P$ is graphically constructed in the figure. Since $0 < \alpha_P < \pi$, its magnitude is always positive and phase-shifted by $\pi/2$ with respect to the v_0 wave. The instantaneous reactive power given by Equation (4.21) is also plotted graphically on Figure 4.26. It can be easily shown that the curve will repeat for the negative converter, which is also indicated in the figure. The expression for average reactive power Q_i contributed by the line can be given as

$$Q_i = \frac{1}{\pi}\int_0^{\frac{1}{2}\text{cycle}} (1.17V_s \sin\alpha_P)i_o\,d\omega_o t \qquad (4.22)$$

or, with the help of Figure 4.26, the full expression is

$$Q_i = \frac{1}{\pi}\left[\int_\phi^{\frac{\pi}{2}}(1.17V_s\cos\omega_o t)(I_m\sin(\omega_o t-\phi))d\omega_o t \right.$$
$$\left. + \int_{\frac{\pi}{2}}^{\pi+\phi}(-1.17V_s\cos\omega_o t)(I_m\sin(\omega_o t+\phi))d\omega_o t\right] \quad (4.23)$$

where φ = load power factor angle.

Equation (4.23) can be solved as

$$Q_i = \frac{2P_o}{\pi}\cos^2\phi + \frac{2Q_o}{\pi}\left(\phi + \frac{1}{2}\sin 2\phi\right) \quad (4.24)$$

where P_0 = output active power per phase and Q_0 = output reactive power per phase. These parameters are given, respectively, by the expressions

$$P_o = \frac{1}{\pi}\int_0^{\frac{1}{2}cycle}(v_o i_o)d\omega_0 t = V_o I_o \cos\phi \quad (4.25)$$

and

$$Q_o = V_o I_o \sin\phi \quad (4.26)$$

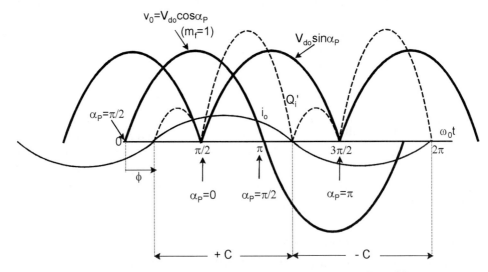

Figure 4.26 Waveforms explaining reactive power loading of line

Phase-Controlled Cycloconverters

Since the active output power is the same as the active input power ($P_0 = P_i$), we can write

$$P_i + jQ_i = P_i + j\frac{2}{\pi}\left[P_i \cos^2\phi + Q_o\left(\phi + \frac{1}{2}\sin 2\phi\right)\right] \quad (4.27)$$

and therefore, the input DPF expression can be written as

$$DPF = \cos\phi_i = \frac{P_i}{|P_i + jQ_i|}$$

$$= \frac{1}{\left|1 + j\frac{2}{\pi}\left[\cos^2\phi + \frac{Q_o}{P_i}\left(\phi + \frac{1}{2}\sin 2\phi\right)\right]\right|}$$

$$= \frac{1}{\left|1 + j\frac{2}{\pi}\left[\cos^2\phi + \tan\phi\left(\phi + \frac{1}{2}\sin 2\phi\right)\right]\right|} \quad (4.28)$$

$$= \frac{1}{\left|1 + j\frac{2}{\pi}(1 + \phi\tan\phi)\right|}$$

where $\tan\varphi = Q_0/P_i = Q_0/P_0$.

If more phase groups are added, parameters P_i and Q_i reflect equally at the input and DPF remains unchanged. It can be shown easily that the DPF expression remains valid also for a 36-thyristor cycloconverter. Note that Equation (4.28) has been derived assuming a modulation factor of $m_f = 1$. Since the modulation factor does not affect the current demanded by the load, but affects the displacement angle directly, we can write the general expression

$$DPF = \frac{m_f}{\left|1 + j\frac{2}{\pi}(1 + \phi\tan\phi)\right|} \quad (4.29)$$

Equation (4.29) indicates that DPF is a function of modulation factor m_f and load power factor angle φ. It has been plotted in Figure 4.27 for different m_f and for both lagging and leading power factors and motoring and regenerative conditions. The curves are symmetrical in all the quadrants, and indicate that if the load DPF = 1 and $m_f = 1$, the best line DPF = 0.843. The line DPF deteriorates as m_f or load DPF decreases. In extreme cases, line DPF = 0 for $m_f = 0$ or load DPF = 0 with either leading or lagging reactive load.

4.2.7 Control of Cycloconverter

The control of a cycloconverter is very complex, and only elementary discussion will be given in this section. Figure 4.28 shows a typical variable-speed, constant-frequency (VSCF)

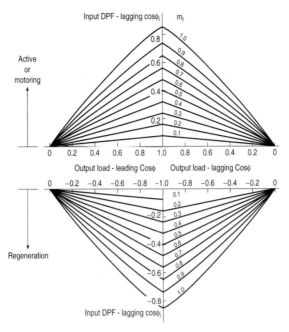

Figure 4.27 Input and output displacement power factor relations with modulation factors

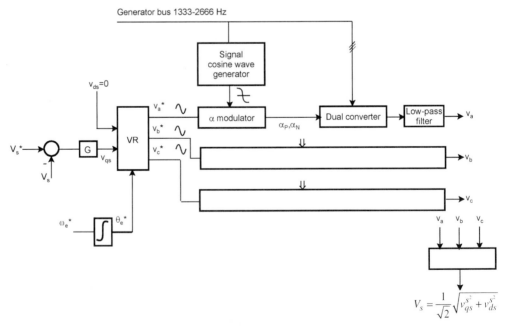

Figure 4.28 Variable-speed, constant-frequency (VSCF) system control block diagram

Phase-Controlled Cycloconverters

system where the power and control elements are shown in block diagram form. The generator bus with regulated voltage but variable frequency (1333 – 2666 Hz) is fed to the cycloconverter phase groups, as shown. A generator speed variation in the range of 2:1 is assumed in this case. The dual converter in each phase group has an output low-pass filter to generate sinusoidal output voltage. The α modulator receives the biased cosine signal wave from the generator bus voltage and sinusoidal control signal voltage to generate the firing angles of the converter. Three-phase sinusoidal control signals can be generated from the primary voltage control loop through the vector rotator (VR), which receives the angular signal θ_e from the frequency signal, as shown. The feedback voltage V_s can be generated from the output phase voltages, as indicated. The details of the α modulator are explained in Figure 4.29. The cosine wave-crossing method with $\alpha_P + \alpha_N = \pi$ assures linear transfer characteristics between the control signal and output voltage, indicating that the cycloconverter is basically a linear amplifier. If current control is desired instead of voltage control, this principle is shown in Figure 4.30. The command current is compared with the feedback current (synthesized from phase currents) and the P-I controller generates the synchronously rotating voltage signal, as shown. Further details of such control methods will be given in Chapter 8.

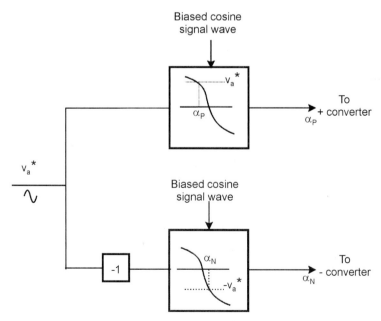

Figure 4.29 Firing angle generation principle for positive and negative converters

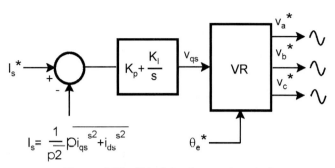

Figure 4.30 Principle of current control

4.2.8 DPF Improvement Methods

As discussed above, the DPF of a phase-controlled cycloconverter is very poor. The DPF can be improved by the following methods:

- Square-wave operation at output
- Asymmetrical firing angle control
- Circulating current control
- Static VAR compensator

A brief review will be given for each of these improvements, except the static VAR compensator, which will be described in Chapter 5.

4.2.8.1 Square-Wave Operation

A simple method of improving the line DPF is to fabricate square-wave voltages at the output instead of sinusoidal voltages. Figure 4.31 shows the fabrication of square-wave voltages for different fundamental magnitudes by controlling the firing angle. If, for example, $\alpha_P = 0$ ($\alpha_P + \alpha_N = \pi$), the peak value of the output fundamental voltage is $4 V_{d0}/\pi$. If α_P is controlled to be $\pi/3$, the fundamental is reduced to 50% amplitude as shown. Without sinusoidal firing angle modulation, the cycloconverter gives phase-controlled, converter-like DPF at the input. The additional advantages of a square-wave cycloconverter are higher output fundamental voltage and simpler control. Of course, the disadvantage is worse voltage ripple at the output. High-power cycloconverter drives sometimes use square-wave cycloconverters [5].

4.2.8.2 Asymmetrical Firing Angle Control

The sequential, or asymmetrical, firing angle control principle, discussed in Chapter 3, can be applied to a cycloconverter to improve the input DPF. Figure 4.32 shows the block diagram of a 72-thyristor cycloconverter [6] where the asymmetrical firing angle control principle has been applied. Each phase group consists of two dual bridges in series, which are supplied by delta-

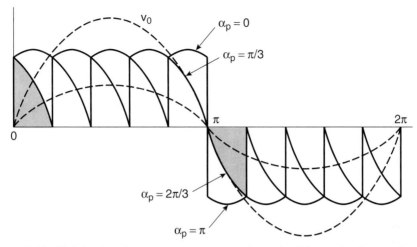

Figure 4.31 Fabrication of output square-wave voltage for different fundamental values

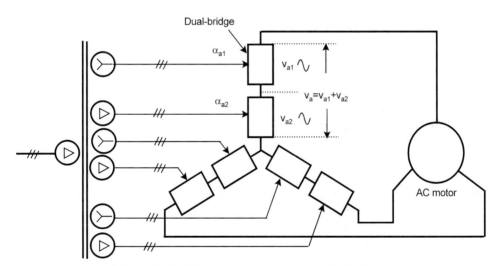

Figure 4.32 72-thyristor cycloconverter block diagram

and wye- connected transformer secondaries, respectively. With concurrent firing angle control, the circuit operates as a 12-pulse cycloconverter, and the corresponding voltage expressions are

$$v_a = v_{a1} + v_{a2} \tag{4.30}$$

$$v_{a1} = v_{a2} = V_{do} \cos\alpha_a \tag{4.31}$$

where $\alpha_{a1} = \alpha_{a2} = \alpha_a$.

With asymmetrical voltage control, the component voltage expressions are given as follows,

$$v_{a1} = V_{do} \cos\alpha_{a1} \qquad (4.32)$$

$$v_{a2} = V_{do} \cos\alpha_{a2} \qquad (4.33)$$

where α_{a1} and α_{a2} are the respective firing angles.

Figure 4.33 explains the fabrication principle of v_{a1} and v_{a2}, such that v_a is always a controllable sine wave. The principle can be summarized as follows:

For positive half-cycle ($v_a \geq 0$):
$v_{a2} = V_{d0}$
$v_{a1} = v_a - V_{d0}$

For negative half-cycle ($v_a < 0$):
$v_{a1} = -V_{d0}$
$v_{a2} = v_a - (-V_{d0})$

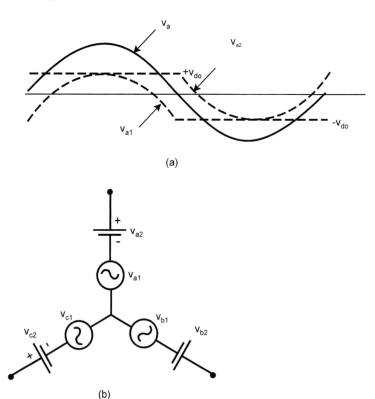

Figure 4.33 (a) Asymmetrical voltage control principle, (b) Equivalent three-phase load voltage generation in positive half-cycle

Phase-Controlled Cycloconverters

The generation of a three-phase output voltage in positive half-cycle is given in Figure 4.33(b). Obviously, the disadvantages are use of large number of thyristors and equivalent six-pulse operation at the output.

4.2.8.3 Circulating Current Control

The circulating current in a cycloconverter with IGR can be intentionally increased and controlled for DPF improvement. This principle was first used in a dual converter-controlled dc motor drive [7], and was later extended to a cycloconverter drive [8]. If an additional dc circulating current I_{cc} is introduced in the circuit of Figure 4.16, the currents i_P and i_N can be given by

$$i_P = I_{CC} + 0.5I_m + 0.5I_m \sin\omega_o t \qquad (4.34)$$

$$i_N = I_{CC} + 0.5I_m - 0.5I_m \sin\omega_o t \qquad (4.35)$$

Since the Thevenin impedance of the circuit is very small, the circulating current can be induced by symmetrically injecting a positive voltage increment of ΔV_d in the positive converter and a negative increment of ΔV_d in the negative converter. The average power due to I_{cc} is practically zero, but it creates equal lagging reactive current loading at both converter inputs.

Figure 4.34 shows a phasor diagram with active (I_P) current, reactive (I_Q) current, and total (I) current. It also includes a variation with circulating current control. In the figure, I_{Q1} =

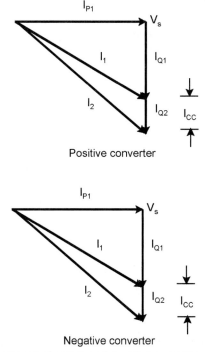

Figure 4.34 Phasor diagrams of active and reactive currents of component converters

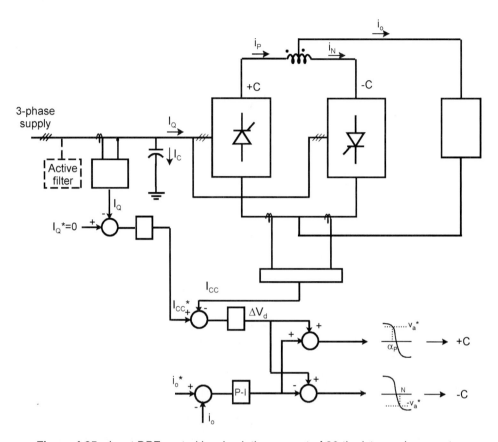

Figure 4.35 Input DPF control by circulating current of 36-thyristor cycloconverter

inherent lagging current due to loading, I_{Q2} = total reactive current, and $I_{Q2} - I_{Q1}$ = reactive current due to I_{cc}. Since the load component fluctuates with loading condition, the I_{Q2} component can be controlled to be constant by I_{cc} control. This means that I_{cc} control permits the cycloconverter to operate at a constant lagging current irrespective of load fluctuation. For such a condition, a fixed capacitor bank at the input can control the line DPF to unity.

The control principle of a phase group is given by the simplified block diagram in Figure 4.35. The load current i_o is controlled in the inner loop, as shown. The next higher loop controls the circulating current I_{cc} by injecting the ΔV_d signal. The outer loop controls the total $I_Q = 0$ by adjusting the I_{cc} magnitude so that the cycloconverter I_Q balances with the constant capacitor bank current I_C.

An active filter is shown connected at the input line to filter the harmonics so that the total PF is unity.

Matrix Converters

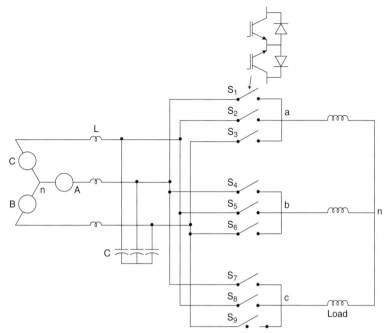

Figure 4.36 Three-phase to three-phase matrix converter

4.3 MATRIX CONVERTERS

So far, we have discussed thyristor cycloconverters that are based on the phase control line commutation principle. In this section and the next, new classes of cycloconverters that use high-frequency, self-controlled ac switches (see Figure 4.3) will be discussed. An example is the three-phase-to-three-phase matrix converter shown in Figure 4.36 [9]. The matrix converter or PWM (pulsewidth modulation) frequency changer was first proposed in 1980, and since then, many papers have been published, but hardly anybody uses the scheme in practical application. It will be described here briefly for completeness. The converter, as the name indicates, is a matrix of nine ac switches that permit the connection of any input phase to any output phase. The switches are controlled by PWM to fabricate output fundamental voltage, which can vary in magnitude and frequency to control an ac motor. PWM techniques are described in the next chapter.

Figure 4.37 explains the output waveform fabrication principle. Consider, for example, the fabrication of an output line voltage v_{ab} wave, as shown in Figure 4.37(c). Figure 4.37(a) shows the input line voltage waves and (b) shows the fictitious diode bridge rectifier output wave. Phases a and b can be connected between the positive and negative envelopes, or they can be shorted to fabricate a v_{ab} wave. For example, if switches S_3 and S_5 are closed, the input line voltage v_{BC} segment will appear at the output line voltage v_{ab} segment. The v_{ab} will be shorted if switches S_3 and S_6 are closed. The same fabrication principle applies to v_{bc} and v_{ca} waves. At

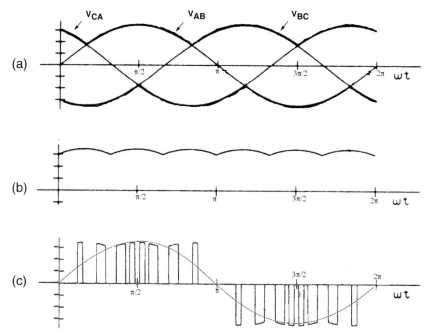

Figure 4.37 Output waveform fabrication principle (a) Three-phase input line voltages, (b) Fictitious diode-bridge rectifier output, (c) PWM fabrication of v_{ab} wave

any instant, one switch in each bank (that is, three switches altogether) is closed, and there are $3^3 = 27$ permissible switching states. Note that lateral line switches should not be closed to prevent line shorting. The line LC filter, as shown, is essential for (1) commutation of ac switches so that load inductive current can be transferred from one line to another, and (2) filtering the line current harmonics. The converter is regenerative, and unlike a thyristor cycloconverter, the line current can be controlled to be sinusoidal with unity PF. Besides the input capacitive filter, the circuit needs 18 IGBTs and 18 diodes, which provide a much higher device count compared to 12 IGBTs and 12 diodes in the conventional double-sided PWM, dc link, voltage-fed frequency conversion scheme (see Figure 5.60).

4.4 HIGH-FREQUENCY CYCLOCONVERTERS

A high-frequency cycloconverter converts single-phase, high-frequency (typically 20 kHz) ac to three-phase, variable-voltage, variable-frequency ac for motor drives using the "soft-switching" principle of ac switches. Soft-switching will be discussed in Chapter 5. Figure 4.38 shows a typical configuration of a high-frequency cycloconverter, where the high-frequency ac is generated by an inverter. Inverters are discussed in Chapter 5. The advantages of a high-frequency link circuit are that the output can have galvanic isolation and the voltage level can change from the input through a light weight, high-frequency transformer. Of course, the penalty

High-Frequency Cycloconverters

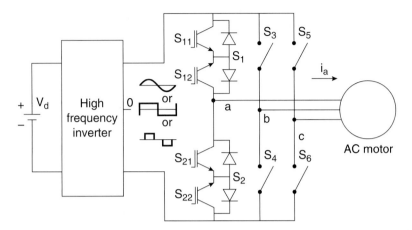

Figure 4.38 Single-phase-to-three-phase high-frequency cycloconverter

is a larger number of devices. The high-frequency ac may be sinusoidal and generated by a resonant link inverter. Or, it can be square- or quasi-square-wave (with a zero voltage gap), as shown, which may be generated by a non-resonant link inverter. With a resonant link, two cycloconverters can be connected back-to-back with a three-phase 60 Hz line supply.

4.4.1 High-Frequency, Phase-Controlled Cycloconverter

With a sinusoidal or square-wave high-frequency link, it is possible to have the phase control principle [10][11] as discussed before to synthesize the output voltage waves. Figure 4.39 illustrates the operation principle with a square-wave for a single-phase, half-bridge circuit where the load is connected between points a and 0. Note that in positive half-cycle, positive load current is carried by device S_{11} of the ac switch S_1, whereas in negative half-cycle, it is carried by device S_{22} of switch S_2. The waveforms show the comparison of a sawtooth carrier wave with a sine-modulating wave to generate the firing angle α. The commutation of positive i_a from the outgoing device to the incoming device is shown in the figure. Note the delay in turning off S_{22} to be sure that current is zero in a self-controlled switch. The phase control provides switching at zero current eliminating the switching loss. The ac switch can be replaced by anti-parallel thyristors [10] if the link frequency is low.

4.4.2 High-Frequency, Integral-Pulse Cycloconverter

4.4.2.1 Sinusoidal Supply

With a sinusoidal, or quasi-square-wave, power supply, it is possible to use the integral half-cycle pulse-width modulation (IPM) principle to synthesize output voltage waves. Figure 4.40 illustrates the operation of a half-bridge circuit with sinusoidal supply voltage. The advantage of IPM is that devices can be switched at zero voltage, which reduces the switching loss; in

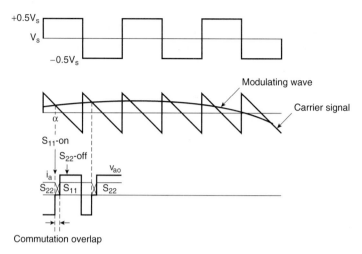

Figure 4.39 Explanatory waveforms of half-bridge, phase-controlled cycloconverter with square-wave for a segment of output voltage

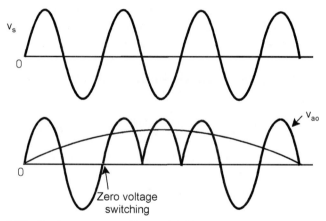

Figure 4.40 Explanatory waveforms of half-bridge integral-pulse cycloconverter with sine wave for a segment of output voltage

other words, it improves the converter's efficiency. Of course, a zero voltage interval is extremely narrow with sinusoidal supply. The additional disadvantages of the scheme are distortion of link voltage due to harmonic loading of the resonant circuit and link frequency drift.

4.4.2.2 Quasi-Square-Wave Supply

A high-frequency, non-resonant link can easily generate a quasi-square-wave by phase-skewing the inverter legs. The resulting zero voltage gap easily permits soft-switching of the cycloconverter devices [12]. Figure 4.41 explains the fabrication principle of output voltage wave, which is essentially similar to Figure 4.40. The advantage in this case is the larger zero

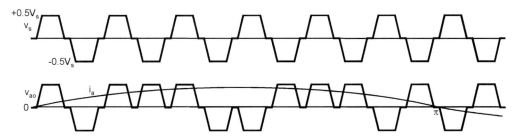

Figure 4.41 Explanatory waveforms of half-bridge, integral-pulse cycloconverter with quasi-square-wave for a segment of output voltage

voltage gap for easy soft-switching. However, extra care should be taken to minimize leakage inductance in the link, which can create large spike voltage at the instant of commutation.

4.5 SUMMARY

This chapter gives a general review of cycloconverter principles and different types of cycloconverter circuits. Phase-controlled cycloconverters with thyristors were discussed in detail because they are widely used in industry. Both 18-thyristor and 36-thyristor circuits with circulating current and blocking modes were covered. The matrix converter and high-frequency cycloconverter were briefly reviewed for completeness. They have extensive literature, but are rarely used in practice. The great disadvantages of phase-controlled cycloconverters are generation of complex harmonics in load and line, and poor line DPF. Different methods of line DPF improvement were discussed. Attempts are being made to replace phase-controlled cycloconverters by dc link frequency changers using GTO/IGBT devices, which will be discussed in Chapter 5.

REFERENCES

1. B. R. Pelly, *Thyristor Phase-Controlled Converters and Cycloconverters*, Wiley, New York, 1971.
2. W. Slabiac and L. J. Lawson, "Precise control of a three-phase squirrel cage induction motor using practical cycloconverter", *IEEE Trans. Ind. Appl.*, vol. 2, pp. 274–280, 1966.
3. T. Nakano, H. Ohsawa, and K. Endoh, "A high performance cycloconverter fed synchronous machine drive system", *IEEE IAS Annu. Meet. Conf. Rec.*, pp. 334–340, 1982.
4. D. L. Plette and H. G. Carlson, "Performance of a variable speed constant frequency electrical system", *IEEE Trans. Aerospace*, vol. 2, pp. 957–970, 1964.
5. J. A. Allan, W. A.Wyeth, G. W. Herzog, and J. I. Young, "Electrical aspects of the 8750 hp gearless ball-mill drive at St. lawrence cement company", *IEEE Trans. Ind. Appl.*, vol. 11, pp. 681–687, Nov./Dec. 1975.
6. R. Kurosawa, T. Shimura, H. Uchino, and K. Sugi, "A microcomputer-based high power cycloconverter-fed induction motor drive", *IEEE IAS Annu. Meet. Conf. Rec.*, pp. 462–467, 1982.
7. J. Rosa and P. Finlayson, "Power factor correction of thyristor dual converter via circulating current control", *IEEE IAS Annu. Meet. Conf. Rec.*, pp.415–422, 1978.
8. Hosoda, Tatura, etc., "A new concept high performance large scale ac drive system – cross current type cycloconverter-fed induction motor with high performance digital vector control", *IEEE IAS Annu. Meet. Conf. Rec.*, pp. 229–234, 1986.

9. P. D. Ziogas, S. I. Khan, and M. H. Rashid, "Some improved force commutated cycloconverter structures", *IEEE IAS Annu. Meet. Conf. Rec.*, pp. 739–748, 1984.
10. P. M. Espalage and B. K. Bose, "High frequency link power conversion", *IEEE Trans. Ind. Appl.*, vol. 13, pp. 387–394, Sept./Oct. 1977.
11. B. Ozpineci and B. K. Bose, "A soft-switched performance enhanced high frequency non-resonant link phase-controlled converter for ac motor drive", *IEEE IECON Conf. Rec.*, pp.733–739, 1998.
12. L. Hui, B. Ozpineci, and B. K. Bose, "A soft-switched high frequency non-resonant link integral pulse modulated DC-AC converter for ac motor drive", *IEEE IECON Conf. Rec.*, pp. 726–732, 1998.

CHAPTER 5

Voltage-Fed Converters

5.1 INTRODUCTION

Voltage-fed converters, as the name indicates, receive dc voltage at one side and convert it to ac voltage on the other side. The ac voltage and frequency may be variable or constant depending on the application. In fact, the general name "converter" is given because the same circuit can operate as an inverter as well as a rectifier. We have seen this dual-mode operation for phase-controlled converters, which were discussed in Chapter 3. A voltage-fed inverter should have a stiff voltage source at the input, that is, its Thevenin impedance should ideally be zero. A large capacitor can be connected at the input if the source is not stiff. The dc voltage may be fixed or variable, and may be obtained from a utility line or rotating ac machine through a rectifier and filter. It can also be obtained from a battery, fuel cell, or solar photovoltaic array. The inverter output can be single-phase or polyphase, and can have square wave, sine wave, PWM wave, stepped wave, or quasi-square wave at the output. Voltage-fed converters are used extensively, and some of their applications may be as follows:

- AC motor drives
- AC uninterruptible power supplies (UPSs)
- Induction heating
- AC power supply from battery, photovoltaic array, or fuel cell
- Static VAR generator (SVG) or compensator (SVC)
- Active harmonic filter (AHF)

In voltage-fed converters, the power semiconductor devices always remain forward-biased due to the dc supply voltage, and therefore, self-controlled forward or asymmetric blocking devices, such as GTOs, BJTs, IGBTs, power MOSFETs, and IGCTs are suitable. Force-commutated thyristor converters were used before, but now they have become obsolete. A feedback diode is always connected across the device to have free reverse current flow. One important characteristic of a voltage-fed converter is that the ac fabricated voltage wave is not affected by the load parameters.

In this chapter, we will study various types of voltage-fed converters, PWM techniques, hard switching vs. soft switching of devices, soft-switched converters, active line current wave shaping and PF correction, static VAR generators, and active filters. Most of the converters will be shown with IGBT devices, and the devices will be considered as ideal on-off switches, as before. The term "inverter" will be used in most cases. This chapter will end with a brief review of the simulation program SIMULINK will be given with examples because of its common usage.

5.2 SINGLE-PHASE INVERTERS

5.2.1 Half-Bridge and Center-Tapped Inverters

One of the simplest possible inverter configurations is the single-phase, half-bridge inverter shown in Figure 5.1 (a); (b) gives its explanatory waves. The circuit consists of a pair of devices Q_1 and Q_2 connected in series across the dc supply, and the load is connected between point a and the centerpoint 0 of a split-capacitor power supply. The snubber across the devices is omitted for simplicity. The devices Q_1 and Q_2 are closed alternately for π angle to generate the square-wave output voltage, as shown. In fact, a short gap, or lock-out time (t_d), is maintained, as indicated, to prevent any short-circuit or "shoot-through" fault due to turn-off switching delay. The load is usually inductive, and assuming perfect filtering, the sinusoidal load current will lag the fundamental voltage by angle φ, as shown. When the supply voltage and load current are of the same polarity, the mode is active, meaning the power is absorbed by the load. On the other hand, when the voltage and current are of opposite polarity (indicated by diode conduction), the

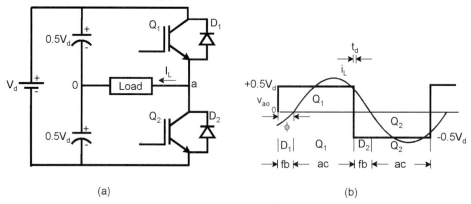

Figure 5.1 (a) Half-bridge inverter, (b) Output voltage and current waves in square-wave mode

Single-Phase Inverters

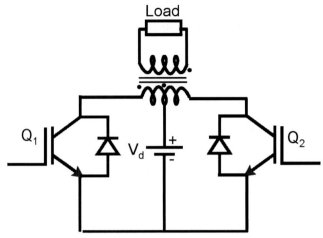

Figure 5.2 Center-tapped inverter

power is fed back to the source. However, the average power will flow from the source to the load. To maintain the centerpoint of the supply voltage V_d, the capacitors should be large. An inverter with a center-tapped transformer is shown in Figure 5.2. The advantages of this transformer are voltage level change and Ohmic isolation from the primary.

5.2.2 Full, or H-Bridge, Inverter

Two half-bridges or phase-legs can be connected to construct a full, or H-bridge, inverter, as shown in Figure 5.3. The split-capacitor power supply is not needed in this case, and the load is connected between the centerpoints, a and b. In the square-wave operation mode shown in Figure 5.4, the device pairs Q_1Q_3 and Q_2Q_4 are switched alternately to generate the square-wave

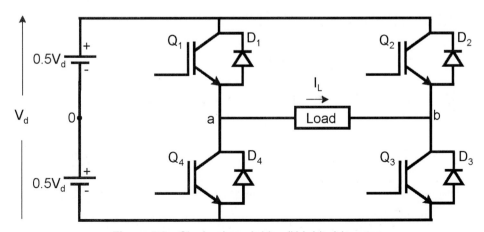

Figure 5.3 Single-phase bridge (H-bridge) inverter

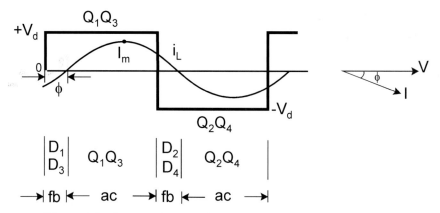

Figure 5.4 Output voltage and current waves in square-wave mode

output voltage of amplitude V_d. Again, assuming inductive and harmonic-free load current at phase angle φ, the load current in active mode will be carried by the Q_1Q_3 or Q_2Q_4 pair, whereas the feedback current will flow through the D_1D_3 or D_2D_4 pair, as shown in the figure. Both the diodes and IGBTs are designed to withstand the supply voltage V_d. With the current waves, it can be easily seen that the peak current in the IGBT is I_m, whereas that in the diode is $I_m \sin \varphi$. H-bridge converters are used in four-quadrant speed control of small dc motors. The converter modes, shown in Figure 5.5, can be summarized as follows:

Quadrant 1: Positive buck (step-down) converter (forward motoring)

Q_1 – On

Q_2 – Chopping

$D_3 Q_1$ – Freewheeling

Quadrant 2: Positive boost (step-up) converter (forward regeneration)

Q_4 – Chopping

$D_2 D_1$ – Freewheeling

Quadrant 3: Negative buck (step-down) converter (reverse motoring)

Q_3 – On

Q_4 – Chopping

$D_1 Q_3$ – Freewheeling

Quadrant 4: Negative boost (step-up) converter (reverse regeneration)

Q_2 – Chopping

$D_3 D_4$ – Freewheeling

Single-Phase Inverters

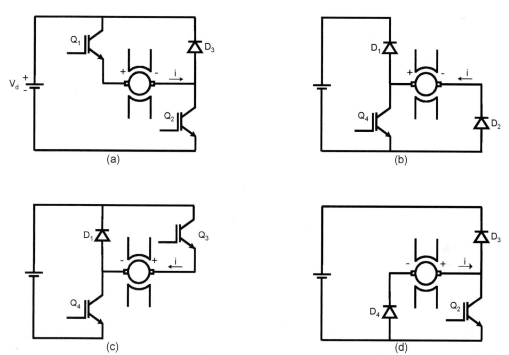

Figure 5.5 H-bridge converter operation for four-quadrant PM dc motor speed control (a) Buck converter (forward motoring), (b) Boost converter (forward regeneration), (c) Buck converter (reverse motoring), (d) Boost converter (reverse regeneration)

5.2.2.1 Phase-Shift Voltage Control

The output voltage wave of the H-bridge inverter, shown in Figure 5.4, can be controlled by phase shifting the control of the component half bridges. Often, this method is known as phase shift-PWM. The waveforms in Figure 5.6 explain this operation. Each half-bridge is operated in square-wave mode, as shown in Figure 5.1, but the right half-bridge is phase shifted by lag angle φ with respect to the other. The voltage waves v_{a0} and v_{b0} are positioned with respect to the fictitious centerpoint of the dc supply V_d. The output line voltage wave v_{ab} ($v_{a0} - v_{b0}$) as shown, is a quasi-square wave of pulse width φ, which can control its fundamental component. The Fourier series of v_{ab} can be given as the following, where n = odd integer (1, 3, 5, etc.):

$$v_{ab} = \sum_{n=1,3,5...} \frac{4V_d}{n\pi} \left[\sin\frac{n\phi}{2} \right] \cos n\omega t \qquad (5.1)$$

The convergent series contains the fundamental component (n = 1) and odd harmonics. The fundamental component peak value is given by

$$a_1 = \frac{4V_d}{\pi} \sin\frac{\phi}{2} \qquad (5.2)$$

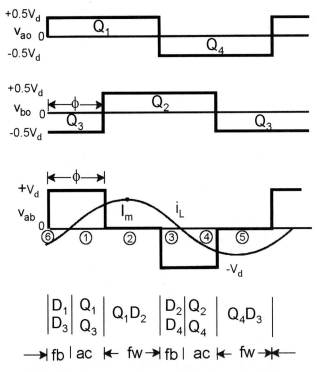

Figure 5.6 Waveforms for phase-shift control of output voltage

Figure 5.7 gives the amplitude (per unit) of the fundamental and harmonic components with the phase-shift angle φ. Note that the fundamental component reaches to the maximum value of $4V_d/\pi$ (1.0 pu) at square wave when $\varphi = \pi$. Assuming again the typical lagging load current with perfect filtering, the different segments, shown in the lower part of Figure 5.6, can be summarized as follows:

Segment 1: Active mode with positive voltage and positive current – Q_1Q_3 conducting
Segment 2: Freewheeling mode with positive current (zero voltage) – Q_1D_2 conducting
Segment 3: Feedback mode with positive current – D_2D_4 conducting
Segment 4: Active mode with negative current and negative voltage – Q_2Q_4 conducting
Segment 5: Freewheeling mode with negative current (zero voltage) – Q_4D_3 conducting
Segment 6: Feedback mode with negative current – D_1D_3 conducting

So far, single-phase inverters have been explained in square-wave mode of operation with average power flow from the dc to ac side only. It is also possible to operate the circuits with multiple PWM and in rectification mode with power flow from the ac to the dc side. These principles will be explained later.

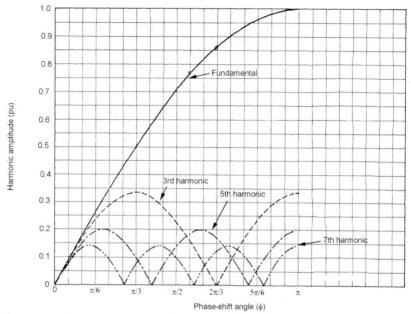

Figure 5.7 Spectrum of output voltage with phase-shift control ($V_{base} = 4V_d/\pi$)

5.3 THREE-PHASE BRIDGE INVERTERS

5.3.1 Square-Wave, or Six-Step, Operation

Three-phase bridge inverters are widely used for ac motor drives and general-purpose ac supplies. Figure 5.8 shows the inverter circuit, and Figure 5.9 explains the fabrication of the output voltage waves in square-wave, or six-step, mode of operation. The circuit consists of three half-bridges, which are mutually phase-shifted by $2\pi/3$ angle to generate the three-phase voltage waves. The input dc supply is usually obtained from a single-phase or three-phase utility power supply through a diode-bridge rectifier and LC or C filter, as shown. The square-wave phase voltages with respect to the fictitious dc center tap can be expressed by Fourier series as

$$v_{ao} = \frac{2V_d}{\pi}\left[\cos\omega t - \frac{1}{3}\cos 3\omega t + \frac{1}{5}\cos 5\omega t - ...\right] \quad (5.3)$$

$$v_{bo} = \frac{2V_d}{\pi}\left[\cos\left(\omega t - \frac{2\pi}{3}\right) - \frac{1}{3}\cos 3\left(\omega t - \frac{2\pi}{3}\right) + \frac{1}{5}\cos 5\left(\omega t - \frac{2\pi}{3}\right) - ...\right] \quad (5.4)$$

$$v_{co} = \frac{2V_d}{\pi}\left[\cos\left(\omega t + \frac{2\pi}{3}\right) - \frac{1}{3}\cos 3\left(\omega t + \frac{2\pi}{3}\right) + \frac{1}{5}\cos 5\left(\omega t + \frac{2\pi}{3}\right) - ...\right] \quad (5.5)$$

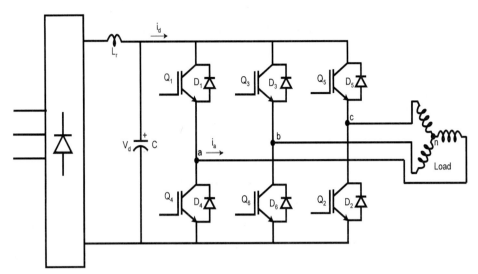

Figure 5.8 Three-phase bridge inverter (shown with a diode rectifier in the front end)

where V_d = dc supply voltage. The line voltages can therefore be constructed from Equations (5.3)–(5.5) as

$$v_{ab} = v_{ao} - v_{bo}$$
$$= \frac{2\sqrt{3}V_d}{\pi}\left[\cos\left(\omega t + \frac{\pi}{6}\right) + 0 - \frac{1}{5}\cos 5\left(\omega t + \frac{\pi}{6}\right) - \frac{1}{7}\cos 7\left(\omega t + \frac{\pi}{6}\right) + ...\right] \quad (5.6)$$

$$v_{bc} = v_{bo} - v_{co}$$
$$= \frac{2\sqrt{3}V_d}{\pi}\left[\cos\left(\omega t - \frac{\pi}{2}\right) + 0 - \frac{1}{5}\cos 5\left(\omega t - \frac{\pi}{2}\right) - \frac{1}{7}\cos 7\left(\omega t - \frac{\pi}{2}\right) + ...\right] \quad (5.7)$$

$$v_{ca} = v_{co} - v_{ao}$$
$$= \frac{2\sqrt{3}V_d}{\pi}\left[\cos\left(\omega t + \frac{5\pi}{6}\right) + 0 - \frac{1}{5}\cos 5\left(\omega t + \frac{5\pi}{6}\right) - \frac{1}{7}\cos 7\left(\omega t + \frac{5\pi}{6}\right) + ...\right] \quad (5.8)$$

Note that the line fundamental voltage amplitude is $\sqrt{3}$ times that of the phase voltage, and there is a leading phase-shift angle of $\pi/6$.

The line voltage waves, shown in Figure 5.9, have the characteristic six-stepped wave shape, and are analogous to line current waves in a phase-controlled bridge rectifier (see Figure 3.10(e)). The characteristic harmonics in the waveform are $6n \pm 1$, where n = integer. The three-phase fundamental as well as the harmonic components are balanced with a mutual phase-shift angle of $2\pi/3$. Because of the characteristic wave shape, this type of inverter is called a square-wave, or six-stepped, inverter.

Three-Phase Bridge Inverters

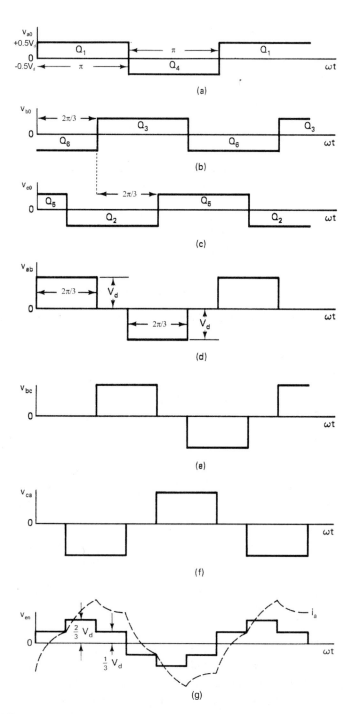

Figure 0.1 Synthesis of output voltage waves in square-wave mode

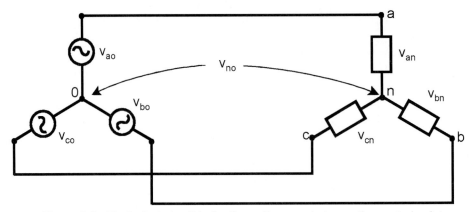

Figure 5.9 Equivalent circuit indicating voltage v_{no} between the neutral points

If the three-phase load neutral n is connected to the center tap of the dc voltage V_d, then the load phase voltages are v_{a0}, v_{b0}, and v_{c0}, as discussed above. With an isolated neutral load (usual for a machine), the equivalent circuit is shown in Figure 5.10. Here, the triplen harmonics that is, the zero-sequence components of the supply, will appear across points n and 0. We can write the following relations:

$$v_{ao} = v_{an} + v_{no} \tag{5.9}$$

$$v_{bo} = v_{bn} + v_{no} \tag{5.10}$$

$$v_{co} = v_{cn} + v_{no} \tag{5.11}$$

Since the load-phase voltages are balanced, in other words $v_{an} + v_{bn} + v_{cn} = 0$, adding these equations, we get

$$3v_{no} + 0 = v_{ao} + v_{bo} + v_{co} \tag{5.12}$$

or

$$v_{no} = \frac{1}{3}(v_{ao} + v_{bo} + v_{co}) \tag{5.13}$$

Therefore, substituting Equation (5.13) in (5.9), (5.10), and (5.11), respectively, we get

$$v_{an} = \frac{2}{3}v_{ao} - \frac{1}{3}v_{bo} - \frac{1}{3}v_{co} \tag{5.14}$$

$$v_{bn} = \frac{2}{3}v_{bo} - \frac{1}{3}v_{ao} - \frac{1}{3}v_{co} \tag{5.15}$$

$$v_{cn} = \frac{2}{3}v_{co} - \frac{1}{3}v_{ao} - \frac{1}{3}v_{bo} \tag{5.16}$$

Three-Phase Bridge Inverters

These isolated neutral-phase voltages can be described by Fourier series, or the waveforms can be constructed graphically, as shown for v_{an} in Figure 5.9(g). It is also a six-stepped wave, but its fundamental component is phase-shifted by $\pi/6$ angle from that of the respective line voltage. Basically, the isolated neutral and connected neutral phase voltages are the same, except in the former case, the triplen harmonics have been suppressed.

For a linear and balanced three-phase load, the line current waves are also balanced. The individual line current components can be solved for each component of the Fourier series voltage and then the resultant can be derived by the superposition principle. A typical line current wave with inductive load is shown on the v_{an} wave in Figure 5.9(g).

5.3.2 Motoring and Regenerative Modes

An inverter can supply average power to the load in the usual inverting or motoring mode shown in Figure 5.11(a). The phase current wave i_a is assumed to have perfect filtering by the load, and it is indicated with lagging phase angle $\varphi = \pi/3$. In the first segment, the phase voltage is positive, but the phase current is negative and flowing through the diode D_1, indicating that power is fed back to the source. In the next segment, the IGBT Q_1 is carrying the active load cur-

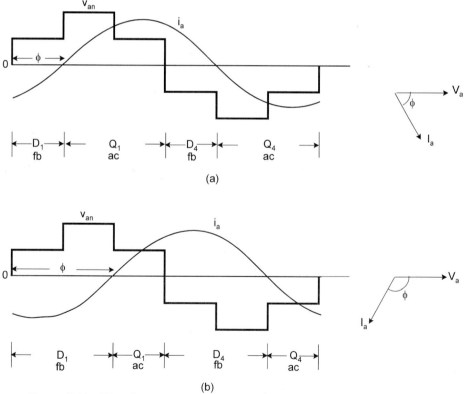

Figure 5.10 Waveforms showing (a) Motoring mode, (b) Regenerative modes

rent. The next half-cycle is symmetrical, and the respective conduction intervals of D_4 and Q_4 are shown in the figure. It may be inferred that if the load is purely resistive or has unity DPF ($\phi = 0$), each IGBT conducts for π angle.

It can be shown that the inverter can also operate in rectification or regeneration mode, pumping average power from the ac to the dc side. Figure 5.11(b) shows waveforms for regeneration mode at angle $\phi = 2\pi/3$, indicating that the feedback interval is considerably larger than the active interval. In the extreme condition, if $\phi = \pi$, the inverter operates as a diode rectifier with only the diodes conducting.

5.3.3 Input Ripple

The inverter input voltage and current waves will deviate from ideal dc values due to the presence of ripple components. Ripple voltage may be introduced by the line-side rectifier due to a finite pulse number and practical filter size, and this ripple voltage will affect the inverter output voltage waves and will correspondingly deteriorate its input current wave. The distortion introduced by the inverter itself can be calculated by considering the instantaneous power balance between the input and the output, since there is no energy storage in the switching elements. Hypothetically, if the inverter output voltage and current waves are considered sinusoidal and balanced, the instantaneous power given by

$$p_i = v_{an}i_a + v_{bn}i_b + v_{cn}i_c \tag{5.17}$$

is always constant, irrespective of load power factor condition. In this case, the reactive currents will circulate through the inverter and the input current will not contain any harmonics. With the ideal stepped voltage and sine current waves at $\phi = \pi/3$ (shown in Figure 5.11(a)) in all the phases and considering instantaneous power balance between the output and the input, the input current wave can be derived graphically, as indicated in Figure 5.12. Considering the duality of voltage and current waves between a phase-controlled converter and a square-wave, voltage-fed inverter, the input current wave i_d is identical to the voltage wave shape at the rectifier output for firing angle $\alpha = \pi - \varphi$. The input ripple current will flow in the filter capacitor and will introduce some additional distortion in the dc voltage. In fact, the electrolytic capacitor size is determined

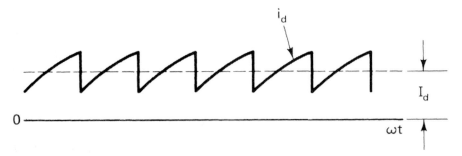

Figure 5.11 Input ripple current with sinusoidal load current at $\pi/3$ lagging angle

on the basis of the worst-case ripple current flowing through it. The accurate determination of the voltage and current waves are somewhat involved and can be determined by computer simulation study.

5.3.4 Device Voltage and Current Ratings

For a specified KVA rating of an inverter, a suitable power switching device is to be selected and then the voltage and current ratings of the devices are to be designed on a worst-case basis. As mentioned before, the devices have to withstand the voltage V_d in the forward direction. A typical 50 percent margin is to be added to withstand the transient overshoot at turn-off switching. The diodes should be fast recovery type to reduce the switching loss. The peak and average current in the devices can be determined from the current waves. If the inverter is required to have a higher power rating on a short-term basis, the design should take this into consideration.

5.3.5 Phase-Shift Voltage Control

The phase-shift voltage control principle, described in Figure 5.6 for a single-phase inverter, can easily be extended to control the output voltages of a three-phase inverter. Figure 5.13 shows the circuit of a three-phase, H-bridge inverter for phase-shift voltage control, and Figure 5.14 explains the fabrication principle of the output voltage waves. The circuit consists of three H-bridge inverters, one for each phase group, where each H-bridge is connected to a transformer's primary winding. The output voltages are derived from the transformer's secondary

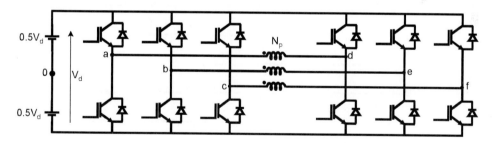

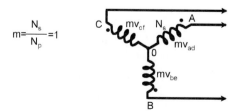

Figure 5.12 Three-phase, H-bridge, square-wave inverter for phase-shift voltage control

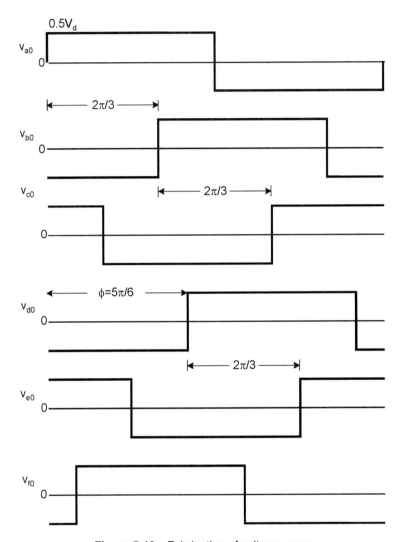

Figure 5.13 Fabrication of voltage waves

windings connected in wye, as shown in the figure. In Figure 5.14, the voltage waves v_{a0}, v_{b0}, and v_{c0}, derived with respect to the dc supply center tap, have an amplitude of 0.5 V_d and are mutually phase-shifted by $2\pi/3$ angle, as shown. The three-phase voltage waves v_{d0}, v_{e0}, and v_{f0} are similar, but these are phase-shifted by angle φ with respect to v_{a0}, v_{b0}, and v_{c0}, respectively. The transformer's secondary phase voltages v_{A0}, v_{B0}, and v_{C0} can be derived from the above waves by the following expressions:

$$v_{A0} = mv_{ad} = m(v_{ao} - v_{do}) \tag{5.18}$$

$$v_{B0} = mv_{be} = m(v_{bo} - v_{eo}) \qquad (5.19)$$

$$v_{C0} = mv_{cf} = m(v_{co} - v_{fo}) \qquad (5.20)$$

where the transformer turns ratio $m = N_s/N_p$ is assumed to be unity for sketching the waves. Note that each of these waves, similar to the v_{ab} wave in Figure 5.6, is a function of φ angle, which can control the fundamental component. The output line voltages are given by

$$v_{AB} = v_{A0} - v_{B0} \qquad (5.21)$$

$$v_{BC} = v_{B0} - v_{C0} \qquad (5.22)$$

$$v_{CA} = v_{C0} - v_{A0} \qquad (5.23)$$

Evidently, the component voltage waves v_{a0}, v_{d0}, v_{A0}, etc. in each three-phase group contain triplen harmonics ($n = 3, 6, 9$, etc.). However, since the triplen harmonics are co-phasal, these will be eliminated from the line voltage v_{AB}, that is, v_{AB} is a six-stepped wave (with an order of harmonics $6n \pm 1$) at any phase-shift angle φ. With cosine symmetry, the Fourier series of v_{A0} and v_{B0} can be given as (see Equation (5.1))

$$v_{A0} = \sum_{n=1,3,5,\ldots} \frac{4mV_d}{n\pi}\left[\sin\frac{n\phi}{2}\right]\cos n\omega t \qquad (5.24)$$

$$v_{B0} = \sum_{n=1,3,5,\ldots} \frac{4mV_d}{n\pi}\left[\sin\frac{n\phi}{2}\right]\cos n\left(\omega t - \frac{2\pi}{3}\right) \qquad (5.25)$$

and that of v_{AB} is

$$v_{AB} = v_{A0} - v_{B0}$$
$$= \sum_{n=1,5,7,11\ldots} \frac{4mV_d}{n\pi}\left[\sin\frac{n\phi}{2}\right]\left[\cos n\omega t - \cos n\left(\omega t - \frac{2\pi}{3}\right)\right] \qquad (5.26)$$

The above equations indicate that although triplen harmonics are present in v_{A0} and v_{B0} waves, they are eliminated from the v_{AB} wave. Note that at $\varphi = \pi$, v_{A0}, v_{B0}, and v_{C0} are square waves, and the waveform of v_{AB} is similar to that of the v_{ab} wave in Figure 5.9.

5.3.6 Voltage and Frequency Control

Normally, the inverter output voltage and frequency are controlled continuously. For machine drive applications, the range of voltage and frequency control is wide. For a regulated ac power supply, the frequency is constant, but the voltage requires control due to supply and load variations. Inverter frequency control is very straightforward. The frequency command may be fixed or variable, and it can be generated from a microprocessor with the help of look-up tables, hardware and software counters, and D/A converters. Alternatively, a stable analog dc voltage may represent the frequency command, which can be converted to a proportional frequency through a voltage-controlled oscillator (VCO) and processed through counters and logic circuits. The stability of inverter frequency is determined by the stability of the reference signal frequency and is not affected by load and source variations. The inverter output voltage can, in general, be controlled by the following two methods:

- Inverter input voltage control (sometimes defined as wave or pulse amplitude modulation (PAM))
- Voltage control within the inverter by PWM

The phase-shift voltage control discussed before falls into the second category (often defined as phase-shift PWM). Other PWM techniques will be described later. If the input ac supply is rectified to dc, there may be two possible schemes of input voltage control: (1) a phase-controlled rectifier and LC filter, and (2) a diode rectifier with an LC or C filter, followed by a dc-dc converter. The dc-dc converter can also be used for voltage control if the primary power is dc, say, from a battery or photovoltaic array. If the ac power is obtained from an engine-driven alternator, it can be converted to dc by a diode rectifer and the voltage can be controlled by the alternator field. All these methods have characteristic advantages and disadvantages. However, the PWM method of inverter voltage control is the most common.

5.4 MULTI-STEPPED INVERTERS

An inverter with a multiple of six steps, for instance, 12, 18, 24, etc., is defined as a multi-stepped inverter. For a large power rating, a six-stepped inverter can be used with a series-parallel connection of devices. With a series-parallel operation, the devices require matching, and some amount of voltage or current derating of the devices is essential. Besides the design complexity, the matched devices may not be easily available. A possible solution for a higher power rating is a parallel connection of three-phase inverters through center-tapped reactors at the output.

For large-power applications, it is desirable that the inverter output wave should be multi-stepped (approaching a sine wave), because the filter size can be reduced on both the dc and ac sides. The lower order harmonics of a six-stepped wave (i.e., 5th and 7th) can be neutralized by synthesizing a 12-stepped waveform as discussed in Chapter 3. It can be shown that the significant harmonics present in multi-stepped waveforms (including the sixth step) are $Kn \pm 1$, where K = number of steps (6, 12, 24, etc.) and n = integer.

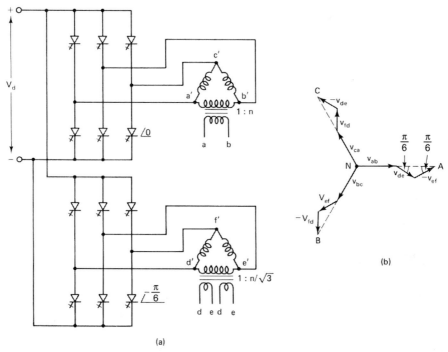

Figure 5.14 Twelve-step inverter indicating synthesis of v_{AN} voltage (bypass diodes are omitted)

5.4.1 12-Step Inverter

Figure 5.15 shows the circuit of a 12-step inverter using two three-phase bridges and a transformer connection. Figure 5.16 shows the phasor diagram of output voltage synthesis and the waveform. The component bridges operate in square-wave, or six-stepped, mode, but the lower bridge is phase-shifted by $\pi/6$ angle and each inverter is connected to the primary delta winding of the respective transformers, as shown. The phasor diagram of the primary fundamental voltages are shown in Figure 5.16(a). The upper transformer has one secondary winding for each phase, whereas the lower transformer has two secondary windings per phase, and the winding ratios are indicated in the figure. The output phase voltages are obtained by the interconnection of three secondary windings to satisfy the phasor diagram in Figure 5.16(b). For example, phase A voltage v_{NA} is given by

$$v_{NA} = v_{ab} + v_{de} - v_{ef} \tag{5.27}$$

Since the lower bridge operates at a lagging angle of $\pi/6$, and considering v_{ab} as the reference, the Fourier series of the component voltages can be given as

$$v_{ab} = \frac{2\sqrt{3}nV_d}{\pi}\left[\cos\omega t + \frac{1}{5}\cos 5\omega t + \frac{1}{7}\cos 7\omega t + \ldots\right] \tag{5.28}$$

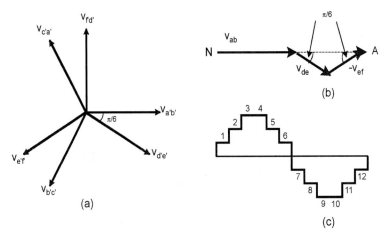

Figure 5.15 Phasor diagram of output voltage synthesis and the waveform (a) Phasor diagram of transformer primary fundamental voltages, (b) Phasor diagram for three-phase output synthesis, (c) 12-step v_{NA} wave

$$v_{de} = \frac{v_{ab}}{\sqrt{3}} \angle -\frac{\pi}{6}$$
$$= \frac{2nV_d}{\pi}\left[\cos\left(\omega t - \frac{\pi}{6}\right) + \frac{1}{5}\cos 5\left(\omega t - \frac{\pi}{6}\right) + \frac{1}{7}\cos\left(\omega t - \frac{\pi}{6}\right) + \ldots\right] \quad (5.29)$$

$$v_{ef} = \frac{v_{ab}}{\sqrt{3}} \angle -\frac{5\pi}{6}$$
$$= \frac{2nV_d}{\pi}\left[\cos\left(\omega t - \frac{5\pi}{6}\right) + \frac{1}{5}\cos 5\left(\omega t - \frac{5\pi}{6}\right) + \frac{1}{7}\cos\left(\omega t - \frac{5\pi}{6}\right) + \ldots\right] \quad (5.30)$$

where n = turns ratio of the upper transformer. The fundamental component of v_{NA} is given as

$$v_{NA(f)} = v_{ab(f)} + v_{de(f)} - v_{ef(f)} = \frac{4\sqrt{3}nV_d}{\pi}\cos\omega t \quad (5.31)$$

which is also evident from the phasor diagram of Figure 5.16(b). The output phase voltage Fourier series can be given as

$$v_{NA} = \frac{4\sqrt{3}nV_d}{\pi}\left[\cos\omega t + \frac{1}{11}\cos 11\omega t + \frac{1}{13}\cos 13\omega t + \ldots\right] \quad (5.32)$$

Note that the 5th and 7th harmonics are eliminated, and the lowest harmonics present are 11, 13, 23, 25, etc. The resulting 12-step wave is sketched in Figure 5.16(c). The principle can be extended to 24- and 48-step inverters, which require proportionally more bridges and transformers. The output voltage in this type of inverter can be controlled by controlling the dc voltage V_d.

Multi-Stepped Inverters

Of course, the lower bridge phase angle can be shifted from $\pi/6$ to control the output voltage, but in this case, 5th and 7th harmonic voltages will appear at the output.

5.4.2 18-Step Inverter by Phase-Shift Control

The phase-shift voltage control principle for a three-phase, H-bridge, square-wave inverter, described in Figures 5.13 and 5.14, can be extended for a multi-stepped inverter. Figure 5.17 shows a GTO-based power circuit of an 18-step inverter with phase-shift voltage control. This 10 MW converter was installed in Southern California Edison's electric grid for battery peaking service [6]. The converter system consists of three groups of H-bridges, where the second and third groups are phase-shifted by 20° and 40°, respectively, with respect to the first group. The transformer of the first group has one secondary winding in each phase, whereas the second and third groups' transformers have two secondary windings in each phase. The secondary connections are self-explanatory in the figure. The output phase and line voltages have 18-step waveforms (i.e., the harmonic orders are 17, 19, 35. 37, etc.), and the output fundamental voltage magnitude and phase can be controlled continuously by a phase-shift angle without introducing any additional harmonics. The converter system acts as a rectifier and charges the battery in off-peak hours of the utility system, but it functions as an inverter to supply the battery-stored energy during peak demand hours. When neither of these functions are

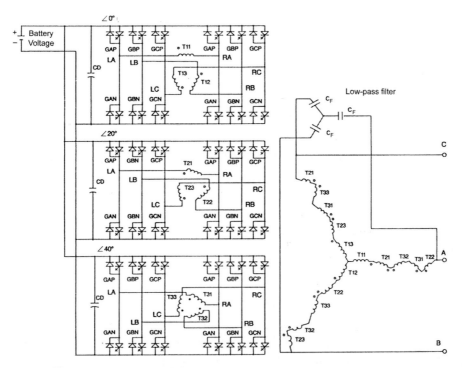

Figure 5.16 18-step GTO inverter for utility battery peaking service [6]

demanded, the converter can operate as a leading or lagging static VAR compensator, which will be explained later in this chapter. An advantage of multi-stepping over the traditional PWM method of control is that the device switching frequency (same as line frequency) is low and the corresponding switching loss is low, which is particularly important for GTO devices. For such high-power applications, GTOs are invariably used.

Multi-stepped inverters with harmonic neutralization, as discussed above, require bulky transformers. The transformers do not saturate if the volt/Hz ratio is maintained constant, but saturation is difficult to avoid near zero frequency operation of the ac drive. However, the advantages are that the transformer permits Ohmic isolation and voltage level change. For an ac machine load, the stator winding can be designed to integrate the transformer function, thus eliminating the requirement of a separate transformer.

5.5 PULSE WIDTH MODULATION TECHNIQUES

The three-phase, six-step inverter discussed before has several advantages and limitations. The inverter control is simple and the switching loss is low because there are only six switchings per cycle of fundamental frequency. Unfortunately, the lower order harmonics of the six-step voltage wave will cause large distortions of the current wave unless filtered by bulky and uneconomical low-pass filters. Besides, the voltage control by the line-side rectifier has the usual disadvantages.

5.5.1 PWM Principle

Because an inverter contains electronic switches, it is possible to control the output voltage as well as optimize the harmonics by performing multiple switching within the inverter with the constant dc input voltage V_d. The PWM principle to control the output voltage is explained in Figure 5.18. The fundamental voltage v_1 has the maximum amplitude ($4V_d/\pi$) at square wave, but by creating two notches as shown, the magnitude can be reduced. If the notch widths are increased, the fundamental voltage will be reduced.

5.5.1.1 PWM Classification

There are many possible PWM techniques proposed in the literature. The classifications of PWM techniques can be given as follows:

- Sinusoidal PWM (SPWM)
- Selected harmonic elimination (SHE) PWM
- Minimum ripple current PWM
- Space-Vector PWM (SVM)
- Random PWM
- Hysteresis band current control PWM
- Sinusoidal PWM with instantaneous current control

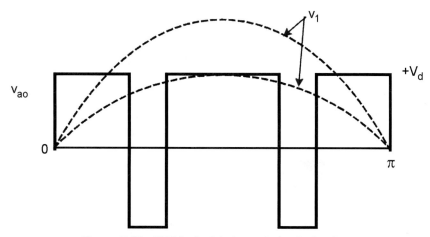

Figure 5.17 PWM principle to control output voltage

- Delta modulation
- Sigma-delta modulation

Often, PWM techniques are classified on the basis of voltage or current control, feedforward or feedback methods, carrier- or non-carrier-based control, etc. Note that the phase-shift PWM discussed before can also be classified as a PWM technique. In this section, we will briefly review the principal PWM techniques.

5.5.1.1.1 Sinusoidal PWM

The sinusoidal PWM technique is very popular for industrial converters and is discussed extensively in the literature. Figure 5.19 explains the general principle of SPWM, where an isosceles triangle carrier wave of frequency f_c is compared with the fundamental frequency f sinusoidal modulating wave, and the points of intersection determine the switching points of power devices. For example, v_{a0} fabrication by switching Q_1 and Q_4 of half-bridge inverter, is shown in the figure. The lock-out time between Q_1 and Q_4 to prevent a shoot-through fault is ignored in the figure. This method is also known as the triangulation, subharmonic, or suboscillation method. The notch and pulse widths of v_{a0} wave vary in a sinusoidal manner so that the average or fundamental component frequency is the same as f and its amplitude is proportional to the command modulating voltage. The same carrier wave can be used for all three phases, as shown. The typical wave shape of line and phase voltages for an isolated neutral load can be plotted graphically as shown in Figure 5.20. The Fourier analysis of the v_{a0} wave is somewhat involved and can be shown to be of the following form:

$$v_{a0} = 0.5mV_d \sin(\omega t + \phi) + \text{high-frequency}(M\omega_c \pm N\omega) \text{ terms} \qquad (5.33)$$

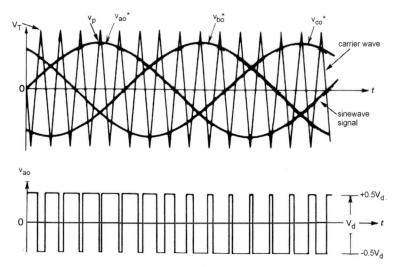

Figure 5.18 Principle of sinusoidal PWM for three-phase bridge inverter

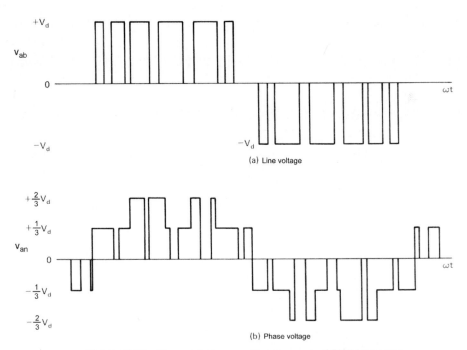

Figure 5.19 Line and phase voltage waves of PWM inverter

where m = modulation index, ω = fundamental frequency in r/s (same as the modulating frequency) and φ = phase shift of output, depending on the position of the modulating wave.

The modulation index m is defined as

$$m = \frac{V_P}{V_T} \quad (5.34)$$

where V_p = peak value of the modulating wave and V_T = peak value of the carrier wave. Ideally, m can be varied between 0 and 1 to give a linear relation between the modulating and output wave. The inverter basically acts as a linear amplifier. Combining Equations (5.33) and (5.34), the amplifier gain G is given as

$$G = \frac{0.5mV_d}{V_P} = \frac{0.5V_d}{V_T} \quad (5.35)$$

At $m = 1$, the maximum value of fundamental peak voltage is $0.5\ V_d$, which is 78.55 percent of the peak voltage ($4V_d/2\pi$) of the square wave. In fact, the maximum value in the linear range can be increased to 90.7 percent of that of the square wave by mixing the appropriate values of triplen harmonics with the modulating wave. At $m = 0$, v_{a0} is a square wave at carrier frequency with symmetrical pulse and notch widths. The PWM output wave contains carrier frequency-related harmonics with modulating frequency-related sidebands in the form $M\omega_c \pm N\omega$, which are shown in Equation (5.33), where M and N are integers and $M + N$ = an odd integer. For a carrier-to-modulating frequency ratio $P = \omega_c/\omega = 15$, Table 5.1 gives a summary of output harmonics.

Table 5.1 Family of Output Harmonics for Sinusoidal PWM with $\omega_c/\omega = 15$

m	Harmonics
1	15ω
	$15\omega \pm 2\omega$
	$15\omega \pm 4\omega$
	$\vdots$
2	$30\omega \pm \omega$
	$30\omega \pm 3\omega$
	$30\omega \pm 5\omega$
	$\vdots$
3	45ω
	$45\omega \pm 2\omega$
	$45\omega \pm 4\omega$
	$\vdots$
	$\vdots$

It can be shown that the amplitude of the harmonics is independent of P and diminishes with higher values of M and N. With higher carrier frequency ratio P, the inverter line current harmonics will be well-filtered by nominal leakage inductance of the machine and will practically approach a sine wave. The selection of a carrier frequency depends on the trade-off between the inverter loss and the machine loss. Higher carrier frequency (same as switching frequency) increases inverter switching loss but decreases machine harmonic loss. An optimal carrier frequency should be selected such that the total system loss is minimal. An important effect of PWM switching frequency is the generation of acoustic noise (known as magnetic noise) by the magnetostriction effect when the inverter supplies power to a machine. The effect can be alleviated by randomly varying the switching frequency (random SPWM), or it can be completely eliminated by increasing the switching frequency above the audio range. Modern high-speed IGBTs easily permit such acoustically noise-free variable-frequency drives. Low-pass line filters can also eliminate this problem.

Overmodulation Region

As the modulation index m approaches 1, the notch and pulse widths near the center of positive and negative half-cycles, respectively, tend to vanish. To complete switching operation of the devices, minimum notch and pulse widths must be maintained. When minimum-width notches and pulses are dropped, there will be some transient jump of load current. The jump may be small for IGBT inverters, but it is substantial for high-power GTO inverters because of the slow switching of the devices. The value of m can be increased beyond the value of 1 to enter into the quasi-PWM region, shown in Figure 5.21 for positive half-cycle only. The v_{a0} wave indicates that the notches near the center part have disappeared, giving a quasi-square-wave output with a higher fundamental component. The transfer characteristics in the overmodulation region are nonlinear, as indicated in Figure 5.22, and the harmonics 5th, 7th, etc. reappear. Ultimately, with a high m value, that is, a large modulating signal, there will be one switching at the leading edge and another switching at the trailing edge, giving square-wave output. At this condition, the fundamental phase voltage peak value is $4(0.5\ V_d)/\pi$, which is 100 percent, as indicated in Figure 5.22.

Frequency Relation

For variable-speed drive applications, the inverter output voltage and frequency are to be varied in the relation shown in Figure 2.14. In the constant power region, the maximum voltage can be obtained by operating the inverter in square-wave mode, but in the constant torque region, the voltage can be controlled using the PWM principle. It is usually desirable to operate the inverter with an integral ratio P of carrier-to-modulating frequency, where the modulating wave remains synchronized with the carrier wave in the entire region. A fixed value of P causes a low carrier frequency as the fundamental frequency goes down, which is not desirable from the machine harmonic loss point of view. A practical carrier-to-fundamental frequency relation of a GTO inverter is shown in Figure 5.23. At a low fundamental frequency, the carrier frequency is maintained constant and the inverter operates in the free-running, or asynchronous, mode. In this region, the ratio P may be nonintegral, and the phase may continually drift. This gives rise to a

Pulse Width Modulation Techniques

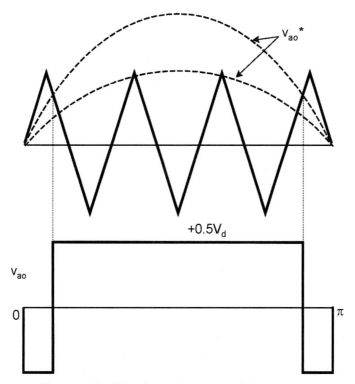

Figure 5.20 Waveforms in overmodulation region

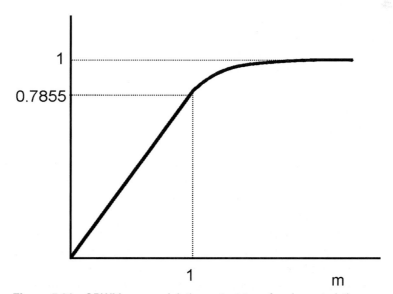

Figure 5.21 SPWM overmodulation output transfer characteristics

subharmonic problem with drifting dc offset (beating effect), which tends to be worse as the f_c/f ratio decreases. It should be mentioned here that the modern IGBT switching frequency is so large compared to the fundamental frequency range, the PWM inverter may operate satisfactorily in the entire asynchronous range. The free-running region is followed by the synchronized region, where P is varied in steps as shown so that the maximum and minimum carrier frequencies remain bounded within a definite zone. The value of P is maintained as a multiple of three because triplen harmonics are of no concern in isolated neutral load. Near the base frequency ($f/f_b = 1$), transition occurs to the square-wave mode, where the carrier frequency is assumed to be the same as the fundamental frequency. The control should be designed carefully so that at the jump of carrier frequency, there is no voltage jump problem, and chattering between adjacent P's should be avoided by providing a narrow hysteresis band at the critical points.

Dead-Time Effect and Compensation

The actual phase voltage (v_{a0}) wave in a PWM inverter deviates to some extent from the ideal wave shown in Figure 5.19 because of the dead-time (or lock-out) effect. This effect is explained in Figure 5.24 for the phase leg a of a three-phase bridge inverter. A fundamental control principle of a voltage-fed inverter is that the incoming device should be delayed by a dead-time t_d (typically a few µs.) from the outgoing device to prevent a shoot-through fault. This is because the turn-on of a device is very fast, but the turn-off is slow. The dead-time effect causes distortion of the output voltage and reduces its magnitude.

Consider the sinusoidal PWM operation in Figure 5.24. The direction of phase a current i_a is positive, as shown. With Q_1 conducting initially, v_{a0} magnitude is $+0.5V_d$, as indicated. When Q_1 is turned off at the ideal transition point, there is a time gap t_d before Q_4 is turned on. During this gap, both Q_1 and Q_4 are off, but $+i_a$ causes switching of v_{a0} to $-0.5V_d$ naturally at the ideal transition point. Consider now the switching from Q_4 to Q_1 with a delay t_d from the ideal transition point. When both devices are off, $+i_a$ continues flowing through D_4, causing a loss of volt-sec. area ($V_d t_d$) pulse given by the shaded area. Consider now that the polarity of current i_a is

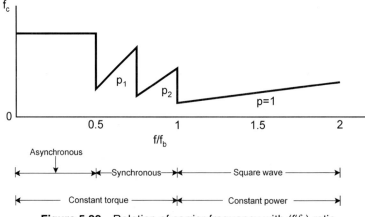

Figure 5.22 Relation of carrier frequency with (f/f_b) ratio

Pulse Width Modulation Techniques

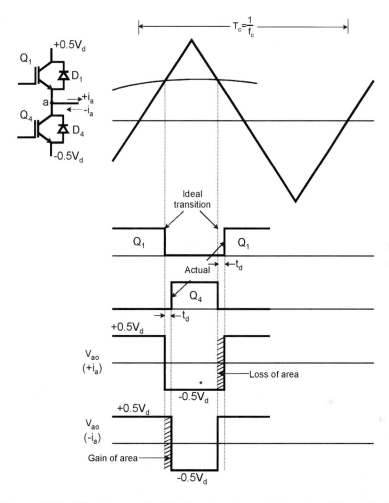

Figure 5.23 Waveforms of half-bridge inverter explaining dead-time effect

negative. Close examination of the waves shows that at the leading edge of Q_4 turn-on, there is a gain of similar volt-sec. area. Note that the loss or gain of the area depends only on the polarity of current, but not its magnitude. The cumulative effect of these "losses" and "gains" of volt-sec. area $V_d t_d$ during every carrier frequency period T_c for $+i_a$ and $-i_a$, respectively, on the fundamental voltage wave is explained in Figure 5.25. The fundamental current i_a is shown to lag the fundamental voltage v_{a0} by phase angle φ. The dead-time effect is indicated in the lowest part of the figure. The areas contributed by $V_d t_d$ can be accumulated and averaged in the half-cycle of fundamental frequency to calculate the square-wave offset V_ε as

$$V_\varepsilon = V_d t_d \left(\frac{P}{2}\right)(2f) = f_c t_d V_d \qquad (5.36)$$

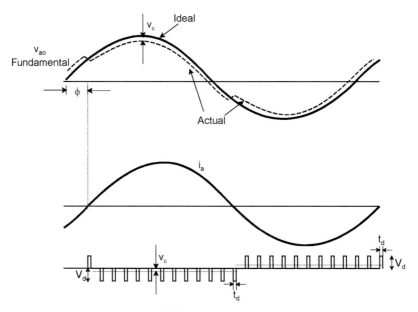

Figure 5.24 Dead-time effect on output phase voltage wave

where $P = f_c/f$ and f = fundamental frequency. The effect of the V_ε wave on the ideal v_{a0} wave is shown at the top of the figure. The loss of fundamental voltage and low-frequency harmonic distortion become serious at low fundamental frequency. The dead-time effect can be compensated easily by the current or voltage feedback method [20]. In the former method, the polarity of the phase current is detected and a fixed amount of compensating bias voltage is added with the modulating wave. In the latter method, the detected output phase voltage is compared with the PWM voltage reference signal and the deviation compensates the reference PWM modulating wave.

5.5.1.1.2 Selected Harmonic Elimination PWM

The undesirable lower order harmonics of a square wave can be eliminated and the fundamental voltage can be controlled as well by what is known as selected harmonic elimination (SHE) PWM. In this method, notches are created on the square wave at predetermined angles, as shown in Figure 5.26. In the figure, positive half-cycle output is shown with quarter-wave symmetry. It can be shown that the four notch angles α_1, α_2, α_3, and α_4 can be controlled to eliminate three significant harmonic components and control the fundamental voltage. A large number of harmonic components can be eliminated if the waveform can accommodate additional notch angles.

The general Fourier series of the wave can be given as

$$v(t) = \sum_{n=1}^{\infty} \left(a_n \cos n\omega t + b_n \sin n\omega t \right) \tag{5.37}$$

Pulse Width Modulation Techniques

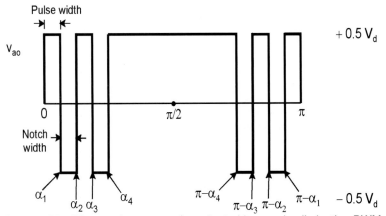

Figure 5.25 Phase voltage wave for selected harmonic elimination PWM

where

$$a_n = \frac{1}{\pi}\int_0^{2\pi} v(t)\cos n\omega t\, d\omega t \tag{5.38}$$

$$b_n = \frac{1}{\pi}\int_0^{2\pi} v(t)\sin n\omega t\, d\omega t \tag{5.39}$$

For a waveform with quarter-cycle symmetry, only the odd harmonics with sine components will be present. Therefore,

$$a_n = 0 \tag{5.40}$$

$$v(t) = \sum_{n=1}^{\infty} b_n \sin n\omega t \tag{5.41}$$

where

$$b_n = \frac{4}{\pi}\int_0^{\frac{\pi}{2}} v(t)\sin n\omega t\, d\omega t \tag{5.42}$$

Assuming that the wave has unit amplitude, that is, $v(t) = +1$, b_n can be expanded as

$$b_n = \frac{4}{\pi}[\int_0^{\alpha_1}(+1)\sin n\omega t\, d\omega t + \int_{\alpha_1}^{\alpha_2}(-1)\sin n\omega t\, d\omega t$$
$$+ \int_{\alpha_2}^{\alpha_3}(+1)\sin n\omega t\, d\omega t + \ldots + \int_{\alpha_{K-1}}^{\alpha_K}(-1)^{K-1}\sin n\omega t\, d\omega t \tag{5.43}$$
$$+ \int_{\alpha_K}^{\frac{\pi}{2}}(+1)\sin n\omega t\, d\omega t]$$

Using the general relation

$$\int_{\theta_1}^{\theta_2} \sin n\omega t \, d\omega t = \frac{1}{n}(\cos n\theta_1 - \cos n\theta_2) \qquad (5.44)$$

the first and last terms are

$$\int_0^{\alpha_1} (+1)\sin n\omega t \, d\omega t = \frac{1}{n}(1 - \cos n\alpha_1) \qquad (5.45)$$

$$\int_{\alpha_K}^{\frac{\pi}{2}} (+1)\sin n\omega t \, d\omega t = \frac{1}{n}\cos n\alpha_K \qquad (5.46)$$

Integrating the other components of Equation (5.43) and substituting (5.45) and (5.46) in it yields

$$b_n = \frac{4}{n\pi}\left[1 + 2(-\cos n\alpha_1 + \cos n\alpha_2 - \ldots + \cos n\alpha_K)\right]$$

$$= \frac{4}{n\pi}\left[1 + 2\sum_{K=1}^{K}(-1)^K \cos n\alpha_K\right] \qquad (5.47)$$

Note that Equation (5.47) contains K number of variables (i.e., $\alpha_1, \alpha_2, \alpha_3, \ldots \alpha_K$), and K number of simultaneous equations are required to solve their values. With K number of α angles, the fundamental voltage can be controlled and $K-1$ harmonics can be eliminated.

Consider, for example, that the 5th and 7th harmonics (lowest significant harmonics) are to be eliminated and the fundamental voltage is to be controlled. The 3rd and other triplen harmonics can be ignored if the machine has an isolated neutral. In this case, $K = 3$ and the simultaneous equations can be written from Equation (5.47) as

Fundamental:

$$b_1 = \frac{4}{\pi}(1 - 2\cos\alpha_1 + 2\cos\alpha_2 - 2\cos\alpha_3) \qquad (5.48)$$

5th harmonic:

$$b_5 = \frac{4}{5\pi}(1 - 2\cos 5\alpha_1 + 2\cos 5\alpha_2 - 2\cos 5\alpha_3) = 0 \qquad (5.49)$$

7th harmonic:

$$b_7 = \frac{4}{7\pi}(1 - 2\cos 7\alpha_1 + 2\cos 7\alpha_2 - 2\cos 7\alpha_3) = 0 \qquad (5.50)$$

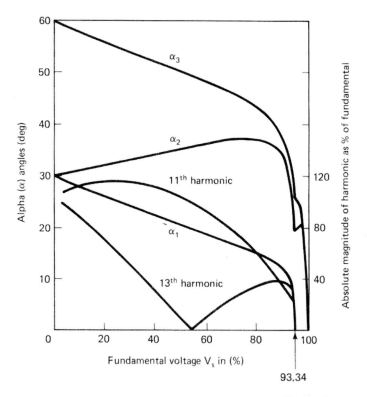

Figure 5.26 Notch angle relation with fundamental output voltage for 5th and 7th harmonic elimination

These nonlinear, transcendental equations can be solved numerically by a computer program for the specified fundamental amplitude and α_1, α_2, and α_3 can be determined, as shown in Figure 5.27. As an example, for a fundamental voltage of 50 percent ($b_1 = 0.5$), the α values are

$\alpha_1 = 20.9°$
$\alpha_2 = 35.8°$
$\alpha_3 = 51.2°$

Figure 5.27 also indicates that the lower order, significant harmonics (i.e., 11th and 13th) have been considerably boosted as a result of lower order harmonics elimination. The effect of these harmonics will possibly be small because of their large separation from the fundamental. Also note that in Figure 5.27, the 5th and 7th harmonics can be eliminated up to a voltage level of 93.34 percent (100 percent corresponds to the square wave) where $\alpha_1 = 0$. The single notch remaining on the outer side of the half-cycle can be narrowed symmetrically by reducing the α_2 angle, and then dropped to attain the full square wave. This segment of the α angle table for the voltage jump within 1percent is illustrated in Table 5.2. The typical waveform at 98 percent volt-

Table 5.2 α Angle Table for Voltage from 93% to 1 0 0 %

V_s	α_1	α_2	α_3
93	0	15.94	22.02
94	0	16.17	21.56
95	0	16.41	20.86
96	0	16.88	20.39
97	0	17.34	19.92
98	0	11.02	13.59
99	0	4.69	7.27
100 (square wave)	0	0	0

age is shown in Figure 5.28. Note that the phase inversion of the fundamental in the whole range of α angles is of no concern. Some amount of 5th and 7th harmonic voltages will reappear in the range of 93.3 percent to 100 percent voltage, but can be ignored in favor of limiting voltage jump.

The selected harmonic elimination method can be conveniently implemented with a microcomputer using a lookup table of notch angles. The simple block diagram in Figure 5.29 indicates the implementation strategy. At a certain command voltage V^*, the angles are retrieved from the lookup table, and correspondingly, the phase voltage pulse widths are generated in the time domain with the help of down-counters, where the counters are clocked at $f_{ck} = Kf$. If, for example, $K = 360$, then 1.0-degree resolution waveforms can be generated.

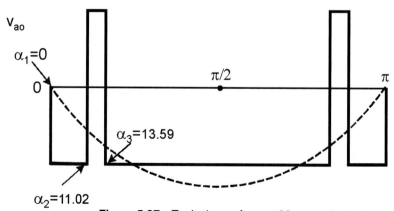

Figure 5.27 Typical waveform at 98 percent output

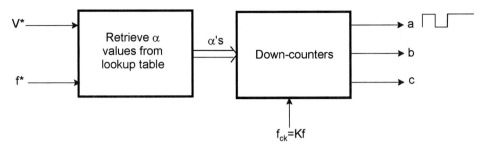

Figure 5.28 Block diagram for SHE implementation

As the fundamental frequency decreases, the number of notch angles can be increased so that a higher number of significant harmonics can be eliminated. Again, the number of notch angles/cycle, or the switching frequency, is determined by the switching losses of the inverter. An obvious disadvantage of the scheme is that the lookup table at low fundamental frequency is unusually large. For this reason, a hybrid PWM scheme where the low-frequency, low-voltage region uses the SPWM method, and the high-frequency region uses the SHE method appears very attractive [8].

5.5.1.1.3 Minimum Ripple Current PWM

One disadvantage of the SHE PWM method is that the elimination of lower order harmonics considerably boosts the next higher level of harmonics, as shown in Figure 5.27. Since the harmonic loss in a machine is dictated by the rms ripple current, it is this parameter that should be minimized instead of emphasizing the individual harmonics. In Chapter 2, it was shown that the effective leakage inductance of a machine essentially determines the harmonic current corresponding to any harmonic voltage. Therefore, the expression of rms ripple current can be given as

$$I_{ripple} = \sqrt{I_5^2 + I_7^2 + I_{11}^2 + \ldots}$$
$$= \sqrt{\frac{\hat{I}_5^2}{2} + \frac{\hat{I}_7^2}{2} + \frac{\hat{I}_{11}^2}{2} + \ldots} \quad (5.51)$$
$$= \sqrt{\frac{1}{2} \sum_{n=5,7,11,\ldots}^{\infty} \left(\frac{\hat{V}_n}{n\omega L}\right)^2}$$

where

$I_5, I_7 \ldots$ = rms harmonic currents
L = effective leakage inductance of the machine per phase
$\hat{I}_5, \hat{I}_7 \ldots$ = peak value of harmonic currents
n = order of harmonic
$\hat{V}_n$ = peak value of nth-order harmonic
ω = fundamental frequency

The corresponding harmonic copper loss is

$$P_L = 3I_{ripple}^2 R \tag{5.52}$$

where R = effective per phase resistance of the machine.

For a given number of notch angles, the expression of $\hat{V}_n$ is given from Equation (5.47). Substituting it in Equation (5.51), I_{ripple} is found as a function of α angles. The α angles can then be iterated in a computer program so as to minimize I_{ripple} for a certain desired fundamental magnitude. The modified lookup of α's based on harmonic loss minimization is more desirable than using the SHE method.

5.5.1.1.4 Space-Vector PWM

The space-vector PWM (SVM) method is an advanced, computation-intensive PWM method and is possibly the best among all the PWM techniques for variable-frequency drive applications. Because of its superior performance characteristics, it has been finding widespread application in recent years.

The PWM methods discussed so far have only considered implementation on a half-bridge of a three-phase bridge inverter. If the load neutral is connected to the center tap of the dc supply, all three half-bridges operate independently, giving satisfactory PWM performance. With a machine load, the load neutral is normally isolated, which causes interaction among the phases. This interaction was not considered before in the PWM discussions. The SVM method considers this interaction of the phases and optimizes the harmonic content of the three-phase isolated neutral load. To understand the SVM theory, the concept of a rotating space vector, as discussed in Chapter 2, is very important. For example, if the three-phase sinusoidal and balanced voltages given by the equations

$$v_a = V_m \cos \omega t \tag{5.53}$$

$$v_b = V_m \cos\left(\omega t - \frac{2\pi}{3}\right) \tag{5.54}$$

$$v_c = V_m \cos\left(\omega t + \frac{2\pi}{3}\right) \tag{5.55}$$

are applied to a three-phase induction motor, using Equation (2.89), it can be shown that the space vector $\overline{V}$ with magnitude V_m rotates in a circular orbit at angular velocity ω where the direction of rotation depends on the phase sequence of the voltages. With the sinusoidal three-phase command voltages, the composite PWM fabrication at the inverter output should be such that the average voltages follow these command voltages with a minimum amount of harmonic distortion.

Converter Switching States

A three-phase bridge inverter, as shown in Figure 5.8, has $2^3 = 8$ permissible switching states. Table 5.3 gives a summary of the switching states and the corresponding phase-to-neutral voltages of an isolated neutral machine. Consider, for example, state 1, when switches Q_1, Q_6, and Q_2 are closed. In this state, phase a is connected to the positive bus and phases b and c are

Table 5.3 Summary of Inverter Switching States

State	On devices	v_{an}	v_{bn}	v_{cn}	Space voltage vector
0	$Q_4Q_6Q_2$	0	0	0	$\overline{V}_0(000)$
1	$Q_1Q_6Q_2$	$2V_d/3$	$-V_d/3$	$-V_d/3$	$\overline{V}_1(100)$
2	$Q_1Q_3Q_2$	$V_d/3$	$V_d/3$	$-2V_d/3$	$\overline{V}_2(110)$
3	$Q_4Q_3Q_2$	.	.	.	$\overline{V}_3(010)$
4	$Q_4Q_3Q_5$	.	.	.	$\overline{V}_4(011)$
5	$Q_4Q_6Q_5$	.	.	.	$\overline{V}_5(001)$
6	$Q_1Q_6Q_5$	.	.	.	$\overline{V}_6(101)$
7	$Q_1Q_3Q_5$	0	0	0	$\overline{V}_7(111)$

connected to the negative bus. The simple circuit solution indicates that $v_{an} = 2/3V_d$, $v_{bn} = -1/3V_d$, and $v_{cn} = -1/3V_d$. The inverter has six active states (1–6) when voltage is impressed across the load, and two zero states (0 and 7) when the machine terminals are shorted through the lower devices or upper devices, respectively. The sets of phase voltages for each switching state can be combined with the help of Equation (2.89) to derive the corresponding space vectors. The graphical derivation of $\overline{V}_1$ (100) in Figure 5.30 indicates that the vector has a magnitude of $2/3V_d$ and is aligned in the horizontal direction as shown. In the same way, all six active vectors and two zero vectors are derived and plotted in Figure 5.31(a). The active vectors are $\pi/3$ angle apart and describe a hexagon boundary (shown as dotted). The two zero vectors $\overline{V}_0(000)$ and $\overline{V}_7(111)$ are at the origin. For three-phase, square-wave operation of the inverter, as shown in Figure 5.9, it can be easily verified that the vector sequence is $\overline{V}_1 \to \overline{V}_2 \to \overline{V}_3 \to \overline{V}_4 \to \overline{V}_5 \to \overline{V}_6$, with each dwelling for an angle of $\pi/3$, and there are no zero vectors.

The question is how to control the inverter space vectors in order to generate harmonically optimum PWM voltage waves at the output.

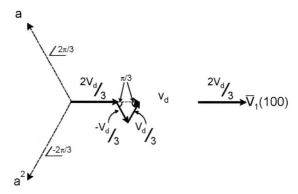

Figure 5.29 Construction of inverter space vector $\overline{V}_1(100)$

Linear or Undermodulation Region

Let us first consider the linear or undermodulation region where the inverter transfer characteristics are naturally linear. The modulating command voltages of a three-phase inverter are always sinusoidal, and therefore, they constitute a rotating space vector $\overline{V}^*$, as shown in Figure 5.31(a). Figure 5.31(b) shows the phase a component of the reference wave on the six-step phase voltage wave profile. For the location of the $\overline{V}^*$ vector shown in Figure 5.31 (a), as an example, a convenient way to generate the PWM output is to use the adjacent vectors $\overline{V}_1$ and $\overline{V}_2$ of sector 1 on a part-time basis to satisfy the average output demand. The $\overline{V}^*$ can be resolved as (dropping bar)

$$V^* \sin\left(\frac{\pi}{3} - \alpha\right) = V_a \sin\frac{\pi}{3} \tag{5.56}$$

$$V^* \sin\alpha = V_b \sin\frac{\pi}{3} \tag{5.57}$$

that is,

$$V_a = \frac{2}{\sqrt{3}} V^* \sin\left(\frac{\pi}{3} - \alpha\right) \tag{5.58}$$

$$V_b = \frac{2}{\sqrt{3}} V^* \sin\alpha \tag{5.59}$$

Pulse Width Modulation Techniques

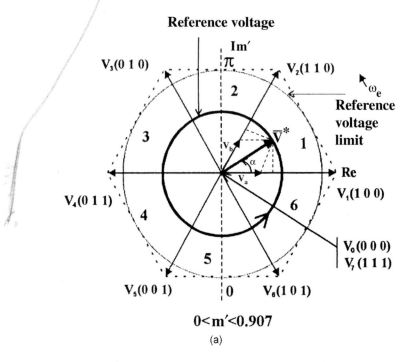

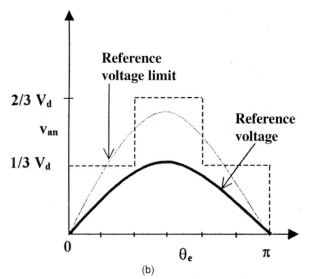

Figure 5.30 (a) Space vectors (bar sign is omitted) of three-phase bridge inverter showing reference voltage trajectory and segments of adjacent voltage vectors, (b) Corresponding reference phase voltage wave

where V_a and V_b are the components of V^* aligned in the directions of V_1 and V_2, respectively. Considering the period T_c during which the average output should match the command, we can write the vector addition

$$V^* = V_a + V_b = V_1 \frac{t_a}{T_c} + V_2 \frac{t_b}{T_c} + (V_o \text{ or } V_7) \frac{t_o}{T_c} \quad (5.60)$$

or

$$V^* T_c = V_1 t_a + V_2 t_b + (V_o \text{ or } V_7) t_o \quad (5.61)$$

where

$$t_a = \frac{V_a}{V_1} T_c \quad (5.62)$$

$$t_b = \frac{V_b}{V_2} T_c \quad (5.63)$$

$$t_o = T_c - (t_a + t_b) \quad (5.64)$$

Note that the time intervals t_a and t_b satisfy the command voltage, but time t_0 fills up the remaining gap for T_c with the zero or null vector. Figure 5.32 shows the construction of the symmetrical pulse pattern for two consecutive T_c intervals that satisfy Equations (5.62)–(5.64). Here, $T_s = 2T_c = 1/f_s$ (f_s = switching frequency) is the sampling time. Note that the null time has been

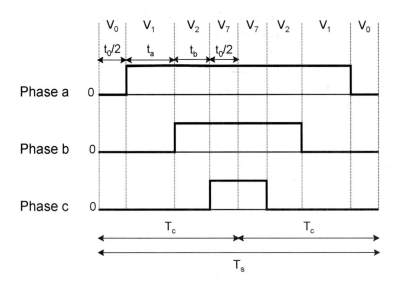

Figure 5.31 Construction of symmetrical pulse pattern for three phases

conveniently distributed between the V_0 and V_7 vectors to describe the symmetrical pulse widths. Studies have shown that a symmetrical pulse pattern gives minimal output harmonics.

In the undermodulation region shown in Figure 5.31(a), the vector V^* always remains within the hexagon. The mode ends in the upper limit when V^* describes the inscribed circle of the hexagon. Let us define a modified modulation factor m' given by

$$m' = \frac{\hat{V}^*}{\hat{V}_{1sw}} \tag{5.65}$$

where $\hat{V}^*$ = vector magnitude, or phase peak value and $\hat{V}_{1sw}$ = fundamental peak value ($2V_d/\pi$) of the square-phase voltage wave. The m' varies from 0 to 1 at the square-wave output. From the geometry of Figure 5.31(a), the maximum possible value of m' at the end of the undermodulation region can be derived. The radius of the inscribed circle can be given as

$$V_m^* = \frac{2}{3} V_d \cos\frac{\pi}{6} = 0.577 V_d \tag{5.66}$$

Therefore, m' at this condition can be derived as

$$m' = \frac{\hat{V}_m^*}{\hat{V}_{1sw}} = \frac{0.577 V_d}{2/\pi \cdot V_d} = 0.907 \tag{5.67}$$

This means that 90.7 percent of the fundamental at the square wave is available in the linear region, compared to 78.55 percent in the sinusoidal PWM.

Overmodulation Region

Overmodulation, or nonlinear, operation [9]–[11] starts when the reference voltage V^* exceeds the hexagon boundary. In overmodulation mode 1, shown in Figure 5.33, V^* crosses the hexagon boundary at two points in each sector. There will be a loss of fundamental voltage in the region where the reference vector exceeds the hexagon boundary. To compensate for this loss, that is, to track the output with the reference voltage, a modified reference voltage trajectory that remains partly on the hexagon and partly on the circle is selected as shown in the figure. The circular part of the modified trajectory has larger radius V_m^* ($V_m^* > V^*$) and crosses the hexagon at angle θ, as shown in the figure. Note that Equations (5.62)–(5.64) remain valid for the circular part of the trajectory, except V^* is replaced by V_m^*. But, on the hexagon trajectory, the time t_0 vanishes, giving only t_a and t_b time intervals. At this condition, t_a and t_b expressions can be derived as

$$t_a = T_c \left[\frac{\sqrt{3}\cos\alpha - \sin\alpha}{\sqrt{3}\cos\alpha + \sin\alpha} \right] \tag{5.68}$$

$$t_b = T_c - t_a \tag{5.69}$$

Substituting $\alpha = 0$ and $\pi/3$ in these equations verifies the corresponding values of $t_a = T_c$, $t_b = 0$ and $t_a = 0$, $t_b = T_c$, respectively. The v_{an} voltage wave is given approximately by linear

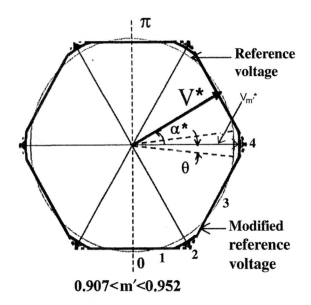

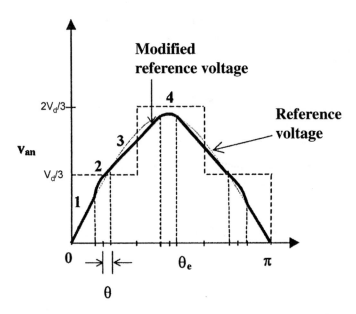

Figure 5.32 Voltage trajectory in overmodulation mode 1 and corresponding average phase voltage wave

segments for the hexagon trajectory and sinusoidal segments for the circular trajectory, as shown in the lower part of Figure 5.33. The equations for the four voltage segments (shown on the trajectory) in the first quarter-cycle [10] can be given as

Segment 1:
$$v_1 = m_1 \theta_e \quad \text{for} \quad 0 < \theta_e < \left(\frac{\pi}{6} - \theta\right) \tag{5.70}$$

Segment 2:
$$v_2 = V_m^* \sin\theta_e \quad \text{for} \quad \left(\frac{\pi}{6} - \theta\right) < \theta_e < \left(\frac{\pi}{6} + \theta\right) \tag{5.71}$$

Segment 3:
$$v_3 = A + \frac{m_1}{2}\theta_e \quad \text{for} \quad \left(\frac{\pi}{6} + \theta\right) < \theta_e < \left(\frac{\pi}{2} - \theta\right) \tag{5.72}$$

Segment 4:
$$v_4 = V_m^* \sin\theta_e \quad \text{for} \quad \left(\frac{\pi}{2} - \theta\right) < \theta_e < \frac{\pi}{2} \tag{5.73}$$

where $\theta_e = \omega_e t$, $m_1 = 2V_d/\pi$ = slope of linear segment 1, $A = V_d/6$, and V_m^* = modified reference voltage. The voltage V_m^* can be determined as a function of crossover angle θ by equating (5.71) and (5.72) at angle $(\pi/6 - \theta)$ as

$$V_m^* = \frac{2V_d\left(\frac{\pi}{6} - \theta\right)}{\pi \sin\left(\frac{\pi}{6} - \theta\right)} \tag{5.74}$$

Because of quarter-cycle symmetry, the fundamental output voltage (peak value) can be written from Equations (5.70)–(5.73) as

$$V_1 = \frac{4}{\pi}[\int_0^{\frac{\pi}{6}-\theta} v_1 \sin\theta_e d\theta_e + \int_{\frac{\pi}{6}-\theta}^{\frac{\pi}{6}+\theta} v_2 \sin\theta_e d\theta_e$$

$$+ \int_{\frac{\pi}{6}+\theta}^{\frac{\pi}{2}-\theta} v_3 \sin\theta_e d\theta_e + \int_{\frac{\pi}{2}-\theta}^{\frac{\pi}{2}} v_4 \sin\theta_e d\theta_e] \tag{5.75}$$

A computer program can be written to solve crossover angle θ as a function of modulation factor m' from Equations (5.65) and (5.70)–(5.75). Figure 5.34 shows the plot of this relation. Mode 1 ends when the angle $\theta = 0$ at $m' = 0.952$, that is, the trajectory is fully on the hexagon, giving only linear segments of v_{an} wave.

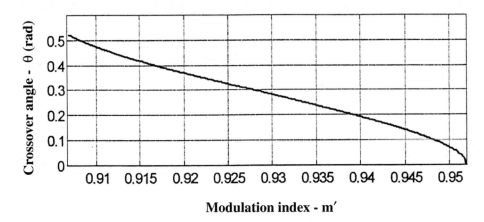

Figure 5.33 Modulation factor m' relation with crossover phase angle θ in overmodulation mode 1

In overmodulation mode 2, the reference voltage V^* increases further, tending toward square-wave mode operation. Again, the actual trajectory is to be modified so that the output fundamental voltage matches with that of the reference voltage. The operation in this region, as explained in Figure 5.35, is characterized by partly holding the modified vector at the hexagon corner for holding angle α_h, and partly by tracking the hexagon sides in every sector. During the holding angle, the magnitude of v_{an} remains constant, whereas during hexagon tracking, the voltage changes approximately in a linear manner, as shown in the lower part of Figure 5.35. At the end of mode 2, the linear segments vanish, giving six-step, or square-wave, operation when the modified vector is held at hexagon corners for $\pi/3$, that is, $\alpha_h = \pi/6$. Linear segments 1 and 3 are given by Equations (5.70) and (5.72), respectively. The expression for modified α_h angle (α_m) in mode 2 can be given as

$$\alpha_m = \begin{cases} 0 & \text{for} \quad 0 < \alpha^* < \alpha_h \\ \dfrac{\alpha^* - \alpha_h}{\dfrac{\pi}{6} - \alpha_h} \cdot \dfrac{\pi}{6} & \text{for} \quad \alpha_h < \alpha^* < \left(\dfrac{\pi}{3} - \alpha_h\right) \\ \dfrac{\pi}{3} & \text{for} \quad \left(\dfrac{\pi}{3} - \alpha_h\right) < \alpha^* < \dfrac{\pi}{3} \end{cases} \tag{5.76}$$

Pulse Width Modulation Techniques

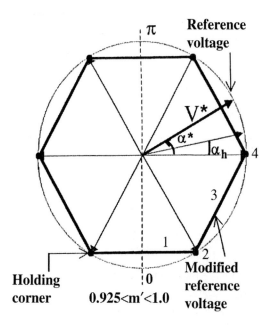

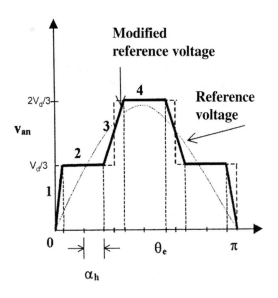

Figure 5.34 Voltage trajectory in overmodulation mode 2 and corresponding average phase voltage wave

For the v_{an} wave, the equations for the four segments in the first quarter-cycle [10] can be given as follows:

Segment 1:
$$v_1 = m_1\theta_e \quad \text{for} \quad 0 < \theta_e < \left(\frac{\pi}{6} - \alpha_h\right) \tag{5.77}$$

Segment 2:
$$v_2 = \frac{1}{3}V_d \quad \text{for} \quad \left(\frac{\pi}{6} - \alpha_h\right) < \theta_e < \left(\frac{\pi}{6} + \alpha_h\right) \tag{5.78}$$

Segment 3:
$$v_3 = A + m_2\theta_e \quad \text{for} \quad \left(\frac{\pi}{6} + \alpha_h\right) < \theta_e < \left(\frac{\pi}{2} - \alpha_h\right) \tag{5.79}$$

Segment 4:
$$v_4 = \frac{2}{3}V_d \quad \text{for} \quad \left(\frac{\pi}{2} - \alpha_h\right) < \theta_e < \frac{\pi}{2} \tag{5.80}$$

where

$$m_1 = \frac{V_d}{3\left(\frac{\pi}{6} - \alpha_h\right)} \tag{5.81}$$

$$m_2 = \frac{V_d}{3\left(\frac{\pi}{3} - 2\alpha_h\right)} \tag{5.82}$$

$$A = \frac{V_d\left(\frac{\pi}{6} - 3\alpha_h\right)}{3\left(\frac{\pi}{3} - 2\alpha_h\right)} \tag{5.83}$$

Again, because of quarter-cycle symmetry, the fundamental (peak) voltage expression can be given as

$$V_1 = \frac{4}{\pi}\left[\int_0^{\frac{\pi}{6}-\alpha_h} v_1 \sin\theta_e \, d\theta_e + \int_{\frac{\pi}{6}-\alpha_h}^{\frac{\pi}{6}+\alpha_h} v_2 \sin\theta_e \, d\theta_e \right.$$
$$\left. + \int_{\frac{\pi}{6}+\alpha_h}^{\frac{\pi}{2}-\alpha_h} v_3 \sin\theta_e \, d\theta_e + \int_{\frac{\pi}{2}-\alpha_h}^{\frac{\pi}{2}} v_4 \sin\theta_e \, d\theta_e\right] \tag{5.84}$$

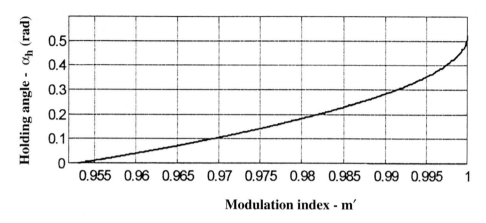

Figure 5.35 Modulation factor m' relation with holding angle α_h in overmodulation mode 2

For the output fundamental voltage to match with the reference voltage, a relation between the holding angle α_h and the modulation factor m' can be derived from Equations (5.65) and (5.77)–(5.84). This relation is plotted in Figure 5.36.

Implementation Step s

A simplified flow diagram for the implementation of the SVM algorithm with the help of DSP is shown in Figure 5.37. The synchronously rotating input voltage components v_{ds}^* and v_{qs}^* are sampled and the corresponding V^* and θ parameters are calculated, as shown. The θ angle is added with the θ_e' (angle between d^s and d^e axes) to obtain the θ_e angle. The θ_e' angle is obtained by the integration of frequency command ω_e^*. The vector parameters V^* and θ_e can be easily derived if the instantaneous phase voltage commands are given instead. The modulation index m' is then calculated to identify undermodulation, mode 1 or mode 2 operation. For undermodulation mode, the adjacent voltage vector segments and time segments t_a, t_b, and t_0 are calculated. The corresponding digital words are then converted to time segments with the help of timers. Mode 1 and mode 2 operations are somewhat similar, except they need the respective crossover angle θ and holding angle α_h.

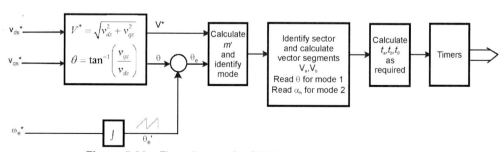

Figure 5.36 Flow diagram for SVM implementation by DSP

The SVM technique is complex and computation-intensive; therefore, the PWM switching frequency is somewhat limited. The frequency can be increased somewhat by simplifying computations and using lookup tables. Recently, artificial neural network (ANN)-based SVM was proposed [11] (see Figure 12.32), which can significantly increase the PWM frequency. The complexity of implementation can be justified because of the larger range of linearity and improved harmonic performance of SVM.

5.5.1.1.5 Sinusoidal PWM with Instantaneous Current Control

So far, we have discussed only feedforward voltage control PWM techniques. In a machine drive system, control of machine current is important because it influences the flux and developed torque directly. High-performance drives invariably require current control. For a voltage-fed inverter with voltage control PWM, a feedback current loop can be applied to control the machine current. In such cases, the inverter operates as a programmable current source.

Figure 5.38 shows an instantaneous current control scheme with sinusoidal voltage PWM in the inner loop. The error in the sinusoidal current control loop is converted to sinusoidal voltage command through a proportional-integral (P-I) controller. The remaining part is the standard SPWM scheme discussed before. For a three-phase inverter, three similar controllers are used. The control is simple, but there are a few problems. Due to limited bandwidth of the control system, the actual current will have a phase lag and magnitude error which will increase with frequency. Such phase deviation is very harmful in high-performance drive systems. The sinusoidal voltage command generated by the current control loop may contain ripple, which may create a multiple zero crossing problem in the SPWM comparator. All of these problems can be solved in a synchronous current controller, which will be discussed in Chapter 8.

5.5.1.1.6 Hysteresis-Band Current Control PWM

Hysteresis-band PWM is basically an instantaneous feedback current control method of PWM where the actual current continually tracks the command current within a hysteresis band. Figure 5.39 explains the operation principle of hysteresis-band PWM for a half-bridge inverter. The control circuit generates the sine reference current wave of desired magnitude and fre-

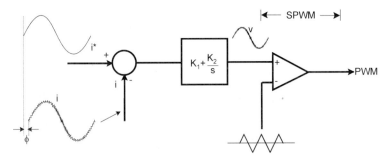

Figure 5.37 Control block diagram for instantaneous current control SPWM

Pulse Width Modulation Techniques

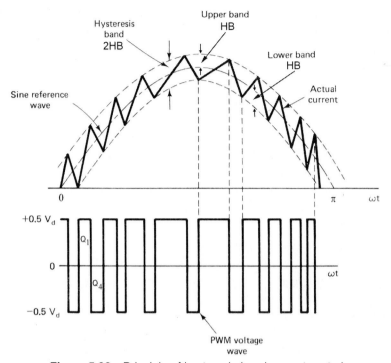

Figure 5.38 Principle of hysteresis-band current control

quency, and it is compared with the actual phase current wave. As the current exceeds a prescribed hysteresis band, the upper switch in the half-bridge is turned off and the lower switch is turned on. As a result, the output voltage transitions from $+0.5V_d$ to $-0.5V_d$, and the current starts to decay. As the current crosses the lower band limit, the lower switch is turned off and the upper switch is turned on. A lock-out time (t_d) is provided at each transition to prevent a shoot-through fault. The actual current wave is thus forced to track the sine reference wave within the hysteresis band by back-and-forth (or bang-bang) switching of the upper and lower switches. The inverter then essentially becomes a current source with peak-to-peak current ripple, which is controlled within the hysteresis band irrespective of V_d fluctuation. When the upper switch is closed, the positive slope of the current is given as

$$\frac{di}{dt} = \frac{0.5V_d - V_{cm}\sin\omega_e t}{L} \qquad (5.85)$$

where $0.5V_d$ is the applied voltage, $V_{cm}\sin\omega_e t$ = instantaneous value of the opposing load CEMF, and L = effective load inductance. The corresponding equation when the lower switch is closed is given as

$$\frac{di}{dt} = \frac{-(0.5V_d + V_{cm}\sin\omega_e t)}{L} \qquad (5.86)$$

The peak-to-peak current ripple and switching frequency are related to the width of the hysteresis band. For example, a smaller band will increase switching frequency and lower the ripple. An optimum band that maintains a balance between the harmonic ripple and inverter switching loss is desirable. The hysteresis-band PWM can be smoothly transitioned to square-wave voltage mode through the quasi-PWM region. In the low-speed region of the machine, when the CEMF is low, there is no difficulty in the current controller tracking. However, at higher speeds, the current controller will saturate in part of the cycle due to a higher CEMF and fundamental frequency-related harmonics will appear. At this condition, the fundamental current will be less and its phase will lag with respect to the command current.

Figure 5.40 shows the simple control block diagram for a hysteresis-band PWM implementation. The error in the current control loop is impressed at the input of a comparator with a hysteresis band, as shown. The bandwidth of HB is given as

$$HB = V \frac{R_2}{R_1 + R_2} \tag{5.87}$$

where V = comparator supply voltage. The conditions for switching the devices are:

Upper switch on:

$$(i^* - i) > HB \tag{5.88}$$

Lower switch on:

$$(i^* - i) < -HB \tag{5.89}$$

For a three-phase inverter, a similar control circuit is used in all phases.

The hysteresis-band PWM has been very popular because of its simple implementation, fast transient response, direct limiting of device peak current, and practical insensitivity of dc link voltage ripple that permits a lower filter capacitor. However, there are a few drawbacks of

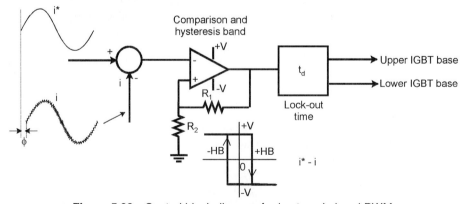

Figure 5.39 Control block diagram for hysteresis-band PWM

Pulse Width Modulation Techniques

this method. It can be shown that the PWM frequency is not constant (varies within a band) and, as a result, non-optimum harmonic ripple is generated in the machine current. An adaptive hysteresis band can alleviate this problem. It can be shown that the fundamental current suffers a phase lag that increases at higher frequency. This phase deviation causes problems in high-performance machine control. Of course, isolated neutral load (which is not discussed here) creates additional distortion of the current wave.

5.5.1.1.7 Sigma-Delta Modulation

A PWM technique known as sigma-delta modulation has often been used in high-frequency link converter systems (see Figures 4.40 and 4.41) to generate variable-frequency, variable-voltage sinusoidal waves by fabrication of integral half-cycle pulses. The principle is explained in Figure 5.41. The modulator receives the command phase voltage v_{ao}^* at variable magnitude and frequency, and it is compared with the actual discrete phase voltage pulses. The resulting error (delta operation) is integrated (sigma operation) to generate the integral error function e given as

$$e = \int v_{ao}^* dt - \int v_{ao} dt \tag{5.90}$$

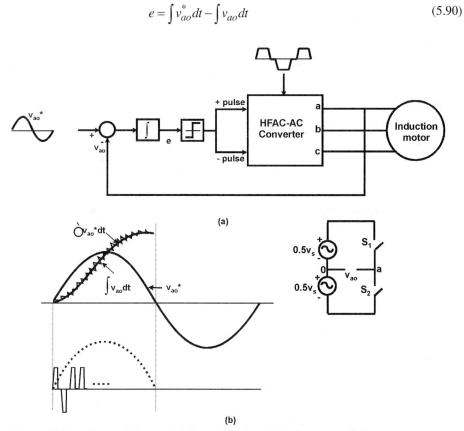

Figure 5.40 Sigma-delta modulation principle of high-frequency link converter

The polarity of the error function is detected by a bipolar comparator. The positive polarity of e selects a positive voltage pulse, whereas the negative polarity selects a negative voltage pulse. The switching is made at zero voltage gap (zero voltage switching) to get the soft-switching advantages of the inverter, which will be explained later. Note, for example, that the positive voltage pulse can be selected by closing the ac switch S_1 in the positive half-cycle or by closing S_2 in the negative half-cycle. It can be easily seen that the tracking accuracy of the command and feedback volt-sec. integrals will improve at a higher converter frequency. If $v_{a0}^* = 0$, the HFAC converter pulses will alternate. The modulator smoothly transitions to the square-wave region when the v_{a0}^* magnitude increases to a high value. At this condition, the polarity of all the pulses is unidirectional during the fundamental half-cycle to get the maximum fundamental voltage. Three such modulators are used in the three phases, and again, the effect of load neutral isolation is ignored.

Instead of fundamental voltage control, if fundamental current control is desired, then the simple delta modulation principle can be used. In such a case, the instantaneous current control loop error polarity can select the appropriate voltage pulse directly at the nearest zero voltage gap.

5.6 THREE-LEVEL INVERTERS

So far, this chapter has discussed the two-level inverter, which can be seen in Figure 5.26. In this type of inverter, the switching of the upper and lower devices in a half-bridge inverter generates a v_{a0} wave with positive and negative levels ($+0.5V_d$ and $-0.5V_d$), respectively. If the fundamental output voltage and corresponding power level of the PWM inverter are to be increased to a high value, the dc link voltage V_d must be increased and the devices must be connected in series. By using matched devices in series, static voltage sharing may be somewhat easy, but dynamic voltage sharing during switching is always difficult. The problem may be solved by using a multi-level, or neutral-point clamped (NPC), inverter.

Figure 5.42 shows the circuit of a three-level, three-phase inverter using GTO devices. In the figure, the dc link capacitor C has been split to create the neutral point 0. Since the operation of all the phase groups is essentially identical, consider only the operation of the half-bridge for phase a. A pair of devices with bypass diodes are connected in series with an additional diode connected between the neutral point and the center of the pair, as shown. The devices Q_{11} and Q_{14} function as main devices (like a two-level inverter), and Q_{12} and Q_{13} function as auxiliary devices which help to clamp the output potential to the neutral point with the help of clamping diodes D_{10} and D_{10}'. All the PWM techniques discussed so far can be applied to this inverter.

To explain the inverter operation, consider a simple selected harmonic elimination PWM method where it is desired to eliminate the lowest two significant harmonics (5th and 7th) and control the fundamental voltage. Figure 5.43 shows the phase voltage v_{a0} waveform with three α angles and the corresponding gate voltage switching waves. Note that the main devices (Q_{11} and Q_{14}) generate the v_{a0} wave, whereas the auxiliary devices (Q_{13} and Q_{12}) are driven complementary to the respective main devices. With such control, each output potential is clamped to the

Three-Level Inverters

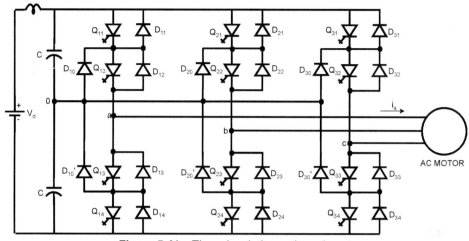

Figure 5.41 Three-level, three-phase inverter

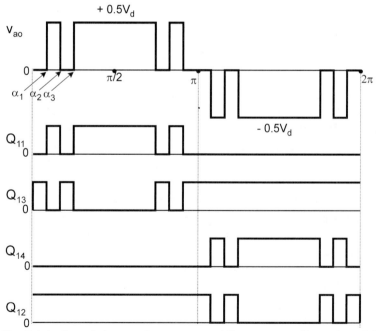

Figure 5.42 Fabrication of phase voltage wave and the corresponding gate switching waves

neutral potential in the off periods of the PWM control, as indicated in Figure 5.43. Evidently, positive phase current $+i_a$ will be carried by devices Q_{11} and Q_{12} when v_{a0} is positive, by devices D_{13} and D_{14} when v_{a0} is negative, and by devices D_{10} and Q_{12} at the neutral clamping condition. On the other hand, negative phase current $-i_a$ will be carried by D_{11} and D_{12} when v_{a0}

is positive, by Q_{13} and Q_{14} when v_{a0} is negative, and by Q_{13} and D_{10}' at the neutral clamping condition. This operation mode gives three voltage levels ($+0.5V_d$, 0, and $-0.5V_d$) at the v_{a0} wave, shown in Figure 5.43, compared to two levels ($+0.5V_d$ and $-0.5V_d$) in a conventional two-level inverter. The levels of line voltage wave v_{ab} are $+V_d$, $-V_d$, $+0.5V_d$, $-0.5V_d$, and 0 compared to levels $+V_d$, $-V_d$, and 0 in a two-level inverter.

Neglecting the fluctuation of neutral point potential, it can be shown that each device has to withstand $0.5V_d$ voltage. When diode D_{10} or D_{10}' is conducting, the voltage across the main device is clamped to $+0.5V_d$. When, for example, the lower devices are conducting, the full bus voltage V_d appears across the upper devices in series, in other words, the devices share $0.5V_d$ statically. However, it can be seen that at any switching, the voltage step size across the series string is only $0.5V_d$, which easily permits series connection of devices without exceeding their $0.5V_d$ rating.

In summary, each half-bridge inverter has the following three switching states:

State A: Upper switches on

State B: Lower switches on

State 0: Auxiliary switches on

Therefore, the three-phase inverter has $3^3 = 27$ switching states. This compares to the eight states of a two-level inverter. Figure 5.44 shows the space-vector diagram of the inverter with all switching states. The larger hexagon states (*ABB*, *AAB*, *BAB*, *BAA*, *BBA*, and *ABA*) correspond to those of a two-level inverter. The three zero states (000, *AAA*, *BBB*) are possible due to the conduction of auxiliary devices, upper devices, or lower devices, respectively. Each space-vector for the inner hexagon has two possible switching states, as indicated in the figure. There are, in addition, six space vectors that correspond to the middle of the outer hexagon sides.

Figure 5.44 also includes a rotating command voltage vector V^*, which should be enclosed within the larger hexagon for the undermodulation region of operation. At any instant, the vector selects the three inverter states corresponding to the apexes of the triangle, which include the V^* for PWM wave generation. At position 7, for example, as shown in the figure, the selected states might be 000/*AAA*/*BBB* – A00/0BB – A0A/0B0. Figure 5.45(a) shows the symmetrical PWM waves for this location using the sequence 0BB – 0B0 – 000 – A00 – 000 – 0B0 – 0BB in a sampling period T_s. Note that the zero inverter state must be included in part of the cycle if V^* lies within the inner hexagon. Also note the difference between the PWM waves corresponding to the states A00 and 0BB. For a larger magnitude of V^*, as shown by location 4, the triangle apexes are A00/0BB – ABB – AB0. The corresponding PWM wave is shown in Figure 5.45(b), which does not include any zero state. Note that switching occurs in one state only at a time with a step size of $0.5V_d$. The fabrication of the line voltage wave from Figure 5.45 is left as an exercise to the reader.

Three-Level Inverters

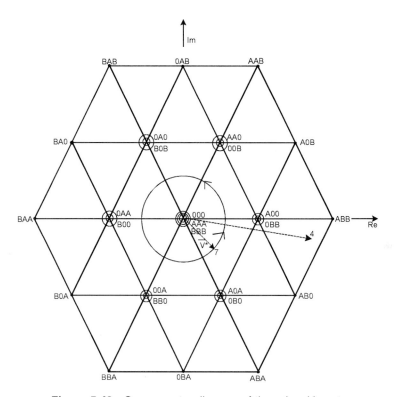

Figure 5.43 Space-vector diagram of three-level inverter

5.6.1 Control of Neutral Point Voltage

The split capacitor bank in the dc link must maintain a constant voltage level ($+0.5V_d$ with respect to negative rail) at the neutral point 0. Otherwise, additional distortion will be contributed to the output voltage. If positive current is drawn from the neutral point, the upper capacitor will charge and the lower capacitor will discharge, which will lower the potential of 0. If, on the other hand, the current is fed to the neutral point, the potential of 0 will go up. Consider, for example in Figure 5.45(b) that i_a is positive whereas i_b and i_c are negative. During the state $0BB$, positive current is drawn from the capacitor bank, lowering the potential of 0. In the state $A00$, negative current will be fed to the neutral point by b and c phases, tending to increase its potential. The neutral point potential can be controlled by manipulating these clamped time intervals, either in open loop manner or by a feedback control loop. Note that there is no neutral current in upper hexagon corner states (ABB, AAB, BAB, BAA, BBA, and ABA) and zero states (000, AAA, and BBB).

Besides the advantage of voltage clamping, a three-level inverter has the additional merit of improved PWM harmonic quality. By preventing the opposite voltage swing during the notch interval of a two-level inverter, the harmonic voltage ripple and corresponding current ripple are reduced. Numerical evaluation of harmonic voltages with selected harmonic elimination PWM

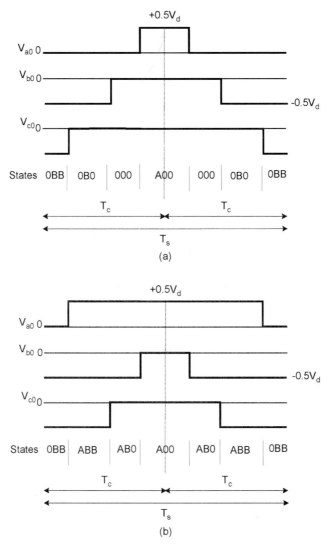

Figure 5.44 PWM voltage waves for vector V* showing the state sequences (a) Vector location 7 (b) Vector location 4

[12] indicates considerable attenuation of lower significant harmonics of a three-level inverter compared to those of a two-level inverter. For an isolated neutral load, the SVM technique with 27 space-vectors evidently gives a considerable improvement of harmonic quality compared to that of a two-level inverter with only 8 states.

For higher voltages and higher power levels, a multi-level inverter with more voltage levels can easily be constructed by extending the same principle. The inverter has some penalty with the extra devices. An additional problem is the fluctuation of neutral point voltages with the finite size of the dc link capacitors. The multiplicity of switching states for inner space vectors (see Figure 5.44) permits the manipulation of PWM signals to compensate for this fluctuation without diminishing PWM quality.

Three-level inverters have recently become popular in high-power applications. Multi-megawatt induction and synchronous motor drives have been built with three-level GTO inverters for industrial applications. Regenerative snubbers are often used in inverters to minimize the high switching losses of GTOs. Three-level IGBT converters are also becoming popular in the lower power ranges of GTO inverters because of improved PWM quality with higher switching frequencies.

5.7 HARD SWITCHING EFFECTS

So far in this chapter, we have discussed hard-switched inverters. The device switching waves of a hard-switched inverter, as shown in Figure 1.21, have a number of detrimental effects, which can be summarized as follows:

Switching Loss – The overlapping of voltage and current waves during each turn-on and turn-off switching cause a large pulse of energy loss, as explained before. With an RC snubber, the turn-off loss can be decreased, but the stored energy in the capacitor is lost at turn-on switching. Therefore, with a snubber, the total switching loss may increase. With higher switching frequency, inverter loss increases, that is, its efficiency decreases. An additional problem is that the cooling system is burdened due to higher loss. In fact, the PWM switching frequency of an inverter is limited because of switching loss. Of course, a regenerative snubber, often used in a GTO inverter, can alleviate this problem.

Device Stress – In hard switching, the switching locus moves through the active region of the volt-ampere area, which stresses the device. The reliability (i.e., MTBF) of the device may be impaired due to prolonged hard switching operation.

EMI Problems – High *dv/dt, di/dt*, and parasitic ringing effect at the switching of a fast device can create severe EMI problems, which may affect the control circuit and nearby apparatus. Parasitic leakage or coupling inductance, although quite small, can be a source of EMI due to large induced (*Ldi/dt*) voltage. Similarly, large *dv/dt* can induce common mode coupling current (*Cdv/dt*) in the control circuit through a parasitic capacitance. EMI problems may be more severe in a snubberless inverter.

Effect on Machine Insulation – High *dv/dt* impressed across the stator winding insulation can create large displacement current (*Cdv/dt*), which can deteriorate machine insulation.

Machine Bearing Current – Recently, it was determined that PWM inverter drives with fast switching IGBT devices are known to cause a machine bearing current problem [13]. Fast switching IGBTs create high *dv/dt*, and inverters can be represented by common mode-equivalent circuits with high *dv/dt* sources. These *dv/dt* sources create circulating current to ground through the machine bearing stray capacitances, as indicated in Figure 5.46. Figure 5.47 shows the stray capacitances linking a machine's stator winding with the rotor and the stator. The common mode *dv/dt* impressed on the stator winding couples to the rotor and creates a circulating

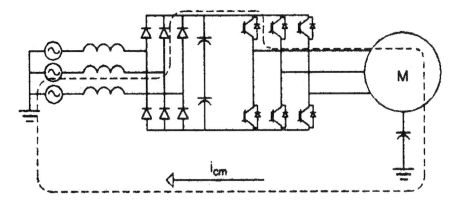

Figure 5.45 Common mode dv/dt induced current flow in a drive through motor bearing

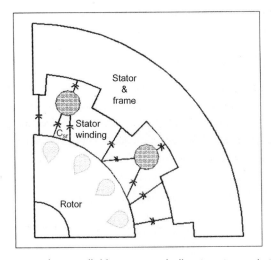

Figure 5.46 Stray capacitances linking stator winding to rotor and stator iron and frame

current to the ground through the machine shaft and stray capacitance of the insulated bearing. There are, of course, additional circulating current paths through the machine stator and grounded frame, and the motor cable to the ground. If the rectifier supply is not grounded (see Figure 5.46), the circulating current can flow through the stray capacitances of the power circuit linking to the ground. The circulating current through the bearing will increase with higher dv/dt and higher PWM switching frequency. The bearing may be insulated, shorted to the frame with a brush, or a low-pass filter may be installed at the machine terminal to alleviate this problem.

Machine Terminal Overvoltage – PWM inverters are often required to link a machine with a long cable, such as in submersible pump drives. The high dv/dt at the inverter output boosts the machine terminal voltage by the reflection of the high-frequency travelling wave. High-frequency ringing occurs at the machine terminal with stray circuit parameters. The result-

Resonant Inverters

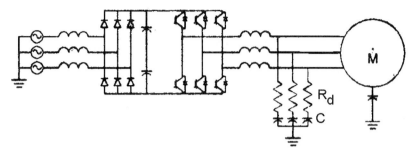

Figure 5.47 Low-pass filter at machine terminal to solve dv/dt-induced machine problems

ing excessive overvoltage threatens the motor insulation. The solution to the problem has been suggested by installing a low-pass *LC* filter at the machine terminal, as shown in Figure 5.48. The damping resistance R_d prevents resonance in the filter, which may be induced by inverter harmonics. The high *dv/dt* effect is essentially shunted to ground through the low-pass filter. The filter also solves bearing current and machine insulation deterioration problems.

5.8 RESONANT INVERTERS

A voltage-fed inverter with a series resonant circuit load can be defined as a resonant inverter. An H-bridge resonant inverter circuit is shown in Figure 5.49. The circuit is not directly useful for motor drives. The principal applications may be high-frequency induction heating, a high-frequency, general-purpose power supply, or as an inverter for a resonant-link, dc-dc converter. The frequency may be in the range of tens of kHz. The power level and frequency dictate the devices to be used.

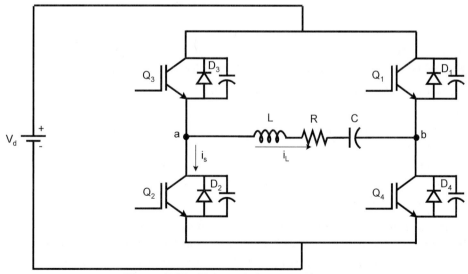

Figure 5.48 Voltage-fed resonant inverter

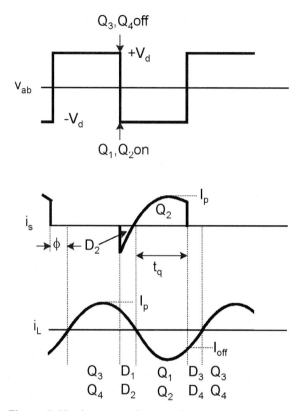

Figure 5.49 Inverter voltage and current waves

Consider, for example, the induction heating application where the load can be represented by an equivalent *RL* circuit. The magnitude of current i_L determines power consumption in the load. Figure 5.50 shows the inverter voltage and current waves for square-wave operation. If the inverter frequency is the same as the natural resonance load circuit frequency, the circuit will be purely resistive and the load current i_L will be maximum. Assume that the harmonic voltages are highly attenuated and the load current wave is nearly sinusoidal. The inverter frequency is intentionally made higher than the resonance frequency so that the load circuit is inductive and the load current lags the fundamental supply voltage, as shown in Figure 5.50. The advantage of lagging power factor operation is that the device switching loss can be practically eliminated. Consider, for example, that initially Q_3Q_4 are conducting and the positive load current with positive load voltage supplies active power to the load. At the trailing edge of a v_{ab} wave, Q_3Q_4 are turned off, and with a short time delay, Q_1Q_2 are turned on. At turn-off of Q_3Q_4, the respective snubber capacitors charge, impressing low *dv/dt* across the device and reducing its turn-off switching loss. When snubber capacitor charging is complete, the voltage v_{ab} becomes negative and the inductive load current is fed back to the source through diodes D_1D_2. Devices Q_1Q_2 are switched on when the bypass diodes are conducting, thus eliminating their turn-on switching loss. This means that the snubber capacitors are lossless, or energy recovery type, because the energy is fed back to the circuit instead of being lost. The elimination of switching loss increases inverter efficiency and the inverter can be operated at higher frequency. Note that the bypass diode need not be a fast-recovery type because it (D_2) gets a long time (t_q) for recovery, when its active device (Q_2) is conducting.

The load current can be regulated by varying the inverter frequency or the input dc voltage. The two half-bridges of the inverter can be phase-shifted to generate a quasi-square wave that can regulate the load current. However, in this case, the benefit of lossless snubbing is lost.

The resonant inverter is popularly used in resonant-link dc-dc converters. The high-frequency capacitor voltage can be rectified by a diode bridge rectifier and filtered to generate the output dc voltage. A high-frequency transformer can be connected across the capacitor that provides isolation and voltage level change. This configuration is usually known as a parallel-loaded resonance converter (PRC). Alternately, the transformer primary winding may be connected in series with the resonance circuit (defined as a series-loaded resonance converter (SRC)). For low-power output, a half-bridge inverter can be used where the output voltage is usually regulated by frequency control.

5.9 SOFT-SWITCHED INVERTERS

The disadvantages of hard switching effects, as discussed before, can be practically eliminated in a soft-switched inverter. In fact, the resonant inverter discussed above is an example of soft-switched inverter. Soft-switched inverters for motor drives were proposed a number of years ago [14], and the literature in this area is very rich. However, hardly any industrial drives use the technology. It will be discussed here for completeness of the subject.

5.9.1 Soft Switching Principle

The principle of soft switching using zero current switching (ZCS) and zero voltage switching (ZVS) is explained in Figure 5.51. It was shown before that in hard switching, the voltage and current overlap to create large switching loss. The main idea in soft switching is to prevent or minimize this overlap so that the switching loss is minimal. In ZCS, when the device turns on, the current build-up can be delayed with series inductance. Similarly, at turn-off, the current may be zero when device voltage builds up. A good example of ZCS is thyristor commutation in a phase-controlled converter. Zero voltage turn-on may occur when the bypass diode is conducting. Zero voltage turn-off occurs with a capacitive snubber which slows down the device voltage build-up. Both zero voltage turn-on and turn-off were explained for the resonant inverter in Figure 5.49. Note that slow voltage and current build-up during turn-off and turn-on, respectively, create less dv/dt- and di/dt-related problems, as discussed before.

5.9.1.1 Inverter Circuits

Varieties of soft-switched inverter circuits for motor drives have been proposed in the literature. In general, they may be dc link or ac link types. The dc link types can be classified as

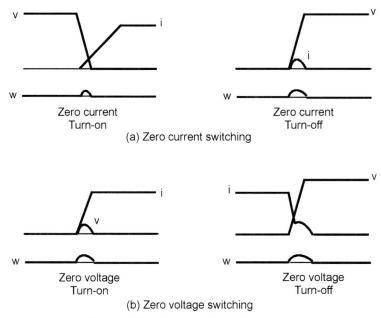

Figure 5.50 Soft switching of devices

resonant link dc (voltage-fed or current-fed) and resonant pole dc inverters. Voltage-fed resonant link inverters can be further classified as free resonance and quasi-resonance types. The ac link types can be classified as resonant link and non-resonant link types. There are further sub-classifications in these groups. In this section, we will briefly review three different topologies.

Resonant Link dc Converter (RLDC) — A voltage-fed inverter that operates on free-running resonance (tens of kilohertz) in the dc link is shown in Figure 5.52 [14]. The dc voltage source V_d, obtained from a battery or through a rectifier, is converted to sinusoidal voltage pulses (v_d) with zero voltage gap on the inverter bus through an $L_r C_r$ resonance circuit, as shown in the figure. The gap permits zero voltage soft switching of the inverter devices. The variable-voltage, variable-frequency waves at the motor terminal can be fabricated by delta modulation principle using the integral voltage pulses, as explained before. To establish the resonant bus voltage pulses, an initial current in the resonant inductor L_r is needed for the compensation of the resonant circuit loss and the reflected inverter input current. This current can be established to the appropriate value during the zero voltage gap by shorting the inverter devices. The bus voltage (higher than V_d) can be controlled either by a passive or active clamping technique. In the passive clamping method, a transformer with a series diode in the secondary is connected across the inductor L_r and, as the voltage tends to exceed a threshold value, the diode conducts pumping the inductor's trapped energy to the source. In this method, the peak dc link voltage v_d can be limited typically to $2.5V_d$. Using the active clamping method, as shown in the figure, the peak value can be limited typically to $1.5 V_d$. At the end of the zero voltage gap, when the desired initial current is reached, the selected devices of the inverter are opened to establish the output

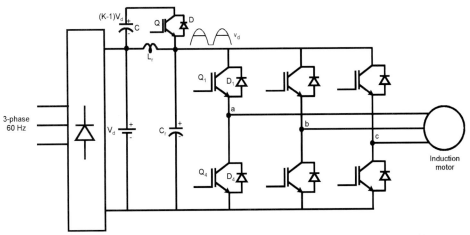

Figure 5.51 Resonant dc link inverter with active voltage clamping

phase voltages as dictated by the PWM algorithm. At the end of the resonant cycle, the inverter diodes provide a path for negative current in the gap interval. The active clamping circuit consists of a precharged capacitor in series with an IGBT-diode pair, as shown, and its operation can be explained as follows: On releasing the dc bus short, the link voltage swings toward its natural peak. However, on reaching the voltage level KV_d, the diode D conducts and clamps bus voltage at this level. With D conducting, Q is turned on in a lossless manner. The trapped inductor current linearly decays to zero and then becomes negative through Q to balance the capacitor charge. At current zero, Q is turned off to initiate the resonance again until the bus voltage falls to zero. The circuit has the disadvantage of voltage penalty on the devices, besides the need of extra components and additional loss in the resonance circuit.

Auxiliary Resonant Commutated Pole (ARCP) Converter — Instead of free-running resonance in the dc link, as discussed above, the inverter shown in Figure 5.53 activates resonance-assisted commutation at the instant of switching inverter devices [15]. The commutation principle of the inverter is somewhat similar to a McMurray inverter. A split capacitor power supply establishes the centerpoint 0 of dc voltage V_d. Each half-bridge inverter (defined as a pole) has a common resonant circuit consisting of the inductor L_r and shunt capacitors C, which are connected to the centerpoint 0 through an auxiliary ac switch S. The resonant snubber capacitor across each device is an energy recovery type. The circuit was originally proposed for a GTO inverter to minimize the switching loss. The inverter in general has three modes of operation. Consider the pole corresponding to phase a, and assume that the phase current i_a is positive, as indicated. Assume that i_a is reasonably large and is initially flowing through device Q_1. Q_1 can be turned off to commutate the current to diode D_4. The charging of bypass capacitor C will provide soft turn-off as well as fast commutation. If the current i_a is low, fast commutation will require the assistance of the resonance circuit. In such a case, activation of the resonance circuit by closing the switch S_1 boosts the current in Q_1, which helps fast commutation. If the phase

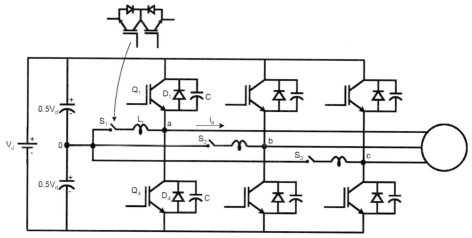

Figure 5.52 Auxiliary resonant commutated pole three-phase inverter

current is initially flowing through diode D_4, then for commutation to Q_1, the resonance circuit is activated to force current through bypass diode D_1 so that Q_1 can be turned on at zero voltage. The inverter operation with negative phase current is similar, except devices Q_4 and D_1 participate. The circuit does not have any voltage or current penalty on devices, but the disadvantages are extra circuit components and the need for a split capacitor power supply. The capacitor size is somewhat large to prevent fluctuation of the centerpoint voltage.

High-Frequency, Non-Resonant Link Converter — The high-frequency link cycloconverter, discussed in Chapter 4 (see Figure 4.38), requires an inverter to convert the input dc to a high-frequency supply (tens of kHz). Figure 5.54 shows the converter system [Ref. 12, Ch. 4] using an H-bridge, soft-switched inverter that operates on the ARCP principle. The inverter poles are phase-shifted to generate a quasi-square wave with zero voltage gap, as shown, for soft switching the HFAC-AC converter. Although ideally zero voltage switching is possible by the

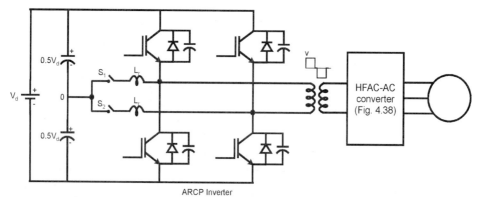

Figure 5.53 High-frequency, non-resonant link converter system with ARCP inverter

Dynamic and Regenerative Drive Braking

gap, the effect of leakage inductance should be carefully considered. The gap can be modulated to control cycloconverter output voltage with some harmonics improvement. Although the converter system uses many devices, there are advantages of using transformer coupling, such as Ohmic isolation from the primary, voltage level change, and availability of auxiliary power supplies coupled to the transformer. Note that the size of the high-frequency transformer is quite small.

5.10 DYNAMIC AND REGENERATIVE DRIVE BRAKING

In a variable-frequency drive, a machine may be subjected to mechanical or electrical braking for speed reduction. The mechanical brake may be applied externally or may be inherent by the load. For example, in a pump or fan-type drive, the load itself may exert mechanical braking torque to stop the motor. Electrical braking is classified into dynamic braking and regenerative braking. In either case, the motor is operated in the generating mode and the kinetic energy stored in the system inertia is converted to electrical energy. An induction motor can operate as a generator in supersynchronous speed, which is made possible by lowering the inverter frequency below the machine speed ($\omega_e < \omega_r$); (see Figure 2.7). The same condition may be attained when the machine drives an overhauling type of load. A synchronous motor can run as a generator if the power angle δ is transitioned from a negative to a positive value.

5.10.1 Dynamic Braking

In dynamic braking, the recovered electrical energy at the machine terminal is converted to dc through the inverter (the inverter acts as a rectifier) and is dissipated in a resistor. The principle of a dynamic brake is illustrated in Figure 5.55. With the machine acting as a generator and the inverter acting as a rectifier, the inverter input dc current reverses. Since the current cannot flow backward through the diode-rectifier, it charges the filter capacitor, raising the dc link voltage. A dynamic braking circuit, as shown, is connected to absorb the excess energy in the resis-

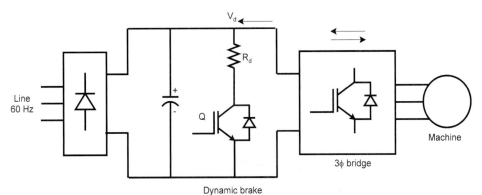

Figure 5.54 Three-phase PWM inverter with diode rectifier showing dynamic brake in the dc link

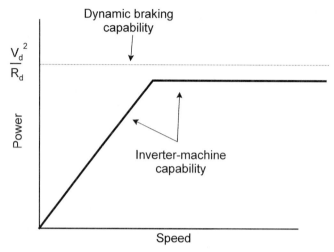

Figure 5.55 Dynamic braking capability of PWM inverter

tor R_d and limit the voltage V_d. Basically, it is a dc-dc buck converter, which is operated on a duty cycle basis to control the bus voltage within a hysteresis band. The dynamic braking capability of a PWM inverter is shown in Figure 5.56.

Since the dc link voltage V_d is constant, the maximum dynamic braking power that can be absorbed by the resistor R_d is given by V_d^2/R_d when the IGBT Q is fully on. The inverter-machine system's power capability is shown in Figure 5.56. Ideally, the curve is identical to that in motoring and is given by a fixed-slope straight line in the constant torque region and a horizontal line in the constant power region. At any speed, the duty cycle of the IGBT is adjusted so that the inverter-machine power matches with the power consumed in R_d. Low-power drives, such as servos, machine tools, and robotics where the recovered power is small, normally use this type of braking.

5.10.2 Regenerative Braking

In regenerative braking, electrical energy is recovered in the source to improve drive efficiency. If the inverter is directly supplied by a dc source, such as in an electric vehicle or subway drive, the braking power instead of dissipating in a resistor can easily flow back to the source. Continuous regenerative operation of a drive is possible if the load machine is a source of power, such as in a wind generation system. In a utility-supplied drive, a PWM rectifier in the front end can recover this energy, which will be explained later. In a four-quadrant speed control, the machine speed is brought down to zero by regenerative or dynamic braking, and then the phase sequence of the inverter is reversed to reverse the direction of machine rotation. High-power four-quadrant industrial drives invariably use regenerative braking.

5.11 PWM RECTIFIERS

In Chapter 3, we discussed diode and phase-controlled rectifiers. These converter circuits are simple, but the disadvantages are large distortion in line current and poor displacement power factor (DPF) (in the latter case), which make the power factor poor. To combat these problems, various power factor correction (PFC) techniques based on active wave shaping of the line current have been proposed. In this section, only a few important methods will be discussed.

5.11.1 Diode Rectifier with Boost Chopper

5.11.1.1 Single-Phase

This type of converter consists of a diode rectifier cascaded with a PWM boost chopper, as shown in Figure 5.57 for a single-phase supply. The boost chopper has essentially two functions: (1) it controls the line current to be sinusoidal at unity power factor, and (2) it regulates the capacitor voltage V_d, which should always be higher than the peak line voltage. Note that the converter system permits power flow to the load, but reverse power flow, that is, regeneration is not possible.

The circuit operation is simple. Assume, for example, the operation in the positive half-cycle of supply voltage v_s, and consider a small time segment when v_s can be represented by a dc

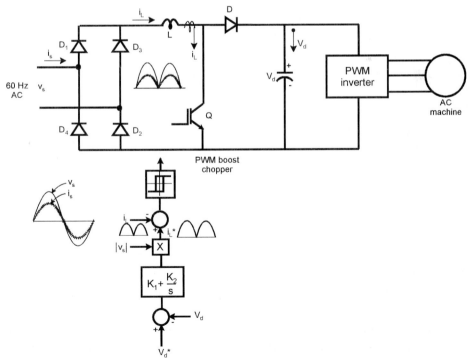

Figure 5.56 Single-phase diode rectifier with boost chopper for line current wave shaping

voltage. When the IGBT Q is turned on, the current will flow in diodes D_1 and D_2 through the boost inductor L. The current increment $+\Delta i_L$ is given as

$$+\Delta i_L = \frac{v_s}{L} t_{on} \quad (5.91)$$

where $+\Delta i_L$ = change in inductor current during the on interval (t_{on}) of Q. When Q is turned off, i_L will charge the capacitor through the diode D and decrease. The expression for the current decrement is

$$-\Delta i_L = \frac{V_d - v_s}{L} t_{off} \quad (5.92)$$

where t_{off} = off time of the chopper. The slope of the current i_L will vary depending on the voltage impressed across the inductance L. Since $|+\Delta i_L| = |-\Delta i_L|$, the switching frequency of the chopper is given as

$$f_{sw} = \frac{1}{t_{on} + t_{off}} = \frac{v_s (V_d - v_s)}{\Delta i_L L V_d} \quad (5.93)$$

where Δi_L = peak-to-peak ripple current. The equation indicates that for the same ripple current, higher switching frequency will require less inductance, but the switching loss will be higher. The additional cost and loss of efficiency are to be considered against the advantages for any application.

The control principle of the circuit is simple, and is included in Figure 5.57. The desired capacitor voltage V_d^* is compared with the actual voltage V_d, and the error through a proportional-integral (P-I) controller is multiplied by the absolute supply voltage wave $|v_s|$ to generate the boost chopper input current command i_L^*. The line is assumed as an ideal voltage source. The command i_L^* is compared with the actual current i_L within a hysteresis band (hysteresis-band PWM) to generate the chopper gate drive signal. The actual line current follows the same profile, but it is an ac wave. Commercial IC (integrated circuit) chips are available for such control.

There are a few additional advantages of the circuit beyond what were mentioned above. In a traditional diode-rectifier capacitor filter, the form factor of line current is very poor because of its pulsating nature. Therefore, the volt-amp requirement of the 60 Hz source should be high compared with that of the sinusoidal line current. This means that a UPS system or household power outlet can be more effectively utilized with this circuit. The utility system has the usual "brownout," or voltage sag, problem. With the help of a boost chopper, the dc link voltage can be maintained constant, thus adding reliability to the system. The PWM harmonic characteristics of the inverter will improve because the dc link ripple is less. Besides, constant and regulated V_d, irrespective of line voltage fluctuation, improves the drive performance by extending its constant torque region. Although the circuit is seldom used in motor drives, it is expected that as the cost of power electronics decreases and utility harmonic standards are strictly enforced, its application will increase.

5.11.1.2 Three-Phase

For a three-phase power supply, three single-phase units, as described above, can be used and their outputs can be paralleled across the output capacitor. However, the scheme shown in Figure 5.58, is much simpler and more economical [16]. Note that the boost chopper inductance, in this case, has been transferred to the line side and is distributed equally in the phases, as shown. The chopper operates at constant switching frequency, but with variable duty cycle to fabricate the phase current wave is as shown. When the IGBT Q is turned on, a symmetrical short circuit occurs at the rectifier input, the phase currents build up linearly, independently of each other, and the magnitude is proportional to the respective phase voltage amplitude. This means that the positive phase voltages cause positive currents through the upper diodes, which return as negative currents through the lower diodes that are caused by the negative phase voltages. When the IGBT turns off, the phase currents flow to the output capacitor until they fall to zero linearly, as shown. The discontinuous phase current pulses at high PWM frequency and the sinusoidal locus of the peak values can be filtered with a small LC filter to ideally obtain a sinusoidal average current with unity power factor. The line-side leakage inductance (not shown) and capacitor C constitute the LC filter. The duty cycle of the chopper can be modulated to control the sine current, and the voltage V_d can be controlled in the outer loop as discussed before. Note that each diode carries high peak current, and a highly pulsating current wave may create serious EMI problem.

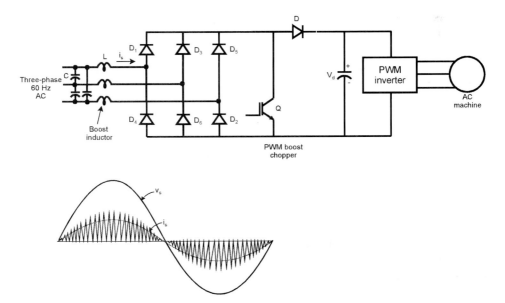

Figure 5.57 Three-phase diode rectifier with boost chopper for line current wave shaping

5.11.2 PWM Converter as Line-Side Rectifier

5.11.2.1 Single-Phase

It was mentioned before that a PWM inverter (either single-phase or three-phase) can be operated as a PWM rectifier; in other words, the power on the ac side can be converted to dc. Therefore, a similar unit can be connected to the line side to function as a PWM rectifier. Figure 5.59 shows a PWM converter system using a single-phase H-bridge converter on the line side as well as on the load side. Normally, the line-side converter functions as a rectifier, but it can also be operated as an inverter. In either case, it is controlled to maintain constant dc link voltage V_d. Note that from the viewpoint of a 60 Hz power balance, the dc link voltage will contain a 120 Hz ripple, which can be reduced by large capacitor size.

The PWM rectifier basically operates as a boost chopper (often called a boost rectifier) with bipolar voltage (ac) at the input, but unipolar voltage (dc) at the output, maintaining sinusoidal line current, as shown. The voltage V_d should be higher than the peak supply voltage ($V_d > \sqrt{2} V_s$). The control principle of a PWM rectifier for sinusoidal line current at unity power factor is shown in the lower part of Figure 5.59. The command dc link voltage V_d^* is compared with

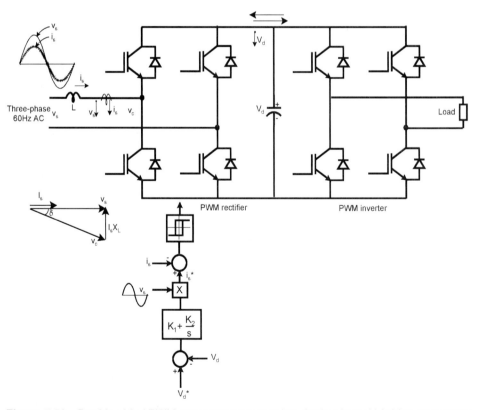

Figure 5.58 Double-sided PWM converter system using single-phase H-bridge converters

the actual V_d and the error signal through a P-I controller is multiplied with the line voltage wave v_s to generate the line current command i_s^*. The rectifier line current i_s can be controlled by the hysteresis-band PWM method so that the actual current tracks the command current. A phasor diagram on the line side is given for unity power factor operation. The CEMF voltage v_c is basically a PWM voltage wave fabricated from the dc link voltage V_d (similar to inverter PWM), and line inductance L helps to smooth the line current. The magnitude and phase of the fundamental component of v_c can be controlled by the rectifier. For V_s and I_s to be co-phasal, as shown in the figure, the CEMF V_c has to be larger and lag in phase to satisfy the phasor diagram. With higher switching frequency, the value of L and the corresponding reactive drop $I_s X_L$ can be decreased so that the V_c phasor approaches the V_s phasor. For unity power factor operation in regenerative mode, the I_s phasor reverses, which causes the reversal of the $I_s X_L$ phasor so that the δ angle reverses. Although unity line power factor operation is discussed here, the power factor can also be leading or lagging, and this will be discussed later for a three-phase system. The H-bridge units can be replaced by half-bridge converters, and the operation principles will remain the same.

5.11.2.2 Three-Phase

The principle of a single-phase, double-sided converter system, as discussed above, can be extended to a three-phase system. Figure 5.60 shows a double-sided PWM converter system for a motor drive application using three-phase, two-level converters.

Most of the discussions for a single-phase system (Figure 5.59) are also valid for a three-phase system. The drive configuration is extremely important for industrial applications. The essential features can be summarized as follows:

- **Four-Quadrant Operation** – In normal motoring mode of the drive, power flows to the motor, and the line-side converter operates as a rectifier, whereas the load-side converter operates as an inverter. In regenerative braking mode, their roles are reversed, that is, the load-side converter operates as a rectifier, whereas the line-side converter operates as an inverter. The direction of rotation of the machine will be determined by the phase

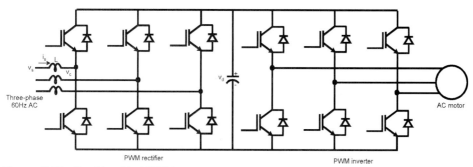

Figure 5.59 Double-sided PWM converter system using three-phase, two-level converters (12-switch)

sequence of the load-side converter. The system can continuously regenerate power if the machine is a generator, such as in a wind generation system.

- **Line Voltage Sag Compensation** – Since the line-side converter operates in boost rectifier mode to maintain the dc link voltage V_d constant irrespective of line voltage, the converter system improves reliability of the drive by line voltage sag compensation.
- **Programmable Line Power Factor** – In addition to unity line power factor operation, as discussed for Figure 5.59, the power factor can be made programmable leading or lagging. This requires further explanation. Figure 5.61 shows the line-side phasor diagram at leading power factor condition. From the figure, the active and reactive power expressions on the line side can be given, respectively, as

$$P = 3V_s I_s \cos\phi \tag{5.94}$$

$$Q = 3V_s I_s \sin\phi \tag{5.95}$$

where V_s and I_s are the line-side phasors, and φ = leading power factor angle, as indicated in Figure 5.61. From the phasor diagram, we can write the expressions

$$V_L = \omega L I_s \tag{5.96}$$

$$V_L \sin\left(\frac{\pi}{2} - \phi\right) = A = V_c \sin\delta \tag{5.97}$$

$$V_L \cos\left(\frac{\pi}{2} - \phi\right) = B = V_c \cos\delta - V_s \tag{5.98}$$

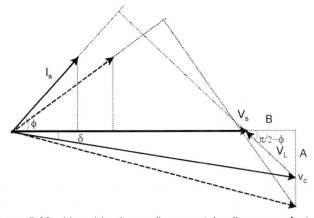

Figure 5.60 Line-side phasor diagram at leading power factor

PWM Rectifiers

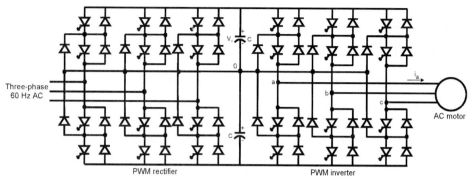

Figure 5.61 Double-sided PWM converter system using three-phase three-level converters (24-switch)

where L = line current smoothing inductor, ω = line frequency, V_c = CEMF phasor, and δ = angle between V_s and V_c phasors. Substituting Equations (5.96)–(5.98) in (5.94) and (5.95) yields

$$P = \frac{3V_s V_c}{\omega L} \sin \delta \tag{5.99}$$

$$Q = \frac{3V_s}{\omega L}(V_c \cos \delta - V_s) \tag{5.100}$$

Equations (5.99) and (5.100) indicate that the inverter CEMF V_c and the δ angle can be controlled to control P and Q. If it is desired to maintain Q as constant but vary the active power P, phasors $I_s \sin \varphi$ and $V_c \cos \delta$ should remain constant. This is indicated by the dotted phasor diagram in Figure 5.61. Note that the rectifier input leading VAR is higher than that of the line side because the inductor L absorbs lagging VAR. The phasor diagram for the lagging power factor condition can be drawn easily. The discussion indicates that the line-side converter can also be operated for static VAR compensation, which will be discussed later.

Note that the dc link voltage V_d should be adequately high for sinusoidal PWM synthesis of a CEMF wave. This means

$$\sqrt{2}V_s \leq 0.5 V_d \tag{5.101}$$

which means

$$1.63 V_L \leq V_d \tag{5.102}$$

where the line voltage $V_L = \sqrt{3} V_s$. For any saturation of PWM, the line current will be heavily distorted with lower order harmonics.

For high-voltage, high-power drives, three-level converters can substitute for the two-level converters in Figure 5.60. Figure 5.62 shows such a converter system using GTOs. As the GTO power level is increasing, multi-megawatt, three-level converter systems are being developed to replace phase-controlled cycloconverters for applications such as rolling mill drives. Double-sided PWM converter systems have also been considered for utility systems asynchronous inter-tie between 60 Hz and 50 Hz systems.

5.12 STATIC VAR COMPENSATORS AND ACTIVE HARMONIC FILTERS

The analysis with the phasor diagram in Figure 5.61 indicates that a PWM rectifier can operate at a programmable leading or lagging power factor. If the active power $P = 0$, then from Equations (5.99) and (5.100), the reactive power expression is given as

$$Q = 3V_s I_s = \frac{3V_s}{\omega L}(V_c - V_s) \tag{5.103}$$

Figure 5.63 shows the phasor diagrams for leading and lagging VAR conditions. This means that in Figure 5.60, if the inverter is disconnected and the rectifier loss is ignored, the PWM rectifier with dc link capacitor voltage can act as a static VAR compensator (SVC). The circuit is shown in Figure 5.64 with a thyristor phase-controlled rectifier (nonlinear load).

As a leading VAR compensator, the circuit acts like a three-phase variable capacitor load, whereas as a lagging VAR compensator, it acts like a three-phase variable inductor load. It is well-known that ideally at no-load operation of a synchronous machine, the excitation current can be controlled to operate it as a leading or lagging VAR compensator. The PWM rectifier is a static version of a synchronous machine VAR compensator. In this mode, large current will circulate in the PWM rectifier circuit, but the capacitor current will remain zero (except harmonics), that is, voltage V_d will remain constant. This is because in a three-phase balanced and sinusoidal system, instantaneous power is zero with reactive load. A finite power loss in the converter will be supplied by the line, making its power factor non-zero. In Figure 5.64, the leading VAR compensator compensates the lagging current I_Q taken by the phase-controlled converter so that the line displacement factor is unity.

Figure 5.65 shows a simplified control block diagram of an SVC for induction motor reactive current compensation. The control is somewhat complex and will be explained in detail in Chapter 8. Basically, the capacitor voltage V_d is controlled in the outer loop by feedback control of the compensator loss component of current ΔI_P. The motor reactive current I_{QM}^* is measured,

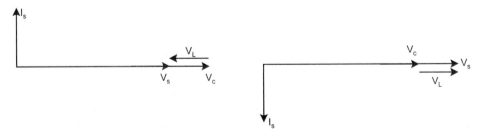

Figure 5.62 Phasor diagrams (a) Leading VAR, (b) Lagging VAR

Static VAR Compensators and Active Harmonic Filters

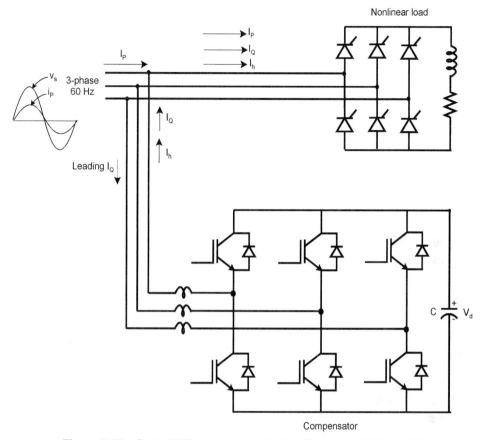

Figure 5.63 Static VAR generator and active filter using PWM rectifier

and it controls the SVC's reactive current I_{QC} by feedback so that line reactive current is zero. Very large (multi-MW) SVCs using GTO multi-stepped converters have been installed in the utility system for voltage control [17]. They are also important elements for a future flexible ac transmission system (FACT).

The SVC in Figure 5.64 can also be operated as an active harmonic filter if the PWM switching frequency is sufficiently high. Consider, for example, that the nonlinear load consumes a 5th harmonic current, which is to be compensated by the active filter. In this case, the current-controlled converter should generate a 5th harmonic current that is equal in magnitude, but opposite in phase. In fact, ideally, the distorted current wave in the nonlinear load can be sensed and compensated entirely by the filter. Figure 5.66 shows a simplified control block diagram of an active filter. The harmonic currents of a nonlinear load (phase-controlled converter, or PCC) are sensed through a high-pass filter and added to the active phase current command generated by the voltage control loop. The resulting command current is controlled in the filter by the hysteresis-band PWM method.

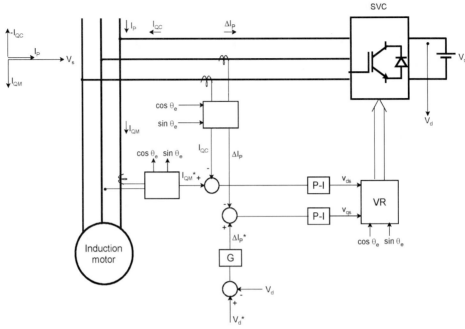

Figure 5.64 Simplified control block diagram of static VAR compensator for induction motor load

In fact, if the load is unbalanced and consumes a negative sequence current, the converter can compensate this unbalance. The functions of VAR compensation, active filtering, and negative sequence compensation can be combined into one compensator to control the line current to be balanced and sinusoidal at unity power factor. Such a compensator is often defined as a universal power line conditioner (UPLC).

Active filters are very attractive for existing installations of PCCs. However, their application is not common because of the high cost of converters. More strict enforcement of harmonic standards will promote their application. Of course, converters with a built-in power factor and harmonic correction, as discussed above, will eliminate their needs.

5.13 INTRODUCTION TO SIMULATION—MATLAB/SIMULINK

When a new converter circuit is developed, or a control strategy of a converter or drive system is formulated, it is often convenient to study the system performance by simulation before building the breadboard or prototype. The simulation not only validates the system's operation, but also permits optimization of the system's performance by iteration of its parameters. Besides control and circuit parameters, the plant parameter variation effect can be studied. For example, the rotor resistance (R_r) variation effect on a detuned vector-controlled drive or accuracy of speed (ω_r) estimation (discussed in Chapter 8) can be studied. Valuable time is thus saved in the development and design of a product, and the failure of components of poorly

Introduction to Simulation—MATLAB/SIMULINK

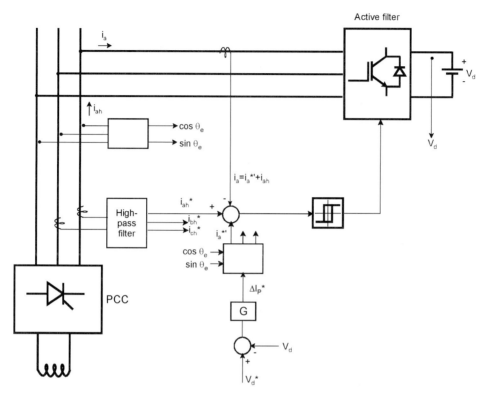

Figure 5.65 Simplified control block diagram of active harmonic filter

designed systems can be avoided. The harmonic spectra of simulated voltage and current waves can be analyzed by FFT to verify their harmful effects and possible design of remedial measures. In some cases, the simulation program helps to generate real-time controller software codes for downloading to a microprocessor or digital signal processor (DSP).

Fortunately, a large number of PC-based, user-friendly digital simulation programs are available for the study of power electronic systems. Examples include SIMULINK, PSPICE, SABER, EMTP, SIMNON, ACSL, MATRIX$_X$, etc. In this section, we will briefly outline the salient features of SIMULINK, which is possibly the most commonly used program for power electronic and drive systems. Then, as examples, a simple voltage-fed inverter and an induction motor dynamic model simulation will be illustrated. In Chapter 11, SIMULINK simulation of a vector-controlled induction motor drive incorporating fuzzy control will be discussed. The reader is directed to reference [18] for details of SIMULINK.

SIMULINK is basically a user-friendly, general digital simulation program of nonlinear dynamical systems which works in the MATLAB environment. A continuous or discrete time system and multi-rate system can also be used. Although basically it is the MATLAB (MathWorks) program, the user has a graphical interface where he or she builds the system with the help of subsystem blocks. In the beginning, the user must define the mathematical model of the system by differential and algebraic equations and express it in state variable or transfer function form. A library of templates or function blocks, such as "sources," "sinks," "discrete," "linear," "nonlinear," and "connections," can be used in the simulation. The power semiconductor device is normally simulated by the element "switch," which is an element in the nonlinear function blocks.

Figure 5.67 shows the simple simulation block diagram for a three-phase, two-level PWM inverter. Each leg of the inverter is represented by a "switch" which has three input terminals and one output terminal. The output of a switch (v_{ao}, v_{bo}, or v_{co}) is connected to the upper input terminal ($+0.5V_d$) if the PWM control signal (middle input) is positive. Otherwise, the out-

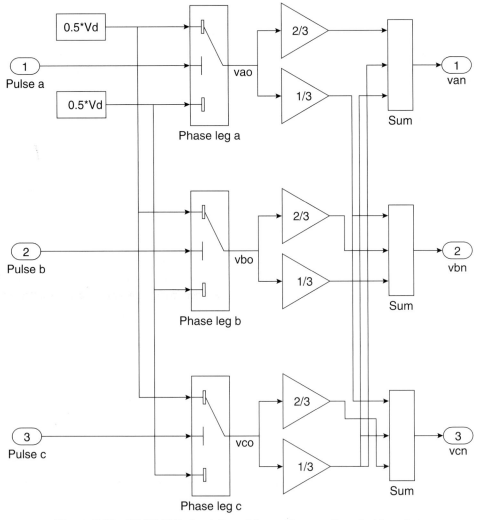

Figure 5.66 SIMULINK simulation of three-phase voltage-fed inverter

put is connected to the lower input terminal ($-0.5 V_d$). The output, or v_{ao}, voltage thus oscillates between $+0.5V_d$ and $-0.5V_d$, which is characteristic of a pole of an inverter. The output phase voltages are constructed by the following equations:

$$v_{an} = \frac{2}{3}v_{ao} - \frac{1}{3}v_{bo} - \frac{1}{3}v_{co} \tag{5.14}$$

$$v_{bn} = \frac{2}{3}v_{bo} - \frac{1}{3}v_{ao} - \frac{1}{3}v_{co} \tag{5.15}$$

$$v_{cn} = \frac{2}{3}v_{co} - \frac{1}{3}v_{ao} - \frac{1}{3}v_{bo} \tag{5.16}$$

The simulation of these equations is evident from the figure.

Figure 5.68 shows the flux model simulation of the d^e-q^e model of the induction motor, which was discussed in Chapter 2. The model receives the input voltages v_{qs} and v_{ds} and frequency ω_e and solves the output currents i_{qs} and i_{ds} with the help of flux linkage equations. At the input, the inverter output voltages v_{an}, v_{bn}, and v_{cn} are converted to v_{qs} and v_{ds} (block "abc-syn") by the following equations:

$$v_{qs}^s = \frac{2}{3}v_{an} - \frac{1}{3}v_{bn} - \frac{1}{3}v_{cn} \tag{2.72}$$

$$v_{ds}^s = -\frac{1}{\sqrt{3}}v_{bn} + \frac{1}{\sqrt{3}}v_{cn} \tag{2.73}$$

$$v_{qs} = v_{qs}^s \cos\omega_e t - v_{ds}^s \sin\omega_e t \tag{2.74}$$

$$v_{ds} = v_{qs}^s \sin\omega_e t + v_{ds}^s \sin\omega_e t \tag{2.75}$$

Similarly, the output block "2_3" uses the inverse equations given by Equations (2.76)–(2.77) and (2.69)–(2.71). The torque and speed are calculated by Equations (2.160) and (2.105), respectively. The implementation of the subsystem block "F_{qs}" in detail is shown in Figure 5.69, which simulates the Equations (2.156).

5.14 SUMMARY

This chapter gave a broad review of different types of voltage-fed converters and their principles of operation. The converter topologies included half-bridge, center-tap, full or H-bridge, and three-phase bridges with two levels and three levels. The three-level, or multi-level, inverter is also known as a neutral-point-clamped inverter, which is used for high-voltage, high-power (multi-MW) applications. Multi-stepped inverters with the help of transformers can have 12, 18, 24, etc. steps. This type of converter operates with devices that switch at line supply fre-

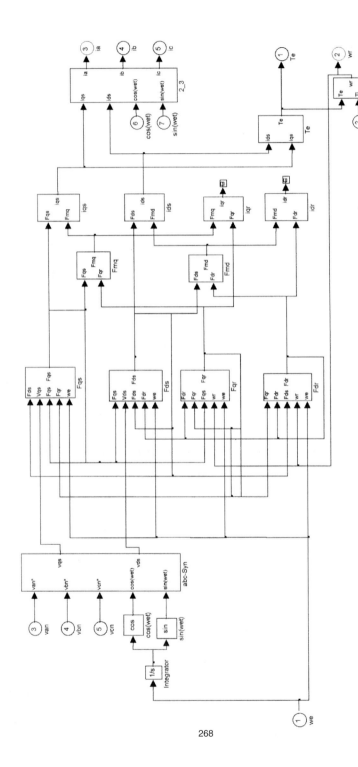

Figure 5.67 SIMULINK simulation of d^e-q^e flux linkage model of induction motor

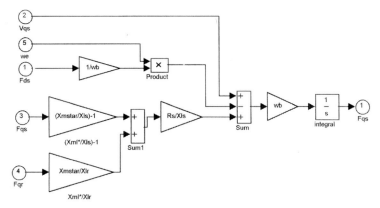

Figure 5.68 SIMULINK simulation details of the block "F_{qs}"

quency and therefore is suitable for GTO-like devices which have high switching loss. A 10 MW application was illustrated with an 18-step inverter.

PWM techniques are extensively used for the control of voltage-fed inverters. Among different PWM methods, sinusoidal and space-vector PWM were emphasized. Both undermodulation and overmodulation regions were discussed. The harmful effects of hard-switched PWM converters were discussed. The principle of soft switching and a few converter topologies with soft switching were covered. Power factor correction with a boost chopper and PWM rectifier was discussed. Both a static VAR compensator and active filter with control principle were covered. Finally, simulation was discussed, with a brief review of MATLAB/SIMULINK, which included two example simulations. Simulation will be further discussed in Chapters 11 and 12. It is needless to mention that voltage-fed converters constitute the most important class in power electronic systems and therefore justify elaborate discussion.

REFERENCES

1. B. D. Bedford and R. G. Hoft, *Principles of Inverter Circuits*, Wiley, New York, 1964.
2. N. Mohan, T. M. Undeland, and W. P. Robbins, *Power Electronics*, John Wiley, New York, 1995.
3. M. P. Kazmierkowski and H. Tunia, *Automatic Control of Converter-Fed Drives*, Elsevier, 1994.
4. B. K. Bose, "Power electronics—a technology review", *Proc. of the IEEE*, vol. 80, pp. 1303–1334, August 1992.
5. B. K. Bose, "Recent advances and trends in power electronics and drives", *Proc. Of NORPIE Workshop*, Helsinki, pp.170–182, 1998.
6. L. H. Walker, "10-MW GTO converter for battery peaking service", *IEEE Trans. Ind. Appl.*, vol. 26, pp. 63–72, Jan./Feb. 1990.
7. J. Holtz, "Pulse width modulation for electric power conversion", *Proc. of the IEEE*, vol. 82, pp. 1194–1214, August 1994.
8. B. K. Bose and H. A. Sutherland, "A high performance pulse-width modulator for an inverter-fed drive system using a microcomputer", *IEEE Trans. Ind. Appl.*, vol. 19, pp. 235–243, Mar./Apr. 1983.
9. J. Holtz, W. Lotzkat, and M. Khambadkone, "On continuous control of PWM inverters in the overmodulation range including the six-step mode", *IEEE Trans. Power Electronics*, , vol. 8, pp. 546–553, October 1993.

10. D. C. Lee and G. M. Lee, "A novel overmodulation technique for space vector PWM inverters", *IEEE Trans. Power lectronics*, vol. 13, pp. 1144–1151, Nov. 1998.
11. J. O. P. Pinto, B. K. Bose, L. E. Borges, and M. P. Kazmierkowski, "A neural network based space vector PWM controller for voltage-fed inverter induction motor drive", *IEEE IAS Annu. Meet. Conf. Rec.*, pp. 2614–2622, 1999.
12. A. Nabae, T. Isao, and A. Hirofumi, "A new neutral-point-clamped PWM inverter", *IEEE Trans. Ind. Appl.*, vol. 17, pp. 512–523, Sept./Oct. 1987.
13. P. J. Link, "Minimizing electric bearing currents in ASD systems", *IEEE Industry Appl. Magazine*, pp. 55–66, July/Aug. 1999.
14. D. M. Divan et al., "Design methodologies for soft switched inverters", *IEEE IAS Annu. Meet. Conf. Rec.*, pp. 758–766, 1988.
15. R. W. De Donker and J. P. Lyons, "The auxiliary resonant commutated pole converter", *IEEE IAS Annu. Meet. Conf. Rec.*, pp. 1228–1235, 1990.
16. A. R. Prasad, P. D. Ziogas, and S. Manias, "An active power factor correction technique for three phase diode rectifiers", *IEEE PESC Rec.*, pp. 58–66, 1989.
17. M. Hirakawa, N. Eguchi, M. Yamamoto, S. Konishi, and Y. Makano, "Self-commutated SVC for electric railways", *Int'l. Conf. on Power Elec. and Drives Syst. Rec.*, Singapore, pp. 732–737, 1995.
18. Math Works, Inc., "SIMULINK User's Guide", Version 2, Jan. 1997.
19. N. Mohan et. al., "Simulation of power electronic and motion control systems—an overview", *Proc. Of the IEEE*, vol. 82, pp. 1287–1302, August 1994.
20. T. Sukegawa, K. Kamiyama, K. Mizuno, T. Matsui, and T. Okuyama, "Fully digital, vector controlled PWM VSI-fed ac drives with an inverter dead-time compensation strategy", *IEEE Trans. Ind. Appl.*, vol. 27, pp. 552–559, May/June 1991.

CHAPTER 6

Current-Fed Converters

6.1 INTRODUCTION

The general principles of current-fed converters using phase control and line commutation techniques were already discussed in Chapter 3. In this chapter, the concept will be expanded further and additional types of current-fed converters will be introduced. Again, the term "inverter" will be used frequently, although the same circuit can be used either as an inverter or a rectifier. A current-fed or current-source inverter (CFI or CSI), as the name indicates, likes to see a stiff dc current source (ideally with infinite Thevenin impedance) at the input, which is in contrast to a stiff voltage source (zero Thevenin impedance), which is desirable in a voltage-fed inverter. A current-fed inverter should not be confused with the current-controlled, voltage-fed inverter discussed in Chapter 5. A variable voltage source can be converted to a variable current source by connecting a large inductance in series and controlling the voltage within a feedback current control loop. The variable dc voltage can be obtained from a utility supply through a phase-controlled rectifier, or from a rotating excitation-controlled ac generator through a diode rectifier, or from a battery-type power supply through a dc-dc converter. With a stiff dc current source, the output ac current waves are not affected by the load condition (just as voltage waves in a voltage-fed inverter are not affected by the load). The power semiconductor devices in a current-fed inverter must withstand reverse voltage, and therefore, standard asymmetric voltage blocking devices, such as power MOSFETs, BJTs, IGBTs, MCTs, IGCTs, and GTOs cannot be used. Symmetric voltage blocking GTOs and thyristor devices should be used. Of course, forward-blocking devices can be used with series diodes. It can be shown that in many respects, current-fed inverters are somewhat dual to voltage-fed inverters.

The application of current-fed converters may include the following:

- Speed control of large power induction and synchronous motors
- Variable-frequency starting of 60 Hz wound-field synchronous motors
- High-frequency induction heating
- Superconducting magnet energy storage (SMES)
- DC motor drives
- Static VAR compensators
- Active harmonic filters

In this chapter, we will study the different types of current-fed converters, forced-commutation and load-commutation principles, PWM techniques, various application considerations, and finally, we will compare current-fed converters with voltage-fed converters. Again, power semiconductor devices will be considered as ideal on-off switches.

6.2 GENERAL OPERATION OF A SIX-STEP THYRISTOR INVERTER

Six-step thyristor inverters are extremely important circuit configurations for high-power, wound-field, synchronous motor drives. First, we will study the general operation principles of this type of inverter with induction and synchronous machine loads. Figure 6.1 shows the power circuit for a current-fed inverter, which is being supplied by a phase-controlled rectifier on the line side. A variable dc link voltage V_r is generated by the phase control of the rectifier, and it is converted to a current source I_d by connecting a series inductance L_d and providing a feedback

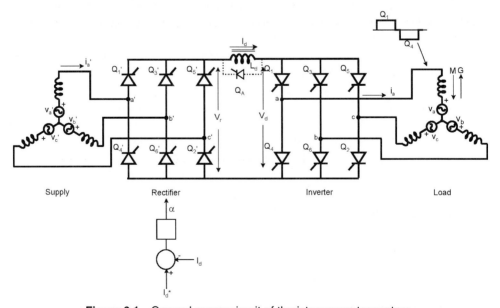

Figure 6.1 General power circuit of thyristor converter system

General Operation of a Six-Step Thyristor Inverter

current control loop as shown. The current magnitude, as desired, can be established by the command current I_d^*. Although an infinite value of L_d is desirable for an ideal ripple-free current source, size and cost constraints limit the practical L_d within a reasonable value, thus permitting some amount of ripple. Ignoring commutation considerations for the present, the inverter circuit appears to be identical to that of the rectifier. The load of the inverter is an induction or synchronous machine, which can be represented by phase CEMF in series with equivalent leakage inductance (see Figures 2.28(b) and 2.35). The power circuit thus appears symmetrical about the dc link. An auxiliary thyristor Q_A is shown connected across the inductor L_d in the reverse direction, which will be explained later. The dc current I_d is switched through the inverter thyristors so as to establish three-phase, six-stepped, symmetrical line current waves as shown in Figure 6.2. The load or line current wave is given by the Fourier series

$$i_a = \frac{2\sqrt{3}I_d}{\pi}\left[\cos\omega t - \frac{1}{5}\cos 5\omega t + \frac{1}{7}\cos 7\omega t + \ldots\right] \tag{6.1}$$

where the peak value of the fundamental component is $2\sqrt{3}I_d/\pi$. Each thyristor conducts for $2\pi/3$ angle, and at any instant, one upper device and one lower device remain in conduction. The operation is similar to the rectifier in the front end. The dc link current is considered harmonic-free ($L_d \to \infty$), and the commutation effect from thyristor-to-thyristor is ignored. The inverter input voltage V_d can be constructed by the amplitude between the two phase voltage envelopes, as indicated in the figure. At steady state, $V_r = V_d$ if the resistance of L_d is neglected. For a variable-speed drive motor, the inverter can be operated at variable frequency and adjustable magnitude of dc current I_d.

The current waves in Figure 6.2 are drawn for maximum power inversion, that is, firing angle $\alpha = \pi$ when the fundamental component of phase current is out of phase with the respec-

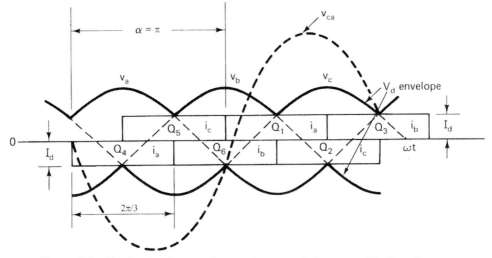

Figure 6.2 Idealized voltage and current waves of six-stepped thyristor inverter

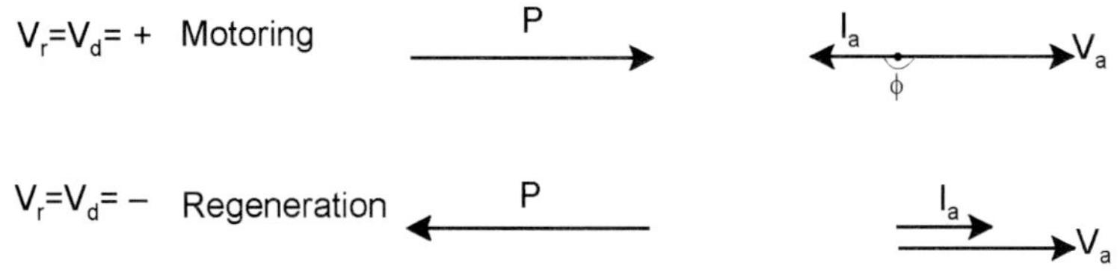

Figure 6.3 Phasor diagrams at unity displacement power factor

tive phase voltage. At this condition, the line-side converter is acting as a rectifier, supplying power to the inverter. Since I_d is always positive, both V_r and V_d are also positive. If firing angle α of the inverter is shifted to zero, it acts as a rectifier, pumping power in the dc link from the machine, which acts as a generator. At this condition, the line-side converter acts as an inverter (making both V_r and V_d negative) and regenerated power is fed to the line.

Figure 6.3 shows the phasor diagrams in motoring and regeneration modes. Note that the line-side converter is always line-commutated irrespective of the rectifier or inverter mode of operation. This line commutation also implies lagging VAR demand at the input of the converter. From the same considerations, we can infer that the load-side converter can be commutated by the load CEMF (load commutation) if it can supply lagging VAR to the converter, or the load power factor is leading. This condition turns off the outgoing thyristor by applying a negative voltage segment of the CEMF wave. Verification of this by sketching waveforms is left as an exercise. The conclusion we can draw is that an over-excited synchronous machine load at a leading power factor can provide load commutation, whereas an induction motor load with a lagging power factor requires some type of forced commutation.

6.2.1 Inverter Operation Modes

The inverter firing angle α can vary ideally in the range 0 to 2π with respect to the CEMF wave. The following general operation modes can be obtained, as explained in Figure 6.4.

6.2.1.1 Mode 1: Load-Commutated Rectifier ($0 \leq \alpha \leq \pi/2$)

This mode corresponds to the familiar line-commutated rectifier mode of operation, except that here, the commutation is performed by the load instead of the line. Figure 6.4(a) shows the phase voltage and current waves for $\alpha = \pi/4$. When the incoming thyristor Q_1 in Figure 6.1 is fired, the outgoing thyristor Q_5 will be impressed with a negative segment of anode voltage v_{ca}, which is shown by the dashed curve (see Figure 6.2), and therefore will turn off by load commutation. The fundamental component of i_a in Figure 6.4(a) lags the voltage wave by $\pi/4$, and the corresponding phasor diagram is shown on the right side. The active power will flow from the load to the dc link, which will be pumped back to the line by the line-side converter acting as an inverter, as explained before. The dc link voltages are given by the following general expressions:

For the inverter:

$$V_d = -V_{do} \cos\alpha \tag{6.2}$$

General Operation of a Six-Step Thyristor Inverter

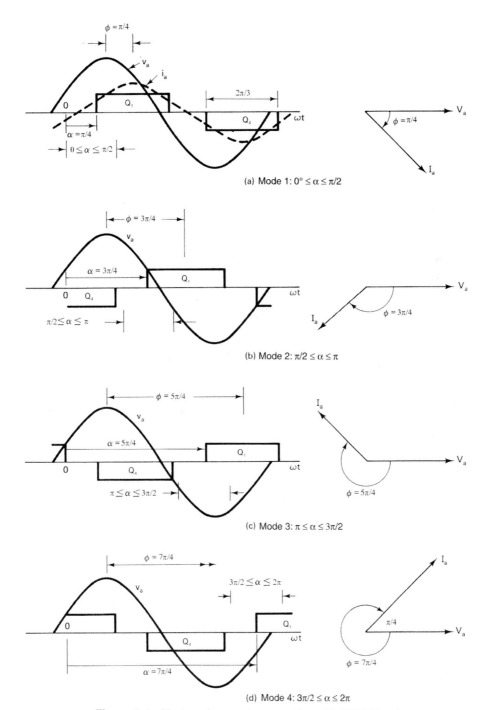

Figure 6.4 Modes of converter operation with CEMF load

For the rectifier:

$$V_r = V_{ro} \cos\alpha' = V_d = V_{do} \cos(\pi - \alpha) \qquad (6.3)$$

where α' = firing angle of the rectifier, and it is assumed that the supply and load emfs are equal. For $\alpha = \pi/4$, angle $\alpha' = 3\pi/4$, giving both V_d and V_r as negative values. At this condition, as explained before, the load will supply lagging VAR to the inverter (i.e., leading VAR will be supplied to the load from the inverter). This mode can be considered as over-excited synchronous machine operation in regenerative mode.

6.2.1.2 Mode 2: Load-Commutated Inverter ($\pi/2 \leq \alpha \leq \pi$)

This mode is explained in Figure 6.4(b) for a typical angle $\alpha = 3\pi/4$. The outgoing thyristor Q_5 is commutated by the load since v_{ca} is negative in this range. The active power flows to the load and the dc link voltages are positive, as determined by Equations (6.2) and (6.3). The load is required to operate at leading power factor, as in the previous mode. This mode can therefore be considered as the motoring mode of a synchronous machine operating at over-excitation.

6.2.1.3 Mode 3: Force-Commutated Inverter ($\pi \leq \alpha \leq 3\pi/2$)

By delaying the inverter firing angle further beyond π, the advantage of load commutation is lost since the outgoing thyristor Q_5 is impressed with a positive v_{ca} voltage. Therefore, for successful operation in this range, some type of forced commutation is required. The phasor diagram at the typical phase angle $\varphi = 5\pi/4$ indicates that the active power flows to the load, giving motoring operation, and lagging VAR is consumed by the load. This mode therefore corresponds to induction motor operation.

6.2.1.4 Mode 4: Force-Commutated Rectifier ($3\pi/4 < \alpha < 2\pi$)

In this mode, as in Mode 3, the inverter requires forced commutation. The phasor diagram in Figure 6.4(d) indicates rectifier operation, with the load demanding lagging VAR. This mode can therefore be identified as induction motor operation with regeneration. All four modes are summarized in Figure 6.5.

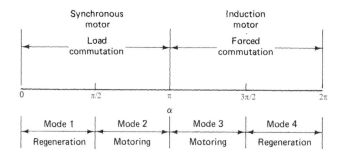

Figure 6.5 Summary of modes for ac machine operation

6.3 LOAD-COMMUTATED INVERTERS

In the previous section, we discussed different operation modes of current-fed inverters with a CEMF-type load. These modes are also generally valid for single-phase inverters with a passive-type load. The concepts are further expanded in this section.

6.3.1 Single-Phase Resonant Inverter

Let us consider a single-phase, H-bridge thyristor inverter with passive *RL* load as shown in Figure 6.6. The inverter is fed by a phase-controlled rectifier (single- or three-phase). A capacitor *C* of sufficiently high value is connected across the load so that at fundamental frequency, the effective load has a leading power factor. The purpose of the capacitor is to have load commutation of the thyristors. The circuit is generally used for high-frequency induction heating applications. Figure 6.7(a) shows the inverter load voltage and current waves, assuming that the dc link inductance is very high and has near perfect filtering of harmonic currents by the capacitor. The thyristor pairs Q_1Q_2 and Q_3Q_4 are switched alternately for π angle to impress a square current wave at the output. The fundamental component of load current leads the nearly sinusoidal load voltage wave by $\beta°$. When thyristor pair Q_1Q_2 is switched on, outgoing pair Q_3Q_4 is impressed with a negative voltage segment for duration $\beta°$, causing load commutation. Since $\beta = \omega t_q$, the minimum value of β should be sufficient to turn off the thyristors during time t_q. The thyristors can be replaced by symmetric blocking GTOs or IGBTs with series

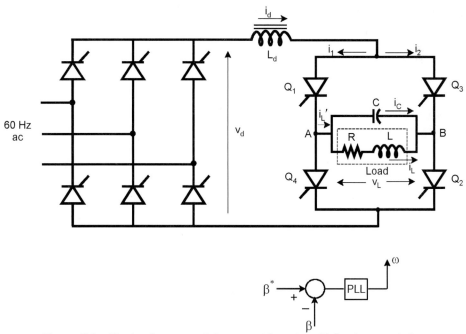

Figure 6.6 Single-phase parallel resonant inverter with load commutation

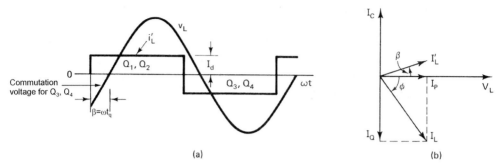

Figure 6.7 (a) Load voltage and current waves, (b) Phasor diagram

diodes. Like a thyristor converter, a load-commutated converter can be considered soft-switched with zero-current switching. Figure 6.7(b) shows the fundamental frequency phasor diagram where the load rms voltage V_L is drawn as the reference phasor. The load current I_L is at lagging power factor angle φ and can be resolved into a reactive component I_Q and an active component I_P. The leading capacitor current I_C overcomes the lagging current I_Q so that the effective DPF $\cos \beta$ becomes leading. The effective load can be considered as a parallel resonant circuit and the inverter frequency is higher than the resonance frequency ($\omega > \omega_r$) so that the effective DPF is leading. Figure 6.6 can be considered as dual to the voltage-fed series resonant inverter shown in Figure 5.49.

6.3.1.1 Circuit Analysis

The general circuit equations of the inverter can be given as

$$v_L = i_L R + L \frac{di_L}{dt} \tag{6.4}$$

$$i_C = C \frac{dv_L}{dt} \tag{6.5}$$

$$i_L' = i_L + i_C \tag{6.6}$$

$$i_L' = i_1 - i_2 \tag{6.7}$$

$$i_d = i_1 + i_2 \tag{6.8}$$

$$v_d - v_L = L_d \frac{di_d}{dt} + i_d R_d \tag{6.9}$$

where R_d is the resistance of inductor L_d. The equations can be expressed in state-variable form and solved for transient and steady-state conditions with the help of a computer program. Such a generalized model-based analysis permits complete design of the inverter, including the effect of

Load-Commutated Inverters

harmonics. The control loops of the converter system can be added for completeness of the design. We will attempt here an approximate steady-state analysis, assuming that L_d is lossless and of infinite value and that the load is highly inductive ($\omega L \gg R$). Again, the operation's near load resonance (small β) will be considered and the harmonic effect on the load will be ignored. The series RL load can be resolved into parallel L_1 and R_1 components so that reactive component I_Q flows through L_1 and active component I_P flows through R_1. The load impedance Z_L can be written in the form

$$Z_L = R + j\omega L = \frac{R_1 j\omega L_1}{R_1 + j\omega L_1} \qquad (6.10)$$

$$= \frac{R_1 \omega^2 L_1^2}{R_1^2 + j\omega^2 L_1^2} + j\frac{R_1^2 \omega L_1}{R_1^2 + j\omega^2 L_1^2}$$

If the circuit is highly inductive, $I_Q \gg I_P$, that is, $R_1 \gg \omega L_1$, then

$$R \approx \frac{\omega^2 L_1^2}{R_1}, \quad L \approx L_1 \qquad (6.11)$$

or

$$R_1 = \frac{\omega^2 L_1}{R}, \quad L_1 = L \qquad (6.12)$$

The total load fundamental component of current (i.e., the fundamental component of the square wave) is given as

$$I_L' = \frac{2\sqrt{2}}{\pi} I_d \qquad (6.13)$$

The active dc power supplied by the source is consumed in the load; in other words,

$$P_d = V_d I_d = \frac{V_L^2}{R_1} \qquad (6.14)$$

Again, the active and reactive components of the total load current can be given as

$$I_P = I_L' \cos\beta = \frac{V_L}{R_1} \qquad (6.15)$$

$$I_Q' = I_C - I_Q = I_L' \sin\beta = \frac{V_L}{X_c'} \qquad (6.16)$$

where $X_c' = j\omega L_1 \| 1/j\omega C$. Combining Equations (6.15) and (6.16) yields

$$V_L = I_L' \frac{R_1 X_c'}{\sqrt{R_1^2 + X_c'^2}} \quad (6.17)$$

From Equations (6.13) and (6.14) we get

$$I_L' = \frac{2\sqrt{2}I_d}{\pi} = \frac{2\sqrt{2}V_L^2}{\pi R_1 V_d} \quad (6.18)$$

Combining Equations (6.17) and (6.18) yields

$$V_L = \frac{\pi}{\sqrt{8}} V_d \frac{\sqrt{R_1^2 + X_c'^2}}{X_c'} \quad (6.19)$$

The load voltage, currents, and commutating angle β can be calculated from the above equations for the given circuit parameters. Or else, for a specified β angle, the value of capacitance can be determined for a given load.

Figure 6.8 shows an effective load equivalent circuit with a phasor diagram. There are basically two control variables of the inverter, the dc link current I_d and frequency ω. With constant frequency, if the current I_d, that is, I_L' is increased, the β angle will remain constant and the output power will increase. If, on the other hand, I_d is constant but frequency ω is increased, the I_L' phasor will rotate to increase the β angle and the output power will be reduced.

In practice, the load may be variable. Therefore, for satisfactory load commutation, one possibility is to vary the capacitance C such that the desired margin angle β is always maintained. It is more convenient to keep the capacitor fixed and vary the inverter frequency, as mentioned above, such that it is always slightly higher than the load resonance frequency. This condition can assure leading power factor operation at a fixed leading angle β. Figure 6.6 shows the inverter frequency control derived from a feedback β angle through a phase-locked-loop (PLL) control. Some frequency variation is of no concern for an induction heating-type load.

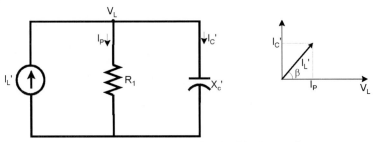

Figure 6.8 Load equivalent circuit with phasor diagram

Instead of a constant margin angle, a constant margin time t_β is desirable in variable-frequency operation. With a constant β angle, as the frequency decreases, the margin time t_β becomes larger than necessary, causing unnecessary VAR loading of the inverter.

6.3.2 Three-Phase Inverter

6.3.2.1 Lagging Power Factor Load

The concept of load commutation, as explained above with a single-phase inverter, can be extended to polyphase inverters. Figure 6.9 shows a current-fed, three-phase bridge inverter with lagging power factor load where the load commutation is achieved with a leading VAR load connected at the load terminal. Again, as discussed before, with a variable load, a fixed capacitor bank can be connected at the terminal and the inverter frequency can be manipulated so that the effective inverter load has a leading power factor and commutation occurs at a fixed advance angle β. If, however, the load is an induction motor, the inverter frequency is dictated by the machine's speed requirement. With a constant volts/Hz requirement at the machine terminal, as the machine speed rises with frequency, the voltage also increases proportionally, and therefore, the capacitor current I_C increases with the following parabolic relation:

$$I_C = V\omega C = K\omega^2 C \tag{6.20}$$

where V = phase voltage, ω = machine frequency, and C = equivalent capacitance/phase. The correct leading VAR requirement at the machine terminal can be met, for example, by an SVC, which was discussed in Chapter 5. The SVC compensates the lagging VAR of the machine, but in addition, consumes the correct leading VAR so that the effective load power factor is leading at angle β.

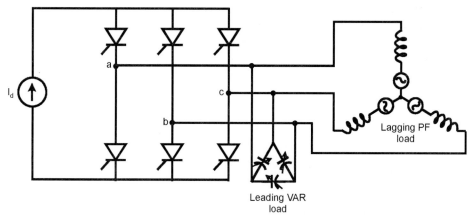

Figure 6.9 Three-phase bridge inverter with load commutation of lagging power factor load

6.3.2.2 Over-Excited Synchronous Machine Load

Thyristor current-fed, load-commutated inverters (LCI) are extremely popular in high-power (multi-MW) wound-field synchronous motor drives where it is easy to maintain the required leading power factor angle by adjusting the field excitation. This type of drive has been used widely in pumps and compressors. The absence of a forced commutation requirement for thyristors makes the inverter simple, reliable, cost-effective, and more efficient. The slow-speed, rectifier-type thyristors are well-suited for the inverter. The torque and speed of the machine are controlled by the dc link current and inverter frequency, respectively, and the machine is operated by self-control mode, which will be explained in Chapter 9.

Figure 6.10 shows the fundamental frequency phasor diagram of a salient pole synchronous machine for load commutation under motoring condition, and Figure 6.11 shows the motor phase voltage and current waves, including the commutation overlap effect. Note the phase reversal of currents in these figures conform with the standard motoring phasor diagram (see Figure 2.31). The phasor diagram uses all the standard notations, and winding resistance and commutation overlap have been neglected. A flux linkage phasor diagram has been added in the figure, where ψ_f = field flux (linkage), ψ_a = armature reaction flux, and ψ_s = resultant stator flux. The machine usually has a damper winding to reduce the commutation overlap angle (μ), but current does not flow in it except during the commutation transient. The d^e and q^e components of ψ_a can be written from the phasor diagram as

$$\psi_{ds} = L_{ds}I_{ds} = \sqrt{2}L_{ds}I_s \sin(\delta + \phi) \tag{6.21}$$

$$\psi_{qs} = L_{qs}I_{qs} = \sqrt{2}L_{qs}I_s \cos(\delta + \phi) \tag{6.22}$$

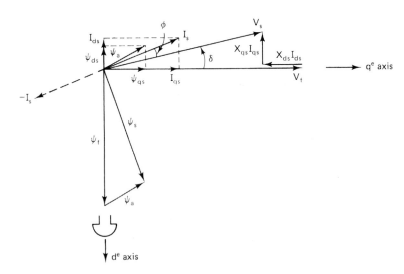

Figure 6.10 Phasor diagram of synchronous motor with load commutation

Load-Commutated Inverters

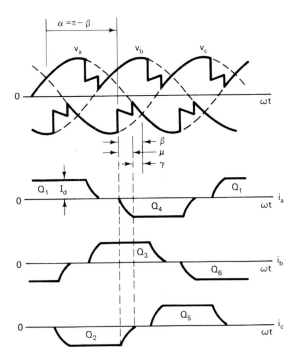

Figure 6.11 Phase voltage and current waves at synchronous motor terminal indicating load commutation

where the variables are in rms values. Since $L_{ds} \neq L_{qs}$ in a salient pole machine, the phasors ψ_a and I_s are not cophasal.

Note that the voltage and current waves of a load-commutated inverter are essentially identical with those of the phase-controlled inverter, and indeed, Figure 6.11 has already been explained without the commutation overlap effect in Chapter 3 (see Figure 3.27). The machine can easily go into regeneration mode by reducing the firing angle α to zero when the inverter operates as a diode rectifier. The machine terminal voltages will deviate slightly from the nominal waves shown in Figure 6.11 due to resistance drop and commutation voltage spikes. The commutating inductance L_c, which causes the voltage spike and overlap angle μ, should be small. An expression of the overlap angle that was derived in Equation (3.56) can be repeated as

$$\cos(\beta - \mu) - \cos\beta = \frac{2\omega L_c I_d}{\sqrt{6} V_s} \tag{3.56}$$

The parameter L_c is basically the subtransient inductance of the machine, which can be approximately given by

$$L_c = 0.5(L_d'' + L_q'') \tag{6.23}$$

where

$$L_d'' = L_{ls} + L_{dm} \| L_{1fr} \| L_{1dr} \quad (6.24)$$

$$L_q'' = L_{ls} + L_{qm} \| L_{1qr} \quad (6.25)$$

These parameters are the Thevenin inductances of d^e and q^e equivalent circuits, respectively, of the machine equivalent circuits shown in Figure 2.36, where the winding resistances were ignored. Although not included in Equations (6.24) and (6.25), the damper winding resistances R_{dr} and R_{qr} play an important part in reducing the effective value of L_c. Thus, during commutation, overlap angle μ a pulse of current flows in the damper winding.

6.3.2.3 Synchronous Motor Starting

The load commutation of an inverter with an over-excited synchronous machine load as discussed above depends on the CEMF, and therefore, the machine should not operate below a critical speed. The CEMF becomes lower at lower speeds and causes the commutation overlap angle μ to be larger. Typically, below 5 percent of the base speed, the load commutation does not work satisfactorily. In this speed range, the inverter needs some type of forced commutation, which is discussed later. A type of forced commutation that does not require any additional power circuit components is known as the pulsed or dc link current interruption method [2], and is explained in Figure 6.12. In this method, the inverter thyristors are commutated by periodically interrupting the dc link current I_d as shown. The three-phase inverter firing pulses are derived from an absolute position encoder mounted on the machine shaft. Assume, for example, that commutation is desired from thyristor Q_2 to Q_4 in Figure 6.1 when Q_3 on the positive side is conducting (see Figure 6.2). At point A, the line-side converter is blocked. The converter will be dragged into inverting mode when negative V_r pumps the energy stored in L_d and the machine phases into the line, and I_d falls to zero. During current interruption, inverter conducting thyristors Q_2 and Q_3 will turn off, but then at point B, Q_3 and Q_4 are fired and the line-side converter is enabled to re-establish current I_d. Thus, conduction is successfully transferred from Q_2 to Q_4. The dc link bypass thyristor Q_A (acting as a diode), shown in Figure 6.1, can help fast current interruption. When the line-side converter is blocked, Q_A is fired to lock the current in L_d through it. This helps faster current interruption in Q_2 and Q_3 because of reduced feedback energy. This mode of control is repeated every $\pi/3$ interval and the machine phase current becomes slightly less than $2\pi/3$ wide per half-cycle with a small gap in the middle. The machine is started from zero speed with field current and the developed torque gradually increases the speed until the control is transitioned to load commutation. Nearly fully developed torque can be obtained in this method, and the machine can have sustained operation in this mode.

Besides variable-speed drives, load-commutated inverters are also used as solid-state, variable-frequency starters for line-synchronized, constant-speed operation of the machine. As a general-purpose starter for a machine, it has the following three modes of operation: (1) at low speed, the motor is started with forced commutation of the inverter as explained above; (2) when

Force-Commutated Inverters

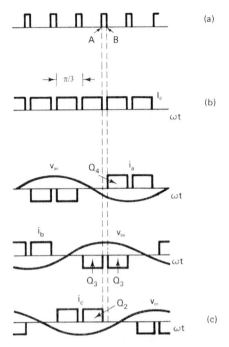

Figure 6.12 Pulse method of starting synchronous motor (a) Pulse train from rotor position encoder, (b) DC link current, (c) Inverter phase CEMF voltage and current waves

adequate speed is developed, the drive enters into load commutation mode; (3) as the machine speeds up and the voltage, frequency, and phase conditions match with the utility line, the machine is switched into the utility system. The solid-state starter, although expensive, gives the machine the additional capability for speed control in four quadrants. This starting scheme becomes especially attractive in a multiple-machine installation where one starter can be time-shared.

6.4 FORCE-COMMUTATED INVERTERS

It was mentioned before that a current-fed inverter requires forced commutation if the load power factor is not leading. Although a number of forced commutation schemes are available in the literature, in this section, only one scheme will be reviewed briefly.

6.4.1 Auto-Sequential Current-Fed Inverter (ASCI)

Figure 6.13 shows a three-phase bridge inverter with an auto-sequential method of forced commutation, which supplies power to an induction motor load. As mentioned before, the induction motor model is approximately represented by a per-phase equivalent circuit that consists of a sinusoidal CEMF in series with an effective leakage inductance L. At the stalled condition of the machine, the CEMF is zero and the motor becomes ideally an inductive load.

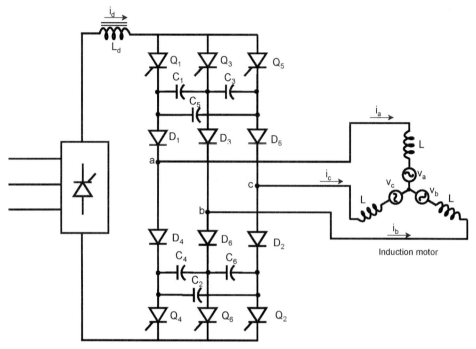

Figure 6.13 Three-phase bridge ASCI inverter with induction motor load

Thyristors Q_1–Q_6 are the principal switching devices of the inverter, where each of them conducts in sequence, ideally for $2\pi/3$ angle, to establish the usual six-stepped current waves in the machine. The series diodes and delta-connected capacitor banks (all with equal values), which are connected to each of the upper and lower groups of thyristors, constitute the forced commutation elements. The capacitors store a charge with the correct polarity for commutation and the series diodes tend to isolate them from the load. In normal inverter operation, the upper group and lower group devices operate independently, and there are six commutations per cycle of fundamental frequency.

Figure 6.14 explains commutation from thyristor Q_2 to Q_4, and shows the corresponding equivalent circuit. All the other commutations are similar. The CEMF magnitudes and their polarity are dictated by motoring mode operation (Mode 3 in Figure 6.5). The discussion on commutation is also valid during the regenerative mode (Mode 4). When incoming thyristor Q_4 is fired, outgoing thyristor Q_2 is impressed with reverse voltage across the capacitor bank, as shown in the figure. It is assumed that the capacitor bank has a stored charge at this correct polarity. Thyristor Q_2 turns off almost instantaneously with the reverse voltage. The dc link current I_d flows through Q_3 and D_3 in the upper group, phases b and c of the machine, device D_2, the delta capacitor bank, and Q_4 to the negative supply. The capacitor bank then charges linearly with the dc current I_d. During this constant current charging of the capacitor bank, there is no drop in the inductance L and diode D_4 remains reverse-biased by the dominating line CEMF

Harmonic Heating and Torque Pulsation

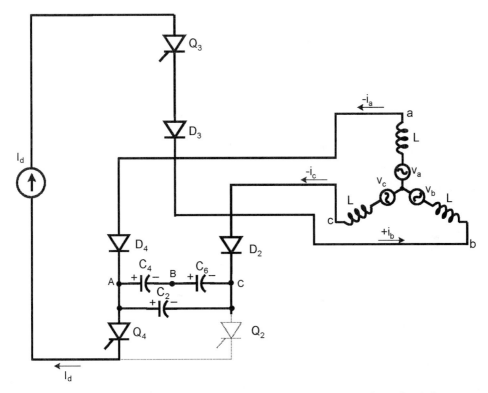

Figure 6.14 Equivalent ASCI circuit during commutation from Q_2 to Q_4

voltage v_{ca}. The linear charging period ends when the capacitor bank voltage equals the line voltage, and then diode D_4 begins to conduct. The current I_d then resonantly transfers to D_4 completely and terminates the commutation process.

During the current transfer interval, a large voltage spike (Ldi/dt) will be induced across each L and will be added with the CEMF. Machine spike voltage is a serious problem, which can be suppressed by a diode bridge at the machine terminal with a zener diode load. The machine can also be designed with low leakage inductance to reduce this voltage.

ASCI inverter-fed induction motor drives of medium to large capacity have been widely used in industry, but recently, they have become obsolete in competition with inverters using self-controlled devices, which will be discussed later.

6.5 HARMONIC HEATING AND TORQUE PULSATION

We have discussed six-step current-fed inverters in a previous section. Six-stepped current waves create harmonic heating and torque pulsation problems like six-stepped voltage waves in voltage-fed inverters, which were discussed in Chapter 2. The pulsating torque problem may be serious for inverter operation below a few Hz of fundamental frequency because the pulsating torque frequency may be near the mechanical resonance frequency of the drive. Similar to voltage-fed inverters, there are essentially the following two methods for harmonic heating and pulsating torque compensation of current-fed inverters:

- Multi-stepped inverters
- PWM inverters

These techniques will be discussed later in detail. If the dc link is not very stiff, there is a possibility of pulsating torque compensation by modulating the dc current I_d with the front-end converter. This principle is explained in Figure 6.15. Figure 6.15(a) shows the torque ripple with the ideal six-stepped phase current wave. Assuming an approximate instantaneous power balance between the inverter input and machine shaft output, we can write

$$p = I_d V_d \approx T_e \omega_m \qquad (6.26)$$

or

$$T_e = \frac{I_d}{\omega_m} V_d \qquad (6.27)$$

Considering current I_d and speed ω_m as constants, the torque ripple follows the inverter dc voltage ripple of a six-pulse, phase-controlled bridge converter, and therefore is given by $\pi/3$ segments of a sine wave, as shown. The dc link current I_d can be controlled by a high-gain instantaneous torque control loop so that it ideally follows the inverse profile of the torque ripple as indicated in Figure 6.15(b). In practice, the inductance L_d in the dc link will not permit such precise control.

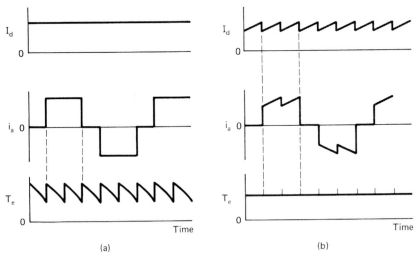

Figure 6.15 (a) Pulsating torque with smooth dc current, (b) Smoothing of pulsating torque by modulation of dc current

6.6 MULTI-STEPPED INVERTERS

Multi-stepping a current-fed inverter above six steps is an obvious method of solving harmonic heating and torque pulsation problems. The principle is essentially the same as that of the multi-stepped voltage-fed inverters discussed in Chapter 5. A 12-pulse, phase-controlled thyristor converter using two six-pulse bridge circuits and phase-shifting transformers was explained in Figure 3.36. The circuit configuration, redrawn in Figure 6.16, can be directly used for a synchronous motor drive with load commutation where the ac source is replaced by the three-phase machine. Here, the line commutation is substituted by the load commutation of the machine. Both six-pulse inverter components operate symmetrically with $\pi/3$ phase shift angle because the machine CEMF waves reflected at the inverter outputs are also shifted by $\pi/3$ angle. A 12-step current wave flows in the machine winding, which has the characteristic harmonic components of 11th, 13th, 23rd, 25th, etc., and the lowest frequency of the pulsating torque corresponds to the 12th harmonic. Thus, higher harmonic frequencies and their lower amplitudes reduce pulsating torque and harmonic heating effects considerably. A similar power circuit configuration can also be used on the line side for feeding the inverter. As usual, during motoring mode, the line-side converter operates as a rectifier and the load-side unit acts as an inverter. However, during regeneration, their roles are reversed. For the induction motor load, the inverter requires forced commutation, and therefore, the load-commutated inverter may be replaced by an ASCI inverter. Of course, to justify the complexity of a 12-pulse power circuit, the drive system power rating should be considerably high.

Twelve-pulse inverters with phase-shifting transformers, as discussed above, become very expensive, although the advantage is that a standard three-phase machine can be used. The transformers can be eliminated by using an asymmetric six-phase machine, as shown in Figure 6.17. This machine has an asymmetric six-phase winding in the sense that the xyz winding group is

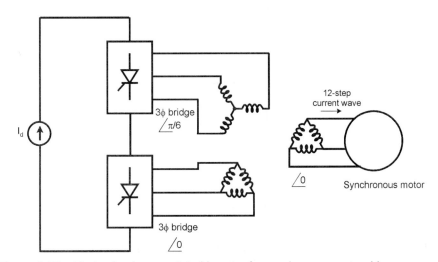

Figure 6.16 12-step load-commutated inverter for synchronous motor drive

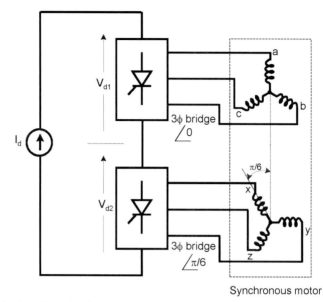

Figure 6.17 Equivalent 12-pulse load-commutated inverter drive for synchronous motor with asymmetric six-phase machine

phase-advanced by $\pi/6$ angle with respect to the abc winding group. In a conventional or symmetric six-phase machine, the three-phase winding groups are displaced by a $\pi/3$ angle. The advantage of the asymmetric connection is that if the component three-phase windings are supplied by the inverters at $\pi/6$ phase shift angle, as shown in the figure, the resultant magnetic field has 12-pulse harmonics. Therefore, an equivalent 12-pulse operation of the machine results as far as the harmonic torque pulsation is concerned. Of course, the windings carry the usual six-stepped current wave. The machine CEMFs are at $\pi/6$ phase-shift angle for each three-phase group, and therefore, the firing pulse train of the lower bridge is phase-shifted by $\pi/6$ with respect to the upper bridge. The inverter input voltages V_{d1} and V_{d2} are equal and their sum, $V_d = V_{d1} + V_{d2}$, will contain a 12-pulse ripple. The same power circuit configuration can be used for an induction motor drive also, but a load-commutated inverter is replaced by a force-commutated inverter.

6.7 INVERTERS WITH SELF-COMMUTATED DEVICES

6.7.1 Six-Step Inverter

So far, we have seen that the speed of an induction motor can not be controlled by a simple six-step thyristor current-fed inverter. The inverter requires either forced commutation or an expensive SVC at the machine terminal. Let us now consider a six-step, current-fed inverter using self-controlled symmetric blocking devices, such as GTOs, as shown in Figure 6.18. Since the GTOs can be turned on and off by gate current pulse, the inverter can easily control the six-

Inverters with Self-Commutated Devices

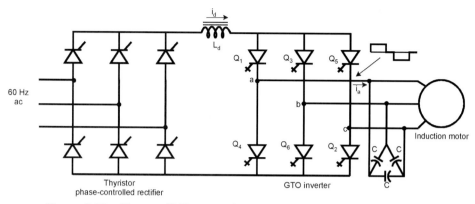

Figure 6.18 Six-step GTO current-fed inverter with induction motor load

step current wave in a lagging power factor induction motor load at an arbitrary phase position. Note that a small delta capacitor bank is yet to be connected at the machine terminal, but its function is entirely different and can be defined as follows: (1) primarily, it permits commutation from the outgoing GTO to the incoming GTO, which will be explained later, and (2) it acts as a load filter for higher harmonics. Reduced harmonics in the machine current reduces machine copper loss and torque pulsation, and attenuates acoustic noise. There can be a harmonic resonance problem with the load, which will be discussed later. All the devices are hard-switched, as usual, but snubbers are not shown for simplicity.

Consider, for example, commutation from device Q_1 to Q_3. Figure 6.19 shows the equivalent circuit during commutation. Initially, device Q_1 in the upper group and Q_2 in the lower group are conducting and the dc link current I_d is flowing through Q_1, phase a, phase c, and Q_2, as indicated. The equivalent capacitance C_{eq} between lines b and a and the polarity of the line

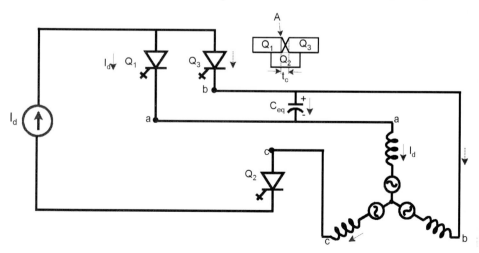

Figure 6.19 Equivalent circuit during commutation from Q_1 to Q_3

voltage v_{ba} across the capacitor are indicated in the figure. The sequence of commutation can be summarized as follows:

- Turn on GTO Q_3 at instant A, which is shown in the upper part of the figure. The polarity of voltage across C_{eq} indicates that Q_1 will not automatically turn off by this voltage polarity as in a load-commutated inverter.
- Turn off Q_1. The current I_d quickly transfers to Q_3 (irrespective of capacitor voltage polarity), which now flows through the equivalent capacitor C_{eq}.
- The capacitor quickly charges and eventually overcomes the line CEMF between phases a and b. Gradually, the current transfers to phase b. The commutation is complete when current I_d is fully transferred to phase b.

The total commutation time t_c is indicated in the figure. Once the commutation is complete, the current can be commutated back to device Q_1, if desired. This back-and-forth commutation can create a PWM current wave in a phase with desirable notch and pulse widths, which will be discussed later. Since the commutation takes a finite time interval t_c, the notch and pulse widths should not fall below a minimum value.

Note that the capacitor bank is not desired to give phase compensation of the lagging load. Figure 6.20 shows a typical phasor diagram at the machine terminal, assuming constant active and reactive currents are consumed by the load. As explained before, with rated flux, the machine terminal voltage and frequency increase proportionally with speed. At low machine speed, both frequency and voltage are low, which make the capacitor current I_C small. Therefore, the inverter's terminal lagging power factor is low. As the machine speed increases, the power factor improves (see Equation (6.20)), as shown, giving less loading on the inverter. At maximum speed, the power factor is unity and the inverter loading is minimum. With a leading power factor, of course, the GTO inverter can be operated with load commutation.

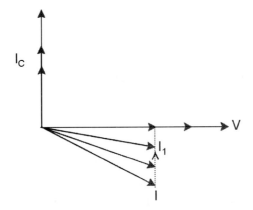

Figure 6.20 Phasor diagram at inverter output terminal at increasing machine speed

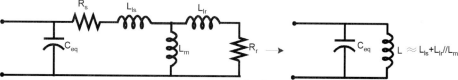

Figure 6.21 Harmonic equivalent circuit of induction motor with commutating capacitor

The converter system of Figure 6.18 can easily be operated in four-quadrant mode. Multi-MW GTO current-fed inverter drives with six-step waves have been built for pump and fan-type drives. For the lower power range, the GTOs can be substituted by IGBTs with series diodes.

6.7.1.1 Load Harmonic Resonance Problem

The current wave supplied to the motor with a capacitor bank in parallel contains harmonics, and one of these harmonics may create a resonance problem. Figure 6.21 shows the per-phase harmonic equivalent circuit of the machine with the equivalent shunt capacitor at the input. In a large machine, resistances R_s and R_r are small and can be neglected. A simplified parallel LC equivalent circuit is shown on the right, where the equivalent inductance L consists of a parallel combination of L_m and L_{lr} in series with L_{ls}. The skin effect will tend to increase the rotor resistance, but it is neglected for simplicity. The resonance frequency f_r of the circuit is given by the expression

$$f_r = \frac{1}{2\pi\sqrt{LC_{eq}}} \tag{6.28}$$

A harmonic component of the inverter output current wave may excite destructive voltage and current in the machine due to the resonance effect. This is also accompanied by excessive copper loss, core loss, and torque pulsation. The typical range of resonance frequency in a practical machine is 100 Hz to 200 Hz. Consider, for example, a resonance frequency of 120 Hz. Figure 6.22 shows the fundamental frequency variation of a six-step inverter current to excite the resonance at different harmonics. The current wave has the characteristic harmonic components of 5th, 7th, 11th, 13th, etc. This means, as shown in the figure, the 13th harmonic of 9.2 Hz, 11th harmonic of 10.9 Hz, 7th harmonic of 17.1 Hz, and 5th harmonic of 24 Hz coincide with the resonance frequency. As the machine starts from zero speed, the order of the harmonic, coinciding with the resonance frequency, will gradually decrease until above 24 Hz fundamental frequency; thus, there is no possibility of resonance effect. If the drive is designed for maximum frequency of 60 Hz, then for operation typically in the range of 30 Hz to 60 Hz, there is no possibility of steady-state resonance. During start-up mode, it is possible to ride through the resonance effect by fast acceleration of the drive. For steady-state operation at any frequency, the characteristic harmonic causing the resonance can be eliminated by the selected harmonic elimination PWM technique, which will be discussed later.

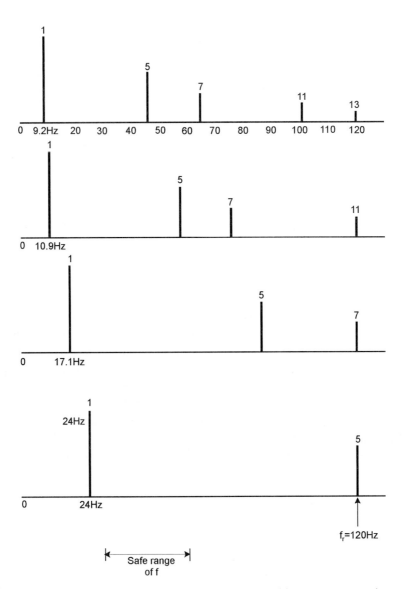

Figure 6.22 Frequency variation of inverter to excite machine resonance at harmonics

6.7.2 PWM Inverters

The disadvantages of a six-step current wave, such as harmonic heating, torque pulsation, and acoustic noise, can be significantly reduced by PWM wave shaping of the inverter's current wave. The PWM current waves with reduced harmonic content are further filtered by the commutating capacitor bank to make the machine current nearly sinusoidal. The PWM techniques of

Inverters with Self-Commutated Devices

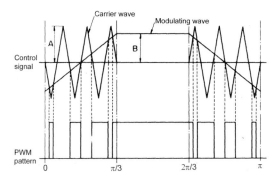

Figure 6.23 Trapezoidal PWM principle

current-fed inverters are somewhat different from those of voltage-fed inverters. In this section, we will briefly discuss a few PWM methods.

6.7.2.1 Trapezoidal PWM

The trapezoidal PWM technique is somewhat analogous to the sinusoidal PWM technique of a voltage-fed inverter, and it was possibly the earliest method proposed in the literature [3]. The principle of this PWM method is explained in Figure 6.23. A trapezoidal modulating wave of maximum amplitude B and at motor fundamental frequency is compared with a triangular carrier wave of peak amplitude A, and the points of intersection generate the PWM pattern, as shown in the figure. The pattern has a quarter-cycle as well as a half-cycle symmetry, and the middle $\pi/3$ segment does not have any modulation. There are essentially two variables for PWM pattern generation: one is the modulation index $m = B/A$, which is the ratio of the modulating wave amplitude to the carrier amplitude, and the other is the pulse number M in the half-cycle of inverter operation. Harmonics in the PWM pattern can be varied by changing these parameters. Figure 6.24 shows the relation between the modulation index and the lower order harmonics in

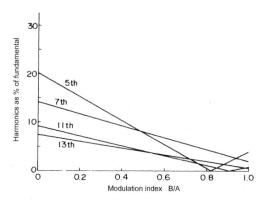

Figure 6.24 Harmonics in PWM current wave with different modulation index and $M = 21$ (pulses per half-cycle)

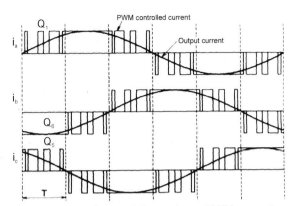

Figure 6.25 Waveforms of three-phase PWM current waves

the PWM current wave for $M = 21$. The harmonic amplitudes decrease with higher modulation index, and the value of 0.82 is considered optimum where the 5th harmonic is zero and the 7th, 11th, and 13th harmonics are given by 4 percent, 1 percent, and 2 percent, respectively. The fundamental component of current is relatively insensitive to the modulation index and M values. Figure 6.25 shows the fabrication of PWM current waves for the three phases at the inverter output using this pattern. The fundamental component of the waves is also indicated in the figure. With this harmonically optimized PWM pattern remaining constant, the magnitude of the sinusoidal current is controlled by the dc link current I_d. Since only two devices conduct at any instant in a current-fed inverter, the PWM pattern in one phase should be made complementary with another phase. In a voltage-fed inverter, on the other hand, the PWM patterns for the three phases are independent since the phase voltages are independently controlled with three devices conducting (one for each phase) at any instant. In Figure 6.25, the middle $\pi/3$ segment does not have any modulation, but the $\pi/3$ segment at the leading and trailing edges of each half-cycle are modulated with an adjacent device in the same group (upper or lower). Consider, for example, the initial $\pi/3$ segment (marked T) in Figure 6.25, where GTO Q_6 is carrying the negative phase b current without modulation, whereas the positive phase a and phase c currents are modulated by devices Q_1 and Q_5. There is the possibility of high transient voltage build-up on commutating capacitors by this back-and-forth switching of devices accompanied by the harmonic resonance effect. This problem has been solved by controlled shoot-through conduction of the series devices.

It is necessary to control the device switching frequency (to limit switching loss) to nearly constant, irrespective of the fundamental frequency of the current. This can be done by making the parameter M constant in many segments of the fundamental frequency, as shown in Figure 6.26, for switching frequency near 1.0 kHz. A look-up table of optimized PWM patterns for different M values can be stored in a microcomputer for implementation. Note that in a multi-MW GTO inverter, the switching frequency hardly exceeds a few hundred Hz.

It can be shown [5] that trapezoidal modulation can give low harmonic components up to harmonic order $n = 1.5 (M + 1)$ for values of $M \geq 9$. Again, although trapezoidal modulation is

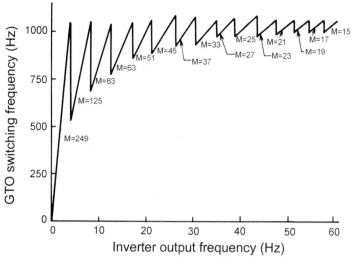

Figure 6.26 Relation between inverter frequency and GTO switching frequency

effective in controlling lower order harmonics, it does produce a pair of harmonics of order $n = 3(M-1) \pm 1$ with large magnitude. For this reason, the importance of trapezoidal modulation is substantially diminished for $M < 9$.

6.7.2.2   Selected Harmonic Elimination PWM (SHE-PWM)

Selected harmonic elimination PWM (SHE-PWM) has the following advantages: it not only can improve the harmonic content of output current, but it can also eliminate the particular harmonic that will tend to cause resonance problem. The principle of SHE-PWM was explained in Chapter 5; however, in a current-fed inverter, there are a number of restrictions. Consider the three-phase PWM current waves with SHE-PWM shown in Figure 6.27 with five pulses per half-cycle ($M = 5$). In addition to quarter-cycle and half-cycle symmetry, the first $\pi/6$ and last $\pi/6$ intervals in each half-cycle are inverse images of each other. All the switching angles of the half-cycle of the i_a wave and the initial part of the i_c wave are indicated in the figure. Note that in the initial $\pi/3$ interval, the positive current is switched between i_a and i_c, whereas i_b is negative, unmodulated current. Angles α_1 and α_2 are the variables, and all the other switching angles are related to them. With two unknown variables, two significant harmonics can be eliminated (such as 5th and 7th) from the current wave. The fundamental current is controlled by the dc link current I_d, as mentioned before. The general relation between the number of harmonics removed (K) and M is given by

$$K = (M-1)/2 \qquad (6.29)$$

where M = number of pulses per half-cycle. Both K and M parameters have odd values. Figure 6.28 shows current waves with $M = 3$ and 7, respectively. With $M = 3$, only one significant harmonic (such as 5th) can be eliminated, and with $M = 7$, three significant harmonics (such as 5th, 7th, and 11th) can be eliminated. The value of M can be increased at low fundamental fre-

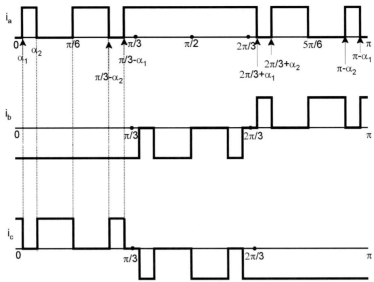

Figure 6.27 Selected harmonic elimnation PWM current wave showing five pulses/half-cycle ($M = 5$)

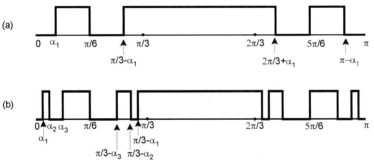

Figure 6.28 Selected harmonic elimnation PWM current wave showing three and seven pulses/half-cycle (a) $M = 3$, (b) $M = 7$

quency to eliminate more significant harmonics. The α angles can be solved from equations discussed in Chapter 5. Table 6.1 gives α angles for the elimination of some significant harmonics [5]. Unfortunately, the elimination of lower order harmonics boosts the higher order significant harmonics, which adversely affects harmonic loss. Note that in a practical implementation of SHE-PWM, the minimum pulse and notch widths are to be maintained for satisfactory commutation of the devices. To eliminate the harmonic resonance problem, as explained in Figure 6.22, the desired α angles given in Table 6.1 should be maintained in the vicinity of fundamental frequency as it varies.

So far, we have considered a harmonically optimized, fixed PWM wave pattern with a variation of the fundamental current magnitude by modulation of the dc link current I_d. It is

Table 6.1 Switching Angles for Elimination of Some Significant Harmonics

Pulses/Half-Cycle (M)	Significant Harmonics to be Eliminated	Switching Angles (α)
3	5th	18.0°
3	7th	21.43°
3	11th	24.55°
3	13th	25.38°
5	5th, 7th	7.93°, 13.75°
7	5th, 7th, 11th	2.24°, 5.6°, 21.26°

possible, for example, to have generalized techniques for realizing PWM patterns which provide selective harmonic elimination as well as current magnitude modulation similar to a voltage-fed inverter [6]. In this scheme, the full-rated dc link current I_d is always maintained and a combination of chops and short-circuit pulses are positioned in such a way that lower order harmonics are eliminated selectively, in addition to current magnitude modulation at a desired switching frequency. This scheme has the advantage of fast transient response, but the losses are excessive.

6.7.3 Double-Sided PWM Converter System

The inverter in Figure 6.18, operating either in six-step or PWM mode, can also be operated in regenerative mode, that is, the variable-frequency electrical energy from the machine side can be converted to dc, which can then be pumped to the line. In this mode, as explained before, the load-side converter acts as a rectifier and the line-side converter acts as an inverter. The polarity of dc link voltages change, while the polarity of dc link current always remains the same. Since the current-fed PWM inverter has the capability to operate as a rectifier, it is apparent that the line-side, phase-controlled thyristor rectifier can be substituted by a self-controlled GTO PWM rectifier. Such a double-sided PWM converter system is shown in Figure 6.29. The configuration is essentially a dual of the double-sided PWM voltage-fed converter system shown in Figure 5.60. The capacitor bank on the line side performs similar functions as those on the load side. The PWM current wave fabrication method remains the same, except the line-side converter has the additional responsibility of controlling the dc link voltage so that the dc link current can be smoothly controlled. The PWM current wave at the rectifier input is filtered by the capacitor bank so that the line current becomes nearly sinusoidal. Additionally, the current wave can be placed at an arbitrary phase position with the phase voltage to make the line power factor unity, leading, or lagging. The line frequency and voltage are nearly constant, and therefore, a fixed leading current will flow through the capacitor bank. As shown by the phasor diagram, the fundamental current at the rectifier input should be lagging slightly to give unity power factor on the line side. Since

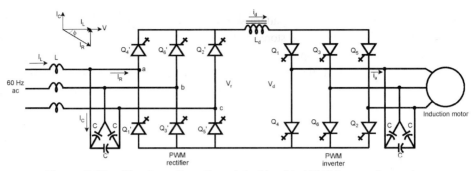

Figure 6.29 Circuit configuration of double-sided PWM converter system

there is PWM control on both sides of the dc link, the inductor L_d can be substantially reduced (typically 10 percent) compared to that of the double-sided, six-step converter system.

As mentioned before, both the inverter and rectifier units require PWM control to produce sinusoidal voltages and currents in the machine and line, respectively. However, the rectifier has the additional duty to control the dc link voltage to regulate the magnitude of motor current, while the inverter controls the frequency corresponding to motor speed. Figure 6.30 shows a simplified control block diagram of a converter system [7] that uses the trapezoidal PWM technique. The main blocks of the inverter section control are the clock generator, counter 2, PWM pattern storage memory, pattern selector, ring counter 2, and PWM waveform synthesizer. The clock generator generates the clock pulse train with a frequency proportional to the inverter frequency command f_s^*. These clock pulses are counted by counter 2 and the counted values are

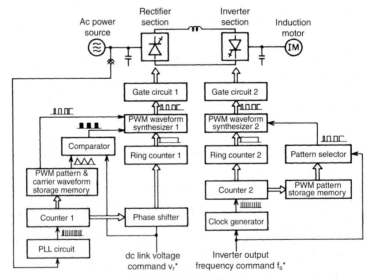

Figure 6.30 Control block diagram of double-sided converter system using trapezoidal PWM method

used as address signals for the PWM pattern storage memory. A number of current wave patterns, as indicated in Figure 6.23, are stored in the PWM pattern storage memory. The pattern selector is used to select an appropriate current pattern from those stored in the PWM pattern storage memory, depending on the fundamental frequency command, as indicated in Figure 6.26. PWM waveform synthesizer 2 synthesizes the output signals of ring counter 2 and the pattern selector to generate the time domain PWM pulse train for gate circuit 2. As the inverter command frequency increases, the pulse pattern for a fixed M value in the range of fundamental frequency shrinks in the time domain (see Figures 6.26 and 6.23) until another pattern is selected for a higher frequency range so that the GTO switching frequency remains nearly constant.

In the PWM rectifier section's control, the PLL (phase-locked loop) circuit generates the clock frequency, which is N times that of the ac supply frequency. These clock pulses are counted by counter 1 and the counted values are used as address signals for the PWM pattern and carrier waveform storage memory. The digital words corresponding to a triangular carrier wave generated from this memory are compared with the dc link voltage command V_d^* to produce the bypass gate pulses which force the rectifier leg to short-circuit to control the dc link voltage. The phase shifter is a circuit that changes the firing angle of the rectifier section (phase-controlled rectifier mode) at a low dc link voltage because of the minimum pulse and notch interval requirements of GTOs. PWM waveform synthesizer 1 synthesizes the output signals of ring counter 1, the PWM pattern, and the comparator output to generate a time domain PWM pulse train for gate circuit 1, as shown.

Figure 6.31 explains the rectifier's operation in one cycle of the power source voltage with the help of gate signals, dc link voltage, dc link current, and input PWM-controlled currents. The

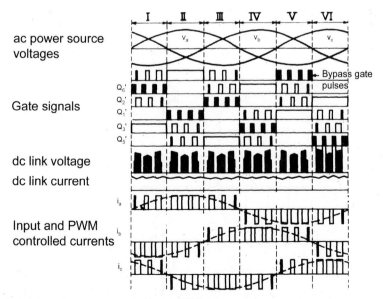

Figure 6.31 Waveforms explaining dc link voltage control by PWM rectifier

gate signals contain the bypass gate pulses, which are shown in a dark color. Consider, for example, the $\pi/3$ segment, marked as V, during which the bypass gate pulses are applied to GTO Q_4'. The gate signals are also applied to Q_1' during the same time. As a result, the phase a rectifier leg is short-circuited. As long as the bypass gate pulses are supplied, the dc link current will flow, bypassing the ac power source. All other segments are similar to the V segment. Consequently, the PWM current waves at the rectifier input become sinusoidally modulated waveforms, as shown in the figure. The capacitor bank, acting as a filter, converts the line-side current waves to nearly sinusoidal form, as discussed before. The gate signals shown in the figure were obtained by combining the bypass gate pulses for controlling the dc link voltage with the basic PWM pulse pattern selected to minimize harmonics. Therefore, the pulse widths in the gate signals can be distributed sinusoidally and the input currents become sinusoidal waveforms, even if the bypass gate pulse widths are changed.

6.7.4 PWM Rectifier Applications

6.7.4.1 Static VAR Compensator/Active Filter

In the double-sided PWM converter system shown in Figure 6.29, it is possible to disconnect the inverter unit and short the inductor output terminal. Such a circuit configuration is shown in Figure 6.32, where the reverse-blocking GTOs are replaced by IGBTs with series diodes. Although both the number of devices and the conduction losses are higher, IGBTs have the following advantages: the PWM frequency can be higher, and the minimum pulse and notch widths are smaller. Neglecting losses in the devices and the inductor for the present, the PWM rectifier can control the desirable dc current I_d with zero active power input. In such a case, the fabricated

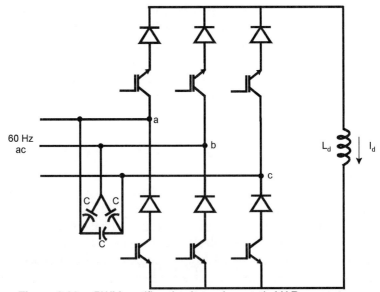

Figure 6.32 PWM rectifier circuit used as static VAR generator

PWM current wave orientation at the input is such that its fundamental component is either leading or lagging at quadrature with respect to the supply voltage. The fundamental current can be controlled by controlling the current I_d. Therefore, the circuit can operate as a controllable leading or lagging SVC. Of course, a small loss component of input current should be added with the input reactive current. This function is similar to a voltage-fed PWM rectifier SVC.

In fact, the circuit of Figure 6.32 can also be used as an active harmonic filter. Consider, for example, that a 5th harmonic current consumed by a parallel thyristor converter is to be compensated. The PWM rectifier can be controlled such that the input current pattern fundamental component is the 5th harmonic, which is placed in phase opposition with the current to be compensated. Although in principle, the current-fed rectifier scheme is a competitor to the voltage-fed SVC/active filter scheme, the former is not preferred because of higher losses, higher cost, and sluggish transient response.

6.7.4.2 Superconducting Magnet Energy Storage (SMES)

Another interesting application of a PWM rectifier is an SMES system. The off-peak energy of the utility system can be rectified to dc by the PWM rectifier and stored as magnetic energy in the superconducting coil L_d. The coil is cooled cryogenically so that it becomes lossless with zero resistance. The stored energy can be inverted and pumped back to the utility system during peak demand period. Compared to the traditional thyristor phase-controlled converter scheme, the line current wave is nearly sinusoidal. In fact, the line current (corresponding to the loss component) can be placed in phase with the supply voltage while the inductor current circulates through the rectifier.

6.7.4.3 DC Motor Speed Control

A PWM rectifier circuit can also be used for two-quadrant speed control of a dc motor where the inductor is replaced by the motor aramature. The motor armature current can be controlled to control the torque at any speed. In regenerative braking mode, the rectifier acts as an inverter and energy is pumped back to the source. Again, the scheme is superior to a phase-controlled converter because the line current is sinusoidal and DPF can be maintained at unity at all operating conditions. A dual-converter system can extend the drive operation into the four-quadrant mode.

6.8 CURRENT-FED VS. VOLTAGE-FED CONVERTERS

So far, we have discussed different types of current-fed converters and their performance characteristics. At this point, it is useful to make a general comparison of their features with those of voltage-fed converters, which were discussed in Chapter 5. A current-fed converter is essentially a dual to the voltage-fed converter. Note that a voltage-fed converter can be operated in the current control mode by adding a current feedback loop. Similarly, a current-fed converter can be operated in the voltage control mode, if desired, by adding a voltage control loop. But, our definitions of voltage-fed and current-fed converters are based on whether the dc supply (or the link) is a voltage source or a current source. A broad summary comparison of the two classes of converters is given below. The comments are generally valid for a six-step, current-fed inverter if not mentioned otherwise.

- In a current-fed inverter, the inverter is more interactive with the load, and therefore, a close match between the inverter and machine is desirable. For example, the inverter likes to see a low leakage inductance, unlike that of a voltage-fed inverter, because this parameter directly influences the inverter commutation process. A large leakage inductance of machine filters harmonics in a voltage-fed inverter, but in a current-fed inverter, it lengthens the current transfer interval which limits the PWM frequency as well as the fundamental frequency of operation. For an ASCI inverter, the inductance causes a voltage spike problem, which may be several times the peak machine CEMF.
- A current-fed converter system has inherent four-quadrant operation capability and does not require any extra power circuit component. On the other hand, a voltage-fed converter system requires a regenerative converter on the line side. In the case of a 60 Hz power failure, regenerated power in the current-fed system cannot be absorbed in the line; therefore, machine speed can be reduced only by a mechanical brake. For a voltage-fed system, however, the dynamic brake in the dc link can be used when line power fails.
- A current-fed converter system is more rugged and reliable, and problems such as shoot-through faults do not exist. Momentary short-circuits in the load and the misfiring of thyristors are acceptable. Fault interruption by gate circuit suppression is simple and straightforward.
- Devices must be symmetric blocking in a current-fed system, unlike the asymmetric blocking devices (with bypass diode) in a voltage-fed system. The devices can be somewhat slow because the commutation time is dominated by the external circuit.
- The control of six-step converters is simple and similar to phase-controlled, line-commutated converters. However, the control of PWM converters may be more complex.
- A multi-machine load on a single inverter or parallel multi-inverter load on a single rectifier is very difficult in a current-fed system. A drive system usually consists of one rectifier, one inverter, and one machine. In applications where multi-machine or multi-inverter capability is required, a voltage-fed system is normally used.
- Current-fed inverters have sluggish dynamic response compared to PWM voltage-fed inverters. A current-fed inverter drive cannot be operated in an open-loop condition, whereas open-loop volts/Hz control of a voltage-fed inverter is very common. This will be further discussed in Chapter 8.
- The successful operation of a current-fed inverter requires that a minimum load always remain connected. The inability to operate at no-load condition invalidates its application in many areas. On the other hand, voltage-fed inverters operate easily at no load.
- From the overall cost, efficiency, and transient response points of view, a voltage-fed PWM converter drive system is definitely superior. This is evident by its widespread application.

6.9 SUMMARY

This chapter gives a broad overview of different types of current-fed converters and their performance characteristics. In the beginning, a six-step thyristor inverter was studied in a generalized manner with machine load to explain its operation with load commutation and forced commutation. A single-phase, load-commutated inverter for induction heating and a three-phase load commutation with an over-excited synchronous motor load were studied in some detail. Current-fed inverter drives are normally used for higher power ranges. Particularly, multi-MW, load-commutated, synchronous motor drives are the most popular application of current-fed inverters. This type of drive will be studied further in Chapter 9. The possibility of induction motor speed control with load commutation was discussed. Force-commutated, thyristor-type, current-fed inverters have been obsoleted recently, and therefore, only a brief description was included.

Principal PWM techniques were reviewed. The double-sided PWM converter system and its control were discussed. A number of important applications of PWM rectifiers were described. Finally, a broad comparison between voltage-fed and current-fed converter systems was given, which should help you to select the topology for a particular application. Although the current-fed converter drive system has a few special merits, overall, the voltage-fed converter drive system is the winner, which is evident by its widespread application.

REFERENCES

1. H. Stemmler, "High-power industrial drives", *Proceedings of the IEEE*, vol. 82, pp. 1266–1286, August 1994.
2. B. Mueller, T. Spinanger, and D. Wallstein, "Static variable frequency starting and drive system for large synchronous motors", *Conf. Rec. IEEE/IAS Annu. Meet.*, pp. 429–438, 1979.
3. M. Hombu, S. Ueda, A. Ueda, and Y. Matsuda, "A new current source GTO inverter with sinusoidal output voltage and current", *IEEE Trans. on Ind. Appl.*, vol. 21, pp. 1192–1198, Sept./Oct. 1985.
4. C. Namuduri and P. C. Sen, "Optimal pulsewidth modulation for current source inverters", *IEEE Trans. on Ind. Appl.*, vol. 22, pp. 1052–1072, Nov./Dec. 1986.
5. B. Wu, S. B. Dewan, and G. R. Slemon, "PWM-CSI inverter for induction motor drives", *IEEE Trans. on Ind. Appl.*, vol. 28, pp. 64–71, Jan./Feb. 1992.
6. H. R. Karshenas, H. A. Kojori, and S. B. Dewan, "Generalized techniques of selective harmonic elimination and current control in current source inverters/converters", *IEEE Trans. on Power Elec.*, vol. 10, pp. 566–573, Sept. 1995.
7. M. Hombu, S. Ueda, and A. Ueda, "A current source GTO inverter with sinusoidal inputs and outputs", *IEEE Trans. on Ind. Appl.*, vol. 23, pp. 247–255, March/April 1987.

CHAPTER 7

Induction Motor Slip-Power Recovery Drives

7.1 INTRODUCTION

Induction motor drives with full-power control on the stator side, as discussed in Chapters 4, 5 and 6, are widely used in industrial applications. Although either a cage-type or wound-rotor machine can be used in the drive, the former is always preferred because a wound-rotor machine is heavier, more expensive, has higher rotor inertia, a higher speed limitation, and maintenance and reliability problems due to brushes and slip rings. However, it is interesting to note that a wound-rotor machine with a mechanically varying rotor circuit rheostat is possibly the simplest and oldest method of ac motor speed control. One feature of this machine is that the slip power becomes easily available from the slip rings, which can be electronically controlled to control speed of the motor. For limited-range speed control applications, where the slip power is only a fraction of the total power rating of the machine, the converter cost reduction can be substantial. This advantage offsets the demerits of the wound-rotor machine to some extent. Slip-power recovery drives have been used in the following applications:

- Large-capacity pumps and fan drives
- Variable-speed wind energy systems
- Shipboard VSCF (variable-speed/constant-frequency) systems
- Variable-speed hydro pumps/generators
- Utility system flywheel energy storage systems

In this chapter, we will study the principles of slip-power control, particularly the popular static Kramer and static Scherbius drives. It should be noted that the nomenclature in these classes of drives is not consistent.

7.2 DOUBLY-FED MACHINE SPEED CONTROL BY ROTOR RHEOSTAT

A simple and primitive method of speed control of a wound-rotor induction motor is by mechanical variation of the rotor circuit resistance, as shown in Figure 7.1.

The torque-slip curves of the motor for varying rotor resistance Rr, as calculated by Equation (2.32), are shown in Figure 7.2. With external resistance $R_1 = 0$, that is, with the slip rings shorted, the inherent torque-slip curve of the machine gives a speed corresponding to point A at the rated load torque. As the resistance is increased, the curve becomes flatter, giving less speed until the speed becomes zero at high resistance ($> R_4$). The maximum or breakdown torque (see Equation (2.35)) remains constant, but the starting torque, given by Equation (2.33), increases with higher resistance. The mechanical variation of resistance has the inherent disadvantage. In addition, this method of speed control is very inefficient because the slip energy is wasted in the rotor circuit resistance. However, several advantages of this method are: absence of in-rush starting current, availability of full-rated torque at starting, high line power factor, absence of line current harmonics, and smooth and wide range of speed control. The scheme is hardly used now-a-days.

Instead of mechanically varying the resistance, the equivalent resistance in the rotor circuit can be varied statically by using a diode bridge rectifier and chopper as shown in Figure 7.3. As usual, the stator of the machine is connected directly to the line power supply, but in the rotor circuit, the slip voltage is rectified to dc by the diode rectifier. The dc voltage is converted to current source I_d by connecting a large series inductor L_d. It is then fed to an IGBT shunt chopper with resistance R as shown. The chopper is pulse width modulated with duty cycle $\delta = t_{on}/T$, where t_{on} = on-time and T = time period. When the IGBT is off, the resistance is connected in the circuit and the dc link current I_d flows through it. On the other hand, if the device is on, the resistance is short-circuited and the current I_d is bypassed through it. It can be shown that the

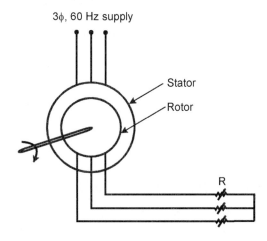

Figure 7.1 Doubly-fed induction motor speed control by rotor rheostat

Static Kramer Drive

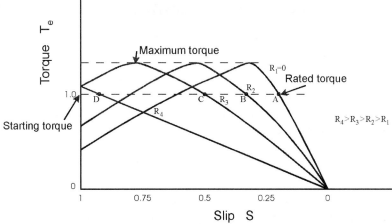

Figure 7.2 Torque-slip curves of motor with variable rotor resistance

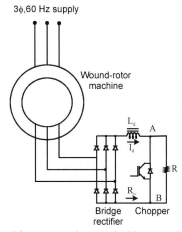

Figure 7.3 Motor speed control with rotor circuit chopper

duty cycle control of the chopper offers an equivalent resistance $R_0 = (1 - \delta)R$ between points A and B. Therefore, the developed torque and speed of the machine can be controlled by the variation of the duty cycle of the chopper. This electronic control of rotor resistance is definitely advantageous compared to rheostatic control, but the problem of poor drive efficiency remains the same. This scheme has been used in intermittent speed control applications in a limited speed range, where the efficiency penalty is not of great concern.

7.3 STATIC KRAMER DRIVE

Instead of wasting the slip power in the rotor circuit resistance, it can be converted to 60 Hz ac and pumped back to the line. The slip power-controlled drive that permits only a subsynchronous range of speed control through a converter cascade is known as a static Kramer

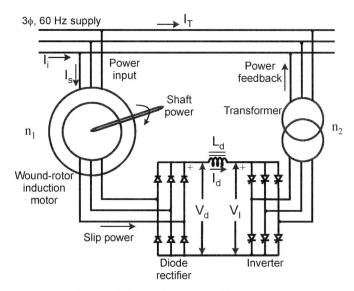

Figure 7.4 Static Kramer drive system

drive, and the scheme is shown in Figure 7.4. It is different from the original Kramer drive, where rotating machines were used for slip energy recovery. The static Kramer drive has been very popular in large power pump and fan-type drives, where the range of speed control is limited near, but below the synchronous speed. The drive system is very efficient and the converter power rating is low, as mentioned before, because it has to handle only the slip power. In fact, the power rating becomes lower with a more restricted range of speed control. The additional advantages, which will be explained later, are that the drive system has dc machine-like characteristics and the control is very simple. These advantages largely offset the disadvantages of the wound-rotor induction machine.

The machine air gap flux is established by the stator supply, and it practically remains constant if stator drops and supply voltage fluctuation are neglected. Ideally, the machine rotor current is a six-stepped wave in phase with the rotor phase voltage if the dc link current I_d is considered harmonic-free, and the commutation overlap angle of the diode rectifier is neglected. The machine fundamental frequency phasor diagram referred to the stator is shown in Figure 7.5, where V_s = phase voltage, I_{rf}' = fundamental frequency rotor current referred to the stator, ψ_g = air gap flux, I_m = magnetizing current, and φ = power factor angle. With constant air gap flux, machine torque becomes directly proportional to current I_{rf}'. Since I_{rf}' is directly proportional to dc link current I_d, the torque is also proportional to I_d. Instead of static resistance control as discussed in the previous section, the scheme here can be considered as CEMF control, where a variable CEMF V_I is being presented by a phase-controlled, line-commutated inverter to control the dc link current I_d. In steady-state operation, the rectified slip voltage V_d and inverter

Static Kramer Drive

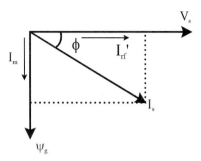

Figure 7.5 Machine phasor diagram referred to the stator

dc voltage V_I will balance, ignoring the resistive drop in the inductance L_d. The voltage V_d is proportional to slip S, whereas the current I_d is proportional to developed torque. At a certain speed, the inverter's firing angle can be decreased to decrease the voltage V_I, which will increase I_d to increase the corresponding developed torque. The simplified speed and torque expressions can be derived as follows. Neglecting the stator and rotor drops, voltage V_d is given as Equation (3.21).

$$V_d = \frac{1.35 S V_L}{n_1} \tag{7.1}$$

where S = per-unit slip, V_L = stator line voltage, and n_1 = stator-to-rotor turns ratio of the machine. Again, the inverter dc voltage V_I is given as Equation (3.57).

$$V_I = \frac{1.35}{n_2} V_L |\cos\alpha| \tag{7.2}$$

where n_2 = transformer line side-to-inverter ac side turns ratio and α = inverter firing angle. For inverter operation, the range of the firing angle is $\pi/2 < \alpha < \pi$. Since in steady state V_d and V_I must balance, Equations (7.1) – (7.2) give

$$S = \frac{n_1}{n_2} |\cos\alpha| \tag{7.3}$$

Therefore, the speed expression ω_r can be given as

$$\begin{aligned}\omega_r &= \omega_e (1-S) \\ &= \omega_e \left\{ 1 - \frac{n_1}{n_2} |\cos\alpha| \right\} \\ &= \omega_e (1 - |\cos\alpha|) \end{aligned} \tag{7.4}$$

where $n_1/n_2 = 1$ has been assumed. Equation (7.4) indicates that ideally, speed can be controlled between zero and synchronous speed ω_e by controlling inverter firing angle α. At zero speed, voltage V_d is maximum, which corresponds to angle $\alpha = \pi$; at synchronous speed, $V_d = 0$ when $\alpha = \pi/2$. In practice, the maximum speed should be slightly less than synchronous speed so that torque (i.e., I_d) can be developed with a finite resistance drop ($V_d = I_d R_d$) of the dc link inductor at $V_I = 0$.

Again, neglecting losses, the following power equations can be written:

$$SP_g = V_I I_d \tag{7.5}$$

$$\begin{aligned} P_m &= (1-S)P_g = T_e \omega_m \\ &= T_e \omega_e (1-S)\frac{2}{P} \end{aligned} \tag{7.6}$$

where P_g = air gap power, P_m = mechanical output power, ω_m = mechanical speed, and P = number of poles. Substituting Equations (7.2), (7.3), and (7.5) in (2.23) gives

$$\begin{aligned} T_e &= (\frac{P}{2})\frac{P_g}{\omega_e} \\ &= (\frac{P}{2})\frac{V_I I_d}{S\omega_e} \\ &= (\frac{P}{2})\frac{\frac{1.35}{n_2}V_L |\cos\alpha| I_d}{\frac{n_1}{n_2}|\cos\alpha|\omega_e} \\ &= (\frac{P}{2})\frac{1.35 V_L}{\omega_e n_1} I_d \end{aligned} \tag{7.7}$$

This equation indicates that the torque is proportional to current I_d. The drive system has nearly the characteristics of a separately excited dc motor, because the air gap flux is nearly constant and the torque is proportional to current I_d. With higher load torque T_L, the machine tends to slow down and current I_d increases so that $T_e = T_L$. In other words, for a fixed firing angle of the inverter, the voltage V_I is fixed. Therefore, to balance the resistance drop of the dc link inductor, V_d must slightly increase, giving speed drop characteristics like a dc machine. Figure 7.6 gives torque-speed curves for different firing angles α. More accurate torque-speed relations will be developed later.

The static Kramer drive has one-quadrant speed control characteristics. The drive cannot have regenerative braking capability, and speed reversal is not possible. Regenerative braking in the subsynchronous speed range will be discussed later. For reversal of speed, a circuit breaker

Static Kramer Drive

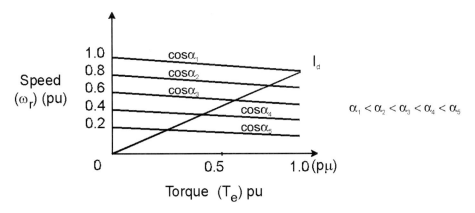

Figure 7.6 Typical torque-speed curves at different inverter firing angles

can be installed on the stator side, which should reverse the phase sequence of the line voltages. For most pump and fan drive applications, simple one-quadrant speed control is acceptable.

7.3.1 Phasor Diagram

A fundamental frequency phasor diagram can be drawn to explain the performance of the drive system. In practice, the rotor current displacement factor will slightly deviate from unity because of the commutation overlap angle shown in Figure 7.7. In fact, the overlap angle μ introduces a lagging angle φ_r to the fundamental current, as indicated in the figure. This current increases as I_d increases with the increase of slip S. Near zero slip, when the rotor voltage is very small, a large current I_d may cause overlap angle μ to exceed the $\pi/3$ angle, causing a short circuit between the upper and lower diodes.

Figure 7.8 shows the approximate phasor diagram of the drive system at the rated torque condition, where all the phasors are referred to the line or stator side. The stator draws a

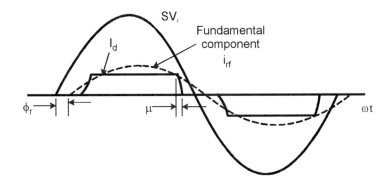

Figure 7.7 Rotor phase voltage and current waves

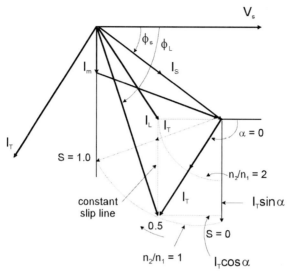

Figure 7.8 Phaser diagram of static Kramer drive system at rated voltage

magnetizing current I_m, which lags $\pi/2$ angle with respect to the stator phase voltage V_s. The total stator current I_s lags the stator voltage by angle φ_s as shown. On the inverter side, although the active power is fed back to the line, it also demands lagging reactive current from the line because of phase control. This additional reactive current drawn by the inverter reduces the overall power factor of the system. Assuming continuous conduction of the inverter and ripple-free current I_d, the inverter output power factor is $|\cos \varphi| = |\cos \alpha|$, that is, its power factor varies linearly with dc voltage V_I. This, of course, neglects the inverter commutation overlap effect. Consider no transformer connection for the present and $n_1 = 1$. The phasor diagram shows the inverter line current I_T at slip $S = 0.5$. Phasors I_T and I_{rf}' are nearly equal in magnitude because of the nearly identical waves of these currents. The active component $I_T \cos \varphi$ opposes the stator active current, whereas the reactive component $I_T \sin \varphi$ adds to the stator magnetizing current I_m. The total line current I_L is the phasor sum of I_s and I_T and it lags at angle φ_L, which is larger than the stator power factor angle φ_s. With constant torque, the magnitude of I_T remains constant, but as the slip varies between 0 and 1, the phasor I_T rotates from $\alpha = 90°$ to $160°$, as shown in the figure. At zero speed ($S = 1$), the machine acts as a transformer, and ignoring losses, all the active power is transferred back to the line through the inverter. The result is that both the machine and inverter consume only reactive power. The inverter margin angle (β) of $20°$ for the inverter covers both commutation (μ) and turn-off (γ) angles. From the phasor diagram, it is evident that at $S = 0$, the system power factor is lagging at low value, which deteriorates as the slip increases. Therefore, with restricted speed range close to synchronous speed, the power factor is comparatively better. For reduced torque condition, the current I_T is proportionally less. The corresponding phasor diagram modification is left as an exercise to the reader. If, for example, the torque is reduced to 50 percent at $S = 0.5$, angle φ remains the same as shown and current I_T is reduced to 50 percent, that is, both $I_T \cos \varphi$ and $I_T \sin \varphi$ are also proportionately reduced.

For a restricted speed range closer to synchronous speed, the system power factor can be further improved by using a step-down transformer. The transformer primary-to-secondary turns ratio n_2 can be adjusted so that at the desired maximum slip, angle $\varphi = \pi$. Of course, the inverter margin angle should always be maintained. Substituting this condition in Equation (7.3) gives

$$S_{max} = n_1 / n_2 \quad (7.8)$$

For example, if $S_{max} = 0.5$ and $n_1 = 1$, then n_2 should ideally be 2. In the phasor diagram of Figure 7.8, this condition corresponds to the transformer line current $I_T' = 0.5 I_T$, which clearly indicates the power factor improvement of line current I_L. As the speed is increased to change the slip from 0.5 to 0, the phasor I_T' rotates anti-clockwise, as shown, until $\varphi = \pi/2$. Equation (7.3) indicates that for constant slip, the variation of n_2/n_1 linearly varies $\cos \alpha$ magnitude. The constant slip line at $S = 0.5$ is indicated in Figure 7.8.

The step-down transformer has essentially two functions: besides improving the line power factor, it also helps to reduce the converter power ratings. Both the rectifier and inverter should be designed to handle the same current I_d as dictated by the torque requirement. The rectifier should be designed for the slip voltage given by SV_L/n_1, whereas the inverter should be designed for the line voltage V_L in the absence of the transformer. The rectifier voltage and corresponding power rating decrease with a smaller speed range, but the inverter must be designed for full power. Installation of the transformer reduces the voltage and corresponding power rating of the inverter, and the criteria for the turns ratio n_2 design is the same as that of Equation (7.8). For the same example (i.e., $S_{max} = 0.5$, $n_1 = 1$, and $n_2 = 2$), both the rectifier and inverter have an equal power rating, which is 50 percent of the full power. It can be shown easily that the converter power rating can be reduced proportionately as S_{max} is reduced. This is an important advantage of the slip-power recovery drive. The discussion above assumes that the machine is not started with the converters in the circuit. Otherwise, the advantage of a reduced converter rating is lost.

A typical starting method of a Kramer drive with resistance switching is shown in Figure 7.9. The motor is started with switch 1 on and switches 2 and 3 off. As the speed builds up, resistances R_1 and R_2 are shorted sequentially until at the desired S_{max} value, switch 1 is opened and the drive controller is brought into action.

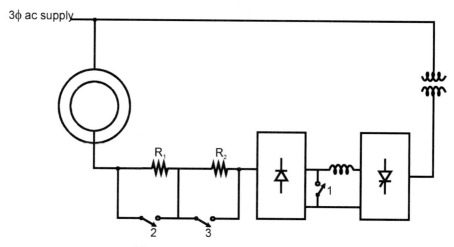

Figure 7.9 Motor starting method

7.3.2 AC Equivalent Circuit

The drive system performance can be analyzed with the help of a dc or ac equivalent circuit of the machine. We will attempt here an approximate ac equivalent circuit with respect to the rotor. Neglecting drops in the semiconductor devices, the slip-power output is partly dissipated in the dc link resistance R_d of the inductor and is partly fed back to the ac line through the transformer.

The respective power components can be given as

$$P_1 = I_d^2 R_d \tag{7.9}$$

$$P_f = \frac{1.35 V_L I_d}{n_2} |\cos\alpha| \tag{7.10}$$

The equivalent rotor circuit power per phase is given as

$$P' = P_1' + P_f' = \frac{1}{3}\left(I_d^2 R_d + \frac{1.35 V_L I_d}{n_2}|\cos\alpha|\right) \tag{7.11}$$

Therefore, the machine air gap power per phase, which includes the rotor copper loss, is given as

$$P_g' = I_r^2 R_r + P' + P_m' \tag{7.12}$$

where I_r = rotor rms current per phase, R_r = rotor resistance, and P_m' = mechanical output power per phase. The torque and corresponding mechanical power P_m' are essentially contributed by the fundamental component of rotor current I_{rf} only. The expression for rotor circuit copper loss per phase is

$$P'_{rl} = I_r^2 R_r + \frac{1}{3} I_d^2 R_d \qquad (7.13)$$
$$= I_r^2 (R_r + 0.5 R_d)$$

where $I_{rf} = \frac{\sqrt{6}}{\pi} I_d$ has been substituted for a six-step wave. Therefore, the expression for P'_m is given as

$$P'_m = (\text{Fund.freq.slip power}) \frac{1-S}{S}$$
$$= \left[I_{rf}^2 (R_r + 0.5 R_d) + \frac{\pi}{3\sqrt{6}} \cdot \frac{1.35 V_L}{n_2} I_{rf} |\cos\alpha| \right] \left(\frac{1-S}{S} \right) \qquad (7.14)$$

where the $I_{rf} = \frac{\sqrt{6}}{\pi} I_d$ relation has been used to replace I_d in Equation (7.10). The air gap power P'_g in Equation (7.12) can be written by substituting Equation (7.14) as follows:

$$P'_g = I_r^2 \frac{R_r}{S}$$
$$= \frac{1}{S} \left[I_{rf}^2 (R_r + 0.5 R_d) + \frac{\pi}{3\sqrt{6}} \cdot \frac{1.35 V_L}{n_2} I_{rf} |\cos\alpha| \right] + (I_r^2 - I_{rf}^2)(R_r + 0.5 R_d) \qquad (7.15)$$
$$= I_{rf}^2 R_X + I_{rf}^2 \frac{R_A}{S}$$

where

$$R_X = \left(\frac{\pi^2}{9} - 1 \right)(R_r + 0.5 R_d) \qquad (7.16)$$

$$R_A = (R_r + 0.5 R_d) + \frac{\pi}{3\sqrt{6}} \cdot \frac{1.35 V_L}{n_2 I_d} |\cos\alpha| \qquad (7.17)$$
$$= (R_r + 0.5 R_d) + \frac{V_s}{n_2 I_{rf}} |\cos\alpha|$$

where the $I_r = \frac{\pi}{3} I_{rf}$ relation has been used in Equation (7.16) to eliminate I_r. Note that the air gap power P'_g consists of two components: one is the fundamental frequency slip power and the other is the ripple power loss. Equation (7.15) indicates that the rotor circuit, which absorbs the active power, can be represented by a per-phase passive ac equivalent circuit where R_A is the equivalent resistance given by Equation (7.17). The resistance R_X represents an additional resis-

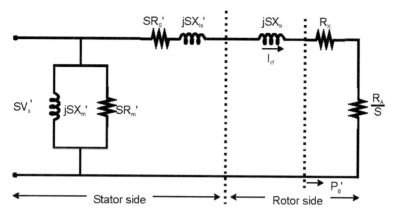

Figure 7.10 Per-phase passive equivalent circuit of the machine (with respect to rotor

tance that consumes harmonic power. The equivalent circuit is shown in Figure 7.10, where R_A is a function of V_s, I_{rf}, and $\cos \alpha$.

Note that all the stator circuit parameters and the supply voltage are multiplied by S to refer to the rotor circuit. The symbol '(prime) indicates rotor-referred parameters with turns ratio n_1; in other words,

$$V_s' = \frac{1}{n_1} V_s \tag{7.18}$$

$$X_m' = \frac{1}{n_1^2} X_m \tag{7.19}$$

$$R_m' = \frac{1}{n_1^2} R_m \tag{7.20}$$

$$R_s' = \frac{1}{n_1^2} R_s \tag{7.21}$$

$$X_{ls}' = \frac{1}{n_1^2} X_{ls} \tag{7.22}$$

It is more convenient to represent the equivalent circuit in terms of the CEMF presented by the inverter. Equation (7.15) can also be written in the form

$$P_g' = I_{rf}^2 (R_X + R_B) + V_c I_{rf} \tag{7.23}$$

Static Kramer Drive

where

$$R_B = \frac{1}{S}(R_r + 0.5R_d) \tag{7.24}$$

$$V_c = \frac{1}{S}\frac{V_s}{n_2}|\cos\alpha| \tag{7.25}$$

Figure 7.11 shows part of the rotor referred equivalent circuit with the CEMF V_c, where torque can be increased by increasing I_{rf}, that is, by decreasing V_c (with the help of S and $\cos\alpha$ parameters).

7.3.3 Torque Expression

The average torque developed by the machine is given by the total fundamental air gap power divided by the synchronous speed ω_e. The expression in terms of a passive ac equivalent circuit is

$$\begin{aligned} T_e &= 3\left(\frac{P}{2}\right)\frac{P'_{gf}}{\omega_e} \\ &= 3\left(\frac{P}{2}\right)\left(\frac{I_{rf}^2 R_A}{S}\right) \end{aligned} \tag{7.26}$$

where P'_{gf} = fundamental frequency per-phase air gap power, and the expression of R_A is given in Equation (7.17). The equation can be solved in terms of circuit parameters as

$$T_e = 3\left(\frac{P}{2}\right)\frac{SV_s'^2 R_A}{\omega_e\left[\left(SR_s' + \frac{R_A}{S}\right)^2 + \left(SX_{ls}' + SX_{lr}\right)^2\right]} \tag{7.27}$$

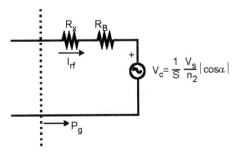

Figure 7.11 Part of rotor referred per-phase equivalent circuit with CEMF V_c

where

$$I_{rf} = \frac{SV_s'}{\sqrt{\left(SR_s' + \frac{R_A}{S}\right)^2 + \left(SX_{ls}' + SX_{lr}'\right)^2}} \quad (7.28)$$

has been substituted. Equation (7.27) relates torque as a function of slip S, rotor current I_{rf}, and inverter firing angle α. An approximate torque expression relating slip and the α angle can be derived more conveniently from the equivalent circuit of Figure 7.11. The torque in terms of fundamental air gap power P_{gf}' is given from Equation (7.23) as

$$T_e = 3\left(\frac{P}{2}\right)\frac{P_{gf}'}{\omega_e}$$
$$T_e = 3\left(\frac{P}{2}\right)\left(V_c I_{rf} + I_{rf}^2 R_B\right) \quad (7.29)$$

An approximate expression of I_{rf} can be written from the equivalent circuit by neglecting the reactances and stator resistance, which are small at small values of slip. Therefore,

$$I_{rf} \approx \frac{SV_s' - V_c}{R_B} = \frac{\frac{SV_s}{n_1} - \frac{V_s}{Sn_2}|\cos\alpha|}{R_B} \quad (7.30)$$

Substituting Equations (7.30) and (7.25) in (7.29) yields

$$T_e \approx 3\left(\frac{P}{2}\right)\frac{1}{\omega_e}\left[\frac{V_s}{Sn_2 R_B}|\cos\alpha|\left\{\frac{SV_s}{n_1} - \frac{V_s}{Sn_2}|\cos\alpha|\right\} + \frac{V_s^2}{R_B^2}\left(\frac{S}{n_1} - \frac{1}{Sn_2}|\cos\alpha|\right)^2\right]$$

$$= 3\left(\frac{P}{2}\right)\frac{V_s^2}{\omega_e}\left[\frac{|\cos\alpha|}{S^2 n_2 R_B}\left(\frac{S^2}{n_1} - \frac{|\cos\alpha|}{n_2}\right) + \frac{1}{R_B}\left(\frac{S}{n_1} - \frac{|\cos\alpha|}{Sn_2}\right)^2\right] \quad (7.31)$$

$$= 3\left(\frac{P}{2}\right)\frac{V_s^2}{\omega_e R_r}\left[\frac{|\cos\alpha|}{Sn_2}\left(\frac{S^2}{n_1} - \frac{|\cos\alpha|}{n_2}\right) + S\left(\frac{S}{n_1} - \frac{|\cos\alpha|}{Sn_2}\right)^2\right]$$

where $R_B = R_r/S$ has been substituted, neglecting the resistance R_d. Equation (7.31) relates torque as a function of both the slip S and α angle approximately. A more accurate torque expression can be derived by a computer program using the equivalent circuit. The results are plotted in Figure 7.12. The shaded area in the figure indicates the normal zone of operation, which can be compared with the approximate curves shown in Figure 7.6.

Static Kramer Drive

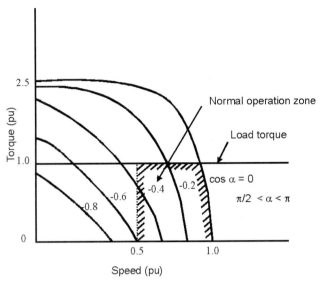

Figure 7.12 Torque-speed curves at different firing angles of inverter

7.3.4 Harmonics

The rectification of slip power causes harmonic currents in the rotor, and these harmonics are reflected to the stator by the transformer action of the machine. The harmonic currents are also injected to the ac line by the inverter. As a result, the machine losses are increased and some amount of harmonic torque is produced. The rotor current wave is ideally six-stepped, which is given by the Fourier series

$$i_r = \frac{2\sqrt{3}}{\pi} I_d \left[\cos\omega_{sl}t - \frac{1}{5}\cos 5\omega_{sl}t + \frac{1}{7}\cos 7\omega_{sl}t - \frac{1}{11}\cos 11\omega_{sl}t + ... \right] \quad (7.32)$$

where the fundamental component contributes the useful torque, but the lower order harmonics, (such as 5th and 7th) have dominating harmful effects. Each harmonic current in the rotor will create a rotating magnetic field and its direction of rotation will depend on the order of the harmonic. The 5th harmonic, for example, at frequency $5\omega_{sl}$, rotates opposite to the direction of the rotor, whereas the 7th harmonic, at frequency $7\omega_{sl}$, rotates in the same direction. The interaction of different harmonics with the air gap flux creates pulsating torque. For example, the 5th and 7th harmonics, interacting with the fundamental ψ_g wave, contribute to the 6th harmonic pulsating torque, which is discussed in Chapter 2. However, it can be shown that the harmonic torque is small compared to the average torque and can be neglected in a practical drive.

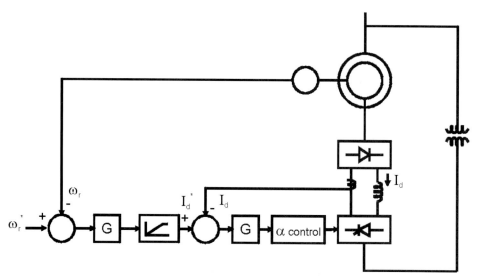

Figure 7.13 Speed control of static Kramer drive

7.3.5 Speed Control of a Kramer Drive

A speed control system of a Kramer drive is shown in Figure 7.13, and Figure 7.14 shows its typical performance. As explained before, the drive has the characteristics of a separately excited dc motor, and therefore, the control strategy is similar to a phase-controlled rectifier dc drive. With essentially constant air gap flux, the torque is proportional to dc link current I_d, which is controlled by an inner feedback loop. If the command speed ω_r^* is increased by a step, as shown in Figure 7.14, the motor accelerates at a constant developed torque corresponding to the I_d^* limit set by the speed control loop. The inverter firing angle α initially decreases with high slope to establish I_d, and then gradually decreases as speed increases. As the actual speed approaches the command speed, the dc link current is reduced to balance the load torque at a certain α angle in steady state. As the speed command is decreased by a step, I_d approaches zero and the machine slows down by the inherent load torque braking effect. During deceleration, the α angle increases continuously so that the inverter voltage V_I balances the rectifier voltage V_d. Then, as the speed error tends to be zero in the steady state, I_d is restored so that the developed torque balances with the load torque. The air gap flux during the whole operation remains approximately constant, as dictated by the stator voltage and frequency. As mentioned before, the maximum speed should be slightly less than the synchronous speed so that the current I_d can be established with finite V_d.

7.3.6 Power Factor Improvement

As discussed above, the static Kramer drive is characterized by poor line power factor because of a phase-controlled inverter. The power factor can be improved by a scheme called

Static Kramer Drive

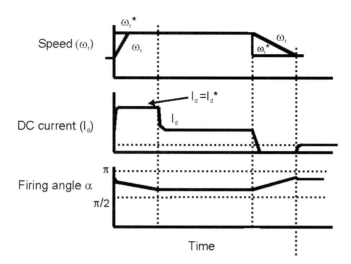

Figure 7.14 Performance characteristics of Kramer drive

commutatorless Kramer drive, which is shown in Figure 7.15. The scheme is somewhat analogous to a primitive Kramer system, where a dc motor is coupled to the induction motor shaft and is fed by the rectified slip power through a diode rectifier. In such a case, the dc motor absorbs the slip power and returns to its shaft coupled to the ac machine. In Figure 7.15, the dc motor is replaced by a synchronous motor with a load-commutated inverter, which acts as a "commutatorless dc machine." The commutatorless dc motor will be explained in Chapter 9. The inverter is fired from signals of an absolute position sensor mounted on the synchronous machine shaft.

The power flow diagram in the commutatorless Kramer drive is shown in Figure 7.16. The air gap power P_g flowing from the stator is split into shaft input power and slip power as in the static Kramer system. But, the slip power drives the synchronous motor and adds to the shaft

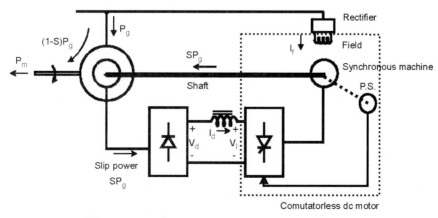

Figure 7.15 Commutatorless Kramer drive system

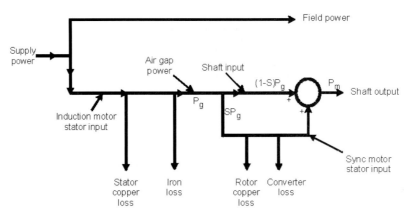

Figure 7.16 Power flow diagram in commutatorless Kramer drive system

input power to constitute the total mechanical power. The synchronous motor field is supplied from the line through a controlled rectifier. The speed and torque of the drive system are controlled by the field current I_f with the inverter firing angle α set to an optimum value for load commutation of the inverter. At any speed, a higher value of I_f will increase the machine CEMF, that is, the inverter dc voltage V_I, which will decrease the developed torque. As the machine speed decreases from synchronous speed, V_d increases linearly, but I_f increases ($I_f = K(V_I/\omega_r)$), reaching saturation soon at a lower speed. Again, as is characteristic to the load-commutated inverter drive, speed control is not possible at low values because of insufficient CEMF. Besides having an improved power factor, the system will operate reliably with short-time power failure, which is not possible in a static Kramer drive. The drive, as usual, has one-quadrant characteristics.

7.4 STATIC SCHERIUS DRIVE

As explained above and indicated by the phasor diagram of Figure 7.5, the Kramer drive has only a forward motoring mode (one quadrant) of operation. For regenerative mode operation, rotor current wave should be reversed and the corresponding phasor I_{rf} should be negative, as indicated in Figure 7.17.

This feature requires that the slip power in the rotor flow in the reverse direction. If the diode rectifier on the machine side is replaced by a thyristor bridge, as shown in Figure 7.18, the slip power can be controlled to flow in either direction. With reverse slip-power flow at subsynchronous speed, the power corresponding to shaft input mechanical power can be pumped out of the stator. It can be shown that such a drive system, with bidirectional slip-power flow, can be controlled for motoring and regenerating in both the subsynchronous and supersynchronous ranges of speed. This scheme is often defined as a static Scherbius drive system. The line commutation of the machine-side converter becomes difficult near synchronous speed (excessive commutation overlap time), when the ac voltage is very small.

Static Scherius Drive

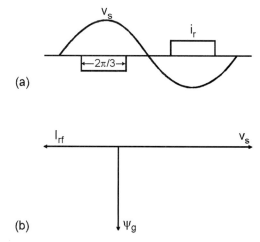

Figure 7.17 (a) Waveforms for regenerative operation of Kramer drive, (b) Phasor diagram

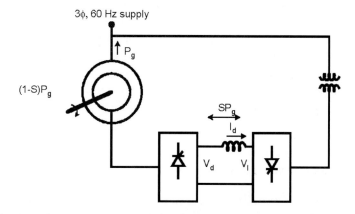

Figure 7.18 Static Scherbius drive system using dc-link thyristor converters

The dual-bridge converter system in Figure 7.18 can be replaced by a single phase-controlled line-commutated cycloconverter, as shown in Figure 7.19. The use of a cycloconverter means additional cost and complexity of control, but the resulting advantages are obvious. The problem of commutation near synchronous speed disappears, and the cycloconverter can easily operate as a phase-controlled rectifier, supplying dc current in the rotor and permitting true synchronous machine operation. The additional advantages are near-sinusoidal current waves in the rotor, which reduce harmonic loss, and a machine over-excitation capability that permits leading power factor operation on the stator side. In fact, the cycloconverter's input lagging power factor can be cancelled by the leading machine's power factor so that the line's power factor is unity. The cycloconverter should be controlled so that its output frequency and phase track precisely with those of the rotor slip frequency voltages. Like a Kramer drive, a Scherbius drive also

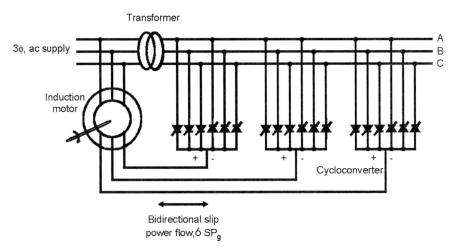

Figure 7.19 Static Scherbius drive using cycloconverter

requires resistive starter starting, which is not shown in Figure 7.19. Of course, like the Kramer drive, speed control is only possible in the forward direction. The Scherbius drive has recently found application in multi-MW, variable-speed pumps/generators [7] and flywheel energy storage [8] systems. The transformer shown on the line side of the cycloconverter is needed, as explained before, to reduce the converter power rating.

7.4.1 Modes of Operation

There are four modes of operation of a Scherbius drive, which can be explained with the help of Figure 7.20. In all the cases, it is assumed that the shaft torque is constant to the rated value and losses in the machine and cycloconverter are neglected.

Mode 1: Subsynchronous Motoring

This mode, shown in Figure 7.20(a), is identical to that of the static Kramer system. The stator input or air gap power P_g is positive and remains constant, and the slip power SP_g, which is proportional to slip (which is positive), is returned back to the line through the cycloconverter, as shown. Therefore, the line supplies the net mechanical power output $P_m = (1 - S)P_g$ consumed by the shaft. The slip frequency current in the rotor creates a rotating magnetic field in the same direction as in the stator, and the rotor speed ω_r corresponds to the difference $(\omega_e - \omega_{sl})$ between these two frequencies. At true synchronous speed ($S = 0$), the cycloconverter supplies dc excitation to the rotor and the machine behaves like a standard synchronous motor.

Mode 3: Subsynchronous Regeneration

In regenerative braking condition, as shown in Figure 7.20(b), the shaft is driven by the load and the mechanical energy is converted into electrical energy and pumped out of the stator. With negative rated torque, the mechanical power input to the shaft $P_m = (1 - S)P_g$ increases with speed and this equals the electrical power fed to the line. In the subsynchronous speed

Static Scherius Drive

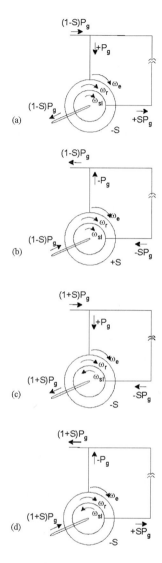

Figure 7.20 Modes of operation of static Scherbius drive (cycloconverter not shown):
  (a) Mode 1: Subsynchronous motoring,
  (b) Mode 3: Subsynchronous regeneration,
  (c) Mode 2: Supersynchronous motoring,
  (d) Mode 4: Supersynchronous regeneration

range, the slip S is positive and the air gap power P_g is negative; correspondingly, negative slip power SP_g is fed to the rotor from the cycloconverter so that the total air gap power is constant. The rotor current has positive phase sequence as before. At synchronous speed, the cycloconverter supplies dc excitation current to the rotor and the machine behaves as a synchronous generator. The drive can have sustained operation in mode 3. The typical application in this is a variable-speed wind generation system.

Mode 2: Supersynchronous Motoring

In this mode, as shown in Figure 7.20(c), the shaft speed increases beyond the synchronous speed, the slip becomes negative, and the slip power is absorbed by the rotor. The slip power supplements the air gap power for the total mechanical power output $(1 + S)P_g$. The line therefore supplies slip power in addition to stator input power. At this condition, the phase sequence of slip frequency is reversed so that the slip current-induced rotating magnetic field is opposite to that of the stator.

Mode 4: Supersynchronous Regeneration

In this mode, indicated in Figure 7.20(d), the stator output power P_g remains constant, but the additional mechanical power input is reflected as slip-power output. The cycloconverter phase sequence is now reversed so that the rotor field rotates in the opposite direction. The variable-speed wind generation mentioned for mode 3 can also be used in this mode.

Power distribution as a function of slip in subsynchronous and supersynchronous speed ranges is summarized for all four modes in Figure 7.21, where the operating speed range of ±50 percent about the synchronous speed is indicated. The control of the Scherbius drive is somewhat complex. It will be discussed in Chapter 8 with the help of vector or field-oriented control (see Figure 8.45).

Bidirectional slip-power flow with a cycloconverter, as discussed above, is also possible if the cycloconverter is replaced by a double-sided, PWM, voltage-fed converter system, as shown in Figure 7.22. For a high power rating, IGBTs can be replaced by GTOs. The dc link voltage V_d should be sufficiently higher than the inverter line voltage to permit PWM operation in the linear or undermodulation region. Both the rectifier and inverter can be operated at a programmable input power factor so that the effective line power factor can be maintained at unity. The rectifier operates satisfactorily at variable voltage and variable slip frequency on the ac side, including the ideal dc condition at synchronous speed.

7.4.2 Modified Scherbius Drive for VSCF Power Generation

A modified Scherbius drive, which has a somewhat similar topology to Figure 7.15, has been used for stand-alone shipboard VSCF power generation. The scheme, shown in Figure 7.23, has some interesting features. Induction generator output power is fed to a stand-alone, 60 Hz, constant voltage bus, which supplies the active and reactive load power as shown. The distribution of active and reactive powers in the supersynchronous and subsynchronous ranges is shown in the figure. The stator active power output P_m of the generator is equal to the turbine shaft power and the slip power fed to the rotor by the cycloconverter. The stator reactive power output Q_L is reflected to the rotor as SQ_L, which is added with the machine magnetizing power

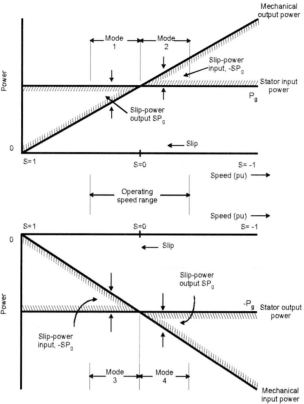

Figure 7.21 Power distribution vs. slip power in sub/supersynchronous speed ranges:
(a) Motoring at constant torque,
(b) Generation at constant torque

requirement to constitute the total reactive power Q_L' of the cycloconverter. The power Q_L' is further increased to Q_L'' at the cycloconverter input, which is supplied by the shaft-mounted synchronous exciter. The slip frequency and its phase sequence are adjusted for varying shaft speed so that the resultant air gap flux rotates at synchronous speed, as explained before. In subsynchronous speed range, the slip power SP_m is supplied to the rotor by the exciter, and therefore, the remaining output power $(1 - S)P_m$ is supplied by the shaft. In supersynchronous speed range, the rotor output power flows in the opposite direction and runs the exciter as a synchronous motor. Therefore, the total shaft power increases to $(1 + S)P_m$. Rotor voltage and frequency vary linearly with deviation from synchronous speed. For example, if the shaft speed varies in the range of 800 to 1600 rpm with 1200 rpm as the synchronous speed ($S = \pm 0.33$), the corresponding range of slip frequency is 0 to 20 Hz for 60 Hz supply frequency.

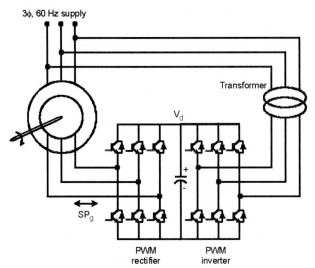

Figure 7.22 Static Scherbius drive using double-sided PWM voltage-fed converter

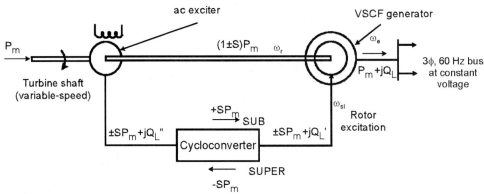

Figure 7.23 Variable-speed, constant frequency generation using modified Scherbius drive

The modified Scherbius system as a VSCF generator has several advantages over the conventional Scherbius system. One principal difference is that the wrap-around, or circulating, KVA demanded by the rotor is supplied from a separate exciter instead of being supplied from the machine's stator terminals. As a result, the main machine is much smaller in size, although in this case, the power is distributed between the two machines. The VSCF bus is much cleaner with regard to harmonics, because the cycloconverter input harmonics are reflected to the exciter. The rotor excitation circuit can be designed with a higher voltage, and the necessity of an input transformer is eliminated. In the case of a supply brown-out or temporary short-circuit fault, the system has improved controllability and reliability of power supply over the standard Scherbius system. Again, the cycloconverter can be replaced by a double-sided, PWM voltage-fed converter system, which will relieve the additional reactive and harmonic power loadings of the exciter.

7.5 SUMMARY

This chapter gives a broad review of different types of slip power-controlled drives. Unlike a standard cage-type induction motor, these drives require a doubly-fed wound-rotor induction motor, which is more expensive and has the disadvantages of slip rings and brushes. In the beginning, the primitive rotor rheostat-type speed control was discussed. Then, the more important slip-power recovery-type drives, such as the static Kramer and static Scherbius drives, were discussed in detail for limited speed range applications. The important advantages of these drives are reduced power rating of the converter (at the expense of a 60 Hz transformer) and a fast dc machine-like transient response. The additional disadvantages of these drives are the need of a separate starting method, low line power factor, and non-reversible speed control. For Scherbius drives with cycloconverters or double-sided voltage-fed PWM converters, the power factor problem does not arise. For large power pump and compressor-type applications within a limited speed range, these drives have been widely used. Scherbius drives have also been used in variable-speed wind generation, hydro/pump storage and utility systems, and flywheel energy storage systems.

REFERENCES

1. A. Lavi and R. L. Polge, "Induction motor speed control with static inverter in the rotor", *IEEE Trans. Power Appar. Syst.*, vol. 85, pp. 76–84, Jan. 1966.
2. T. Hori, H. Nagase, and M. Hombu, "Induction Motor Control Systems", *Industrial Electronics Handbook*, J. D. Irwin, pp. 310–315, CRC Press, 1997.
3. T. Wakabayashi, T. Hori, K. Shimzu, and T. Yoshioka, "Commutatorless Kramer control system for large capacity induction motors for driving water service pumps", *IEEE/IAS Annu. Meet. Conf. Rec.*, pp. 822–828, 1976.
4. H. W. Weiss, "Adjustable speed ac drive systems for pump and compressor applications", *IEEE Trans. on Ind. Appl.*, vol. 10, pp. 162–157, Jan./Feb. 1975.
5. P. Zimmermann, "Super synchronous static converter cascade", *Conf. Rec. IFAC Symp On Control in Power Elec. and Electrical Drives*, pp. 559–574, 1977.
6. G. A. Smith, "Static Scherbius system of induction motor speed control", *Proc. IEE*, vol. 124, pp. 557-565, 1977.
7. S. Mori et al., "Commissioning of 400 MW adjustable speed pumped storage system for Ohkawachi hydro power plant", *Proc. Cigre Symp.*, No. 520–04, 1995.
8. T. Nohara, H. Senaha, T. Kageyama, and T. Tsukada, "Successful commercial operation of doubly-fed adjustable-speed flywheel generating station," *Proc. of CIGRE/IEE Japan Colloquium on Rotating Electric Machinery Life Extn.*, pp. 1–6, 1997.
9. R. Peana, J. C. Clare, and G. M. Asher, "Doubly fed induction generator using back-to-back PWM converters and its application to varriable speed wind energy generation," *IEE Proc. on Elec. Power App.*, vol. 143. pp. 231–24, May 1996.
10. R. Datta and V. T. Ranganathan, "Decoupled control of active and reactive power for a grid-connected doubly-fed wound rotor induction machine without position sensors," *IEEE IAS Annu. Meet. Conf. Rec.*, pp. 2623–2630, 1999.

CHAPTER 8

Control and Estimation of Induction Motor Drives

8.1 INTRODUCTION

The control and estimation of induction motor drives constitute a vast subject, and the technology has further advanced in recent years. Induction motor drives with cage-type machines have been the workhorses in industry for variable-speed applications in a wide power range that covers from fractional horsepower to multi-megawatts. These applications include pumps and fans, paper and textile mills, subway and locomotive propulsions, electric and hybrid vehicles, machine tools and robotics, home appliances, heat pumps and air conditioners, rolling mills, wind generation systems, etc. In addition to process control, the energy-saving aspect of variable-frequency drives is getting a lot of attention nowadays.

The control and estimation of ac drives in general are considerably more complex than those of dc drives, and this complexity increases substantially if high performances are demanded. The main reasons for this complexity are the need of variable-frequency, harmonically optimum converter power supplies, the complex dynamics of ac machines, machine parameter variations, and the difficulties of processing feedback signals in the presence of harmonics. While considering drive applications, we need to address the following questions:

- One-, two- or four-quadrant drive?
- Torque, speed, or position control in the primary or outer loop?
- Single- or multi-motor drive?
- Range of speed control? Does it include zero speed and field-weakening regions?
- Accuracy and response time?
- Robustness with load torque and parameter variations?
- Control with speed sensor or sensorless control?
- Type of front-end converter?

- Efficiency, cost, reliability, and maintainability considerations?
- Line power supply, harmonics, and power factor considerations?

In this chapter, we will study different control techniques of induction motor drives, including scalar control, vector or field-oriented control, direct torque and flux control, and adaptive control. Intelligent controls with expert systems, fuzzy logic, and the neural network will be covered in Chapters 10, 11, and 12, respectively. The estimation of feedback signals, particularly speed estimation in sensorless vector controls will be discussed. Variable-frequency power supplies will be considered with voltage-fed inverters, current-fed inverters, and cycloconverters. However, emphasis will be given to voltage-fed converters because of their popularity in industrial drives. The reader should thoroughly review the fundamentals of induction machines in Chapter 2 before embarking on this chapter.

8.2 INDUCTION MOTOR CONTROL WITH SMALL SIGNAL MODEL

The general control block diagram for variable-frequency speed control of an induction motor drive is shown in Figure 8.1. It consists of a converter-machine system with hierarchy of control loops added to it. The converter-machine unit is shown with voltage (V_s^*) and frequency (ω_e^*) as control inputs. The outputs are shown as speed (ω_r), developed torque (T_e), stator current (I_s), and rotor flux (ψ_r). Instead of voltage control, the converter may be current-controlled with direct or indirect voltage control in the inner loop. The machine's dynamic model given by Equations (2.104), (2.105), and (2.115) is nonlinear and multivariable. Besides, there are coupling effects between the input and output variables. For example, both the torque and flux of a machine are functions of voltage and frequency. Machine parameters may vary with saturation, temperature, and skin effect, adding further nonlinearity to the machine model. The converter can be described by a simplified model, which consists of an amplifier gain and a dead-time lag due to the PWM sampling delay. The system becomes discrete-time because of the converter and digital control sampling effects. All the control and feedback signals can be considered as dc

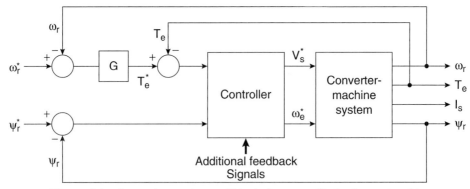

Figure 8.1 General speed control block diagram of induction motor drive

and proportional to actual variables. The speed control is shown with an inner torque control loop, which may be optional. Adding a high-gain inner loop control provides the advantages of linearization, improved bandwidth, and the ability to control the signals within safe limits. Like a dc machine, the flux of an ac machine is normally controlled to be constant at the rated value because it gives fast response and high developed torque per ampere of current. In fact, the flux under consideration may be stator flux (ψ_s), rotor flux (ψ_r), or air gap flux (ψ_m or ψ_g). However, the rotor flux control is considered in the present case. The inner control loops have faster response (i.e., higher bandwidth) than the outer loop. The "controller" block shown in the figure may have different structures, which will be discussed later.

Since an ac drive system is multivariable, nonlinear with internal coupling effect, and discrete-time in nature, its stability analysis is very complex. Computer simulation study becomes very useful for investigating the performance of the drive, particularly when a new control strategy is developed. Once the control structure and parameters of the control system are determined by the simulation study for acceptable performance, a prototype system can be designed and tested with further iteration of the controller parameters.

8.2.1 Small-Signal Model

Neglecting the discrete-time nature of the converter, the converter-machine system in Figure 8.1 can be linearized on a small-signal perturbation basis at a steady-state operating point and a transfer function model can be derived between a pair of input/output signals. The advantage of such a transfer function model is that the stability analysis of the drive system at the quiescent point is now possible using classical control theory, such as the Bode, Nyquist, or root-locus technique. Since the system is nonlinear, the poles, zeros, and gain of the transfer functions will vary as the steady-state operating point shifts. The close loop control system can then be designed with controller parameters such that at the worst operating point, the system is adequately stable and the performances are acceptable.

A dynamic model of an induction motor can be given by a fifth-order system, which is formed by combining Equations (2.104), (2.105), and (2.115). Assembling these equations in matrix form and applying a small-signal perturbation about a steady-state operating point, we get

$$\begin{bmatrix} v_{qso} + \Delta v_{qs} \\ v_{dso} + \Delta v_{ds} \\ v_{qro} + \Delta v_{qr} \\ v_{dro} + \Delta v_{dr} \\ T_{L0} + \Delta T_L \end{bmatrix} = \begin{bmatrix} R_s + SL_s & (\omega_{eo} + \Delta\omega_e)L_s & SL_m & (\omega_{eo} + \Delta\omega_e)L_m & 0 \\ -(\omega_{eo} + \Delta\omega_e)L_s & R_s + SL_s & -(\omega_{eo} + \Delta\omega_e)L_m & SL_m & 0 \\ SL_m & (\omega_{eo} + \Delta\omega_e)L_m & R_r + SL_r & (\omega_{eo} + \Delta\omega_e)L_r & -L_m(i_{dso} + \Delta i_{ds}) - L_r(i_{dro} + \Delta i_{dr}) \\ -(\omega_{eo} + \Delta\omega_e)L_m & SL_m & -(\omega_{eo} + \Delta\omega_e)L_r & R_r + SL_r & L_m(i_{qso} + \Delta i_{qs}) + L_r(i_{qro} + \Delta i_{qr}) \\ \frac{3}{2}\frac{P}{2}L_m(i_{dro} + \Delta i_{dr}) & \frac{-3}{2}\frac{P}{2}L_m(i_{qro} + \Delta i_{qr}) & 0 & 0 & \frac{-2}{P}JS \end{bmatrix} \begin{bmatrix} i_{qso} + \Delta i_{qs} \\ i_{dso} + \Delta i_{ds} \\ i_{qro} + \Delta i_{qr} \\ i_{dro} + \Delta i_{dr} \\ \omega_{ro} + \Delta\omega_r \end{bmatrix} \quad (8.1)$$

where load torque disturbance T_L is considered an input signal. Parameters v_{qso}, v_{dso}, v_{qro}, v_{dro}, T_{Lo}, ω_{eo}, i_{qso}, i_{dso}, i_{qro}, i_{dro}, and ω_{ro} describe the steady-state operating point and can be determined by solving the equations with all time derivatives (terms with S) set equal to zero. Linearizing Equation (8.1) by neglecting the Δ^2 terms and eliminating steady-state terms, we get the small-signal linear state-space equation in the form

$$\frac{dX}{dt} = AX + BU \tag{8.2}$$

where

$$X = \begin{bmatrix} \Delta i_{qs} & \Delta i_{ds} & \Delta i_{qr} & \Delta i_{dr} & \Delta \omega_r \end{bmatrix}^T \tag{8.3}$$

$$U = \begin{bmatrix} \Delta V_s & 0 & 0 & 0 & \Delta \omega_e & \Delta T_L \end{bmatrix}^T \tag{8.4}$$

$$A = \frac{-1}{L_s L_r - L_m^2} \begin{bmatrix} R_s L_r & (L_s L_r - L_m^2)\omega_{eo} + L_m^2 \omega_{ro} & -R_r L_m & L_m L_r \omega_{ro} & L_m^2 i_{dso} + L_m L_r i_{dro} \\ -(L_s L_r - L_m^2)\omega_{eo} - L_m^2 \omega_{ro} & R_s L_r & -L_m L_r \omega_{ro} & -R_r L_m & -L_m^2 i_{qso} - L_m L_r i_{qro} \\ -R_s L_m & -L_m L_s \omega_{ro} & R_r L_s & (L_s L_r - L_m^2)\omega_{eo} - L_s L_r \omega_{ro} & -L_m L_s i_{dso} - L_s L_r i_{dro} \\ L_m L_s \omega_{ro} & -R_s L_m & -(L_s L_r - L_m^2)\omega_{eo} + L_s L_r \omega_{ro} & R_r L_s & L_m L_s i_{qso} + L_s L_r i_{qro} \\ \frac{-3P^2}{8J} L_m(L_s L_r - L_m^2)i_{dro} & \frac{3P^2}{8J} L_m(L_s L_r - L_m^2)i_{qro} & \frac{3P^2}{8J} L_m(L_s L_r - L_m^2)i_{dso} & \frac{-3P^2}{8J} L_m(L_s L_r - L_m^2)i_{qso} & 0 \end{bmatrix} \tag{8.5}$$

$$B = \frac{1}{L_s L_r - L_m^2} \begin{bmatrix} L_r & 0 & -L_m & 0 & -(L_s L_r - L_m^2)i_{dso} & 0 \\ 0 & L_r & 0 & -L_m & (L_s L_r - L_m^2)i_{qso} & 0 \\ -L_m & 0 & L_s & 0 & -(L_s L_r - L_m^2)i_{dro} & 0 \\ 0 & -L_m & 0 & L_s & (L_s L_r - L_m^2)i_{qro} & 0 \\ 0 & 0 & 0 & 0 & 0 & -\frac{P}{2J}(L_s L_r - L_m^2) \end{bmatrix} \tag{8.6}$$

In the expressions above, the machine is considered stator-fed only (i.e., $\Delta v_{qr} = \Delta v_{dr} = 0$) and the stator voltage ΔV_s is aligned to the q^e-axis so that $\Delta v_{qs} = \Delta V_s$ and $\Delta v_{ds} = 0$, leaving only ΔV_s, $\Delta \omega_e$, and ΔT_L as input variables. The small-signal block diagram is shown in Figure 8.2, where the electrical and mechanical responses have been separated. The converter gain can be easily merged with the input voltage signal ΔV_S. The converter-machine model generates currents from input control signals ΔV_s and $\Delta \omega_e$, and feedback speed signal $\Delta \omega_r$ acts to generate the speed-induced CEMF. The developed torque ΔT_e is synthesized from the currents by the equation

$$\Delta T_e = \frac{3}{2}\left(\frac{P}{2}\right) L_m \left[(i_{dro}\Delta i_{qs} + i_{qso}\Delta i_{dr}) - (i_{dso}\Delta i_{qr} + i_{qro}\Delta i_{ds}) \right] \tag{8.7}$$

The other small-signal outputs, ΔI_s and $\Delta \psi_r$ in Figure 8.1, can be synthesized from the current signals as follows:

The stator current $\hat{I}_s$ is given as

$$|\bar{I}_s| = |I_{qds}| = \sqrt{i_{qs}^2 + i_{ds}^2} \tag{8.8}$$

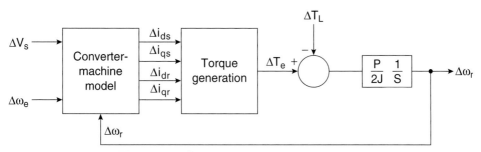

Figure 8.2 Small-signal control block diagram

which can be small-signal linearized as

$$\left|\Delta \bar{I}_s\right| = \frac{i_{qso}}{\sqrt{i_{qso}^2 + i_{dso}^2}} \Delta i_{qs} + \frac{i_{dso}}{\sqrt{i_{qso}^2 + i_{dso}^2}} \Delta i_{ds} \tag{8.9}$$

Similarly, the rotor flux is given as

$$\left|\bar{\psi}_r\right| = \left|\psi_{qdr}\right| = \sqrt{\psi_{qr}^2 + \psi_{dr}^2} \tag{8.10}$$

where

$$\psi_{qr} = L_r i_{qr} + L_m i_{qs} \tag{8.11}$$

$$\psi_{dr} = L_r i_{dr} + L_m i_{ds} \tag{8.12}$$

can be easily derived from the d^e-q^e equivalent circuits of Figure 2.23. The linearized equation for the rotor flux can be derived as

$$\begin{aligned}\left|\Delta \bar{\psi}_r\right| &= \frac{\psi_{qro}}{\psi_{ro}} \Delta \psi_{qr} + \frac{\psi_{dro}}{\psi_{ro}} \Delta \psi_{dr} \\ &= \frac{L_m(L_r i_{qro} + L_m i_{qso})}{\psi_{ro}} \Delta i_{qs} + \frac{L_m(L_r i_{dro} + L_m i_{dso})}{\psi_{ro}} \Delta i_{ds} \\ &+ \frac{L_r(L_r i_{qro} + L_m i_{qso})}{\psi_{ro}} \Delta i_{qr} + \frac{L_r(L_r i_{dro} + L_m i_{dso})}{\psi_{ro}} \Delta i_{dr}\end{aligned} \tag{8.13}$$

where

$$\psi_{r0} = \sqrt{(L_r i_{qro} + L_m i_{qs0})^2 + (L_r i_{dr0} + L_m i_{ds0})^2} \tag{8.14}$$

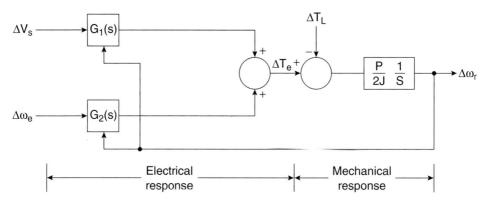

Figure 8.3 Small-signal transfer function block diagram

The small-signal transfer function block diagram derived from Figure 8.2 is shown in Figure 8.3. Here, the transfer functions $G_1(S)$ and $G_2(S)$ are defined as

$$G_1(S) = \frac{\Delta T_e}{\Delta V_s}\bigg|_{\Delta\omega_r=0, \Delta\omega_e=0} \tag{8.15}$$

$$G_2(S) = \frac{\Delta T_e}{\Delta\omega_e}\bigg|_{\Delta\omega_r=0, \Delta V_s=0} \tag{8.16}$$

In these equations, the speed is treated as a constant parameter ($\Delta\omega_r = 0$), assuming the system inertia J is very large (i.e., mechanical response is more sluggish compared to electrical response). In such a case, the response of speed $\Delta\omega_r$ is essentially dictated by the J parameter. For any finite inertia system, transfer functions $\Delta\omega_r/\Delta V_s$, $\Delta\omega_r/\Delta\omega_e$, and $\Delta\omega_r/\Delta T_L$ can be derived from the state-space form of Equation (8.2).

8.3 SCALAR CONTROL

Scalar control, as the name indicates, is due to magnitude variation of the control variables only, and disregards the coupling effect in the machine. For example, the voltage of a machine can be controlled to control the flux, and frequency or slip can be controlled to control the torque. However, flux and torque are also functions of frequency and voltage, respectively. Scalar control is in contrast to vector or field-oriented control (will be discussed later), where both the magnitude and phase alignment of vector variables are controlled. Scalar-controlled drives give somewhat inferior performance, but they are easy to implement. Scalar-controlled drives have been widely used in industry. However, their importance has diminished recently because of the superior performance of vector-controlled drives, which is demanded in many applications. In this section, a few selected scalar control techniques with voltage-fed and current-fed inverters will be discussed.

8.3.1 Voltage-Fed Inverter Control

8.3.1.1 Open Loop Volts/Hz Control

The open loop volts/Hz control of an induction motor is by far the most popular method of speed control because of its simplicity, and these types of motors are widely used in industry. Traditionally, induction motors have been used with open loop 60 Hz power supplies for constant speed applications. For adjustable speed applications, frequency control is natural. However, voltage is required to be proportional to frequency so that the flux ($\psi_s = V_s/\omega_e$) remains constant (see Figure 2.14), neglecting the stator resistance R_s drop. Figure 8.4 shows the block diagram of the volts/Hz speed control method. The power circuit consists of a diode rectifier with a single- or three-phase ac supply, LC filter, and PWM voltage-fed inverter. Ideally, no feedback signals are needed for the control. The frequency ω_e^* is the primary control variable because it is approximately equal to speed ω_r, neglecting the small slip frequency ω_{sl} of the machine. The phase voltage V_s^* command is directly generated from the frequency command by the gain factor G, as shown, so that the flux ψ_s remains constant. If the stator resistance and leakage inductance of the machine are neglected, the flux will also correspond to the air gap flux ψ_m or rotor flux ψ_r. As the frequency becomes small at low speed, the stator resistance tends to absorb the major amount of the stator voltage, thus weakening the flux. The boost voltage V_0 is

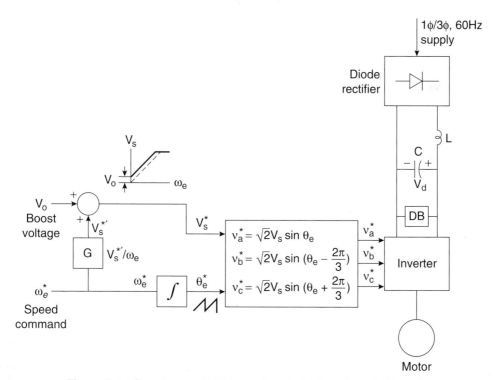

Figure 8.4 Open loop volts/Hz speed control with voltage-fed inverter

added so that the rated flux and corresponding full torque become available down to zero speed. Note that the effect of boost voltage becomes negligible at higher frequencies. The ω_e^* signal is integrated to generate the angle signal θ_e^*, and the corresponding sinusoidal phase voltages (v_a^*, v_b^*, and v_c^* signals) are generated by the expressions shown in the figure. The PWM controller is merged with the inverter block.

Figure 8.5 shows the drive's steady-state performance on a torque-speed plane with a fan or pump-type load ($T_L = K\omega_r^2$). As the frequency is gradually increased, the speed also increases almost proportionally, as indicated by points 1, 2, 3, 4, etc. The operation can be continued smoothly in the field-weakening region where the supply voltage V_s saturates.

Now consider the load torque and line voltage variation effects. If the initial operating point is 3 and the load torque is increased to T_L' for the same frequency command, the speed will droop from ω_r to ω_r'. This droop is small, particularly with a high-efficiency (i.e., low slip) machine, and is easily tolerated for a pump or fan-type drive where precision speed control is not necessary. Assume now that the operation is at point a in another torque-speed curve. If the ac line voltage decreases, that lowers the machine terminal voltage. The speed will also droop corresponding to point b, as shown in the figure. As the literature suggests, a speed droop correction in an open loop control can be achieved by adding an estimated slip signal with the frequency command.

If the frequency command ω_e^* is changed abruptly by a small increment, the slip will change to change the developed torque (within the safe limit), but the speed will tend to remain constant because of machine inertia. However, if it is desired to change the speed by a large step of the frequency command (positive or negative), the drive will easily become unstable. The satisfactory acceleration/deceleration characteristics of the machine are explained in Figure 8.6.

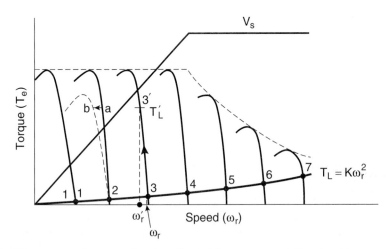

Figure 8.5 Torque-speed curves showing effect of frequency variation, load torque, and supply voltage changes

Scalar Control

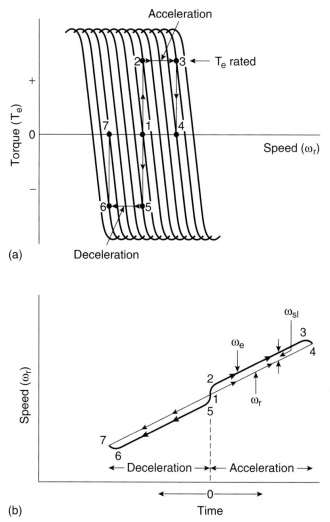

Figure 8.6 Acceleration/deceleration characteristics with volts/Hz control:
(a) On torque-speed curves,
(b) As function of time

Assume a pure inertia-type load for simplicity and that the machine is initially operating at point 1. The command frequency is now increased abruptly by a small step so that with an increment of slip frequency, the operating point shifts to point 2, which corresponds to the rated, developed torque. The drive now goes through constant acceleration with the ramping of the frequency command when the speed tracks the frequency within the limit of slip frequency so that the stability and safe stator current limit are maintained. At operating point 3, the frequency command can be decremented to attain the steady-state operating point, which is 4. The machine torque and speed are related by the following equation:

$$\omega_r = \int \frac{(T_e - T_L)}{J} dt \qquad (8.17)$$

where J = moment of inertia, T_e = developed torque, and T_L = load torque (zero in this case). With the rated T_e, the slope of acceleration $d\omega_r/dt$, as indicated in Figure 8.6(b), is dictated by parameter J, that is, a higher J will permit slow acceleration, and vice versa. If it is possible to estimate J on-line for a variable inertia load, the acceleration of the drive can be predetermined. The deceleration performance is similar, and it is also explained in Figure 8.6. With a diode rectifier on the line side, the inverter will need a dynamic brake, as indicated in Figure 8.4. By decrementing the frequency command in a step, the operating point will shift from point 1 to 5 due to a negative developed torque. It will then decelerate with constant slope given by Equation (8.17) until the steady-state operating point, 7, can be reached again.

Typical drive performance with open loop volts/Hz control in both the accelerating and decelerating conditions with load torque $T_L = K \omega_r^2$ is given by the simulation results shown in Figure 8.7. The inherent machine coupling effect slows down the torque response, as explained before. In addition, there is some amount of underdamping in the torque and flux responses, which increases at lower frequencies but does not affect the speed response because of inertia filtering. Drift in the flux signal for varying torque (i.e., stator current) is also evident. Small-signal analysis and simulation study indicate that at certain regions of operation, the system tends to be unstable [3].

8.3.1.2   Energy Conservation Effect by Variable Frequency Drive

Numerous ac motor drives are used with a pump or fan-type load, where fluid flow control is needed. The traditional method of flow control is by operating the machine at a constant speed with a 60 Hz power supply, and then controlling the flow by throttle opening. The efficiency of this method of flow control is poor, as shown in Figure 8.8, where power consumption is plotted with the loading factor. Variable-frequency speed control of the drive with a fully open throttle reduces power consumption, which is indicated in the figure. For example, with 60 percent loading, the efficiency improvement can be as high as 35 percent. Since drives operate most of the time at light loads, the accumulated energy savings for a prolonged time period can be substantial. The payback period for investment cost of power electronics is small, particularly where the energy cost is high. In addition to the economic factor, efficient utilization of energy reduces the energy demand and correspondingly helps solve the environmental pollution problem where it is generated by fossil fuels. Further improvement of drive efficiency by programming the flux at light loads will be discussed later.

8.3.1.3   Speed Control with Slip Regulation

An improvement of open loop volts/Hz control is close loop speed control by slip regulation as shown in Figure 8.9. Here, the speed loop error generates the slip command ω_{sl}^* through

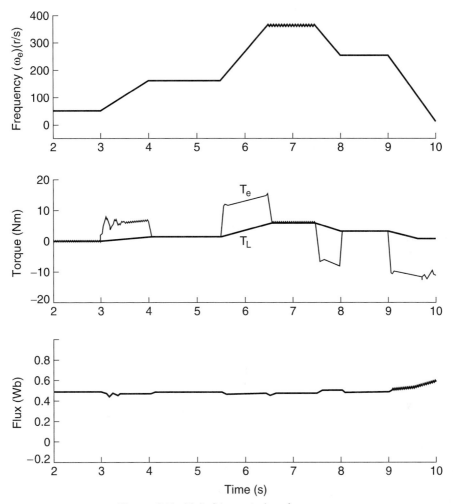

Figure 8.7 Volts/Hz control performance

a proportional-integral (P-I) controller and limiter. The slip is added to the feedback speed signal to generate the frequency command as shown. The frequency command ω_e^* also generates the voltage command through a volts/Hz function generator, which incorporates the low-frequency stator drop compensation. Since the slip is proportional to the developed torque at constant flux (Equation (2.41)), the scheme can be considered as an open loop torque control within a speed control loop. The feedback current signal is not used anywhere in the loop. With a step-up speed command, the machine accelerates freely with a slip limit that corresponds to the stator current or torque limit, and then settles down to the slip value at steady state as dictated by the load torque. If the command speed ω_r^* is reduced by a step, the drive goes into regenerative or dynamic braking mode and decelerates with constant negative slip $-\omega_{sl}^*$, as indicated in the

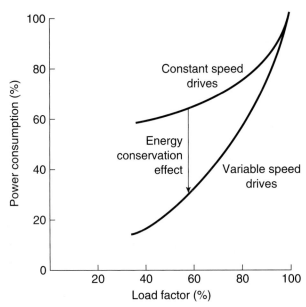

Figure 8.8 Energy-saving characteristics with variable-frequency speed control

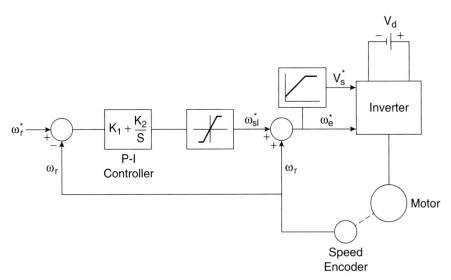

Figure 8.9 Close loop speed control with volts/Hz control and slip regulation

figure. The effects of load torque and line voltage variation are explained in Figure 8.10. If the initial operating point is 1 and the load torque is increased from T_L to T_L', the speed will tend to drop corresponding to point 2. However, the speed control loop will increase the frequency until the original speed is restored at point 3. Since there is no close loop flux control, the line voltage variation will cause some flux drift. Again, if the initial operating point is 1 on curve a of Figure

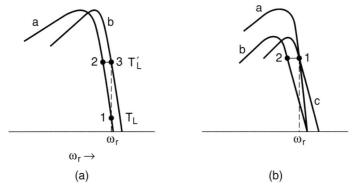

Figure 8.10 (a) Effect of load torque variation, (b) Effect of supply voltage variation

8.10(b), the decrease of line voltage will reduce the flux, tending to shift the operating point to 2. The resulting speed drop will act on the speed loop and raise the frequency to restore the original speed at point 1 on curve c. The system works well in the field-weakening mode also.

8.3.1.4 Speed Control with Torque and Flux Control

As discussed above, the volts/Hz control has the disadvantage that the flux may drift, and as a result, the torque sensitivity with slip (Equation (2.41)) will vary. In addition, line voltage variation, incorrect volts/Hz ratio, stator drop variation by line current, and machine parameter variation may cause weaker flux or the flux may saturate. In Figure 8.9, if the flux becomes weak, the developed torque will decrease with the slip limit and the machine's acceleration/deceleration capability will decrease.

A speed control system with close loop torque and flux control is shown in Figure 8.11. Additional feedback loops mean complexity of additional feedback signal synthesis and potential stability problems. A torque loop within the speed loop improves the speed loop's response. The flux control loop controls the voltage V_s^* as shown. Both the torque and flux feedback signals can be estimated from the machine terminal voltages and currents, as indicated. The feedback signal estimation will be discussed later. With constant $\hat{\psi}_r^*$ command, as the speed increases, the voltage increases proportionally until square-wave mode is reached and field-weakening mode starts. However, if PWM operation is desired in field-weakening mode, the flux command must be decreased to vary inversely with the speed signal so that the PWM controller does not saturate. The flux control loop is usually slower than the torque control loop. The drive can operate in regenerative (or dynamic) braking mode, but the reversal of speed requires a reversal of the phase sequence of the inverter.

With scalar control, as the frequency command ω_e^* is increased by the torque loop, the flux temporarily decreases until it is compensated by the sluggish flux control loop. This inherent coupling effect slows down the torque response.

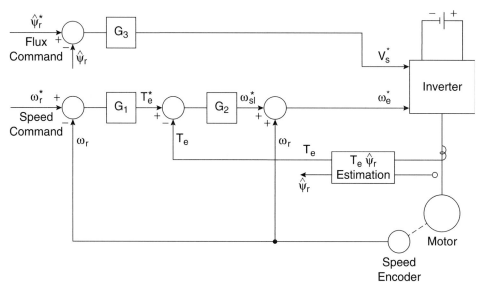

Figure 8.11 Close loop speed control with torque and flux control

8.3.1.5 Current-Controlled Voltage-Fed Inverter Drive

Instead of controlling inverter voltage by the flux loop, the stator current can also be controlled. Close loop current control is beneficial to power semiconductors because of the inherent protection from overcurrent. Besides, the torque and flux of the machine are directly sensitive to currents. A voltage-fed inverter drive with outer loop torque and flux control and hysteresis-band current control in the inner loop is shown in Figure 8.12. Instead of constant rated flux, the flux can be programmed with torque as shown for a light-load efficiency improvement, which will be discussed later. In the figure, the flux control loop generates the stator current magnitude, and its frequency command is generated by the torque loop. The three-phase command currents are then generated by the relations shown in the figure. The feedback phase currents i_a, i_b, and i_c can be sensed by two current sensors because for an isolated neutral load, $i_a + i_b + i_c = 0$.

The performance of the drive for subway traction [5] in both acceleration and regenerative braking modes is explained in Figure 8.13. The figure is essentially the same as Figure 2.14 except that a braking region has been added. In the constant torque region, the inverter operates in the PWM current control mode and has the features of a current-fed inverter. Beyond the base speed, the inverter operates in square-wave mode, because with the constant flux command, the PWM controller fully saturates and the current commands are converted to square-wave voltage commands. In field-weakening mode, only frequency variation by slip control is possible to control the torque. To operate on the torque envelope, the slip is increased to the maximum value in a preprogrammed manner so that the stator current remains constant within the limit value. Beyond the constant power region, the slip remains constant, but the stator current decreases as indicated. The drive may operate at reduced torque at any speed by the reduction of slip. The

Scalar Control

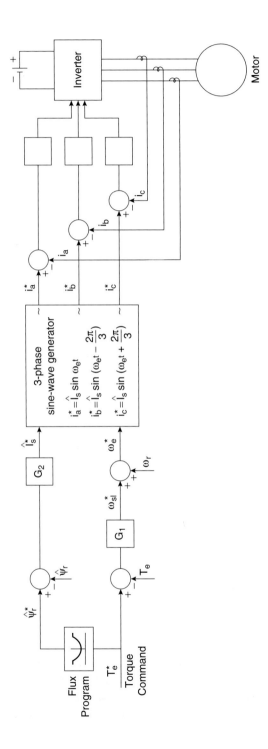

Figure 8.12 Voltage-fed, current-regulated inverter drive with torque and flux control

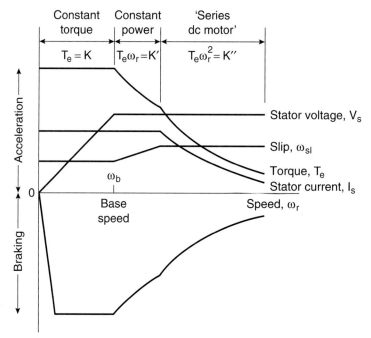

Figure 8.13 Drive characteristics in acceleration and braking modes

available regenerative braking torque is diminished as the speed decreases and then vanishes at zero speed because of the circuit drops.

8.3.1.6 Traction Drives with Parallel Machines

For a voltage-fed inverter system, multiple inverters can be operated in parallel with a single rectifier, or multiple machines can be operated in parallel with a single inverter. In many applications, such as conveyer lines, extruder mills, and subway and locomotive tractions, several identical induction motors are required to operate in parallel with one inverter. Figure 8.14 shows a typical IGBT converter locomotive drive where two identical machines of equal power rating operate in parallel with a single voltage-fed inverter, and each machine drives an axle of the locomotive. If the machines have matched torque-speed characteristics and their speeds are equal, each will offer identical impedance on the variable frequency supply line and their torque sharing will be equal at all operating conditions. In practice, there will be some amount of mismatch between machine characteristics, and speeds may not be identical because of mismatch between the wheel diameters.

Consider first some mismatch in machine characteristics, that is, machine 1 has lower slip than machine 2, as shown in Figure 8.15(a). The wheel diameters for each axle are identical so that the speed ω_r is the same. With supply frequency ω_e, the slip will be the same in both motoring and regenerative conditions. Machine 1, with lower slip characteristics, will share more torque than machine 2. With high-efficiency, low-slip machines, this torque-sharing inequality

Scalar Control

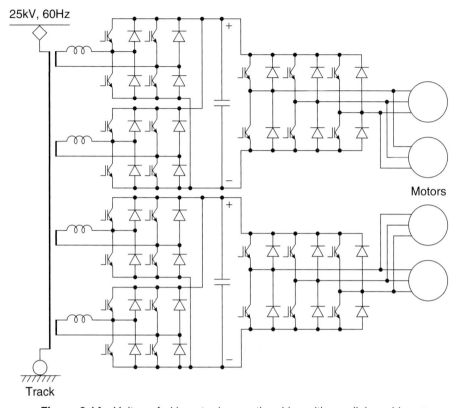

Figure 8.14 Voltage-fed inverter locomotive drive with parallel machines

will be greater. The torque sharing of machine 1 in motoring mode may be excessive to induce slippage of the axle wheels. The incrementation of speed as a result of slippage will decrease its torque sharing, and therefore, will tend to have self-correction. In the regenerative mode, if there is a wheel slippage of machine 1, its speed incrementation will induce larger torque sharing, thus worsening the condition.

Consider now that the machines are exactly matched, but the wheel diameter of machine 2 is slightly shorter than that of machine 1, which will cause $\omega_{r2} > \omega_{r1}$ for the same vehicle speed. This condition, as indicated in Figure 8.15(b), will cause higher torque sharing (T_{e1}) for machine 1 in motoring condition, but lower torque sharing (T_{b1}) in regenerative mode. The wheel slippage characteristics will be the same as before. Again, in practice, both machine mismatch and unequal wheel diameter problems may coexist. The composite problem is left as an exercise to the reader. Open loop volts/Hz speed control is very convenient in these applications. However, close loop high-performance control, such as vector control, is also possible by considering mean feedback signals in closely matched installations.

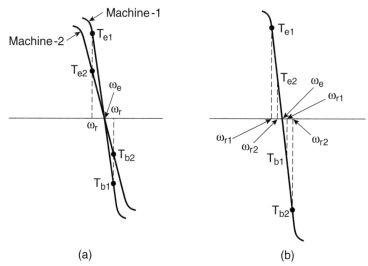

Figure 8.15 (a) Characteristics of mismatched machines but equal wheel diameters, (b) Characteristics of matched machines but unequal wheel diameters

8.3.2 Current-Fed Inverter Control

Some of the control principles of voltage-fed inverter drives, as discussed above, are also applicable for current-fed inverter drives.

8.3.2.1 Independent Current and Frequency Control

In a current-fed inverter drive, the dc link current and inverter frequency are the two control variables where the current can be controlled by varying the firing angle of the front-end thyristor rectifier. Unfortunately, a current-fed inverter system cannot be operated in an open loop like a voltage-fed inverter drive. A minimal close loop control system of a current-fed inverter system, where the dc link current I_d and slip ω_{sl} are controlled independently, is shown in Figure 8.16(a). Figure 8.16(b) shows its performance characteristics [6]. The current I_d is controlled by a feedback loop that controls the rectifier output voltage V_d. The commanded slip is added with the speed signal ω_r to generate the frequency command ω_e^*. In acceleration mode, the slip is positive, but regeneration will be effective with a negative slip command when both V_d and V_I voltages become negative and power is fed back to the source. The main disadvantage of the system is that it has no flux control. The developed torque of the drive can be controlled either by the current I_d or the slip signal ω_{sl}. Machine acceleration in constant torque from point 1 to point 2 with constant I_d and constant ω_{sl} is shown in Figure 8.16(b). As explained in Figure 2.16, the machine is intentionally operated in a statically unstable region of the torque-speed curve so that the flux remains below saturation. At steady-state point 3, if the slip is reduced to balance developed torque with the load torque, saturation will occur. If, on the other hand, the current I_d is reduced at constant slip, the flux may be too low.

Scalar Control

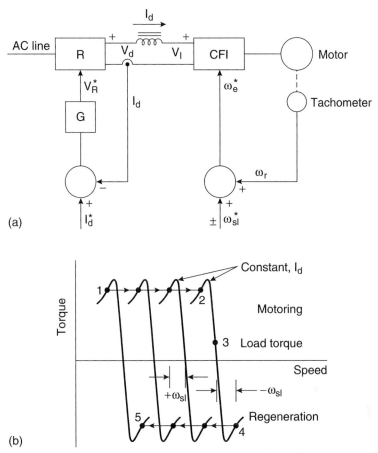

Figure 8.16 (a) Independent current and slip control, (b) Torque-speed characteristics for acceleration and deceleration

8.3.2.2 Speed and Flux Control in Current-Fed Inverter Drive

A practical and much improved scalar control technique of a current-fed inverter drive is shown in Figure 8.17. Here, the speed control loop controls the torque by slip control as usual, but it also controls the current I_d^* through a pre-computed function generator to maintain a constant flux. At zero slip, the developed torque is zero, but the current has a minimum value that corresponds to the magnetizing current of the machine. The slip becomes negative in regenerative mode, but the I_d^* - ω_{sl}^* relation is symmetrical with the motoring mode. The open loop, pre-programmed I_d^* control is satisfactory, but it has one disadvantage: machine flux may vary with parameter variation. An independent flux control loop outside the current control loop, as indicated in Figure 8.12, can be provided for tighter flux control.

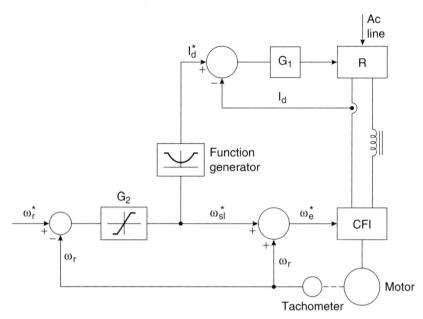

Figure 8.17 Current-fed inverter drive with speed and flux control

8.3.2.3 Volts/Hz Control of Current-Fed Inverter Drive

As the current control mode of a voltage-fed inverter is possible, similarly, a current-fed inverter can also be controlled in voltage-fed mode. Figure 8.18 shows a volts/Hz-controlled drive with a current-fed inverter. In such a case, the features of a current-fed inverter, such as easy regeneration to the line and immunity to the shoot-through fault, can be retained. The command frequency ω_e^* operates directly on the inverter as usual. The voltage command V_s^*, generated from ω_e^* through the volts/Hz function generator, is compared with the machine terminal voltage. The resulting error controls the dc link current I_d in the inner loop, as shown. As the machine speed is increased by ramping up the frequency command, the current I_d changes to match the desired voltage. The drive goes to regenerative mode when the frequency command is ramped down. A particular advantage in the control strategy is that the fluctuation of the line voltage does not affect the machine flux because the current loop makes the desired correction.

8.3.3 Efficiency Optimization Control by Flux Program

Machines are normally operated at rated flux, as mentioned before, so that the developed torque/amp is high and transient response is fast. Industrial drives usually operate at light loads most of the time. If rated flux is maintained at a light load, it can be shown that the core loss (function of flux and frequency) is excessive, giving poor efficiency of the drive. For a certain steady-state, light-load torque condition and at a certain speed, the typical distribution of losses in a converter-machine system and its variation with a variation of flux is shown in Figure 8.19. As the flux ψ_r is reduced from the rated value, the core loss decreases, but the machine copper loss

Scalar Control

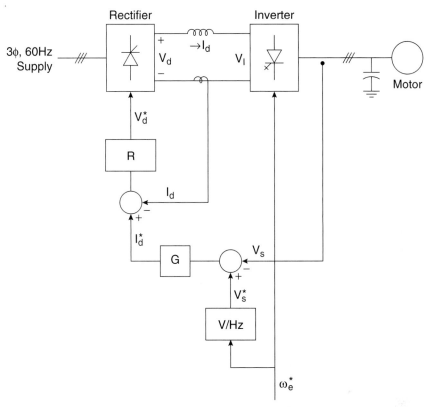

Figure 8.18 Volts/Hz control of current-fed GTO PWM inverter drive

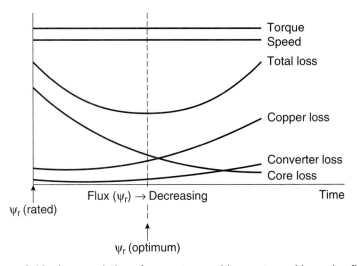

Figure 8.19 Loss variation of converter-machine system with varying flux

and converter losses increase. However, the total loss decreases to a minimum value and then increases again. It is desirable to set the rotor flux at ψ_r (optimum) so that the converter-machine system efficiency is optimum. Figure 8.20 shows the typical optimum flux program for variable torque and constant speed, and compares the corresponding efficiency with that of the rated flux. The flux program is symmetrical for motoring and regenerative torque conditions. Note that at the rated torque, the flux should be at rated value and there is no efficiency improvement. The incremental efficiency improvement becomes larger as the torque is reduced. Figure 8.12 incorporates a similar flux program for efficiency optimization. With full commanded torque T_e^*, the rated flux is established for fast acceleration, but at steady-state, light load, the flux is reduced for efficiency improvement. Since core loss is also influenced by speed (i.e., frequency), the flux program should vary with the variation of speed. Figure 8.21 shows the simulation result of a 5 hp drive's efficiency improvement by using flux programs at three different speeds. Normally, a machine is designed to operate at best efficiency at the rated torque and speed (corner point). The efficiency decreases as speed and/or torque decrease. However, at a certain speed, the incremental efficiency improvement at light load can be substantial by flux programming. Note that Figure 8.21 shows efficiency curves for the total converter-machine system.

Figure 8.22 shows the open loop volts/Hz program for the optimum flux programming control of a pump-type drive, where the load torque as a function of speed is given in the figure. At each speed, the load torque is known and the corresponding optimum flux is shown in the figure. The flux determines the applied voltage. The drive starts with a rated flux and then settles down in the operating range where the optimum flux program is valid. Commercial drives with volts/Hz control are often available with a selectable volts/Hz program for efficiency optimization control. A more advanced method of efficiency optimization control by an on-line search method will be discussed in Chapter 11.

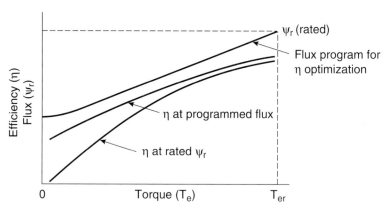

Figure 8.20 Efficiency improvement by flux program at variable torque but constant speed

Scalar Control

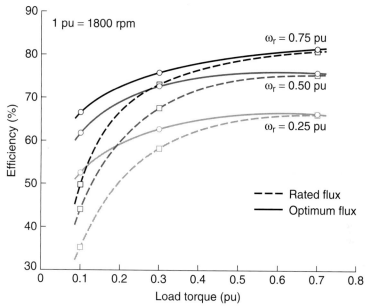

Figure 8.21 Efficiency improvement by flux program at variable torque and different speeds

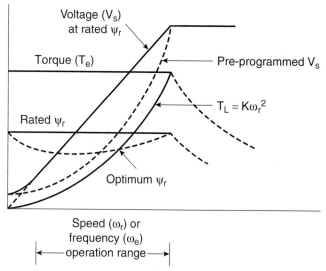

Figure 8.22 Open loop volts/Hz program for optimal efficiency for pump load

8.4 VECTOR OR FIELD-ORIENTED CONTROL

So far, we have discussed scalar control techniques of voltage-fed and current-fed inverter drives. Scalar control is somewhat simple to implement, but the inherent coupling effect (i.e., both torque and flux are functions of voltage or current and frequency) gives sluggish response and the system is easily prone to instability because of a high-order (fifth-order) system effect. To make it more clear, if, for example, the torque is increased by incrementing the slip (i.e., the frequency), the flux tends to decrease. Note that the flux variation is always sluggish. The flux decrease is then compensated by the sluggish flux control loop feeding in additional voltage. This temporary dipping of flux reduces the torque sensitivity with slip and lengthens the response time. This explanation is also valid for current-fed inverter drives.

The foregoing problems can be solved by vector or field-oriented control. The invention of vector control in the beginning of 1970s, and the demonstration that an induction motor can be controlled like a separately excited dc motor, brought a renaissance in the high-performance control of ac drives. Because of dc machine-like performance, vector control is also known as decoupling, orthogonal, or transvector control. Vector control is applicable to both induction and synchronous motor drives, and the latter will be discussed in Chapter 9. Undoubtedly, vector control and the corresponding feedback signal processing, particularly for modern sensorless vector control, are complex and the use of powerful microcomputer or DSP is mandatory. It appears that eventually, vector control will oust scalar control, and will be accepted as the industry-standard control for ac drives.

8.4.1 DC Drive Analogy

Ideally, a vector-controlled induction motor drive operates like a separately excited dc motor drive, as mentioned above. Figure 8.23 explains this analogy. In a dc machine, neglecting the armature reaction effect and field saturation, the developed torque is given by

$$T_e = K_t' I_a I_f \tag{8.18}$$

where I_a = armature current and I_f = field current. The construction of a dc machine is such that the field flux ψ_f produced by the current I_f is perpendicular to the armature flux ψ_a, which is produced by the armature current I_a. These space vectors, which are stationary in space, are orthogonal or decoupled in nature. This means that when torque is controlled by controlling the current I_a, the flux ψ_f is not affected and we get the fast transient response and high torque/ampere ratio with the rated ψ_f. Because of decoupling, when the field current I_f is controlled, it affects the field flux ψ_f only, but not the ψ_a flux. Because of the inherent coupling problem, an induction motor cannot generally give such fast response.

DC machine-like performance can also be extended to an induction motor if the machine control is considered in a synchronously rotating reference frame (d^e–q^e), where the sinusoidal variables appear as dc quantities in steady state. In Figure 8.23(b), the induction motor with the inverter and vector control in the front end is shown with two control current inputs, i_{ds}^* and i_{qs}^*.

Vector or Field-Oriented Control

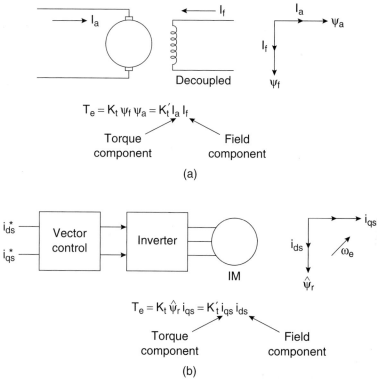

Figure 8.23 (a) Separately excited dc motor, (b) Vector-controlled induction motor

These currents are the direct axis component and quadrature axis component of the stator current, respectively, in a synchronously rotating reference frame. With vector control, i_{ds} is analogous to field current I_f and i_{qs} is analogous to armature current I_a of a dc machine. Therefore, the torque can be expressed as

$$T_e = K_t \hat{\psi}_r i_{qs} \qquad (8.19)$$

or

$$T_e = K_t' i_{ds} i_{qs} \qquad (8.20)$$

where $\hat{\psi}_r$ = absolute $\bar{\psi}_r$ is the peak value of the sinusoidal space vector. This dc machine-like performance is only possible if i_{ds} is oriented (or aligned) in the direction of flux $\hat{\psi}_r$ and i_{qs} is established perpendicular to it, as shown by the space-vector diagram on the right of Figure 8.23(b). This means that when i_{qs}^* is controlled, it affects the actual i_{qs} current only, but does not affect the flux $\hat{\psi}_r$. Similarly, when i_{ds}^* is controlled, it controls the flux only and does not affect the i_{qs} component of current. This vector or field orientation of currents is essential under all operating conditions in a vector-controlled drive. Note that when compared to dc machine space vectors, induction machine space vectors rotate synchronously at frequency ω_e, as indicated in the figure. In summary, vector control should assure the correct orientation and equality of command and actual currents.

8.4.2 Equivalent Circuit and Phasor Diagram

Let us now study the concept of vector orientation further with the help of equivalent circuit and phasor diagrams. Figure 8.24 shows the complex form of d^e-q^e equivalent circuits in steady-state condition derived from Equations (2.109) and (2.110), where rms values V_s and I_s are replaced by corresponding peak values (sinusoidal vector variables), as shown. The rotor leakage inductance L_{lr} has been neglected for simplicity, which makes the rotor flux $\hat{\psi}_r$ the same as the air gap flux $\hat{\psi}_m$. The stator current $\hat{I}_s$ can be expressed as

$$\hat{I}_s = \sqrt{i_{ds}^2 + i_{qs}^2} \qquad (8.21)$$

where i_{ds} = magnetizing component of stator current flowing through the inductance L_m and i_{qs} = torque component of stator current flowing in the rotor circuit. Figure 8.25 shows the phasor (or vector) diagrams in d^e-q^e frame with peak values of sinusoids and air gap voltage $\hat{V}_m$ aligned on the q^e axis. The phase position of the currents and flux is shown in the figure, and the corresponding developed torque expression is given by Equation (8.19). The terminal voltage $\hat{V}_s$ is slightly leading because of the stator impedance drop. The in-phase or torque component of current i_{qs} contributes active power across the air gap, whereas the reactive or flux component of current i_{ds} contributes only reactive power. Figure 8.25(a) indicates an increase of the i_{qs} component of the stator current to increase the torque while maintaining the flux ψ_r constant, whereas (b) indicates a weakening of the flux by reducing the i_{ds} component. Note that although the operation is explained for the steady-state condition, the explanation is also valid for the transient condition in the qds equivalent circuit of Figure 2.24. After understanding the vector orientation principle, the next question is how to control the i_{ds} and i_{qs} components of stator current $\hat{I}_s$ independently with the desired orientation. Instead of considering a cartesian form (i_{ds} and i_{qs}) of control, it is also possible to consider control in a polar form ($|I_s|$ and θ).

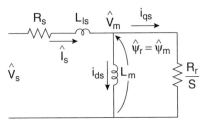

Figure 8.24 Complex (qds) equivalent circuit in steady state (rotor leakage inductance neglected)

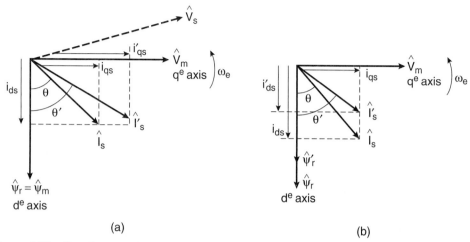

Figure 8.25 Steady-state phasors (in terms of peak values) (a) Increase of torque component of current, (b) Increase of flux component of current

8.4.3 Principles of Vector Control

The fundamentals of vector control implementation can be explained with the help of Figure 8.26, where the machine model is represented in a synchronously rotating reference frame. The inverter is omitted from the figure, assuming that it has unity current gain, that is, it generates currents i_a, i_b, and i_c as dictated by the corresponding command currents i_a^*, i_b^*, and i_c^* from the controller. A machine model with internal conversions is shown on the right. The machine terminal phase currents i_a, i_b, and i_c are converted to i_{ds}^s and i_{qs}^s components by $3\varphi/2\varphi$ transformation. These are then converted to synchronously rotating frame by the unit vector

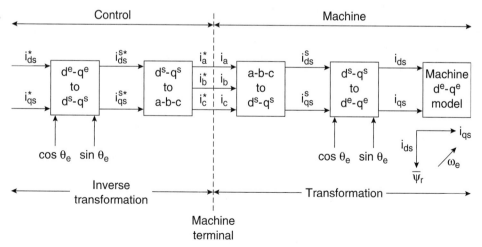

Figure 8.26 Vector control implementation principle with machine d^e-q^e model

components $\cos \theta_e$ and $\sin \theta_e$ before applying them to the d^e–q^e machine model as shown. The transformation equations are given in Chapter 2. The controller makes two stages of inverse transformation, as shown, so that the control currents i_{ds}^* and i_{qs}^* correspond to the machine currents i_{ds} and i_{qs}, respectively. In addition, the unit vector assures correct alignment of i_{ds} current with the flux vector $\bar{\psi}_r$ and i_{qs} perpendicular to it, as shown. Note that the transformation and inverse transformation including the inverter ideally do not incorporate any dynamics, and therefore, the response to i_{ds} and i_{qs} is instantaneous (neglecting computational and sampling delays).

There are essentially two general methods of vector control. One, called the direct or feedback method, was invented by Blaschke [9], and the other, known as the indirect or feedforward method, was invented by Hasse [10]. The methods are different essentially by how the unit vector ($\cos \theta_e$ and $\sin \theta_e$) is generated for the control. It should be mentioned here that the orientation of i_{ds} with rotor flux ψ_r, air gap flux ψ_m, or stator flux (ψ_s) is possible in vector control [11]. However, rotor flux orientation gives natural decoupling control, whereas air gap or stator flux orientation gives a coupling effect which has to be compensated by a decoupling compensation current. This is discussed later.

8.4.4 Direct or Feedback Vector Control

The basic block diagram of the direct vector control method for a PWM voltage-fed inverter drive is shown in Figure 8.27. The principal vector control parameters, i_{ds}^* and i_{qs}^*, which are dc values in synchronously rotating frame, are converted to stationary frame (defined as vector rotation (VR)) with the help of a unit vector ($\cos \theta_e$ and $\sin \theta_e$) generated from flux vector signals ψ_{dr}^s and ψ_{qr}^s. The resulting stationary frame signals are then converted to phase current commands for the inverter. The flux signals ψ_{dr}^s and ψ_{qr}^s are generated from the machine terminal voltages and currents with the help of the voltage model estimator, which will be discussed later. A flux control loop has been added for precision control of flux. The torque component of current i_{qs}^* is generated from the speed control loop through a bipolar limiter (not shown). The torque, proportional to i_{qs} (with constant flux), can be bipolar. It is negative with negative i_{qs}, and correspondingly, the phase position of i_{qs} becomes negative in Figure 8.25. An additional torque control loop can be added within the speed loop, if desired. Figure 8.27 can be extended to field-weakening mode by programming the flux command as a function of speed so that the inverter remains in PWM mode. Vector control by current regulation is lost if the inverter attains the square-wave mode of operation.

The correct alignment of current i_{ds} in the direction of flux $\hat{\psi}_r$ and the current i_{qs} perpendicular to it are crucial in vector control. This alignment, with the help of stationary frame rotor flux vectors ψ_{dr}^s and ψ_{qr}^s, is explained in Figure 8.28. In this figure, the d^e–q^e frame is rotating at synchronous speed ω_e with respect to stationary frame d^s–q^s, and at any instant, the angular position of the d^e-axis with respect to the d^s-axis is θ_e, where $\theta_e = \omega_e t$. From the figure, we can write the following equations:

Vector or Field-Oriented Control

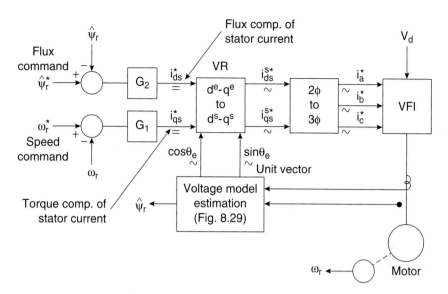

Figure 8.27 Direct vector control block diagram with rotor flux orientation

$$\psi_{dr}{}^s = \hat{\psi}_r \cos\theta_e \tag{8.22}$$

$$\psi_{qr}{}^s = \hat{\psi}_r \sin\theta_e \tag{8.23}$$

In other words,

$$\cos\theta_e = \frac{\psi_{dr}{}^s}{\hat{\psi}_r} \tag{8.24}$$

$$\sin\theta_e = \frac{\psi_{qr}{}^s}{\hat{\psi}_r} \tag{8.25}$$

$$\hat{\psi}_r = \sqrt{\psi_{dr}{}^{s^2} + \psi_{qr}{}^{s^2}} \tag{8.26}$$

where vector $\bar{\psi}_r$ is represented by magnitude $\hat{\psi}_r$. Signals $\cos\theta_e$ and $\sin\theta_e$ have been plotted in correct phase position in Figure 8.28(b). These unit vector signals, when used for vector rotation in Figure 8.27, give a ride of current i_{ds} on the d^e-axis (direction of $\hat{\psi}_r$) and current i_{qs} on the q^e-axis as shown. At this condition, $\psi_{qr} = 0$ and $\psi_{dr} = \hat{\psi}_r$, as indicated in the figure, and the corresponding torque expression is given by Equation (8.19) like a dc machine. When the i_{qs} polarity is reversed by the speed loop, the i_{qs} position in Figure 8.28(a) also reverses, giving negative torque. The generation of a unit vector signal from feedback flux vectors gives the name "direct vector control."

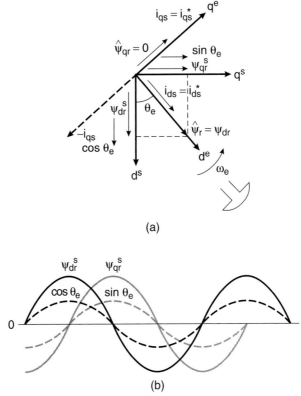

Figure 8.28 (a) d^s-q^s and d^e-q^e phasors showing correct rotor flux orientation, (b) Plot of unit vector signals in correct phase position

Let us now summarize a few salient features of vector control:

- The frequency ω_e of the drive is not directly controlled as in scalar control. The machine is essentially "self-controlled," where the frequency as well as the phase are controlled indirectly with the help of the unit vector.
- There is no fear of an instability problem by crossing the operating point beyond the breakdown torque T_{em} as in a scalar control. Limiting the total $\hat{I}_s(\sqrt{i_{ds}^2 + i_{qs}^2})$ within the safe limit automatically limits operation within the stable region.
- The transient response will be fast and dc machine-like because torque control by i_{qs} does not affect the flux. However, note that ideal vector control is impossible in practice because of delays in converter and signal processing and the parameter variation effect, which will be discussed later.
- Like a dc machine, speed control is possible in four quadrants without any additional control elements (like phase sequence reversing). In forward motoring condition, if the torque T_e is negative, the drive initially goes into regenerative braking mode, which slows down

Vector or Field-Oriented Control

the speed. At zero speed, the phase sequence of the unit vector automatically reverses, giving reverse motoring operation.

8.4.5 Flux Vector Estimation

In the direct vector control method, as discussed above, it is necessary to estimate the rotor flux components ψ_{dr}^s and ψ_{qr}^s so that the unit vector and rotor flux can be calculated by Equations (8.24)–(8.26). Two commonly used methods of flux estimation are discussed below.

8.4.5.1 Voltage Model

In this method, the machine terminal voltages and currents are sensed and the fluxes are computed from the stationary frame (d^s-q^s) equivalent circuit shown in Figure 2.27. These equations are:

$$i_{qs}^s = \frac{2}{3}i_a - \frac{1}{3}i_b - \frac{1}{3}i_c = i_a \tag{8.27}$$

$$i_{ds}^s = -\frac{1}{\sqrt{3}}i_b + \frac{1}{\sqrt{3}}i_c \tag{8.28}$$

$$= -\frac{1}{\sqrt{3}}(i_a + 2i_b) \tag{8.29}$$

since $i_c = -(i_a + i_b)$ for isolated neutral load.

$$v_{qs}^s = \frac{2}{3}v_a - \frac{1}{3}v_b - \frac{1}{3}v_c \tag{2.72}$$

$$= \frac{1}{3}(v_{ab} + v_{ac}) \tag{8.30}$$

$$v_{ds}^s = -\frac{1}{\sqrt{3}}v_b + \frac{1}{\sqrt{3}}v_c \tag{2.73}$$

$$= -\frac{1}{\sqrt{3}}v_{bc} \tag{8.31}$$

$$\psi_{ds}^s = \int (v_{ds}^s - R_s i_{ds}^s)dt \tag{8.32}$$

$$\psi_{qs}^s = \int (v_{qs}^s - R_s i_{qs}^s)dt \tag{8.33}$$

$$\hat{\psi}_s = \sqrt{\psi_{ds}^{s^2} + \psi_{qs}^{s^2}} \tag{8.34}$$

$$\psi_{dm}{}^s = \psi_{ds}{}^s - L_{ls}i_{ds}{}^s = L_m(i_{ds}{}^s + i_{dr}{}^s) \tag{8.35}$$

$$\psi_{qm}{}^s = \psi_{qs}{}^s - L_{ls}i_{qs}{}^s = L_m(i_{qs}{}^s + i_{qr}{}^s) \tag{8.36}$$

$$\psi_{dr}{}^s = L_m i_{ds}{}^s + L_r i_{dr}{}^s \tag{8.37}$$

$$\psi_{qr}{}^s = L_m i_{qs}{}^s + L_r i_{qr}{}^s \tag{8.38}$$

Eliminating $i_{dr}{}^s$ and $i_{qr}{}^s$ from Equations (8.37)–(8.38) with the help of Equations (8.35)–(8.36), respectively, gives the following:

$$\psi_{dr}{}^s = \frac{L_r}{L_m}\psi_{dm}{}^s - L_{lr}i_{ds}{}^s \tag{8.39}$$

$$\psi_{qr}{}^s = \frac{L_r}{L_m}\psi_{qm}{}^s - L_{lr}i_{qs}{}^s \tag{8.40}$$

which can also be written in the following form with the help of Equations (8.35) and (8.36):

$$\psi_{dr}{}^s = \frac{L_r}{L_m}(\psi_{ds}{}^s - \sigma L_s i_{ds}{}^s) \tag{8.41}$$

$$\psi_{qr}{}^s = \frac{L_r}{L_m}(\psi_{qs}{}^s - \sigma L_s i_{qs}{}^s) \tag{8.42}$$

where $\sigma = 1 - L_m^2/L_r L_s$.

Substituting Equations (8.39)–(8.40) in the torque equation (2.113), in stationary frame and simplifying, we get

$$T_e = \frac{3}{2}(\frac{P}{2})\frac{L_m}{L_r}(\psi_{dr}{}^s i_{qs}{}^s - \psi_{qr}{}^s i_{ds}{}^s) \tag{8.43}$$

Figure 8.29 shows the block diagram for feedback signal estimation with the help of a microcomputer, where the estimation of additional signals, such as stator fluxes, air gap fluxes, and torque, are also shown. In the front end, there is some hardware low-pass filtering and 3φ/2φ conversion with the help of op amps before conversion by the A/D converter, which is not shown in detail. Note that machines are normally isolated neutral load, and therefore, only two current sensors are needed. The vector drive uses a current-controlled PWM inverter. The current control is logical, as mentioned before, because both the flux and torque are directly related to currents. The inverter can have hysteresis-band current control, or some type of voltage control within the current control loop (such as synchronous current control, as discussed in Figure

Vector or Field-Oriented Control

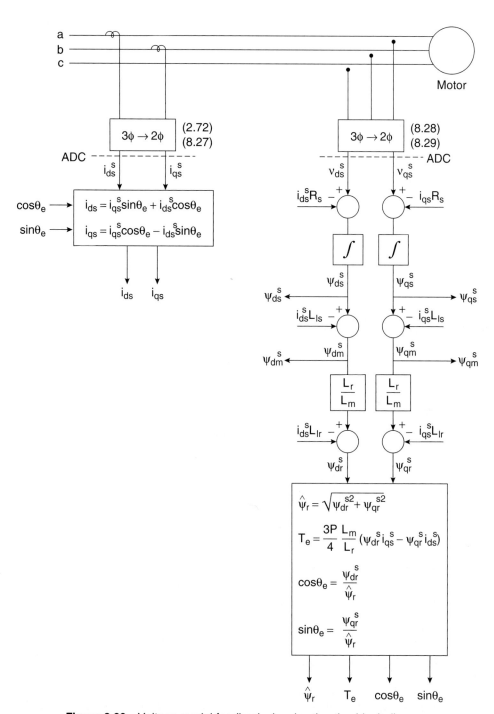

Figure 8.29 Voltage model feedback signal estimation block diagram

8.34). The current estimation equations for i_{ds} and i_{qs} using Equations (2.74)–(2.75) are also included in Figure 8.29. Note that any error in the unit vector or distortion associated with the feedback signals will affect the performance of the drive.

The direct method of vector control discussed so far is difficult to operate successfully at very low frequency (including zero speed) because of the following problems:

- At low frequency, voltage signals v_{ds}^s and v_{qs}^s are very low. In addition, ideal integration becomes difficult because dc offset tends to build up at the integrator output.
- The parameter variation effect of resistance R_s and inductances L_{ls}, L_{lr}, and L_m tend to reduce accuracy of the estimated signals. Particularly, temperature variation of R_s becomes more dominant. However, compensation of R_s is somewhat easier, which will be discussed later. At higher voltage, the effect of parameter variation can be neglected.

In industrial applications, vector drives are often required to operate from zero speed (including zero speed start-up). Here, direct vector control with voltage model signal estimation cannot be used.

8.4.5.2 Current Model

In the low-speed region, the rotor flux components can be synthesized more easily with the help of speed and current signals. The rotor circuit equations of d^s-q^s equivalent circuits (Figure 2.27) can be given as

$$\frac{d\psi_{dr}^s}{dt} + R_r i_{dr}^s + \omega_r \psi_{qr}^s = 0 \tag{8.44}$$

$$\frac{d\psi_{qr}^s}{dt} + R_r i_{qr}^s - \omega_r \psi_{dr}^s = 0 \tag{8.45}$$

Adding terms $(L_m R_r/L_r)i_{ds}^s$ and $(L_m R_r/L_r)i_{qs}^s$, respectively, on both sides of the above equations, we get

$$\frac{d\psi_{dr}^s}{dt} + \frac{R_r}{L_r}(L_m i_{ds}^s + L_r i_{dr}^s) + \omega_r \psi_{qr}^s = \frac{L_m R_r}{L_r} i_{ds}^s \tag{8.46}$$

$$\frac{d\psi_{qr}^s}{dt} + \frac{R_r}{L_r}(L_m i_{qs}^s + L_r i_{qr}^s) - \omega_r \psi_{dr}^s = \frac{L_m R_r}{L_r} i_{qs}^s \tag{8.47}$$

Substituting Equations (8.37) and (8.38), respectively, and simplifying, we get

$$\frac{d\psi_{dr}^s}{dt} = \frac{L_m}{T_r} i_{ds}^s - \omega_r \psi_{qr}^s - \frac{1}{T_r} \psi_{dr}^s \tag{8.48}$$

Vector or Field-Oriented Control

$$\frac{d\psi_{qr}^{s}}{dt} = \frac{L_m}{T_r}i_{qs}^{s} + \omega_r \psi_{dr}^{s} - \frac{1}{T_r}\psi_{qr}^{s} \tag{8.49}$$

where $T_r = L_r/R_r$ is the rotor circuit time constant. Equations (8.48) and (8.49) give rotor fluxes as functions of stator currents and speed. Therefore, knowing these signals, the fluxes and corresponding unit vector signals can be estimated. These equations are defined as the current model for flux estimation, which was originally formulated by Blaschke (often called Blaschke equation). It is shown in block diagram form in Figure 8.30, where the estimation of the $\hat{\psi}_r$, cos θ_e, and sin θ_e signals are shown on the right. Other feedback signals, such as T_e, i_{ds}, and i_{qs}, as well as stator and air gap fluxes, indicated in Figure 8.29, can easily be estimated from the current model. Flux estimation by this model requires a speed encoder, but the advantage is that the drive operation can be extended down to zero speed. However, note that the estimation accuracy is affected by the variation of machine parameters. Particularly, the rotor resistance variation (may be more than 50 percent) becomes dominant by temperature and skin effect. Compensation of this parameter is difficult because of inaccessability.

Since the voltage model flux estimation is better at higher speed ranges, whereas the current model estimation can be made at any speed, it is possible to have a hybrid model [14],

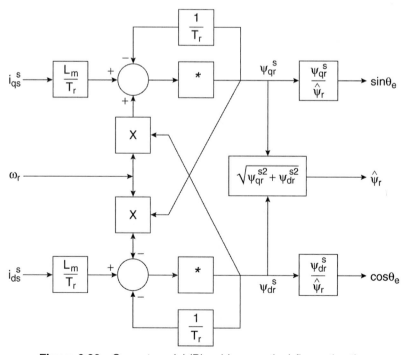

Figure 8.30 Current model (Blaschke equation) flux estimation

where the voltage model becomes effective at higher speed ranges, but transitions smoothly to the current model at lower speed ranges.

8.4.6 Indirect or Feedforward Vector Control

The indirect vector control method is essentially the same as direct vector control, except the unit vector signals ($\cos \theta_e$ and $\sin \theta_e$) are generated in feedforward manner. Indirect vector control is very popular in industrial applications. Figure 8.31 explains the fundamental principle of indirect vector control with the help of a phasor diagram. The d^s-q^s axes are fixed on the stator, but the d^r-q^r axes, which are fixed on the rotor, are moving at speed ω_r as shown. Synchronously rotating axes d^e-q^e are rotating ahead of the d^r-q^r axes by the positive slip angle θ_{sl} corresponding to slip frequency ω_{sl}. Since the rotor pole is directed on the d^e axis and $\omega_e = \omega_r + \omega_{sl}$, we can write

$$\theta_e = \int \omega_e dt = \int (\omega_r + \omega_{sl}) dt = \theta_r + \theta_{sl} \tag{8.50}$$

Note that the rotor pole position is not absolute, but is slipping with respect to the rotor at frequency ω_{sl}. The phasor diagram suggests that for decoupling control, the stator flux component of current i_{ds} should be aligned on the d^e axis, and the torque component of current i_{qs} should be on the q^e axis, as shown.

For decoupling control, we can now make a derivation of control equations of indirect vector control with the help of d^e-q^e equivalent circuits (Figure 2.23). The rotor circuit equations can be written as

$$\frac{d\psi_{dr}}{dt} + R_r i_{dr} - (\omega_e - \omega_r)\psi_{qr} = 0 \tag{8.51}$$

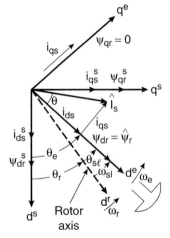

Figure 8.31 Phasor diagram explaining indirect vector control

Vector or Field-Oriented Control

$$\frac{d\psi_{qr}}{dt} + R_r i_{qr} + (\omega_e - \omega_r)\psi_{dr} = 0 \tag{8.52}$$

The rotor flux linkage expressions can be given as

$$\psi_{dr} = L_r i_{dr} + L_m i_{ds} \tag{8.53}$$

$$\psi_{qr} = L_r i_{qr} + L_m i_{qs} \tag{8.54}$$

From the above equations, we can write

$$i_{dr} = \frac{1}{L_r}\psi_{dr} - \frac{L_m}{L_r} i_{ds} \tag{8.55}$$

$$i_{qr} = \frac{1}{L_r}\psi_{qr} - \frac{L_m}{L_r} i_{qs} \tag{8.56}$$

The rotor currents in Equations (8.51) and (8.52), which are inaccessible, can be eliminated with the help of Equations (8.55) and (8.56) as

$$\frac{d\psi_{dr}}{dt} + \frac{R_r}{L_r}\psi_{dr} - \frac{L_m}{L_r} R_r i_{ds} - \omega_{sl}\psi_{qr} = 0 \tag{8.57}$$

$$\frac{d\psi_{qr}}{dt} + \frac{R_r}{L_r}\psi_{qr} - \frac{L_m}{L_r} R_r i_{qs} + \omega_{sl}\psi_{dr} = 0 \tag{8.58}$$

where $\omega_{sl} = \omega_e - \omega_r$ has been substituted.

For decoupling control, it is desirable that

$$\psi_{qr} = 0 \tag{8.59}$$

that is,

$$\frac{d\psi_{qr}}{dt} = 0 \tag{8.60}$$

so that the total rotor flux $\hat{\psi}_r$ is directed on the d^e axis.

Substituting the above conditions in Equations (8.57) and (8.58), we get

$$\frac{L_r}{R_r}\frac{d\hat{\psi}_r}{dt} + \hat{\psi}_r = L_m i_{ds} \tag{8.61}$$

$$\omega_{sl} = \frac{L_m R_r}{\hat{\psi}_r L_r} i_{qs} \tag{8.62}$$

where $\hat{\psi}_r = \psi_{dr}$ has been substituted.

If rotor flux $\hat{\psi}_r$ = constant, which is usually the case, then from Equation (8.61),

$$\hat{\psi}_r = L_m i_{ds} \tag{8.63}$$

In other words, the rotor flux is directly proportional to current i_{ds} in steady state.

To implement the indirect vector control strategy, it is necessary to take Equations (8.50), (8.61), and (8.62) into consideration. Figure 8.32 shows a four-quadrant position servo system using the indirect vector control method. The power circuit consists of a front-end diode rectifier and a PWM inverter with a dynamic brake in the dc link. A hysteresis-band current control PWM is shown, but synchronous current control voltage PWM (see Figure 8.34) can also be used. The speed control loop generates the torque component of current i_{qs}^*, as usual. The flux component of current i_{ds}^* for the desired rotor flux $\hat{\psi}_r$ is determined from Equation (8.63), and is maintained constant here in the open loop manner for simplicity. The variation of magnetizing inductance L_m will cause some drift in the flux. The slip frequency ω_{sl}^* is generated from i_{qs}^* in feedforward manner from Equation (8.62) to satisfy the phasor diagram in Figure 8.31. The corresponding expression of slip gain K_s is given as

$$K_s = \frac{\omega_{sl}^*}{i_{qs}^*} = \frac{L_m R_r}{L_r \hat{\psi}_r} \tag{8.64}$$

Signal ω_{sl}^* is added with speed signal ω_r to generate frequency signal ω_e. The unit vector signals $\cos\theta_e$ and $\sin\theta_e$ are then generated from ω_e by integration and look-up tables, as indicated in the figure. The VR and $2\phi/3\phi$ transformation are the same as in Figure 8.27. The speed signal from an incremental-position encoder is mandatory in indirect vector control because the slip signal locates the pole with respect to the rotor d^r axis in feedforward manner, which is moving at speed ω_r. An absolute pole position on the rotor is not required in this case like a synchronous motor. If the polarity of i_{qs}^* becomes negative for negative torque, the phasor i_{qs} in Figure 8.31 will be reversed, and correspondingly, ω_{sl} will be negative (i.e., θ_{sl} is negative), which will shift the rotor pole position (d^e axis) below the d^r axis. The speed control range in indirect vector control can easily be extended from stand-still (zero speed) to the field-weakening region. The addition of field-weakening control is shown by the dotted block diagram and the corresponding operation is explained in Figure 8.33. In this case, close loop flux control is necessary. In the constant torque region, the flux is constant. However, in the field-weakening region, the flux is programmed such that the inverter always operates in PWM mode, as explained before. The loss of torque and power for field-weakening vector control are indicated in the figure. The same principle of field-weakening control is also valid for direct vector control in Figure 8.27.

Vector or Field-Oriented Control

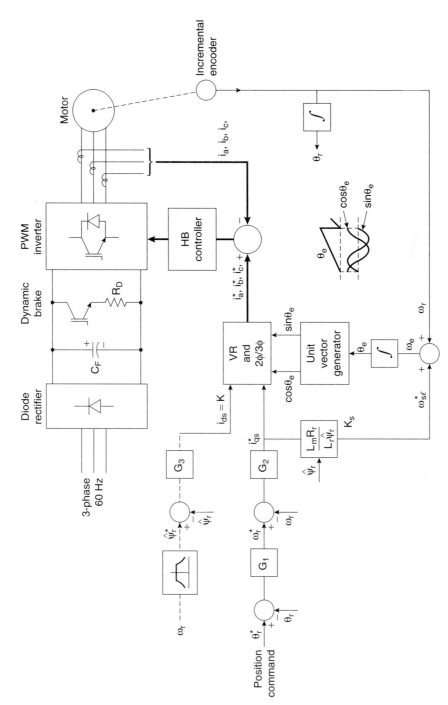

Figure 8.32 Indirect vector control block diagram with open loop flux control

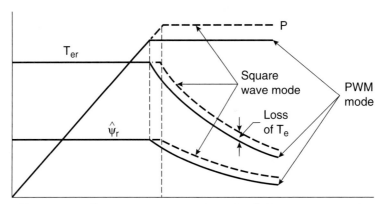

Figure 8.33 Torque-speed curves including field-weakening mode

In both the direct and indirect vector control methods discussed so far, instantaneous current control of the inverter is necessary. Hysteresis-band PWM current control can be used, as indicated in Figure 8.32, but its harmonic content is not optimum. Besides, at higher speeds, the current controller will tend to saturate in part of the cycle (quasi-PWM) because of higher CEMF. In this condition, the fundamental current magnitude and its phase will lose tracking with the command current, and thus, vector control will not be valid. These problems can be solved by synchronous current control (often called dc current control), which is shown in Figure 8.34. Command currents i_{ds}^* and i_{qs}^* in the vector control are compared with the respective i_{ds} and i_{qs} currents generated by the transformation of phase currents ($3\varphi/2\varphi$ conversion and inverse vector rotation VR^{-1}) with the help of the unit vector. The respective errors generate the voltage command signals v_{ds}^* and v_{qs}^* through the P-I compensators, as shown. These voltage commands are then converted into stationary frame phase voltages. The synchronous frame current control with a P-I controller assure amplitude and phase tracking of currents, even when the PWM controller goes into overmodulation range.

Of course, the introduction of feedback loops brings with it a small amount of coupling effect. To enhance the loop response, feedforward CEMF signals (see Figure 2.23) are injected in the respective loops. Signal $\omega_e \psi_{ds}$ is added in the i_{qs} loop, whereas signal $\omega_e \psi_{qs}$ subtracts from the i_{ds} loop signal. Often, the later signal is deleted. The block diagram for estimating the CEMF signals is added in the figure. The estimations of stator fluxes ψ_{ds}^s and ψ_{qs}^s are shown in Figure 8.29. The frequency signal ω_e can be estimated as follows:

$$\cos\theta_e = \frac{\psi_{ds}^s}{\sqrt{\psi_{ds}^{s2} + \psi_{qs}^{s2}}} = \frac{\psi_{ds}^s}{\hat{\psi}_s} \tag{8.65}$$

$$\sin\theta_e^* = \frac{\psi_{qs}^s}{\hat{\psi}_s} \tag{8.66}$$

Vector or Field-Oriented Control

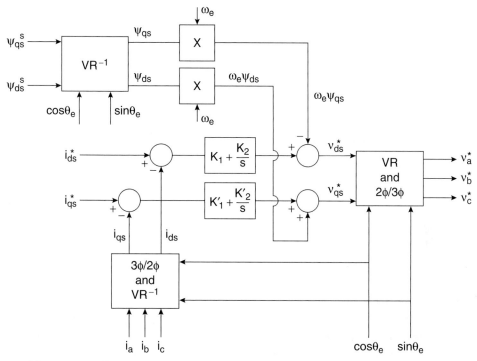

Figure 8.34 Synchronous current control with feedforward CEMF compensation

$$\tan\theta_e^* = \frac{\psi_{qs}^s}{\psi_{ds}^s} \tag{8.67}$$

Differentiating Equation (8.67), we get

$$\sec^2\theta_e \cdot \frac{d\theta_e^*}{dt} = \frac{\psi_{ds}^s \dot{\psi}_{qs}^s - \psi_{qs}^s \dot{\psi}_{ds}^s}{\psi_{ds}^{s\,2}} \tag{8.68}$$

where

$$\sec^2\theta_e = \frac{\hat{\psi}_s^2}{\psi_{ds}^{s\,2}} \tag{8.69}$$

Substituting Equation (8.69) in Equation (8.68), we get

$$\omega_e = \frac{d\theta_e}{dt} = \frac{(v_{qs}^s - i_{qs}^s R_s)\psi_{ds}^s - (v_{ds}^s - i_{ds}^s R_s)\psi_{qs}^s}{\hat{\psi}_s^2} \qquad (8.70)$$

where the voltage expressions behind the stator resistance drops have been substituted for the flux derivatives. Frequency ω_e can also be derived as a function of the rotor fluxes by following a similar procedure.

A dc machine-like electro-mechanical model of an ideal vector-controlled drive can be derived using Equation (8.61) and the following equations:

$$T_e = \frac{3}{2}\left(\frac{P}{2}\right)\frac{L_m}{L_r}\hat{\psi}_r i_{qs} \qquad (8.71)$$

$$T_e - T_L = \left(\frac{2}{P}\right)J\frac{d\omega_r}{dt} \qquad (2.105)$$

Figure 8.35 shows the corresponding transfer function block diagram, where the delay between the command and actual currents has been neglected. The developed torque T_e responds instantaneously with the current i_{qs}, but the flux response has first-order delay (with time constant L_r/R_r), similar to a dc machine. It can be shown that direct vector control also has a similar transfer function model.

The physical principle of vector control can be understood more clearly with the help of the d^e-q^e circuits shown in Figure 8.36. Since currents i_{ds} and i_{qs} are being controlled, ideally, the stator-side Thevenin impedance is infinity, that is, the stator-side circuit parameters and EMFs are of no consequence. With $\psi_{qr} = 0$ under all conditions, EMF $\omega_{sl} \psi_{qr} = 0$ in the d^e circuit. This indicates that at steady state, current i_{ds} will flow through the magnetizing branch only to establish the rotor flux ψ_r, but transiently, the current will be shared by the rotor circuit also and the time constant can be easily seen as L_r/R_r. In the q^e circuit, when torque is controlled

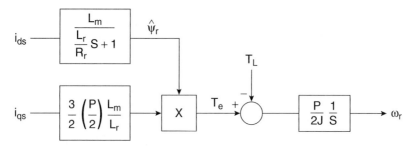

Figure 8.35 Transfer function block diagram of vector-controlled drive

Vector or Field-Oriented Control

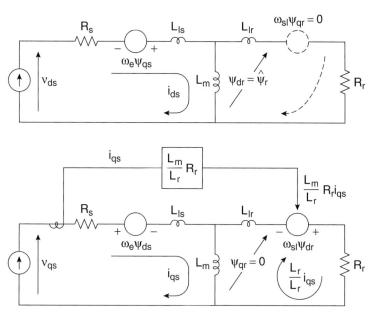

Figure 8.36 Explanation of vector control with the help of d^e-q^e circuits

by i_{qs}, EMF $\omega_{sl} \psi_{dr}$ in the rotor circuit is modified instantly because $\omega_{sl} \psi_{dr} = L_m R_r i_{qs}/L_r$, as dictated by Equation (8.62). Since $\psi_{qr} = 0$, this EMF establishes the current $(L_m/L_r)i_{qs}$ through the resistance R_r. To verify that this current satisfies $\psi_{qr} = 0$, we can write

$$\begin{aligned}\psi_{qr} &= \psi_m + \psi_{lr} \\ &= L_m(\frac{L_m}{L_r}i_{qs} - i_{qs}) + (\frac{L_m}{L_r}i_{qs})L_{lr} \\ &= i_{qs}(\frac{L_m^2}{L_r} - L_m + \frac{L_m L_{lr}}{L_r}) = 0\end{aligned} \quad (8.72)$$

If L_{lr} is neglected for simplicity and flux $\hat{\psi}_r$ is considered as constant, it can easily be seen that magnetizing current component i_{ds} flows through L_m only, whereas the torque component of current i_{qs} is constrained to flow to the rotor side only. This was the original hypothesis in explaining vector control with the help of Figs. 8.24 and 8.25.

8.4.6.1 Indirect Vector Control Slip Gain (K_s) Tuning

The slip gain K_s in indirect vector control (Figure 8.32) is a function of the machine's parameters. It is desirable that these parameters match the actual parameters of the machine at all operating conditions to achieve decoupling control of the machine. The slip gain detuning problem is a serious disadvantage of indirect vector control. Of course, the problem is similar to direct vector control at low speeds, where the current model flux estimation is used.

With the close loop flux control shown in Figure 8.32, the estimated value of $\hat{\psi}_r$ (input to K_s) is known; therefore, the variation of three parameters, (L_m, L_r, and R_r), is of concern. The saturation effect of magnetizing inductance L_m almost cancels the variation of L_m/L_r, thus leaving the dominant effect of rotor resistance variation on K_s. With open loop flux control at steady-state condition, $\hat{\psi}_r = i_{ds} L_m$. Therefore, K_s becomes a function of rotor time constant $T_r = L_r/R_r$ only.

The effect of rotor resistance detuning and the corresponding coupling effect are explained in Figure 8.37, where R_r = actual resistance of the machine and $\hat{R}_r$ = estimated resistance used in the K_s parameter. If $\hat{R}_r$ is lower than R_r, the slip frequency ω_{sl}^* will be lower than the actual (see Figs. 8.32 and 8.36), giving backward alignment of the $i_{ds}' - i_{qs}'$ current pair, as shown in Figure 8.37(a). With such misalignment, if torque is increased by a step of i_{qs}', a component of this current on the d^e axis will increase the flux (overfluxing). The resulting torque and flux responses in both the transient and steady-state conditions are shown in Figure 8.37(b). On the other hand, the effect of higher than actual resistance ($\hat{R}_r/R_r = 1.3$) will cause underfluxing, and the corresponding performance is indicated in Figure 8.37(b).

The initial tuning of slip gain is straightforward if the parameters of the machine are known a-priori. Conversely, an automated measurement of parameters can be made initially by injecting signals in the machine through the inverter and then estimating the parameters with the help of a DSP. This will be discussed later in self-commissioning of the drive. The initial tuning of K_s can also be done by giving a square-wave torque (or i_{qs}^*) command and then matching the actual torque wave with the predicted torque wave under a tuned condition.

The continuous on-line tuning of K_s is very complex and demands rigorous computation by a DSP. A number of methods for slip gain tuning have been suggested in the literature. Unfortunately, most of these algorithms are also dependent on machine parameters. Fortunately, however, the temperature variation of R_r is somewhat slow, and this permits adequate computation time required by the DSP. The extended Kalman filter (EKF) method of parameter (or state) estimation based on the full-order dynamic machine model is an elegant and powerful method, and it will be discussed later for speed estimation.

Another method that is more acceptable is based on the model referencing adaptive control (MRAC) and is shown in general block diagram form in Figure 8.38 [16]. The MRAC principle will be discussed later in this chapter. Here, the reference model output signal X^* that satisfies the tuned vector control is usually a function of command currents i_{ds}^* and i_{qs}^*, machine inductances, and operating frequency. The adaptive model X is usually estimated by the machine feedback voltages and currents, as shown. The reference model output is compared with that of the adaptive model and the resulting error generates the estimated slip gain $\hat{K}_s$

Vector or Field-Oriented Control

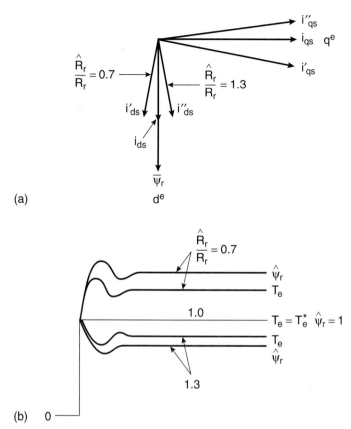

Figure 8.37 (a) Vector control detuning due to rotor resistance mismatch, (b) Resulting response

through a P-I compensator. Thus, slip gain tuning will occur when X matches with X^*. Consider, for example, on-line tuning based on the torque model. Figure 8.37(b) shows the deviation of actual torque from the commanded value because of detuned slip gain. The K_s parameter can be tuned so that $T_e = T_e^*$ at all operating conditions. In Figure 8.38, the reference model output for ideal vector control is given as

$$X^* = T_e^* = \frac{3}{2}(\frac{P}{2})\frac{L_m}{L_r}\hat{\psi}_r^* i_{qs}^* \tag{8.73}$$

Substituting $\hat{\psi}_r = L_m i_{ds}^*$, we get

$$X^* = \frac{3}{2}(\frac{P}{2})\frac{L_m^2}{L_r} i_{ds}^* i_{qs}^* \tag{8.74}$$

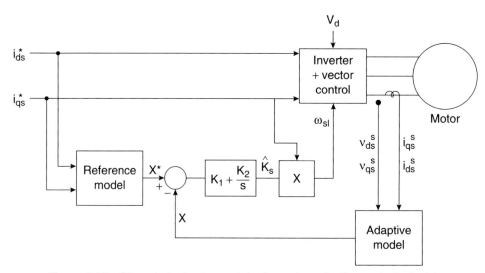

Figure 8.38 Slip gain tuning by model referencing adaptive control principle

The actual torque T_e (X) can be estimated from stationary frame variables as follows:

$$X = T_e = \frac{3}{2}\left(\frac{P}{2}\right)(\psi_{ds}^s i_{qs}^s - \psi_{qs}^s i_{ds}^s) \tag{2.121}$$

Note that the L_m and L_r parameter variations affect the estimation accuracy of X^*. Additionally, at low speeds, the estimation accuracy of X will be affected by the variation of stator resistance R_s. Other MRAC-based slip gain tuning methods with fuzzy logic will be discussed in Chapter 11.

8.4.7 Vector Control of Line-Side PWM Rectifier

The line-side converter in a double-sided PWM converter system (see Figure 5.60) can be vector-controlled to regulate the active I_P and reactive I_Q currents independently. In this case, the unit vector can be generated from the line voltage vector to orient I_P in phase and I_Q in quadrature, respectively. Figure 8.39 shows the control block diagram for the unity line power factor operation, and Figure 8.40 explains the operation's principle. The line voltage vector $\bar{V}_s$ (of magnitude $\hat{V}_s$) is oriented on the d^e axis so that

$$v_{ds}^s = \hat{V}_s \cos\theta_e \tag{8.75}$$

$$v_{qs}^s = \hat{V}_s \sin\theta_e \tag{8.76}$$

$$\cos\theta_e = \frac{v_{ds}^s}{\hat{V}_s} \tag{8.77}$$

Vector or Field-Oriented Control

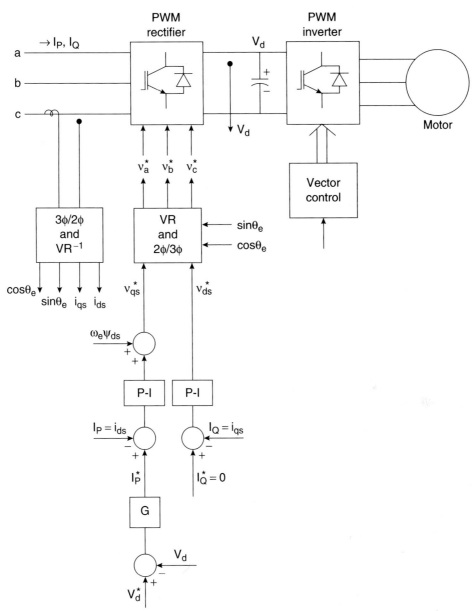

Figure 8.39 Direct vector control of line-side PWM rectifier

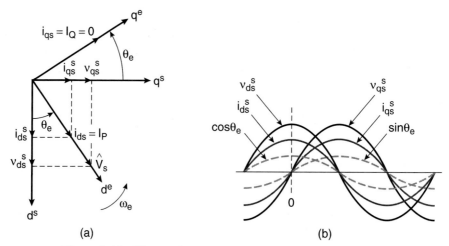

Figure 8.40 Phasor diagram and waveforms for vector control

$$\sin\theta_e = \frac{v_{qs}^s}{\hat{V}_s} \tag{8.78}$$

where

$$\hat{V}_s = \sqrt{v_{ds}^{s^2} + v_{qs}^{s^2}} \tag{2.88}$$

Since $i_{ds} = I_P$ and $i_{qs} = I_Q = 0$,

$$i_{ds}^s = I_P \cos\theta_e \tag{8.79}$$

$$i_{qs}^s = I_P \sin\theta_e \tag{8.80}$$

Figure 8.40(b) shows the time-domain plot of the voltage and current waves, indicating the unity line power factor condition. In the control block diagram, the phase voltages are sensed, filtered, and converted to v_{ds}^s and v_{qs}^s variables. The unit vector signals $\cos\theta_e$ and $\sin\theta_e$ are then calculated by Equations (8.77)–(8.78). These are also used to generate i_{ds} and i_{qs} currents as shown in Figure 8.29. The line-side converter controls the dc link voltage V_d by a feedback loop. The loop error generates the active current command I_P^* with $I_Q^* = 0$. A synchronous current control loop (with CEMF injection in the I_P loop) and vector rotator then generate the phase voltage commands as shown. Note that if I_Q^* is positive, the line current will be at a leading power factor.

8.4.8 Stator Flux-Oriented Vector Control

So far, we have discussed rotor flux-oriented vector control only. It was mentioned before that vector control is also possible with air gap flux or stator flux orientation, but at the cost of a coupling effect that demands decoupling compensation. Stator flux-oriented direct vector control has the advantage that flux vector estimation accuracy is affected by the stator resistance (R_s) variation only. In this section, we will develop a strategy for stator flux-oriented direct vector control by manipulating equations derived from d^e-q^e equivalent circuits.

Multiplying Equations (8.57) and (8.58) by $T_r = L_r/R_r$, we get

$$(1+ST_r)\psi_{dr} - L_m i_{ds} - T_r \omega_{sl} \psi_{qr} = 0 \tag{8.81}$$

$$(1+ST_r)\psi_{qr} - L_m i_{qs} + T_r \omega_{sl} \psi_{dr} = 0 \tag{8.82}$$

In these equations, ψ_{dr} and ψ_{qr} are to be eliminated and replaced by ψ_{qs} and ψ_{ds}. The stator flux expressions can be written from d^e-q^e equivalent circuits as

$$\psi_{ds} = L_s i_{ds} + L_m i_{dr} \tag{8.83}$$

$$\psi_{qs} = L_s i_{qs} + L_m i_{qr} \tag{8.84}$$

or

$$i_{dr} = \frac{\psi_{ds}}{L_m} - \frac{L_s}{L_m} i_{ds} \tag{8.85}$$

$$i_{qr} = \frac{\psi_{qs}}{L_m} - \frac{L_s}{L_m} i_{qs} \tag{8.86}$$

Substituting Equations (8.85) and (8.86) in Equations (8.53) and (8.54), respectively, we get

$$\psi_{dr} = \frac{L_r}{L_m} \psi_{ds} + (L_m - \frac{L_r L_s}{L_m}) i_{ds} \tag{8.87}$$

$$\psi_{qr} = \frac{L_r}{L_m} \psi_{qs} + (L_m - \frac{L_r L_s}{L_m}) i_{qs} \tag{8.88}$$

These equations relate stator and rotor fluxes with stator currents. Substituting Equations (8.87) and (8.88) in Equations (8.81) and (8.82), respectively, then multiplying both sides by L_m/L_r, and simplifying, we get

$$(1+ST_r)\psi_{ds} = (1+\sigma ST_r)L_s i_{ds} + \omega_{sl} T_r \left[\psi_{qs} - \sigma L_s i_{qs}\right] \tag{8.89}$$

$$(1+ST_r)\psi_{qs} = (1+\sigma ST_r)L_s i_{qs} - \omega_{sl} T_r \left[\psi_{ds} - \sigma L_s i_{ds}\right] \tag{8.90}$$

where $\sigma = 1 - L_m^2/L_s L_r$.

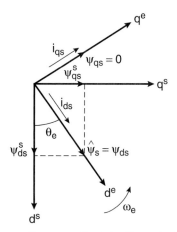

Figure 8.41 Phasor diagram with stator flux-oriented vector control

With stator flux-oriented vector control, as shown in the phasor diagram of Figure 8.41, $\psi_{qs} = 0$, $\psi_{ds} = \hat{\psi}_s$. Therefore, Equations (8.89) and (8.90) can be written, respectively, as

$$(1+ST_r)\psi_{ds} = (1+\sigma ST_r)L_s i_{ds} - \sigma L_s T_r \omega_{sl} i_{qs} \tag{8.91}$$

$$(1+\sigma ST_r)L_s i_{qs} = \omega_{sl} T_r [\psi_{ds} - \sigma L_s i_{ds}] \tag{8.92}$$

These equations indicate that stator flux ψ_{ds} is a function of both the i_{ds} and i_{qs} currents; in other words, there is coupling effect. This means that if the torque is changed by i_{qs}, it will also change the flux. Therefore, this coupling effect must be eliminated by the feedforward control method.

Consider the decoupler block diagram in Figure 8.42, where the decoupling signal i_{dq} as shown is being added in the flux control loop to generate the i_{ds}^* command signal. From Figure 8.42, we can write

$$i_{ds}^* = G(\psi_{ds}^* - \psi_{ds}) + i_{dq} \tag{8.93}$$

where $G = K_1 + K_2/S$. Substituting Equation (8.93) in (8.91), we get

$$(1+ST_r)\psi_{ds} = L_s\left[(1+\sigma ST_r).G(\psi_{ds}^* - \psi_{ds}) + (1+\sigma ST_r)i_{dq} - \sigma T_r \omega_{sl} i_{qs}\right] \tag{8.94}$$

For decoupling control of ψ_{ds} with the help of i_{dq}, terms $(1 + \sigma S T_r) i_{dq} - \sigma T_r \omega_{sl} i_{qs} = 0$, that is,

$$i_{dq} = \frac{\sigma T_r \omega_{sl} i_{qs}}{(1+\sigma S T_r)} \quad (8.95)$$

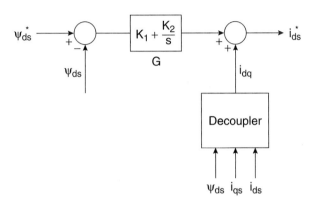

Figure 8.42 Feedforward decoupling signal injection in stator flux-oriented vector control

where ω_{sl} can be written from Equation (8.92) as

$$\omega_{sl} = \frac{(1+\sigma S T_r) L_s i_{qs}}{T_r (\psi_{ds} - \sigma L_s i_{ds})} \quad (8.96)$$

Combining Equations (8.95) and (8.96), we get

$$i_{dq} = \frac{\sigma L_s i_{qs}^2}{(\psi_{ds} - \sigma L_s i_{ds})} \quad (8.97)$$

which indicates that decoupling current i_{dq} is a function of ψ_{ds}, i_{qs}, and i_{ds}. This functionality is indicated in the block diagram of Figure 8.42. The general expression of developed torque is

$$T_e = \frac{3}{2}(\frac{P}{2})(\psi_{ds} i_{qs} - \psi_{qs} i_{ds}) \quad (2.114)$$

With vector control, $\psi_{qs} = 0$, or

$$T_e = \frac{3}{2}(\frac{P}{2})\psi_{ds} i_{qs} \quad (8.98)$$

Figure 8.43 shows a block diagram with a stator flux-oriented vector control, where the feedback estimation block computes Equations (8.32), (8.33), (8.34), (8.65), (8.66), (8.70), (8.99), and (8.100). Since the stator resistance R_s is a function of the stator winding temperature only, it can easily be compensated. Note that i_{dq} accuracy can be affected by parameter variation,

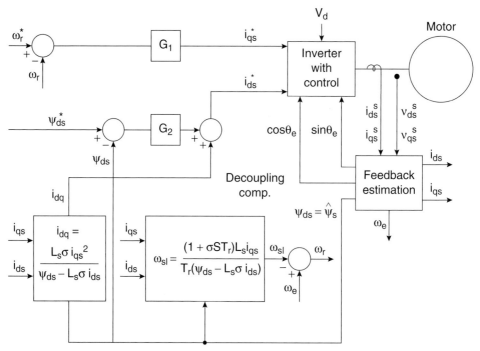

Figure 8.43 Stator flux-oriented vector-controlled drive

but being within a feedback loop, this effect can be neglected. The slip signal ω_{sl} can be estimated by Equation (8.96) and then subtracted from the frequency ω_e estimation to estimate the speed signal, as shown. The speed estimation will be discussed in detail later.

8.4.9 Vector Control of Current-Fed Inverter Drive

The direct and indirect vector control methods have so far been described for voltage-fed, current-controlled converters. These principles can easily be extended to other types of converters. Figure 8.44 shows rotor flux-oriented direct vector control of a current-fed inverter drive. The inverter, which can be a six-step or PWM-type, is fed by a phase-controlled thyristor rectifier in the front end. The drive operates with regulated rotor flux, and speed is being controlled in the outer loop. The speed loop generates the torque command T_e^*, which is divided by rotor flux $\hat{\psi}_r$ to generate the command current i_{qs}^*. The flux loop generates the flux component of current i_{ds}^*. These currents are then converted to polar form as shown. The machine stator current magnitude $\hat{I}_s$ (peak value of sinusoidal component) is related to the dc link current, which is generated by close loop current control of the phase-controlled rectifier. The inverter frequency is controlled by phase-locked loop (PLL) so that the machine stator current $\hat{I}_s$ is positioned at the desired torque angle θ with respect to the rotor flux (see Figure 8.31). This scheme is also known as angle control in the literature [5]. The rotor flux can be estimated from current model equations discussed previously. Torque angle θ can be calculated from the following equations:

Vector or Field-Oriented Control

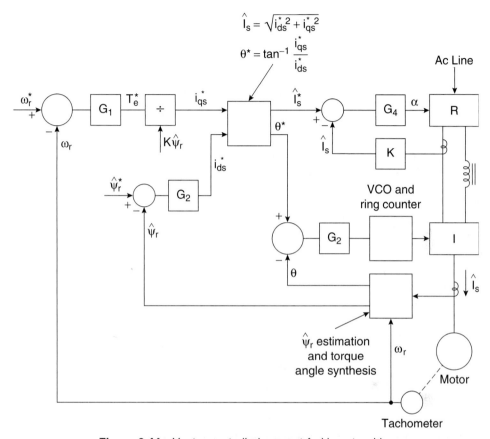

Figure 8.44 Vector-controlled, current-fed inverter drive

$$i_{qs} = i_{qs}{}^s \cos\theta_e - i_{ds}{}^s \sin\theta_e \qquad (8.99)$$

$$i_{ds} = i_{qs}{}^s \sin\theta_e + i_{ds}{}^s \cos\theta_e \qquad (8.100)$$

$$\angle\theta = \tan^{-1}\frac{i_{qs}}{i_{ds}} \qquad (8.101)$$

The control scheme remains valid from zero speed and in all quadrants.

8.4.10 Vector Control of Cycloconverter Drive

The direct and indirect methods of vector control which have been discussed so far can easily be applied to cycloconverter-controlled induction motor drives, where the inverter is replaced by a cycloconverter, and all other signal processing remains the same. With direct vector control using an instantaneous current-controlled cycloconverter, either the voltage model (Figure 8.27) or current model (Figure 8.30) flux vector (ψ_{dr}^s, ψ_{qr}^s) estimation can be used. The indirect vector control shown in Figure 8.32 can also be used. In either type of vector control, the instantaneous phase current control can be replaced by the voltage control within synchronous current control loops (Figure 8.34). The signals can also be processed in polar form instead of cartesian form. Vector control of a cycloconverter-fed synchronous motor drive will be discussed in Chapter 9.

In this section, we will review vector control of a Scherbius drive, which was discussed in Chapter 7. The cycloconverter is placed in the rotor circuit of a wound-rotor induction motor as shown in Figure 8.45. In this figure, the cyloconverter is required to send slip energy ($+SP_g$) to the line in subsynchronous motoring and supersynchronous regeneration, and to feed slip energy to the rotor ($-SP_g$) in subsynchronous regeneration and supersynchronous motoring. Currents I_P and I_Q are the in-phase and quadrature components of the rotor current, respectively, with respect to slip voltage V_r. The positive direction of I_P is shown in the figure. The error from the speed control loop generates the current command $I_P^{*\prime}$. The current I_Q^* can be set to zero or to an arbitrary value. The polarity of $I_P^{*\prime}$ depends on the error polarity. The signal goes through a polarity reverser, as shown, where the signal polarity is reversed if the speed is in supersynchronous range, but otherwise remains the same. The unit vector signals $\cos\theta_{sl}$ and $\sin\theta_{sl}$ are obtained from the following relations:

$$\cos\theta_{sl} = \cos(\theta_e - \theta_r) = \cos\theta_e \cos\theta_r + \sin\theta_e \sin\theta_r \qquad (8.102)$$

$$\sin\theta_{sl} = \sin(\theta_e - \theta_r) = \sin\theta_e \cos\theta_r - \cos\theta_e \sin\theta_r \qquad (8.103)$$

where θ_e = line unit vector obtained from Equations (8.77)–(8.78) (see Figure 8.40(a)), and θ_r = rotor position signal obtained from the speed encoder. The cycloconverter phase current commands at slip frequency are then obtained by standard vector rotation of I_P^* and I_Q^* and $2\varphi/3\varphi$ transformation, as usual. If the machine speed changes from a subsynchronous to supersynchronous ($\omega_r > \omega_e$) range, the slip voltage phase sequence will change from positive to negative. The corresponding reversal of the phase sequence of the $\cos\theta_{sl}$ and $\sin\theta_{sl}$ signals will also reverse the phase sequence of the rotor currents. At true synchronous speed, the machine operates as a synchronous motor or generator, and in this condition, the cycloconverter operates as a rectifier, supplying dc excitation current to the machine rotor winding. At this condition, signals $\cos\theta_{sl}$ and $\sin\theta_{sl}$ freeze to dc values, and correspondingly, dc phase current commands are generated for the cycloconverter.

To understand drive operation clearly, consider, for example, a drive that is accelerating from subsynchrounous speed with the command speed at a supersynchronous value. At subsynchronous motoring, I_P^* is positive because $I_P^{*\prime}$ is positive and the polarity reverser is in a non-reversing mode. With $+\omega_{sl}$, the phase sequence of rotor currents is positive and I_P is in phase with V_r so that slip power SP_g is positive. At synchronous speed, $\theta_{sl} = 0$, that is, I_P is dc current. At supersynchronous speed motoring, I_P^* becomes negative due to the negative polarity of the

Vector or Field-Oriented Control

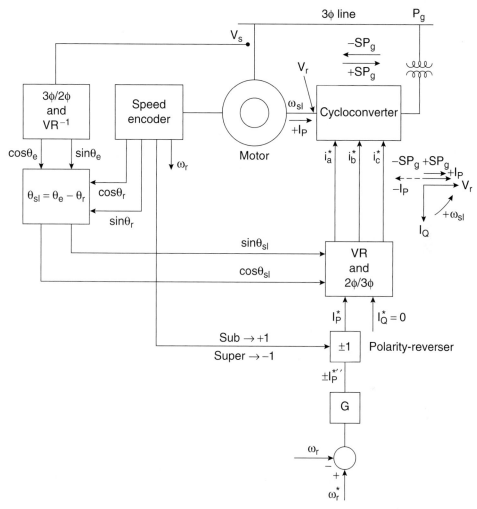

Figure 8.45 Vector control of Scherbius drive with cycloconverter

polarity reverser. This reverses the polarity of I_P, making the slip power negative. Similarly, the sequences can be derived as the speed is reduced by regenerative braking from supersynchronous to subsychrounous range.

If a Scherbius system is used for continuous regeneration (e.g., a VSCF wind generation system), the control strategy essentially remains the same except that the active and reactive currents, I_P and I_Q of the cycloconverter, are controlled to control the generator's real and reactive power, respectively, at the stator terminals by the feedback method. Vector control of the Scherbius drive with double-sided PWM converters is left as an exercise to the reader.

8.5 SENSORLESS VECTOR CONTROL

Sensorless vector control of an induction motor drive essentially means vector control without any speed sensor. An incremental shaft-mounted speed encoder (usually an optical type) is required for close loop speed or position control in both vector- and scalar-controlled drives. A speed signal is also required in indirect vector control in the whole speed range, and in direct vector control for the low-speed range, including the zero speed start-up operation. A speed encoder is undesirable in a drive because it adds cost and reliability problems, besides the need for a shaft extension and mounting arrangement. It is possible to estimate the speed signal from machine terminal voltages and currents with the help of a DSP. However, the estimation is normally complex and heavily dependent on machine parameters. Although sensorless vector-controlled drives are commercially available at this time, the parameter variation problem, particularly near zero speed, imposes a challenge in the accuracy of speed estimation.

8.5.1 Speed Estimation Methods

The induction motor speed estimation techniques proposed in the literature can generally be classified as follows:

- Slip calculation
- Direct synthesis from state equations
- Model referencing adaptive system (MRAS)
- Speed adaptive flux observer (Luenberger observer)
- Extended Kalman filter (EKF)
- Slot harmonics
- Injection of auxiliary signal on salient rotor

Detailed descriptions and historical reviews of all these methods [18] are beyond the scope of this book. However, the methods will be reviewed briefly and a few selected methods will be described in some detail.

8.5.1.1 Slip Calculation

Speed can be calculated from slip frequency ω_{sl} from the relation $\omega_r = \omega_e - \omega_{sl}$, where ω_e = stator frequency (r/s). The ω_{sl} signal was calculated before in stator flux-oriented direct vector control (Figure 8.43) as

$$\omega_{sl} = \frac{(1+\sigma S T_r) L_s i_{qs}}{T_r(\psi_{ds} - \sigma L_s i_{ds})} \tag{8.96}$$

where $\sigma = 1 - L_m^2/L_s L_r$, $T_r = L_r/R_r$, and i_{ds}, i_{qs}, and ψ_{ds} are the signals corresponding to stator flux orientation. Equation (8.96) can also be expressed with variables for rotor flux-oriented vector control. The expression for stator frequency is given as

Sensorless Vector Control

$$\omega_e = \frac{(v_{qs}^s - i_{qs}^s R_s)\psi_{ds}^s - (v_{ds}^s - i_{ds}^s R_s)\psi_{qs}^s}{\hat{\psi}_s^2} \tag{8.70}$$

The speed ω_r can be calculated from Equations (8.96) and (8.70). For scalar and indirect vector control, the ω_e signal can be available directly as a control variable. The signal can also be calculated by processing the waveform. An accurate calculation of slip frequency for a high-efficiency machine, particularly near synchronous speed, is difficult because the signal magnitude is small and highly dependent on machine parameters. Besides, there is the problem of direct integration of machine terminal voltages (see Figure 8.29) at low speeds to synthesize the ω_{sl} and ω_e signals.

8.5.1.2 Direct Synthesis from State Equations

The dynamic d^s-q^s frame state equations of a machine can be manipulated to compute the speed signal directly. The method described here is essentially similar to the slip calculation method discussed previously. The stator voltage equation for v_{ds}^s in a d^s-q^s equivalent circuit can be written as

$$v_{ds}^s = i_{ds}^s R_s + L_{ls}\frac{d}{dt}(i_{ds}^s) + \frac{d}{dt}(\psi_{dm}^s) \tag{8.104}$$

Substituting Equation (8.39) in (8.104) gives

$$v_{ds}^s = \frac{L_m}{L_r}\frac{d}{dt}(\psi_{dr}^s) + (R_s + \sigma L_s S)i_{ds}^s \tag{8.105}$$

where $\sigma = 1 - L_m^2/L_r L_s$
or

$$\frac{d}{dt}(\psi_{dr}^s) = \frac{L_r}{L_m}v_{ds}^s - \frac{L_r}{L_m}(R_s + \sigma L_s S)i_{ds}^s \tag{8.106}$$

Similarly, the ψ_{qr}^s expression can be derived as

$$\frac{d}{dt}(\psi_{qr}^s) = \frac{L_r}{L_m}v_{qs}^s - \frac{L_r}{L_m}(R_s + \sigma L_s S)i_{qs}^s \tag{8.107}$$

The rotor flux equations in a d^s-q^s frame can be given as

$$\frac{d}{dt}(\psi_{dr}^s) = \frac{L_m}{T_r}i_{ds}^s - \omega_r\psi_{qr}^s - \frac{1}{T_r}\psi_{dr}^s \tag{8.48}$$

$$\frac{d}{dt}(\psi_{qr}{}^s) = \frac{L_m}{T_r} i_{qs}{}^s + \omega_r \psi_{dr}{}^s - \frac{1}{T_r} \psi_{qr}{}^s \qquad (8.49)$$

From Figure 8.31, we can write

$$\angle \theta_e = \tan^{-1} \frac{\psi_{qr}{}^s}{\psi_{dr}{}^s} \qquad (8.108)$$

Differentiating Equation (8.108), we get

$$\frac{d\theta_e}{dt} = \frac{\psi_{dr}{}^s \dot{\psi}_{qr}{}^s - \psi_{qr}{}^s \dot{\psi}_{dr}{}^s}{\hat{\psi}_r{}^2} \qquad (8.109)$$

Combining Equations (8.46), (8.47), and (8.109) [19] and simplifying, we get

$$\omega_r = \frac{d\theta_e}{dt} - \frac{L_m}{T_r} \left[\frac{\psi_{dr}{}^s i_{qs}{}^s - \psi_{qr}{}^s i_{ds}{}^s}{\hat{\psi}_r{}^2} \right] \qquad (8.110)$$

or

$$\omega_r = \frac{1}{\hat{\psi}_r{}^2} \left[(\psi_{dr}{}^s \dot{\psi}_{qr}{}^s - \psi_{qr}{}^s \dot{\psi}_{dr}{}^s) - \frac{L_m}{T_r}(\psi_{dr}{}^s i_{qs}{}^s - \psi_{qr}{}^s i_{ds}{}^s) \right] \qquad (8.111)$$

Figure 8.46 gives a block diagram for speed estimation where the voltage model equations, (8.106) and (8.107), have been used to estimate the rotor fluxes. Evidently, the synthesis is highly machine parameter-sensitive and will tend to give poor accuracy of estimation.

8.5.1.3   Model Referencing Adaptive System (MRAS)

The speed can be calculated by the model referencing adaptive system (MRAS), where the output of a reference model is compared with the output of an adjustable or adaptive model until the errors between the two models vanish to zero. A block diagram for speed estimation by the MRAS technique is shown in Figure 8.47. Consider the voltage model's stator-side equations, (8.106) and (8.107), which are defined as a reference model in Figure 8.47. The model receives the machine stator voltage and current signals and calculates the rotor flux vector signals, as indicated. The current model flux equations, (8.46) and (8.47), are defined as an adaptive model in Figure 8.47. This model can calculate fluxes from the input stator currents only if the speed signal ω_r (shown in the coefficient matrix) is known. With the correct speed signal, ideally, the fluxes calculated from the reference model and those calculated from the adaptive model will match, that is, $\psi_{dr}{}^s = \hat{\psi}_{dr}{}^s$ and $\psi_{qr}{}^s = \hat{\psi}_{qr}{}^s$, where $\hat{\psi}_{dr}{}^s$ and $\hat{\psi}_{qr}{}^s$ are the adaptive model out-

Sensorless Vector Control

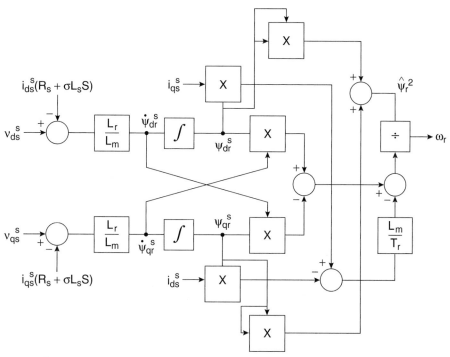

Figure 8.46 Speed estimation by direct synthesis from state equations

puts. An adaptation algorithm with P-I control, as indicated, can be used to tune the speed $\hat{\omega}_r$ so that the error $\xi = 0$.

In designing the adaptation algorithm for the MRAS, it is important to take account of the overall stability of the system and to ensure that the estimated speed will converge to the desired value with satisfactory dynamic characteristics. Using Popov's criteria for hyperstability [29] for a globally asymptotically stable system, we can derive the following relation for speed estimation:

$$\hat{\omega}_r = \xi(K_P + \frac{K_I}{S}) \tag{8.112}$$

where

$$\xi = X - Y = \hat{\psi}_{dr}^s \psi_{qr}^s - \psi_{dr}^s \hat{\psi}_{qr}^s \tag{8.113}$$

In steady state, $\xi = 0$, balancing the fluxes; in other words, $\psi_{dr}^s = \hat{\psi}_{dr}^s$ and $\psi_{qr}^s = \hat{\psi}_{qr}^s$. The MRAS in Figure 8.47 can be interpreted as a vector PLL in which the output flux vector from the reference model is the reference vector and the adjustable model is a vector phase shifter controlled by $\hat{\omega}_r$.

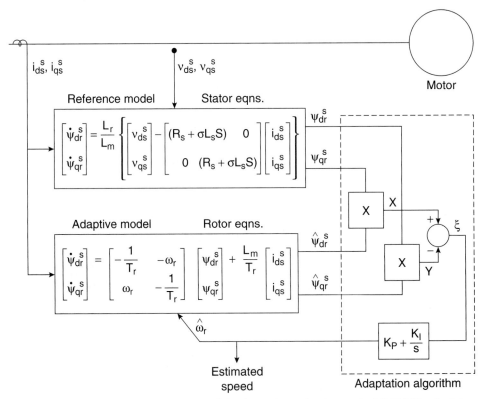

Figure 8.47 Speed estimation by model referencing adaptive control (MRAC) principle

In practice, the rotor flux synthesis based on the reference model is difficult to implement, particularly at low speeds, because of the pure integration of the voltage signals. The MRAS speed estimation algorithm remains valid if, instead of integration, the corresponding CEMF signals are compared directly through some low-pass filters. Estimation accuracy can be good if machine parameters are considered as constant. However, accuracy, particularly at low speeds, deteriorates due to parameter variation.

8.5.1.4   Speed Adaptive Flux Observer (Luenberger Observer)

An improved method of speed estimation that operates on the principle of a speed adaptive flux observer has been proposed in [20]. An observer is basically an estimator that uses a plant model (full or partial) and a feedback loop with measured plant variables. Here, the full-order observer uses the machine electrical model in d^s-q^s frame, where the state variables are stator currents $i_{ds}{}^s$ and $i_{qs}{}^s$ and the rotor fluxes are $\psi_{dr}{}^s$ and $\psi_{qr}{}^s$. Let us first derive this model from the $d^s - q^s$ equivalent circuits shown in Figure 2.27.

Sensorless Vector Control

The rotor voltage equations can be written as

$$i_{dr}^s R_r + \frac{d}{dt}(\psi_{dr}^s) + \omega_r \psi_{qr}^s = 0 \tag{8.114}$$

$$i_{qr}^s R_r + \frac{d}{dt}(\psi_{qr}^s) - \omega_r \psi_{dr}^s = 0 \tag{8.115}$$

Eliminating i_{dr}^s from Equation (8.114) with the help of Equation (8.37) yields

$$\frac{d}{dt}(\psi_{dr}^s) = -\frac{R_r}{L_r}\psi_{dr}^s - \omega_r \psi_{qr}^s + \frac{L_m R_r}{L_r} i_{ds}^s \tag{8.116}$$

Similarly, from the q^s circuit,

$$\frac{d}{dt}(\psi_{qr}^s) = -\frac{R_r}{L_r}\psi_{qr}^s + \omega_r \psi_{dr}^s + \frac{L_m R_r}{L_r} i_{qs}^s \tag{8.117}$$

Substituting Equations (8.116) and (8.117) in (8.106) and (8.107), respectively, and simplifying, we get

$$\frac{d}{dt}(i_{ds}^s) = -\frac{(L_m^2 R_r + L_r^2 R_s)}{\sigma L_s L_r^2} i_{ds}^s + \frac{L_m R_r}{\sigma L_s L_r^2}\psi_{dr}^s + \frac{L_m \omega_r}{\sigma L_s L_r}\psi_{qr}^s + \frac{1}{\sigma L_s} v_{ds}^s \tag{8.118}$$

$$\frac{d}{dt}(i_{qs}^s) = -\frac{(L_m^2 R_r + L_r^2 R_s)}{\sigma L_s L_r^2} i_{qs}^s - \frac{L_m \omega_r}{\sigma L_s L_r}\psi_{dr}^s + \frac{L_m R_r}{\sigma L_s L_r^2}\psi_{qr}^s + \frac{1}{\sigma L_s} v_{qs}^s \tag{8.119}$$

where $\sigma = 1 - L_m^2/L_s L_r$, and L_{ls} and L_{lr} have been substituted by $L_s - L_m$ and $L_r - L_m$, respectively.

Equations (8.116)–(8.119) constitute the desired state equations, which can be written in the form

$$\frac{d}{dt}(X) = AX + BV_s \tag{8.120}$$

where

$$X = \begin{bmatrix} i_{ds}^s & i_{qs}^s & \psi_{dr}^s & \psi_{qr}^s \end{bmatrix}^T \tag{8.121}$$

$$V_s = \begin{bmatrix} v_{ds}^s & v_{qs}^s & 0 & 0 \end{bmatrix}^T \tag{8.122}$$

$$A = \begin{bmatrix} -\dfrac{(L_m^2 R_r + L_r^2 R_s)}{\sigma L_s L_r^2} & 0 & \dfrac{L_m R_r}{\sigma L_s L_r^2} & \dfrac{L_m \omega_r}{\sigma L_s L_r} \\ 0 & -\dfrac{(L_m^2 R_r + L_r^2 R_s)}{\sigma L_s L_r^2} & -\dfrac{L_m \omega_r}{\sigma L_s L_r} & \dfrac{L_m R_r}{\sigma L_s L_r^2} \\ \dfrac{L_m R_r}{L_r} & 0 & -\dfrac{R_r}{L_r} & -\omega_r \\ 0 & \dfrac{L_m R_r}{L_r} & \omega_r & \dfrac{R_r}{L_r} \end{bmatrix} \quad (8.123)$$

$$B = \begin{bmatrix} \dfrac{1}{\sigma L_s} & 0 \\ 0 & \dfrac{1}{\sigma L_s} \\ 0 & 0 \\ 0 & 0 \end{bmatrix} \quad (8.124)$$

Note that parameter matrix A contains the speed signal ω_r, which is to be estimated.

Figure 8.48 shows the block diagram of the speed adaptive flux observer using the above machine model, where the symbol "^" means the estimated value. The output current signals $\hat{i}_{ds}^s$ and $\hat{i}_{qs}^s$) are derived through the following C matrix:

$$C = \begin{bmatrix} 1 & 0 & 0 & 0 \\ 0 & 1 & 0 & 0 \end{bmatrix} \quad (8.125)$$

Input voltage signals v_{ds}^s and v_{qs}^s are measured from the machine terminal. If speed signal ω_r in parameter matrix A is known, the fluxes and currents can be solved from the state equations. However, if the ω_r signal is not correct, there will be a deviation between the estimated states and the actual states. In the figure, the estimated currents are compared with the actual machine terminal currents, and the errors inject the auxiliary corrective signals eG through gain matrix G, as shown, so that matrix e tends to vanish. The observer equation can be given as

$$\frac{d}{dt}(\hat{X}) = \hat{A}\hat{X} + BV_s + G(\hat{i}_s - i_s) \quad (8.126)$$

where $\hat{i}_s = [i_{ds}^s \ i_{qs}^s]$ and G = observer gain matrix. The observer also gives an estimation of the flux vector, as shown in the figure.

The speed adaptive flux observer permits estimation of the unknown speed ω_r (which is considered a parameter) in matrix A. To derive the speed adaptation algorithm, Lyapunov's

Sensorless Vector Control

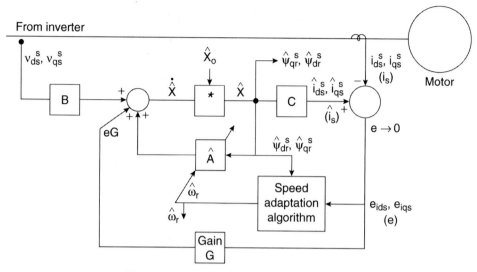

Figure 8.48 Speed adaptive flux observer

theorem is utilized. In general, the estimation error of the stator currents and rotor fluxes is described by the following equation:

$$\frac{d}{dt}(e) = (A+GC)e - \Delta A \hat{X} \tag{8.127}$$

where

$$e = X - \hat{X} \tag{8.128}$$

$$\Delta A = \hat{A} - A = \begin{bmatrix} 0 & -\dfrac{\Delta\omega_r j}{c} \\ 0 & \Delta\omega_r j \end{bmatrix} \tag{8.129}$$

$$j = \begin{bmatrix} 0 & -1 \\ 1 & 0 \end{bmatrix} \tag{8.130}$$

$$\Delta\omega_r = \hat{\omega}_r - \omega_r \tag{8.131}$$

$$c = \frac{\sigma L_s L_r}{L_m} \tag{8.132}$$

Let us define the following Lyapunov function candidate:

$$V = e^T e + \frac{(\hat{\omega}_r - \omega_r)^2}{\lambda} \tag{8.133}$$

where λ is a positive constant.

The time derivative of V becomes

$$\frac{dV}{dt} = e^T \left[(A+GC)^T + (A+GC) \right] e - 2\Delta\omega_r (e_{ids}\hat{\psi}_{qr}^s - e_{iqs}\hat{\psi}_{dr}^s)/c + \frac{2\Delta\omega_r}{\lambda} \frac{d\hat{\omega}_r}{dt} \tag{8.134}$$

where $e_{ids}^s = i_{ds}^s - \hat{i}_{ds}^s$ and $e_{iqs}^s = i_{qs}^s - \hat{i}_{qs}^s$.

From the above equation, we can derive the following adaptation scheme for speed estimation by equalizing the second term with the third term, that is,

$$\frac{d\hat{\omega}_r}{dt} = \lambda(e_{ids}\hat{\psi}_{qr}^s - e_{iqs}\hat{\psi}_{dr}^s)/c \tag{8.135}$$

If the observer gain matrix G is chosen such that the first term of Equation (8.134) is negative-semidefinite, the speed adaptive flux observer is stable. Since the speed ω_r can change quickly, the following proportional and integral adaptive scheme is used in the speed adaptation algorithm to improve the response of the speed estimation:

$$\hat{\omega}_r = K_P(e_{ids}\hat{\psi}_{qr}^s - e_{iqs}\psi_{dr}^s) + K_I \int (e_{ids}\hat{\psi}_{qr}^s - e_{iqs}\psi_{dr}^s) dt \tag{8.136}$$

where K_P and K_I are arbitrary positive gains.

Although the speed estimation accuracy is improved by the observer, there is a finite parameter variation (particularly in the stator and rotor resistances) effect. The estimation error tends to be more dominant as the speed approaches zero.

8.5.1.5 Extended Kalman Filter (EKF)

The extended Kalman filter (EKF) is basically a full-order stochastic observer for the recursive optimum state estimation of a nonlinear dynamical system in real time by using signals that are corrupted by noise. The EKF can also be used for unknown parameter estimation (such as rotor resistance R_r) or joint state and parameter estimation. The Luenberger observer, as discussed previously, is a deterministic observer (without noise) in comparison with the EKF, and is applicable to linear time-invariant systems. The noise sources in EKF take into account mea-

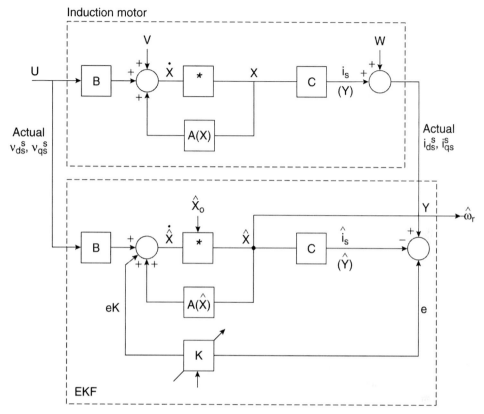

Figure 8.49 Extended Kalman filter for estimation of speed

surement and modeling inaccuracies. The original Kalman filter is applicable only to a linear system. Figure 8.49 shows the block diagram of the EKF algorithm, where the machine model is indicated on the top. The EKF algorithm uses the full machine dynamic model, where the speed ω_r is considered a parameter as well as a state. The augmented machine model can be given as [see Equations (8.120)–(8.124)]

$$\frac{dX}{dt} = AX + BV_s \qquad (8.137)$$

$$Y = CX \qquad (8.138)$$

where

$$A = \begin{bmatrix} -\dfrac{(L_m^2 R_r + L_r^2 R_s)}{\sigma L_s L_r^2} & 0 & \dfrac{L_m R_r}{\sigma L_s L_r^2} & \dfrac{L_m \omega_r}{\sigma L_s L_r} & 0 \\ 0 & -\dfrac{(L_m^2 R_r + L_r^2 R_s)}{\sigma L_s L_r^2} & -\dfrac{L_m \omega_r}{\sigma L_s L_r} & \dfrac{L_m R_r}{\sigma L_s L_r^2} & 0 \\ \dfrac{L_m R_r}{L_r} & 0 & -\dfrac{R_r}{L_r} & -\omega_r & 0 \\ 0 & \dfrac{L_m R_r}{L_r} & \omega_r & -\dfrac{R_r}{L_r} & 0 \\ 0 & 0 & 0 & 0 & 0 \end{bmatrix} \qquad (8.139)$$

$$X = \begin{bmatrix} i_{ds}^s & i_{qs}^s & \psi_{dr}^s & \psi_{qr}^s & \omega_r \end{bmatrix}^T \qquad (8.140)$$

$$B = \begin{bmatrix} \dfrac{1}{\sigma L_s} & 0 \\ 0 & \dfrac{1}{\sigma L_s} \\ 0 & 0 \\ 0 & 0 \\ 0 & 0 \end{bmatrix} \qquad (8.141)$$

$$C = \begin{bmatrix} 1 & 0 & 0 & 0 & 0 \\ 0 & 1 & 0 & 0 & 0 \end{bmatrix} \qquad (8.142)$$

$$Y = \begin{bmatrix} i_{ds}^s & i_{qs}^s \end{bmatrix}^T = i_s \qquad (8.143)$$

and $V_s = [v_{ds}^s \, v_{qs}^s]^T$ is the input vector. Equation (8.137) is of the fifth order, where speed ω_r is a state as well as a parameter. If speed variation is considered negligible, then $d\omega_r/dt = 0$. This is a valid consideration if the computational sampling time is small or load inertia is high. With speed ω_r as a constant parameter, the machine model used in the EKF is linear. For digital implementation of an EKF, the model must be discretized in the following form:

$$X(k+1) = A_d X(k) + B_d U(k) \qquad (8.144)$$

$$Y(k) = C_d X(k) \qquad (8.145)$$

Sensorless Vector Control

The induction motor model in Figure 8.49 is shown with two noise sources, V and W, which correspondingly give the model equations in discrete form as

$$X(k+1) = A_d X(k) + B_d U(k) + V(k) \tag{8.146}$$

$$Y(k) = C_d X(k) + W(k) \tag{8.147}$$

where $V(k)$ and $W(k)$ are zero-mean, white Gaussian noise vectors of $X(k)$ and $Y(k)$, respectively. Both $V(k)$ and $W(k)$ are independant of $X(k)$ and $Y(k)$, respectively. The statistics of noise and measurements are given by three covariance matrices, Q, R, and P, where Q = system noise vector covariance matrix (5X5), R = measurement noise vector covariance matrix (2X2), and P = system state vector covariance matrix (5X5).

The sequence of the EKF algorithm implementation by a flow diagram is shown in Figure 8.50, which also includes the basic computational expressions. Basically, it has two main stages: prediction stage and filtering stage. In prediction stage, the next predicted values of states $X^*(k+1)$ are calculated by the machine model and the previous values of the estimated states. In addition, the predicted state covariance matrix $P^*(k+1)$ is also calculated using the covariance vector Q. In the filtering stage, the next estimated states $\hat{X}(k+1)$ are obtained from the predicted states $X^*(k+1)$ by adding the correction term eK where $e = Y(k+1) - \hat{Y}(k+1)$ and K = Kalman gain. The Kalman gain is optimized for the state estimation errors. The EKF computations are done in recursive manner so that e approaches 0.

Due to high complexity, the computation time, even with a powerful DSP, is long and difficult to apply in real time, particularly for fast-speed variations. Besides, there is also a parameter variation problem, which makes accuracy poor at low speeds.

8.5.1.6 Slot Harmonics

This is one of the simplest methods for the estimation of rotor speed. In an induction motor, the slots on the surface of the rotor provide reluctance modulation, which produces space harmonics in the air gap flux. The slot-induced ripple flux is superimposed on the fundamental flux wave. Therefore, induced stator voltage waves will contain a ripple voltage component, the frequency and magnitude of which are proportional to the rotor speed. The speed can be estimated by identifying the ripple frequency through a signal processing circuit. Due to the finite number of rotor slots and small reluctance variation, the ripple frequency and voltage magnitude become very low at low motor speeds and speed estimation becomes difficult.

8.5.1.7 Injection of Auxiliary Signal on Salient Rotor

The speed estimation techniques discussed so far become impossible at zero stator frequency, that is, the dc condition when the machine speed is zero at zero torque. At pure dc condition, the rotor conditions become unobservable. However, recently, the position and speed estimation of the induction motor, which are based on the injection of an auxiliary carrier frequency signal from the stator side of a custom-designed rotor have been proposed [21]. In the proposed scheme, the machine rotor slots are to be specially designed to get spatial variation of

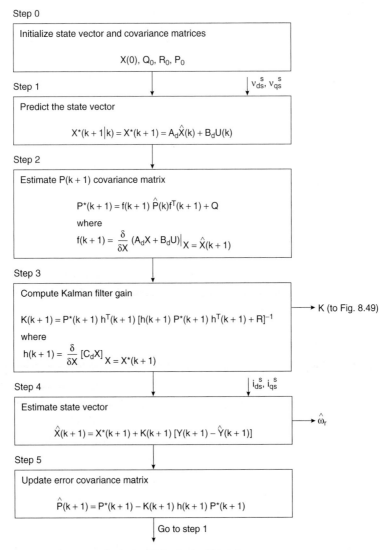

Figure 8.50 EKF algorithm flow diagram

magnetic saliency. This is possible by a number of methods, such as variation of resistance of outer conductors, variation of depth of rotor bars, or variation of conductor heights in slot openings. The saliency of the rotor position is to be tracked by the position estimation algorithm. In this method, a carrier frequency signal (typically 250 Hz) is mixed with the usual three-phase modulating voltage signals of the inverter. The carrier does not essentially affect the drive's

Sensorless Vector Control

performance. The high-frequency machine model with modulating leakage inductance (due to rotor anisotrophy) indicates the generation of a negative sequence current component at carrier frequency that is modulated by the machine position. The machine stator currents are sensed, passed through a band-pass filter, and are then fed to a PLL system to estimate the incremental position and speed signals. The PLL system consists of a salient machine model, low-pass filter, P-I-D controller, and mechanical subsystem model with estimated load torque. The estimation algorithm is obviously very complex. The accuracy of estimation is affected by the parameter variation of the machine model, the load torque estimation error, and the inertia variation error. There is also a skin effect problem in the rotor bars due to carrier frequency. In addition, the finite phase shift in the low-pass filter to retrieve the fundamental frequency phase currents will tend to degrade the vector control. Custom designing of the rotor for speed and position estimation may not be acceptable by drive manufacturers.

8.5.2 Direct Vector Control without Speed Signal

In a torque-controlled drive with direct vector control, such as in electric/hybrid vehicle applications, it is possible to control the drive from zero speed (including the start-up) without using the speed signal. Such control will be described [23] with stator flux orientation.

8.5.2.1 Programmable Cascaded Low-Pass Filter (PCLPF) Stator Flux Estimation

The use of the voltage model for rotor flux vector estimation was discussed before. The method uses an ideal integrator, where the dc offset voltage tends to build up at the output. The problem becomes particularly serious at the low-frequency condition. If, on the other hand, the single-stage integrator is replaced by a number of cascaded low-pass filters with short time constants for integration, the dc offset can be sharply attenuated. Consider a first-order, low-pass filter with transfer characteristics as follows:

$$\frac{Y}{X} = \frac{1}{(1 + j\tau\omega_e)} \qquad (8.148)$$

where τ = filter time constant and ω_e = frequency. The phase lag and gain (or attenuation) of the filter at frequency ω_e can be given, respectively, as

$$\varphi = \tan^{-1}(\tau\omega_e) \qquad (8.149)$$

$$K = \left|\frac{Y}{X}\right| = \frac{1}{\sqrt{1 + (\tau\omega_e)^2}} \qquad (8.150)$$

If n number of filters are cascaded, the total phase shift angle and gain are given, respectively, as

$$\varphi_T = \varphi_1 + \varphi_2 + ... + \varphi_n = \tan^{-1}(\tau_1 \omega_e) + \tan^{-1}(\tau_2 \omega_e) + ... + \tan^{-1}(\tau_n \omega_e) \tag{8.151}$$

$$K_T = K_1 K_2 ... K_n = \frac{1}{\sqrt{\left[1+(\tau_1 \omega_e)^2\right]\left[1+(\tau_2 \omega_e)^2\right]...\left[1+(\tau_n \omega_e)^2\right]}} \tag{8.152}$$

If all the filter stages are identical, the corresponding expressions are

$$\varphi_T = n\varphi = n.\tan^{-1}(\tau \omega_e) \tag{8.153}$$

$$K_T = nK = \frac{1}{\sqrt{\left[1+(\tau \omega_e)^2\right]^n}} \tag{8.154}$$

If the cascaded filter is required to perform integration of a sinusoidal voltage at frequency ω_e, then $\varphi_T = \pi/2$ and $G.K_T = 1/\omega_e$, where G = gain compensation needed for the integration. Substituting these conditions in Equations (8.153) and (8.154), respectively,

$$\tau = (\frac{1}{\omega_e}) \tan(\frac{\pi}{2n}) \tag{8.155}$$

$$G = (\frac{1}{\omega_e}) \sqrt{\left[1+(\tau \omega_e)^2\right]^n} \tag{8.156}$$

The above equations give the parameters τ and G as functions of frequency. Figure 8.51 gives the equivalent op amp representation of the cascaded low-pass filter for integration, where $\varphi_1 = \varphi_2 = \varphi_n = \pi/2n$.

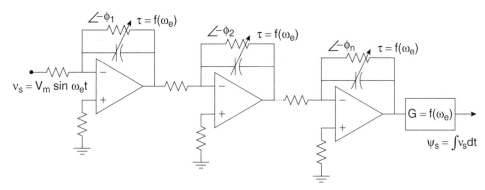

Figure 8.51 Op amp representation of PCLPF for integration

A large number for n is desirable, but in digital implementation, the software computation burden becomes heavy. Figure 8.52 shows a three-stage ($n = 3$) PCLPF for the synthesis of stator fluxes $\psi_{ds}{}^s$ and $\psi_{qs}{}^s$ by solving the following equations:

$$\psi_{ds}{}^s = \int (v_{ds}{}^s - i_{ds}{}^s R_s) dt \tag{8.32}$$

$$\psi_{qs}{}^s = \int (v_{qs}{}^s - i_{qs}{}^s R_s) dt \tag{8.33}$$

Note that identical low-pass analog hardware filters are used in the front end for both the current and voltage signals (not shown here), but their effects are compensated in the PCLPF. The correct time constant and gain expressions can be derived as

$$\tau = (\frac{1}{\omega_e}) \tan \left[\frac{1}{n} \left[\tan^{-1}(\tau_h \omega_e) + \frac{\pi}{2} \right] \right] = f(.)\omega_e \tag{8.157}$$

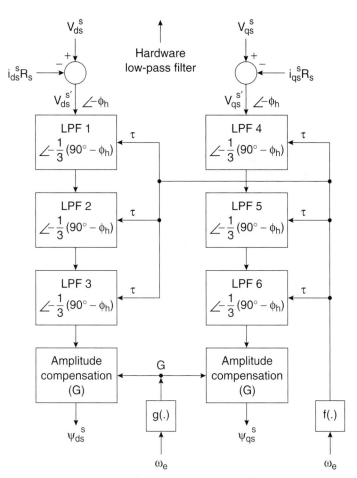

Figure 8.52 Three-stage PCLPF based stator flux vector synthesis

$$G = (\frac{1}{\omega_e})\sqrt{\left[1+(\tau\omega_e)^2\right]^n \left[1+(\tau_h\omega_e)^2\right]} = g(.)\omega_e \qquad (8.158)$$

where τ_h = hardware filter time constant and $\tan^{-1}(\tau_h \omega_e) = -\varphi_h$ (phase lag due to hardware filter) is shown in the figure. If, for example, $\varphi_h = 3°$ at a certain frequency, each low-pass filter (LPF) gives a phase shift angle of 29°, irrespective of the stator frequency variation. Since the hardware filter phase shift angle and attenuation are compensated at all frequencies, it can be designed conservatively to clean the harmonics without fear of phase drift and attenuation of signals.

Figure 8.53 shows the plot of τ and G signals from Equations (8.157) and (8.158) for a particular value of τ_h (in this case, $\tau_h = 0.16$ ms). In Figure 8.52, the frequency ω_e is solved in a circulatory manner from the equation

$$\omega_e = \frac{\left[(v_{qs}^s - i_{qs}^s R_s)\psi_{ds}^{s^2} - (v_{ds}^s - i_{ds}^s R_s)\psi_{qs}^s\right]}{\sqrt{\psi_{ds}^{s^2} + \psi_{qs}^{s^2}}} \qquad (8.70)$$

and is applied to the PCLPFs for solving estimated fluxes. Note that stator resistance R_s will tend to vary with temperature and will cause an error in the estimation. The parameter can be estimated reasonably accurately using stator-mounted thermistor probes. A fuzzy logic method of R_s estimation is discussed in Chapter 11. The PCLPF method of flux vector estimation is satisfactory, typically down to a fraction of a hertz.

8.5.2.2   Drive Machine Start-up with Current Model Equations

The stator flux-oriented direct vector control using PCLPF, as discussed above, is satisfactory from a fraction of a hertz of stator frequency up to the field-weakening region. However, a separate start-up control is required from zero speed stand-still condition. This is possible with the help of a current model or Blaschke equations without any speed signal as discussed below.

The stationary frame current model equations with rotor fluxes are given with standard symbols as

$$\frac{d}{dt}(\psi_{dr}^s) = \frac{L_m}{T_r}i_{ds}^2 - \omega_r\psi_{qr}^s - \frac{1}{T_r}\psi_{dr}^s \qquad (8.48)$$

$$\frac{d}{dt}(\psi_{qr}^s) = \frac{L_m}{T_r}i_{qs}^s + \omega_r\psi_{dr}^s - \frac{1}{T_r}\psi_{qr}^s \qquad (8.49)$$

These equations are to be transformed for stator flux vector estimation so that the stator flux-oriented vector control can be used at start-up condition. Substituting the following equation pairs:

$$\psi_{dr}^s = \frac{L_r}{L_m}\psi_{dm}^s - L_{lr}i_{ds}^s \qquad (8.39)$$

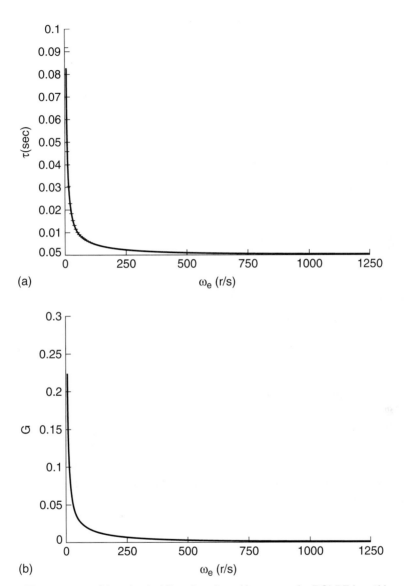

Figure 8.53 Time constant (τ) and gain (G) as function of frequency for PCLPF ($n = 3$)($\tau_h = 0.16$ ms)

$$\psi_{ds}^{s} = \psi_{dm}^{s} + L_{ls}i_{ds}^{s} \tag{8.159}$$

and

$$\psi_{qr}^{s} = \frac{L_r}{L_m}\psi_{qm}^{s} - L_{lr}i_{qs}^{s} \tag{8.40}$$

$$\psi_{qs}^{s} = \psi_{qm}^{s} + L_{ls}i_{qs}^{s} \tag{8.160}$$

in Equations (8.48) and (8.49) and simplifying, we get

$$\frac{d}{dt}(\psi_{ds}^{\ s}) = -\frac{1}{T_r}\psi_{ds}^{\ s} + A\frac{d}{dt}(i_{ds}^{\ s}) + Bi_{ds}^{\ s} - \omega_r(\psi_{qs}^{\ s} - Ai_{qs}^{\ s}) \quad (8.161)$$

$$\frac{d}{dt}(\psi_{qs}^{\ s}) = -\frac{1}{T_r}\psi_{qs}^{\ s} + A\frac{d}{dt}(i_{qs}^{\ s}) + Bi_{qs}^{\ s} - \omega_r(\psi_{ds}^{\ s} - Ai_{ds}^{\ s}) \quad (8.162)$$

where $A = L_{ls} + (L_m L_{lr})/L_r$ and $B = L_s R_r/L_r$.

Equations (8.161) and (8.162) give Blaschke equations for stator flux vector estimation. These equations can be used without the speed signal at zero speed to start the machine. Substituting $\omega_r = 0$ in the above equations, we get

$$\frac{d}{dt}(\psi_{ds}^{\ s}) = -\frac{1}{T_r}\psi_{ds}^{\ s} + A\frac{d}{dt}(i_{ds}^{\ s}) + Bi_{ds}^{\ s} \quad (8.163)$$

$$\frac{d}{dt}(\psi_{qs}^{\ s}) = -\frac{1}{T_r}\psi_{qs}^{\ s} + A\frac{d}{dt}(i_{qs}^{\ s}) + Bi_{qs}^{\ s} \quad (8.164)$$

A block diagram to solve the stator flux vectors $\psi_{ds}^{\ s}$ and $\psi_{qs}^{\ s}$ is shown in Figure 8.54.

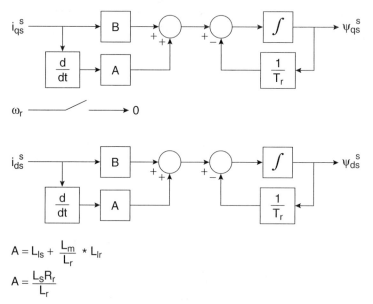

$$A = L_{ls} + \frac{L_m}{L_r} * L_{lr}$$

$$A = \frac{L_s R_r}{L_r}$$

Figure 8.54 Block diagram to solve $\psi_{ds}^{\ s}$ and $\psi_{qs}^{\ s}$ from Blaschke equations for start-up of drive without speed sensor

Sensorless Vector Control

The current derivatives shown can be actually implemented with the DSP in real time, where the derivative is basically considered an increment (or decrement) in a sampling time interval for a heavily filtered current signal. Note that the models in Figure 8.54 are strictly valid at $\omega_r = 0$. Figure 8.55 shows the complete control block diagram, incorporating the PCLPFs and Blaschke equation start-up, as discussed above. The additional unit vector signals $\cos \theta_e'$ and $\sin \theta_e'$ used for the inverse rotation of stator currents are constructed with hardware filter (τ_h) compensation as follows:

$$\psi_{ds}^{s'} = \frac{1}{(1+\tau_h S)} \psi_{ds}^{s} \tag{8.165}$$

$$\psi_{qs}^{s'} = \frac{1}{(1+\tau_h S)} \psi_{qs}^{s} \tag{8.166}$$

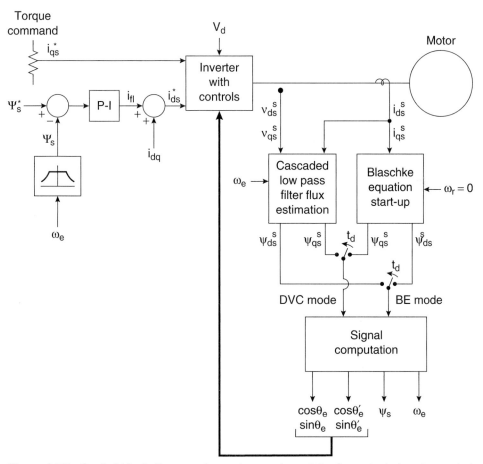

Figure 8.55 Control block diagram of speed sensorless stator flux-oriented vector control incorporating PCLPF flux estimator and Blaschke equation start-up

$$\cos\theta_e' = \frac{\psi_{ds}^{s'}}{\sqrt{\psi_{ds}^{s'2} + \psi_{qs}^{s'2}}} \tag{8.167}$$

$$\sin\theta_e' = \frac{\psi_{qs}^{s'}}{\sqrt{\psi_{ds}^{s'2} + \psi_{qs}^{s'2}}} \tag{8.168}$$

In the beginning, the drive is in Blaschke equation (BE) mode at stand-still condition when the rated stator fluxes are established with dc stator current before applying any command torque (i_{qs}^*). As the torque command is established at stand-still condition, the slip frequency is developed and the speed tends to develop with finite torque. The drive is transitioned to direct vector control (DVC) mode at slip frequency with a time delay (depending on the command torque) before any finite speed is developed. Figure 8.55 can also be modified for rotor flux-oriented vector control.

8.6 DIRECT TORQUE AND FLUX CONTROL (DTC)

In the mid-1980s, an advanced scalar control technique, known as direct torque and flux control (DTFC or DTC) or direct self-control (DSC) [24], was introduced for voltage-fed PWM inverter drives. This technique was claimed to have nearly comparable performance with vector-controlled drives. Recently, the scheme was introduced in commercial products by a major company and therefore created wide interest. The scheme, as the name indicates, is the direct control of the torque and stator flux of a drive by inverter voltage space vector selection through a lookup table. Before explaining the control principle, we will first develop a torque expression as a function of the stator and rotor fluxes.

8.6.1 Torque Expression with Stator and Rotor Fluxes

The torque expression given in Equation (2.121) can be expressed in the vector form as

$$\bar{T}_e = \frac{3}{2}\left(\frac{P}{2}\right)\bar{\psi}_s X \bar{I}_s \tag{8.169}$$

where $\bar{\psi}_s = \psi_{qs}^s - j\psi_{ds}^s$ and $\bar{I}_s = i_{qs}^s - j i_{ds}^s$. In this equation, $\bar{I}_s$ is to be replaced by rotor flux $\bar{\psi}_r$. In the complex form, $\bar{\psi}_s$ and $\bar{\psi}_r$ can be expressed as functions of currents (see Figure 2.28) as

$$\bar{\psi}_s = L_s \bar{I}_s + L_m \bar{I}_r \tag{8.170}$$

$$\bar{\psi}_r = L_r \bar{I}_r + L_m \bar{I}_s \tag{8.171}$$

Direct Torque and Flux Control (DTC)

Eliminating $\bar{I}_r$ from Equation (8.170), we get

$$\bar{\psi}_s = \frac{L_m}{L_r}\bar{\psi}_r + L_s'\bar{I}_s \qquad (8.172)$$

where $L_s' = L_s L_r - L_m^2$. The corresponding expression of $\bar{I}_s$ is

$$\bar{I}_s = \frac{1}{L_s'}\bar{\psi}_s - \frac{L_m}{L_r L_s'}\bar{\psi}_r \qquad (8.173)$$

Substituting Equation (8.173) in (8.169) and simplifying yields

$$\bar{T}_e = \frac{3}{2}(\frac{P}{2})\frac{L_m}{L_r L_s'}\bar{\psi}_r X \bar{\psi}_s \qquad (8.174)$$

that is, the magnitude of torque is

$$T_e = \frac{3}{2}(\frac{P}{2})\frac{L_m}{L_r L_s'}|\psi_r||\psi_s|\sin\gamma \qquad (8.175)$$

where γ is the angle between the fluxes. Figure 8.56 shows the phasor (or vector) diagram for Equation (8.174), indicating the vectors $\bar{\psi}_s$, $\bar{\psi}_r$, and $\bar{I}_s$ for positive developed torque. If the rotor flux remains constant and the stator flux is changed incrementally by stator voltage $\bar{V}_s$ as shown, and the corresponding change of γ angle is $\Delta\gamma$, the incremental torque ΔT_e expression is given as

$$\Delta T_e = \frac{3}{2}(\frac{P}{2})\frac{L_m}{L_r L_s'}|\psi_r||\bar{\psi}_s + \Delta\bar{\psi}_s|\sin\Delta\gamma \qquad (8.176)$$

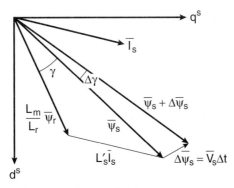

Figure 8.56 Stator flux, rotor flux, and stator current vectors on $d^s - q^s$ plane (stator resistance is neglected)

8.6.2 Control Strategy of DTC

The block diagram for direct torque and flux control is shown in Figure 8.57, and Figure 8.58 explains the control strategy. The speed control loop and the flux program as a function of speed are shown as usual and will not be discussed. The command stator flux $\hat{\psi}_s^*$ and torque T_e^* magnitudes are compared with the respective estimated values, and the errors are processed through hysteresis-band controllers, as shown. The flux loop controller has two levels of digital output according to the following relations:

$$H_\psi = 1 \quad \text{for} \quad E_\psi > +HB_\psi \tag{8.177}$$

$$H_\psi = -1 \quad \text{for} \quad E_\psi < -HB_\psi \tag{8.178}$$

where $2HB_\psi$ = total hysteresis-band width of the flux controller. The circular trajectory of the command flux vector $\bar{\psi}_s^*$ with the hysteresis band rotates in an anti-clockwise direction, as shown in Figure 8.58(a). The actual stator flux $\bar{\psi}_s$ is constrained within the hysteresis band and

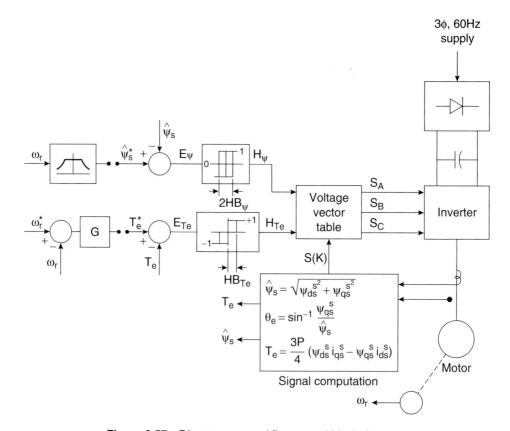

Figure 8.57 Direct torque and flux control block diagram

Direct Torque and Flux Control (DTC)

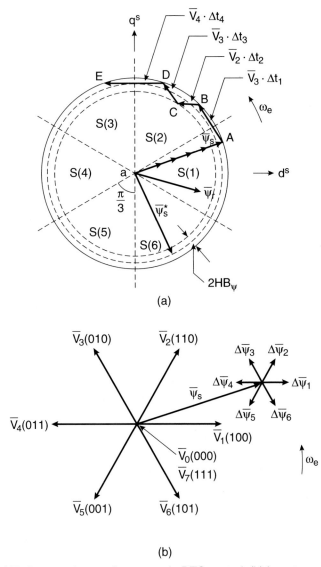

Figure 8.58 (a) Trajectory of stator flux vector in DTC control, (b) Inverter voltage vectors and corresponding stator flux variation in time Δt

it tracks the command flux in a zigzag path. The torque control loop has three levels of digital output, which have the following relations:

$$H_{Te} = 1 \quad \text{for} \quad E_{T_e} > +HB_{T_e} \tag{8.179}$$

$$H_{Te} = -1 \quad \text{for} \quad E_{T_e} < -HB_{T_e} \tag{8.180}$$

$$H_{Te} = 0 \quad \text{for} \quad -HB_{T_e} < E_{T_e} < +HB_{T_e} \qquad (8.181)$$

The feedback flux and torque are calculated from the machine terminal voltages and currents. The signal computation block also calculates the sector number $S(k)$ in which the flux vector $\bar{\psi}_s$ lies. There are six sectors (each $\pi/3$ angle wide), as indicated in Figure 8.58(a). The voltage vector table block in Figure 8.57 receives the input signals H_ψ, H_{Te}, and $S(k)$ and generates the appropriate control voltage vector (switching states) for the inverter by a lookup table, which is shown in Table 8.1 (the vector sign is deleted). The inverter voltage vectors (six active and two zero states) and a typical $\bar{\psi}_s$ are shown in Figure 8.58(b). Neglecting the stator resistance R_s of the machine, we can write

$$\bar{V}_s = \frac{d}{dt}(\bar{\psi}_s) \qquad (8.182)$$

or

$$\Delta\bar{\psi}_s = \bar{V}_s \cdot \Delta t \qquad (8.183)$$

which means that $\bar{\psi}_s$ can be changed incrementally by applying stator voltage vector $\bar{V}_s$ for time increment Δt. The flux increment vector corresponding to each of the six inverter voltage vectors is shown in Figure 8.58(b). The flux in the machine is initially established at zero frequency (dc) along the radial trajectory OA shown in Figure 8.58(a). With the rated flux, the command torque is applied and the $\bar{\psi}_s^*$ vector starts rotating. Table 8.1 applies the selected voltage vector, which essentially affects both the torque and flux simultaneously. The flux trajectory segments AB, BC, CD, and DE by the respective voltage vectors $\bar{V}_3$, $\bar{V}_4$, $\bar{V}_3$, and $\bar{V}_4$ are shown in Figure 8.58(a). The total and incremental torque due to $\Delta\bar{\psi}_s$ are explained in Figure 8.56. Note that the stator flux vector changes quickly by $\bar{V}_s$, but the $\bar{\psi}_r$ change is very sluggish due to large time constant T_r (see Figure 8.35). Since $\bar{\psi}_r$ is more filtered, it moves uniformly at frequency ω_e, whereas ψ_s movement is jerky. The average speed of both, however, remains the same in the steady-state condition. Table 8.2 summarizes the flux and torque change (magnitude and direction) for applying the voltage vectors for the location of $\bar{\psi}_s$ shown in Figure 8.58(b). The flux can be increased by the V_1, V_2, and V_6 vectors (vector sign is deleted), whereas it can be decreased by the V_3, V_4, and V_5 vectors. Similarly, torque is increased by the V_2, V_3, and V_4 vectors, but decreased by the V_1, V_5, and V_6 vectors. The zero vector (V_0 or V_7) short-circuits the machine terminals and keeps the flux and torque unaltered. Due to finite resistance (R_s) drop, the torque and flux will slightly decrease during the short-circuit condition.

Consider, for example, an operation in sector $S(2)$ as shown in Figure 8.58(a), where at point B, the flux is too high and the torque is too low; that is, $H_\psi = -1$ and $H_{Te} = +1$. From Table 8.1, voltage V_4 is applied to the inverter, which will generate the trajectory BC. At point C, $H_\psi = +1$ and $H_{Te} = +1$ and this will generate the V_3 vector from the table. The drive can easily operate in the four quadrants, and speed loop and field-weakening control can be added, if desired. The torque

Adaptive Control

H_ψ	H_{Te}	S(1)	S(2)	S(3)	S(4)	S(5)	S(6)
1	1	V_2	V_3	V_4	V_5	V_6	V_1
	0	V_0	V_7	V_0	V_7	V_0	V_7
	−1	V_6	V_1	V_2	V_3	V_4	V_5
−1	1	(V_3)	(V_4)	V_5	V_6	V_1	V_2
	0	V_7	V_0	V_7	V_0	V_7	V_0
	−1	V_5	V_6	V_1	V_2	V_3	V_4

Table 8.1 Switching Table of Inverter Voltage Vectors

Voltage vector	V_1	V_2	V_3	V_4	V_5	V_6	V_0 or V_7
ψ_s	↑	↑	↓	↓	↓	↑	0
T_e	↓	↑	↑	↑	↓	↓	↓

Table 8.2 Flux and Torque Variations Due to Applied Voltage Vector in Figure 8.58(b) (Arrow indicates magnitude and direction)

response of the drive is claimed to be comparable with that of a vector-controlled drive. There are a few special features of DTC control that can be summarized as follows:

- No feedback current control
- No traditional PWM algorithm is applied
- No vector transformation as in vector control
- Feedback signal processing is somewhat similar to stator flux-oriented vector control
- Hysteresis-band control generates flux and torque ripple and switching frequency is not constant (like hysteresis-band current control)

8.7 ADAPTIVE CONTROL

A linear control system with invariant plant parameters can be designed easily with the classic design techniques, such as Nyquist and Bode plots. Ideally, a vector-controlled ac drive can be considered as linear, like a dc drive system. However, in industrial applications, the electrical and mechanical parameters of the drive hardly remain constant. Besides, there is a load torque disturbance effect. For example, the inertia of an electric vehicle or subway drive will vary with passenger load. In a robotic drive, on the other hand, the inertia will change, depending on the length of the arm and the load it carries. In a rolling mill drive, the drive load torque will change abruptly when a metal slab is introduced within the rolls. Figure 8.59 shows a block diagram of a speed-controlled vector drive indicating the moment of inertia J and load torque T_L variation. For the fixed control parameters in G, an increase of the J parameter will reduce the loop gain, deteriorating the system's performance. Similarly, a sudden increase of load torque T_L or J will temporarily reduce the speed until it is compensated by the sluggish speed loop. The effect of the parameter variation can be compensated to some extent by a high-gain negative feedback loop. But, excessive gain may cause an underdamping or instability problem in extreme cases. The problems discussed above require adaptation of the controller G in real time, depending on the plant parameter variation and load torque disturbance, so that the system response is not affected. Adaptive control techniques can be generally classified as

- Self-tuning control
- MRAC
- Sliding mode or variable structure control
- Expert system control
- Fuzzy control
- Neural control

In this section, the principles of adaptive control will be reviewed. Expert system, fuzzy logic, and neural network-based intelligent control techniques will be discussed in Chapters 10, 11, and 12, respectively. The application of sliding mode control in a vector drive will be discussed in somewhat detail.

8.7.1 Self-Tuning Control

In this method, as the name indicates, the controller parameters are tuned on-line to adapt to the plant parameter variation. A simple example of adaptive control is the loop gain scheduling in Figure 8.59, where the gain is proportional to parameter J, provided the information for the J variation is available. Another example is the slip gain (K_s) tuning of an indirect vector-controlled drive with on-line identification of the rotor time constant T_r. Figure 8.60 gives the

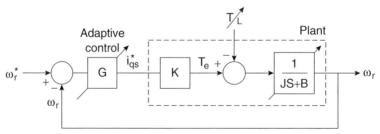

Figure 8.59 Adaptive control block diagram for plant parameter variation and load torque disturbance

Adaptive Control

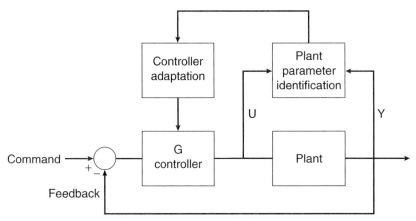

Figure 8.60 Block diagram of self-tuning control

general block diagram for a self-tuning control. Basically, it consists of a plant parameter identification algorithm and controller adaptation algorithm. The identification algorithm estimates the varying plant parameter(s) on-line and updates the control parameters for the desired performance of the system. The controller can be P-I, P-I-D, dead-beat, pole assignment control, or pole-zero cancellation with the plant. The problems arising with P-I or P-I-D control are that no unique mathematical relation between the transfer function of the system to be controlled and the controller parameters exists. Updating must be done with the help of a search algorithm, which may be too slow for on-line adaptation. All the other controllers allow direct computation of the new controller settings based on the identified system parameters. The computational complexity and corresponding time delay may not impose much problem in a slow process control drive system, but in a fast system, rapid on-line computation becomes difficult.

8.7.1.1 Load Torque Disturbance (T_L) Compensation

As mentioned above, a sudden load or disturbance torque T_L can cause a droop in the speed in a speed-controlled drive system, which may not be desirable. The speed droop can be compensated with the help of a disturbance torque observer.

The speed and torque are given by the following relation:

$$J\frac{d\omega_m}{dt} + B\omega_m = T_e - T_L \tag{8.184}$$

where B = viscous friction coefficient. Therefore, T_L can be estimated by the following equation:

$$T_L = T_e - (JS + B)\omega_m \tag{8.185}$$

Figure 8.61 shows the estimation of the load torque and its compensation in a feedforward manner [27]. The actual speed ω_m is measured with the measurement delay time T_d. The signal is processed through the inverse mechanical model $(\hat{J}S + \hat{B})$ and then subtracted from the effective torque T_e' to generate the estimated torque signal $\hat{T}_L$. A low-pass filter is added to clean

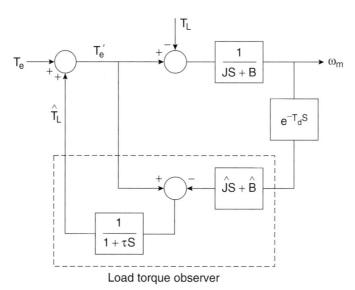

Figure 8.61 Disturbance torque estimation and its compensation in the system

the noise generated by the inverse model (due to differentiation). The compensation will not be accurate because of the speed measurement delay, the inaccuracy of the inverse model, and the delay due to the low-pass filter. However, the compensation will be reasonably good, particularly in a high-inertia system.

8.7.2 Model Referencing Adaptive Control (MRAC)

The MRAC was already discussed for indirect vector control slip gain (K_s) tuning (Figure 8.38) and speed estimation (Figure 8.47). In an MRAC, as the name indicates, the plant's response is forced to track the response of a reference model, irrespective of the plant's parameter variation and load disturbance effect. Such a system is defined as a robust system. The reference model may be fixed or adaptive and is stored in the DSP's memory.

Consider, for example, an indirect vector-controlled induction motor servo drive (Figure 8.32) and assume that the slip gain K_s has been tuned to achieve perfect decoupling. Ideally, the model of such a drive is identical to that of a dc machine. The position loop response of the system can be given by a second-order transfer function. A typical response for such a drive system with variable inertia ($J_2 > J_1$) load is shown in Figure 8.62. Such a response variation may be undesirable in some applications. The problem can be solved by the MRAC system shown in Figure 8.63. The speed command ω_r^*, generated by the position control loop, is applied in parallel to the reference model and plant controller as shown. The reference model output speed ω_{rm} is compared with the measured plant speed ω_r, and the resulting error signal e along with the ω_r signal actuates the adaptation algorithm. The feedforward and feedback gains K_F and K_B, respectively, of the plant controller are iterated by the adaptation algorithm dynamically so as to reduce the error e to zero. The algorithm contains P-I-type control law so that the desired K_F and K_B parameter values are locked in the integrator when the error vanishes to zero. The plant will be capable of tracking the reference model without saturation provided that the J parameter in the reference model is defined on the worst-case (maximum) basis (i.e., slowest response) for a variable-inertia plant load.

The MRAC profile for maximum inertia J_2 is indicated in Figure 8.62. With a fixed machine-converter power rating, the drive will fail to meet the desired acceleration-deceleration profile if actual inertia exceeds J_2. If the plant inertia is $J_1 (J_1 < J_2)$, the drive will have suboptimal transient response. Thus, the robustness of an MRAC system is obtained at the sacrifice of optimum response. Besides, there is the chattering problem on the speed signal, which is somewhat smoothed for the position signal because of integration. If chattering can be tolerated, an MRAC is applicable to a speed control system.

The adaptation algorithm can be defined as [29]

$$K_F = K_{FO} + FV\omega_r^* + \int_0^t GV\omega_r^* \, dt \tag{8.186}$$

$$K_B = K_{BO} + LV\omega_r + \int_0^t MV\omega_r \, dt \tag{8.187}$$

$$V = De \tag{8.188}$$

where K_{FO} and K_{BO} are the initial gain values and F, G, L, M, and D are the adaptation law constants. In general, the structure of the reference model and the plant should be the same and the parameters should be compatible for satisfactory adaptation. In Figure 8.63, the state equations for the reference model and the plant, respectively, can be given as

$$\frac{d}{dt}(\omega_{rm}) = A_m \omega_{rm} + B_m \omega_r^* \tag{8.189}$$

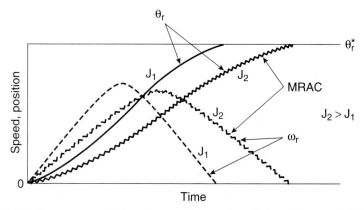

Figure 8.62 Response variation of servo drive with variation of moment of inertia J

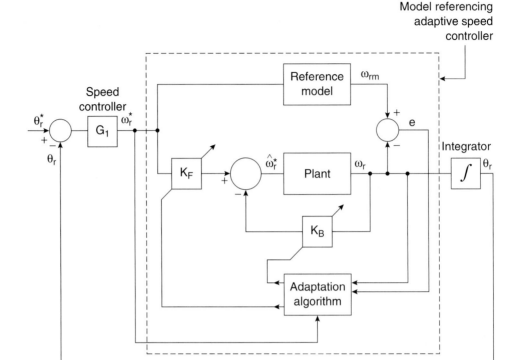

Figure 8.63 Servo drive with model referencing adaptive speed control

$$\frac{d}{dt}(\omega_r) = A_p \omega_r + B_p \hat{\omega}_r \tag{8.190}$$

and the other control loop equations are

$$\omega_r^* = G_1(\theta_r^* - \theta_r) \tag{8.191}$$

$$\hat{\omega}_r^* = K_F \omega_r^* - K_B \omega_r \tag{8.192}$$

$$e = \omega_{rm} - \omega_r \tag{8.193}$$

Note that the control parameters K_F and K_B are time-varying. The speed control system within the dashed line in Figure 8.63 can be represented by an equivalent feedforward time-invariant linear system with a feedback, nonlinear, time-varying block. The global stability of the system can be analyzed by Popov's hyperstability theorem, and correspondingly, parameters F, G, L, M, and D can be determined. An MRAC with a neural network will be discussed in Chapter 12.

8.7.3 Sliding Mode Control

A sliding mode control (SMC) with a variable control structure is basically an adaptive control that gives robust performance of a drive with parameter variation and load torque disturbance. The control is nonlinear and can be applied to a linear or nonlinear plant. In an SMC, as the name indicates, the drive response is forced to tract or "slide" along a predefined trajectory or "reference model" in a phase plane by a switching control algorithm, irrespective of the plant's parameter variation and load disturbance. The control DSP detects the deviation of the actual trajectory from the reference trajectory and correspondingly changes the switching strategy to restore the tracking. In performance, it is somewhat similar to an MRAC, as described before, but the design and implementation of an SMC are somewhat simpler. SMCs can be applied to servo drives with dc motors, induction motors, and synchronous motors for applications such as robot drives, machine tool control, etc. In this section, we will review the principles of the SMC without going into rigorous theoretical aspects [31]. Then, we will apply it to a vector-controlled induction motor servo drive.

8.7.3.1 Control Principle

An SMC is basically a variable structure control system (VSS) where the structure or topology of the control is intentionally varied to stabilize the control and make its response robust. Consider a simple second-order undamped linear system, as shown in Figure 8.64, with variable plant gain K. It can easily be seen that the system is unstable in either negative or positive feedback mode. However, by switching back and forth between the negative and positive feedback modes, the system cannot only be made stable, but its response can be made independent to the plant parameter K.

Consider Figure 8.64 in negative feedback mode with switch 1 closed. We can write

$$X_1 = R - C \tag{8.194}$$

or

$$R - X_1 = C \tag{8.195}$$

where X_1 = loop error.

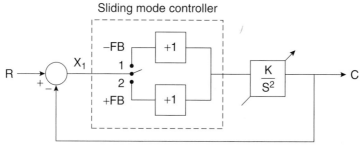

Figure 8.64 Variable structure control of second-order system

Differentiating Equation (8.195),

$$\frac{d}{dt}(R - X_1) = \frac{dC}{dt} = -X_2 \tag{8.196}$$

or

$$\frac{dX_1}{dt} = X_2 \tag{8.197}$$

where R = step input with constant value at time $t+$ and $-X_2 = dC/dt$.
We can also write the derivative of X_2 in the following form to satisfy the loop relation:

$$\frac{dX_2}{dt} = -KX_1 \tag{8.198}$$

Combining Equations (8.197) and (8.198),

$$\frac{d^2 X_1}{dt^2} + KX_1 = 0 \tag{8.199}$$

which gives the second-order system model in terms of the loop error X_1 and its derivative X_2. The general solution of the undamped equation, (8.199), is

$$X_1 = A \sin(\sqrt{K}t + \theta) \tag{8.200}$$

$$X_2 = \frac{dX_1}{dt} = \sqrt{K} A \cos(\sqrt{K}t + \theta) \tag{8.201}$$

where A and θ are arbitrary constants. Combining them, we get

$$\frac{X_1^2}{A^2} + \frac{X_2^2}{(\sqrt{K}A)^2} = 1 \tag{8.202}$$

Equation (8.202) describes an ellipse with semi-axes A and $\sqrt{K}A$. Its phase plane trajectory is plotted in Figure 8.65, which shows concentric ellipses with arbitrary A. The shape of the ellipses will vary with a variation of gain K.

In the positive feedback mode (switch 2 closed) of Figure 8.64, we can write the following equations:

$$\frac{dX_1}{dt} = X_2 \tag{8.203}$$

$$\frac{dX_2}{dt} = KX_1 \tag{8.204}$$

Adaptive Control

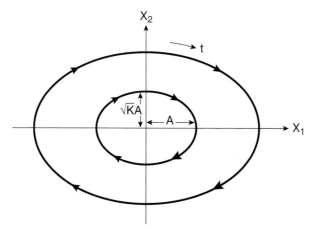

Figure 8.65 Phase plane portrait of Figure 8.64 with negative feedback

Combining these equations,

$$\frac{d^2 X_1}{dt^2} - KX_1 = 0 \tag{8.205}$$

The general solution of Equation (8.205) is

$$X_1 = B_1 e^{\sqrt{K}t} + B_2 e^{-\sqrt{K}t} \tag{8.206}$$

$$X_2 = \frac{dX_1}{dt} = \sqrt{K} B_1 e^{\sqrt{K}t} - \sqrt{K} B_2 e^{-\sqrt{K}t} \tag{8.207}$$

where B_1 and B_2 are arbitrary constants. Squaring each equation and combining them gives us

$$\frac{X_1^2}{4B_1 B_2} - \frac{X_2^2}{4KB_1 B_2} = 1 \tag{8.208}$$

where term $B_1 B_2$ can be positive, negative, or zero. Equation (8.208) describes hyperbolas that are plotted in the phase plane of Figure 8.66. The straight line asymptote equations can be derived by substituting $B_1 B_2 = 0$ as

$$KX_1^2 - X_2^2 = 4KB_1 B_2 = 0 \tag{8.209}$$

i.e.

$$X_2 = \pm \sqrt{K} X_1 \tag{8.210}$$

The families of hyperbolas are plotted for $B_1 B_2 > 0$ and $B_1 B_2 < 0$. The system can be switched back and forth between the positive and negative feedback modes for SMC, as

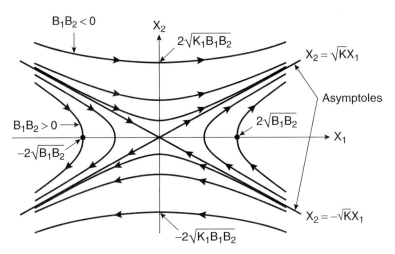

Figure 8.66 Phase plane portrait of Figure 8.64 with positive feedback

explained in Figure 8.67. The operation in Figure 8.67 can be understood by the superposition of Figures 8.65 and 8.66, where the hyperbolic asymptote line is described by the equation

$$\sigma = \sqrt{K}X_1 + X_2 = 0 \tag{8.211}$$

where $\sigma = 0$ is on the line.

Assume that initially, the system is in negative feedback mode and the operating point is at $X_1 = X_{10}$ on the ellipse. As the operating point moves on an elliptic trajectory and touches point B as shown, the positive feedback mode is invoked. Ideally, it will then move along the straight line BO and settle at steady-state point O, where error X_1 and error velocity X_2 are zero. The slope of line BO may vary with a variation of K. Even with a constant K, precision switching at point B is practically impossible to reach the steady-state point O. Let us define the sliding line equation as

$$\sigma = CX_1 + X_2 = 0 \tag{8.212}$$

where $C < \sqrt{K}$ so that the line slope is lower and beyond the range of the K variation. Note carefully that on the sliding line, defined as the "reference trajectory," the positive and negative feedback trajectories, described by hyperbolas and ellipses, respectively, cross in opposite directions. This means that at point B', the control can be switched to positive feedback mode, and then at point D, it can be switched back to negative feedback mode, and so on. The operating point will thus track the sliding line in a zigzag path until the steady-state point is reached at the origin. The time-domain solution of the sliding line is basically a deceleration with exponentially decaying X_1, as indicated by the following equation:

$$X_1(t) = X_1(t_0)e^{-C(t-t_o)} \tag{8.213}$$

Adaptive Control

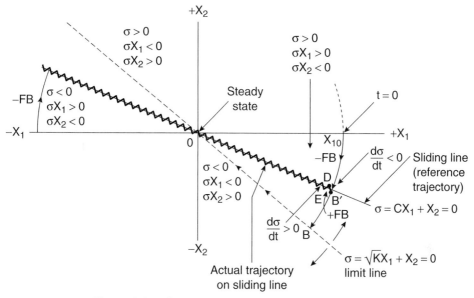

Figure 8.67 Sliding line control in phase plane $X_1 - X_2$

where t_0 = time at which the trajectory reaches the sliding line. The time-domain response for the sliding line control of two different values of C is shown in Figure 8.68, which reflects the characteristic chattering effect. Note that once the operation reaches the sliding line, the response is strictly dictated by slope C, but it is not affected by a variation of parameter K or any load disturbance (robust). The operation of the sliding line control in the second quadrant for $-X_1$ (reverse position error) is similar to that for the fourth quadrant, and is shown in Figure 8.67.

The polarities of parameters σ, σX_1, and σX_2 above and below the sliding line for both $+X_1$ and $-X_1$ are summarized in Figure 8.67. The strategy of switching control is defined by

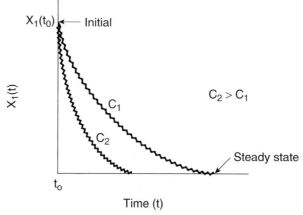

Figure 8.68 Time-domain response in sliding line control for two different values of C

these polarities, which will be described later. Evidently, $\sigma > 0$ and $d\sigma/dt < 0$, as the trajectory tends to cross the sliding line from above, whereas $\sigma < 0$ and $d\sigma/dt > 0$, as the trajectory tends to cross the line from below. Mathematically, we can write

$$\operatorname{Lim} \frac{d\sigma}{dt} < 0 \quad \text{for} \quad \sigma \to +0 \tag{8.214}$$

and

$$\operatorname{Lim} \frac{d\sigma}{dt} > 0 \quad \text{for} \quad \sigma \to -0 \tag{8.215}$$

Combining Equations (8.214) and (8.215), we get

$$\operatorname{Lim} \sigma \frac{d\sigma}{dt} < 0 \quad \text{for} \quad \sigma \to 0 \tag{8.216}$$

Equation (8.216) is defined as an existence or reaching equation, which must be satisfied for the SMC. In other words, the validity of the reaching equation guarantees that the response will cross the trajectory in each switching transition, and is essential for a system to be controllable by the sliding mode. In practice, the parameters in the sliding mode controller are designed with the reaching equation, which will be discussed later.

8.7.3.2 Sliding Trajectory Control of a Vector Drive

We will now apply an SMC to a vector-controlled induction motor servo drive and develop design criteria for the controller's parameters. In addition, the sliding line control will be extended to a full sliding trajectory control, incorporating acceleration, constant speed, and deceleration segments. Figure 8.69 shows the block diagram of an ideal dc machine-like transfer function model of a vector drive that incorporates a sliding mode control. The idea is to make the response insensitive to the plant parameters, that is, the torque constant K_t, moment of inertia J, friction damping coefficient B, and load torque disturbance T_L. Assuming a step command of θ_r^*, we can write the following equations:

$$T_e = K_t i_{qs} = K_t K_1 U \tag{8.217}$$

$$X_1 = \theta_r^* - \theta_r \tag{8.218}$$

$$\frac{dX_1}{dt} = \frac{d\theta_r^*}{dt} - \frac{d\theta_r}{dt} = -\omega_m = X_2 \tag{8.219}$$

$$(T_e - T_L) \frac{1}{JS + B} = -X_2 \tag{8.220}$$

where U = SMC output and K_1 = control gain for active current i_{qs}^*.

Adaptive Control

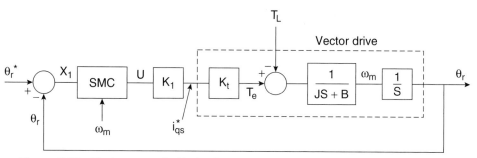

Figure 8.69 Vector-controlled induction motor servo drive with sliding mode control

The second-order plant model is expressed in state-space equations in terms of state variables X_1 and X_2 by the following steps:

$$JSX_2 + BX_2 = -K_t K_1 U + T_L \tag{8.221}$$

$$\frac{dX_2}{dt} = -\frac{B}{J}X_2 - \frac{K_t K_1}{J}U + \frac{1}{J}T_L \tag{8.222}$$

$$\begin{bmatrix} \frac{dX_1}{dt} \\ \frac{dX_2}{dt} \end{bmatrix} = \begin{bmatrix} 0 & 1 \\ 0 & -b \end{bmatrix}\begin{bmatrix} X_1 \\ X_2 \end{bmatrix} + \begin{bmatrix} 0 \\ -a \end{bmatrix}U + \begin{bmatrix} 0 \\ d \end{bmatrix}T_L \tag{8.223}$$

where $b = B/J$, $a = K_t K_1/J$, and $d = 1/J$. Figure 8.70 shows the proposed sliding mode control topology in detail, and Figure 8.71 shows the corresponding trajectory control for the acceleration, constant speed, and deceleration segments for both the $+X_1$ and $-X_1$ regions. Note that the X_2 signal is directly derived from speed signal ω_m.

There are essentially three control loops in Figure 8.70. The main or primary loop receives the position loop error X_1 signal and generates U_1 output through the switching controller with respective gains α_i and β_i. The secondary control loop with derivative input $dX_1/dt = X_2$ generates signal U_2. The input in this loop is derived directly from the motor speed signal ω_m, as indicated. In addition to these loops, there is an auxiliary loop where constant A is injected to eliminate the steady-state error due to coulomb friction and load torque T_L. In a sliding mode controller, all the input signals are transmitted through single-pole double-throw (SPDT) switches, and the criteria for controlling each switch are indicated on the figure. All the loops contribute to the respective signals and the resultant signal U is

$$U = U_0 + U_1 + U_2 \tag{8.224}$$

In Figure 8.71, the outer curve is determined by the limiting values of the acceleration, speed, and deceleration of the drive system. Normally, the variation of plant parameters will cause drift within a band as shown by the dashed lines. For example, if the inertia J is increased, the max-

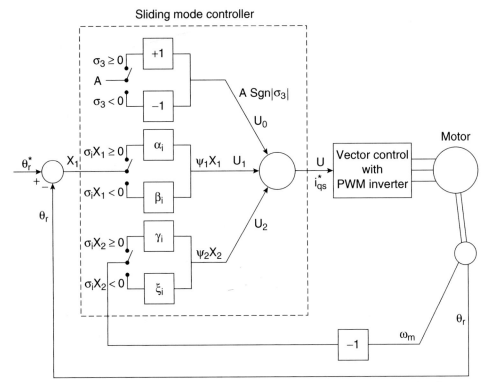

Figure 8.70 Induction motor drive showing sliding mode controller in detail

imum acceleration and deceleration limits will shrink. The sliding trajectory or reference contour in phase plane $X_1 - X_2$, which the drive system will be forced to track, must be described beyond the drift band so that the system becomes controllable and the response is not affected by the drift. The sliding trajectory defined here consists of three segments, and their equations in the fourth quadrant (forward position error) are described as follows:

1. **Acceleration segment:**

$$\sigma_1 = \alpha X_2^2 + (X_1 - X_{10}) \tag{8.225}$$

where X_{10} = initial position error

2. **Constant speed segment:**

$$\sigma_2 = X_2 - X_{20} \tag{8.226}$$

where $-X_2$ corresponds to positive speed and $-X_{20}$ = maximum positive speed ($|X_{20}| < |V_m|$)

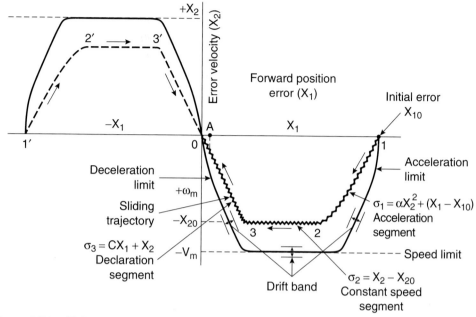

Figure 8.71 Sliding trajectory of drive in acceleration, constant speed, and deceleration segments

3. Deceleration segment:

$$\sigma_3 = CX_1 + X_2 \tag{8.227}$$

which corresponds to the sliding line in Figure 8.67. All the sliding control parameters generally vary with the selection of the segment. Note that in each case, $\sigma = 0$ means the reference trajectory. The actual trajectory that follows the reference trajectory is given by the zig-zag profile in the direction of the arrow, as shown. At steady state, the operating point oscillates at the origin of the phase plane. The trajectory should be defined as close as possible to the limit envelope, but beyond the drift band so as to get the best suboptimal transient response. It can be shown [31] that a second-order system theoretically requires only error signal X_1 and its derivative X_2 as control inputs, as shown in Figure 8.70.

The SMC law can be defined mathematically as

$$U = A.Sgn|\sigma_3| + \psi_1 X_1 + \psi_2 X_2 \tag{8.228}$$

where

$$\begin{aligned} Sgn\,|\sigma_3| &= +1 \quad \text{if} \quad \sigma_3 \geq 0 \\ &= -1 \quad \text{if} \quad \sigma_3 < 0 \end{aligned} \tag{8.229}$$

$$\psi_1 = \alpha_i \quad \text{if} \quad \sigma_i X_1 \geq 0$$
$$= \beta_i \quad \text{if} \quad \sigma_i X_1 < 0 \tag{8.230}$$

$$\psi_2 = \gamma_i \quad \text{if} \quad \sigma_i X_2 \geq 0$$
$$= \xi_i \quad \text{if} \quad \sigma_i X_2 < 0 \tag{8.231}$$

as indicated in Figure 8.70.

Let us now take the individual sliding trajectory segments in Figure 8.71 and derive the relations of the control parameters of the primary and secondary control loops. In each segment, the validity of reaching Equation (8.216) will guarantee the success of the SMC.

Consider first the deceleration segment given by Equation (8.227). Differentiating this equation and substituting in (8.216), we get

$$\sigma_3 (C\frac{dX_1}{dt} + \frac{dX_2}{dt}) < 0 \tag{8.232}$$

But, the state-space equations of the system from (8.223) can be written as

$$\frac{dX_1}{dt} = X_2 \tag{8.233}$$

$$\frac{dX_2}{dt} = -bX_2 - aU + dT_L \tag{8.234}$$

Substituting Equations (8.233) and (8.234) in (8.232) and replacing U by the relation

$$U = \psi_1 X_1 + \psi_2 X_2 \tag{8.235}$$

we get

$$-\sigma_3 X_1 (a\psi_1) - \sigma_3 X_2 (b + a\psi_2 - C) + \sigma_3 dT_L < 0 \tag{8.236}$$

For validity of this equation, the following control relations should be satisfied:

1. Primary loop:

If $\sigma_3 X_1 > 0$, then $a\alpha_3 > 0$, that is, $\alpha_3 > 0$ \hfill (8.237)

If $\sigma_3 X_1 < 0$, then $a\beta_3 < 0$, that is, $\beta_3 < 0$ \hfill (8.238)

Adaptive Control

2. Secondary loop:

If $\sigma_3 X_2 > 0$, then $(b + a\gamma_3 - C) > 0$, that is, $\gamma_3 > \dfrac{(C-b)}{a}$ (8.239)

If $\sigma_3 X_2 < 0$, then $(b + a\xi_3 - C) < 0$, that is, $\xi_3 < \dfrac{(C-b)}{a}$ (8.240)

These equations indicate that control parameter selection in an SMC is very flexible. It appears safe to consider α_3 and γ_3 as positive and β_3 and ξ_3 as negative values. It is better to design the preliminary values of control parameters and then optimize by simulation and experiment. For a wide variation of plant parameters, the control parameters can be adapted, if necessary.

Now, consider the constant speed segment of the trajectory given by Equation (8.226). Differentiating this equation and substituting (8.234), we get

$$\frac{d\sigma_2}{dt} = \frac{dX_2}{dt} = -bX_2 - aU + dT_L$$ (8.241)

Substituting equations (8.241) and (8.235) in (8.216)

$$-\sigma_2 X_1(a\psi_1) - \sigma_2 X_2(b + a\psi_2 - dT_L) < 0$$ (8.242)

1. Primary loop:

If $\sigma_2 X_1 > 0$, then $a\alpha_2 > 0$, that is, $\alpha_2 > 0$ (8.243)

If $\sigma_2 X_1 < 0$, then $a\beta_2 < 0$, that is, $\beta_2 < 0$ (8.244)

2. Secondary loop:

If $\sigma_2 X_2 > 0$, then $(b + a\gamma_2 - dT_L) > 0$, that is, $\gamma_2 > \dfrac{(dT_L - b)}{a}$ (8.245)

If $\sigma_2 X_2 < 0$, then $(b + a\xi_2 - dT_L) < 0$, that is, $\xi_2 < \dfrac{(dT_L - b)}{a}$ (8.246)

Again, it appears safe to select α_2 and γ_2 as positive and β_2 and ξ_2 as negative values.

Now, consider the acceleration segment in the fourth quadrant. Substituting the derivative of Equation (8.225) in (8.216),

$$\sigma_1 \frac{d\sigma_1}{dt} = \sigma_1 (2\alpha X_2 \frac{dX_2}{dt} + \frac{dX_1}{dt}) < 0$$ (8.247)

Substituting Equations (8.233)–(8.235) in (8.247),

$$-\sigma_1 X_1(2\alpha a\psi_1 X_2) + \sigma_1 X_2(-2\alpha b X_2 - 2\alpha a\psi_2 X_2 + 2\alpha dT_L + 1) < 0 \quad (8.248)$$

Since X_2 is always negative in the fourth quadrant, we can write the following equations:

1. Primary loop:

$$\text{If } \sigma_1 X_1 > 0, \text{ then } 2\alpha a\alpha_1 < 0, \text{ that is, } \alpha_1 < 0 \quad (8.249)$$

$$\text{If } \sigma_1 X_1 < 0, \text{ then } 2\alpha a\beta_1 > 0, \text{ that is, } \beta_1 > 0 \quad (8.250)$$

2. Secondary loop:

$$\text{If } \sigma_1 X_2 > 0, \text{ then } \gamma_1 > -\frac{(1+2\alpha dT_L + 2\alpha b|X_2|)}{2\alpha a|X_2|} \quad (8.251)$$

$$\text{If } \sigma_1 X_2 < 0, \text{ then } \xi_1 < -\frac{(1+2\alpha dT_L + 2\alpha b|X_2|)}{2\alpha a|X_2|} \quad (8.252)$$

It can be shown that an SMC becomes valid using the primary loop only. However, the secondary loop with a derivative input improves system response and permits a wider variation of plant parameters.

The "dither signal" $U_0 = A \operatorname{Sgn} |\sigma_3|$ is activated in the deceleration segment only. The signal is bipolar with a mean average value and strengthens the primary loop output in steady-state condition. Without this loop, the drive will have a steady-state position error (X_1) with load torque T_L and coulomb friction because the developed torque with i_{qs} current is supplied by the primary loop error X_1 only. The dither input permits $X_1 = 0$ at steady state when the operating point oscillates in the origin. However, the disadvantage of the dither signal is that it enhances the chattering effect.

The chattering effect of an SMC may not be acceptable in many applications. Chattering in torque and speed may be large, but its effect is small on position because of inertia filtering. Chattering can be improved by a small computation sampling time, higher PWM frequency, and by minimizing any additional delay in feedback signal computation.

8.8 SELF-COMMISSIONING OF DRIVE

Self-commissioning of a drive involves the initial measurement of machine parameters for feedback signal estimation and tuning the control system. Traditionally, machine equivalent circuit parameters are determined by no-load and blocked rotor (or short-circuit) tests with 60 Hz voltage injection in the stator. Information about the parameters is important for the estimation of feedback signals (see Figs. 8.29 and 8.30) and slip gain tuning (see Figure 8.32) of vector-controlled drives. The proportional (*P*) and integral (*I*) gains of feedback control loops

Self-Commissioning of Drive

can also be tuned with knowledge of the machine's parameters. This means that if the transfer function model of a plant is known with the plant parameters, the optimum P and I gains can be determined.

Note that in self-commissioning, we are concerned with initial plant parameters only, not the parameters during operating condition, which might change. Instead of P-I tuning of the loops with the knowledge of plant parameters, the tuning can also be done by observing the loop's response in real time. Expert system and fuzzy logic-based P-I tuning will be discussed in Chapters 10 and 11, respectively. For a digitally controlled drive with an unknown machine, a software routine can automate the whole parameter measurement procedure saving many man-hours.

Let us consider, for example, a direct vector-controlled drive with space vector PWM which has speed loop, flux loop, and synchronous current control loops, as shown in Figure 8.72. The motor has a speed encoder that helps to estimate the rotor flux vector starting from zero speed (current model estimation). Of course, the flux vector can also be estimated with the voltage model. Let us discuss the whole procedure of self-commissioning step-by-step [34].

Step 1: Feed name plate machine parameters

Initially, with the machine at rest, dial into the microcomputer's memory the rated voltage V_s, rated current I_s, rated frequency ω_e, and number of poles P of the machine.

Step 2: Measure stator resistance (R_s)

- Set up the i_{ds} control loop (i_{qs} loop deactivated) with 100 percent dc stator current.
- Select the inverter voltage vector (say) V_1 and PWM modulation index m_1 so that the rated stator current is established. Note the current value of I_{s1}.

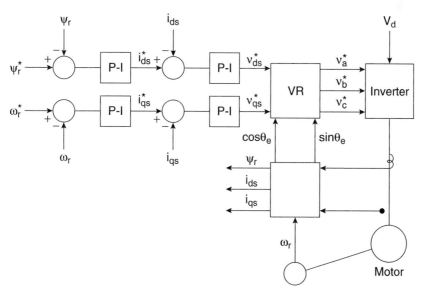

Figure 8.72 Direct vector controlled drive with speed, flux and synchronous current control

- Repeat the above step for 50 percent stator current with modulation index m_2. Note the value of I_{s2}.
- Calculate $R_s = (V_1.m_1 - V_1.m_2)/(I_{s1} - I_{s2})$. The effect of distortion and inverter dead time is cancelled by the two-step measurement.

Step 3: Measure stator transient parameters

- Select voltage vector V_1. Disable the PWM modulator. Apply the voltage for a short time t_1 (in μs), as shown in Figure 8.73. The resulting current response i_s is approximately linear.

Referring to the equivalent circuit Figure 2. 28(a), the transient loop equation is

$$\overline{V}_s = R\overline{i}_s + L\frac{di_s}{dt} \qquad (8.253)$$

where $R = R_s + R_r$ and $L = L_{ls} + L_{lr}$

If the resistance drop is neglected, that is, the current rise is linear, the transient inductance L is given as

$$L = \frac{\overline{V}_s \Delta t}{\Delta \overline{I}_s} \qquad (8.254)$$

Both R and L can be determined from the following solution of Equation (8.253) by measuring i_s for two different times t_1 and t_2. Make a few tests and calculate the average.

$$i_s(t) = \frac{\overline{V}_s}{R}\left[1 - e^{-\frac{t}{\tau}}\right] \qquad (8.255)$$

where $\tau = L/R$.

Step 4: Tune the current loops

- Tune the i_{ds} and i_{qs} loops in Figure 8.72 one at a time. Note that the currents are perpendicular but do not have an orientation with rotor flux. Figure 8.74 shows the i_{qs} loop, where K_{IN} = inverter voltage gain and $1/(R + LS)$ is the transient equivalent circuit transfer function. Determine the optimum P and I gains.
- Repeat the same for the i_{ds} loop.

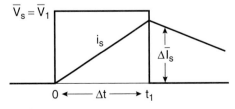

Figure 8.73 Stator transient parameters measurement

Self-Commissioning of Drive

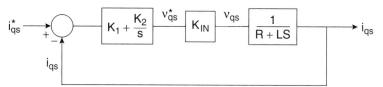

Figure 8.74 Stator current loop with P-I control

Step 5: Measure rotor time constant T_r

- Set up the i_{ds} loop and inject a dc current pulse for several seconds to establish the rotor flux, as shown in Figure 8.75. Turn off the inverter at time t_1, open-circuiting the machine. Measure the induced voltage $v_m(t)$ on the stator side. The rotor flux in the equivalent circuit will decay exponentially with the time constant, which can be given by the equation

$$i_r R_r + L_r \frac{di_r}{dt} = 0 \qquad (8.256)$$

The $v_m(t)$ expression is given as

$$v_m(t) = L_m \frac{di_r}{dt} \qquad (8.257)$$

which is plotted in Figure 8.75. The time constant T_r can be calculated as

$$\frac{T_r}{v_m(t_1)} = \tan\theta = \frac{t_2 - t_1}{v_m(t_2) - v_m(t_1)} \qquad (8.258)$$

or

$$T_r = \frac{v_m(t_1)}{v_m(t_2) - v_m(t_1)} (t_2 - t_1) \qquad (8.259)$$

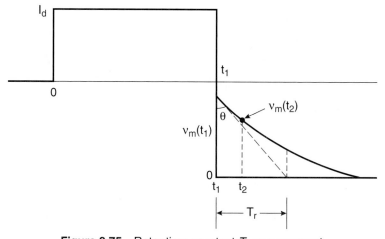

Figure 8.75 Rotor time constant T_r measurement

Step 6: Tune the current model flux vector estimation (see Figure 8.30)
- Operate the drive in no-load volts/Hz control mode to establish a certain speed with a proper V_s/ω_e ratio to establish rated flux ψ_r.
- Plug the T_r value in Figure 8.30 and change L_m until the calculated $i_{qs} = 0$ so that the entire current is i_{ds} (magnetizing current). This i_{ds} is the rated magnetizing current for the rated ψ_r. The vector control currents are now tuned with ψ_r orientation.

Step 7: Tune the flux control loop

The flux control loop transfer function block diagram is given in Figure 8.76, where τ_1 corresponds to the i_{ds} loop response delay. Parameters L_m and T_r are known.
- Select the optimum P and I gains of the flux loop.

Step 8: Measure mechanical inertia J and friction coefficient B
- With the vector control active, establish torque at no-load, as shown in Figure 8.77.

The speed equation can be given as

$$T_e = J(\frac{2}{P})\frac{d\omega_r}{dt} + B\omega_r \tag{8.260}$$

If B is neglected, J can be given as

$$J = \frac{T_e \Delta t}{\Delta \omega_r} \cdot \frac{P}{2} \tag{8.261}$$

where T_e is easily calculated from feedback signals (see Figure 8.29).

Considering B, the exact solution of Equation (8.260) is

$$\omega_r = \frac{T_e}{B}(1 - e^{-\frac{t}{\tau_m}}) \tag{8.262}$$

where $\tau_m = \frac{2J}{PB}$ is the mechanical time constant. Both B and J can be solved from Equation (8.262), which measures ω_r at two different instants.

Although we considered direct vector control with current model estimation as the example, it is possible to tune an indirect vector control and voltage model estimation with these parameter measurements. Finally, it should be noted that with step signal injection, some amount of skin effect in the rotor bars will reduce accuracy.

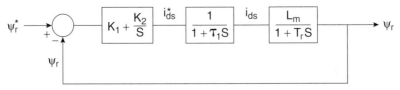

Figure 8.76 Flux control loop with P-I control

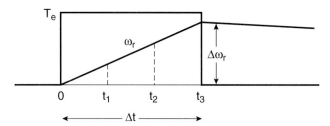

Figure 8.77 Rotor inertia J and friction coefficient B measurement

8.9 SUMMARY

In this chapter, we extensively reviewed the different control and feedback signal estimation techniques for cage-type induction motor drives. This class of drives is widely used in various industrial applications, and the technology is continuously expanding. Therefore, a somewhat lengthy discussion on the subject was justified.

Both scalar and vector control techniques with appropriate feedback signal estimation were covered. In scalar control, we emphasized open loop volts/Hz control because of its popularity in industrial applications. A few more scalar control techniques were reviewed briefly for completeness. Vector control with rotor and stator flux orientation was discussed extensively because of its importance in high-performance drive applications. The vector control implementation with corresponding feedback signal estimation is complex, and therefore, digital control with high-speed, powerful microprocessor or DSP is essential. It is expected that vector control will eventually emerge as the industry-standard control method for induction motor drives.

Speed sensorless vector control is an emerging technology. A number of speed estimation techniques have been reviewed. However, very low-speed operation, including start-up at zero frequency, remains a challenge. DTC control was discussed in detail. It has been accepted commercially, and it will be interesting to see how it competes with vector-controlled drives. The adaptive control methods, particularly the sliding mode control were discussed in some detail for a vector-controlled drive. Fuzzy logic and neural network-based adaptive controls constitute emerging technologies and will be covered in Chapters 11 and 12, respectively. Most of the control and estimation concepts for induction motor drives are also applicable for synchronous motor drives. Self-commissioning of drives was covered at the end of this chapter. Finally, the reader should note the difference between a space vector and rms phasor and the corresponding vector or phasor diagrams.

REFERENCES

1. B. K. Bose (Ed.), *Power Electronics and Variable Frequency Drives,* IEEE Press, NY, 1996.
2. I. Boldea and S. A. Nasar, *Vector Control of AC Drives*, CRC Press, NY, 1992.
3. R. Ueda, T. Sonada, K. Koga, and M. Ichikawa, "Stability analysis in induction motor driven by V/f controlled general purpose inverter", *IEEE Trans. Ind. Appl.*, vol. 28, pp. 472–481, March/April 1992.

4. A. B. Plunkett, "A current-controlled PWM transistor inverter drive", *IEEE IAS Annu. Meet. Conf. Rec.*, pp. 785–792, 1979.
5. A. B. Plunkett and D. L. Plette, "Inverter-induction motor drive for transit cars", *IEEE Trans. Ind. Appl.*, vol. 18, pp. 26–37, 1977.
6. E. P. Cornell and T. A. Lipo, "Modeling and design of controlled current induction motor drive system", *IEEE Trans. Ind. Appl.*, vol. 13, pp. 321–330, July/Aug. 1977.
7. B. K. Bose, "Variable frequency drives – technology and applications", *PEMC Conf. Rec.,* Poland, 1994.
8. W. Leonhard, "Adjustable speed ac drives", *Proc. Of the IEEE*, vol. 76, pp. 455–471, 1988.
9. F. Blaschke, "The principle of field orientation as applied to the new transvector closed loop control system for rotating field machines", *Siemens Review*, vol. 34, pp. 217– 220, May 1972.
10. K. Hasse, "Zur dynamik drehzahlgeregelter antriebe mit stromrichtergespeisten asynchron-kurzschlublaufermaschinen", Darmstadt, *Techn. Hochsch., Diss.*, 1969.
11. R. W. De Doncker and D. W. ovotny, "The universal field oriented controller", *IEEE IAS Annu. Meet. Conf. Rec.*, pp. 450–456, 1988.
12. B. K. Bose, "Variable frequency drives–technology and applications", *Proc. Int'l. Symp. Ind. Elec.*, Budapest, Hungary, pp. 1–18, 1993.
13. B. K. Bose, "High performance control and estimation in ac drives", *IEEE IECON Conf. Rec.*, pp. 377–385, 1997.
14. P. Jansen and R. D. Lorenz, "A physically insightful approach to the design and accuracy assessment of flux observers for field oriented induction machine drives", *IEEE IAS Annu. Meet. Conf. Rec.*, pp. 570–577, 1992.
15. G. Kaufman, L. Garces, and G. Gallagher, "High performance servo drives for machine tool applications using ac motors", *IEEE IAS Annu. Meet. Conf. Rec.*, pp. 604–609, 1982.
16. T. M. Rowan, R. J. Kerkman, and D. Leggate, "A simple on-line adaption for indirect field orientation of an induction machine", *IEEE IAS Annu. Meet. Conf. Rec.*, pp. 579–587, 1989.
17. X. Xu, R. De Doncker, and D. W. Novotny, "A stator flux oriented induction machine drive", *IEEE Power Elec. Spec. Conf.*, pp. 870–876, 1988.
18. K. Rajashekara, A. Kawamura, and K. Matsuse (Ed.), *Sensorless Control of AC Drives*, IEEE Press, NY, 1996.
19. C. Schauder, "Adaptive speed identification for vector control of induction motors without rotational transducers", *IEEE Trans. Indus. Appl*, vol. 28, pp. 1054–1061, Sept./Oct. 1992.
20. H. Kubota, K. Matsuse, and T. Nakano, "DSP-based speed adaptive flux observer of induction motor", *IEEE Trans. Ind. Appl.*, vol. 29, pp. 344–348, March/April 1993.
21. J. Holtz, "Sensorless position control of induction motor – an emerging technology", *IEEE IECON Conf. Rec.*, pp. I1 – I12, 1998.
22. Y. R. Kim, S. K. Sul, and M. H. Park, "Speed sensorless vector control of induction motor using extented Kalman filter", *IEEE Trans. Ind. Appl.*, vol. 30, pp. 1225–1233, Sept./Oct. 1994.
23. B. K. Bose and N. R. Patel, "A sensorless stator flux oriented vector controlled induction motor drive with neuro-fuzzy based performance enhancement", *IEEE IAS Annu. Meet. Conf. Rec.*, pp. 393–400, 1997.
24. I. Takahashi and T. Noguchi, "A new quick response and high efficiency control strategy of an induction motor", *IEEE Trans. Ind. Appl.*, vol. 22, pp. 820–827, Sept./Oct. 1986.
25. G. Buja et al., "Direct torque control of induction motor drives", *ISIE Conf. Rec.*, pp. TU2–TU8, 1997.
26. P. Vas, *Sensorless Vector and Direct Torque Control*, Oxford, NY, 1998.
27. K. Hong and K. Nam, "A disturbance torque compensation scheme considring the speed measurement delay", *IEEE IAS Annu. Meet. Conf. Rec.*, pp. 403–409, 1996.
28. K. J. Astrom, "Theory and applications of adaptive control – a survey", Automata, vol. 19, pp. 471–486, Sept. 1983.
29. Y. D. Landau, *Adaptive Control – The Model Referencing Approach*, Marcel Dekker, 1979.
30. A. Brickwedde, "Microprocessor-based adaptive speed and position control for electrical drives", *IEEE Trans. on Ind. Appl.*, vol. 21, pp. 1154–1161, Sept./Oct. 1985.
31. U. Itkis, *Control Systems of Variable Structures*, Wiley, NY, 1976.

32. B. K. Bose, "Sliding mode control of induction motor", *IEEE IAS Annu. Meet. Conf. Rec.,* pp. 479–486, 1985.
33. F. Harashima, H. Hashimoto, and S. Kondo, "MOSFET converter-fed position servo system with sliding mode control", *IEEE Trans. Ind. Electron.*, vol. 32, pp. 238–244, Mar. 1985.
34. A. M. Khambadkone and J. Holtz, "Vector controlled induction motor drive with a self-commissioning scheme", *IEEE Trans. Ind. Elec.*, vol. 38, pp. 322–327, Oct. 1991.

CHAPTER 9

Control and Estimation of Synchronous Motor Drives

9.1 INTRODUCTION

Synchronous motor drives are close competitors to induction motor drives in many industrial applications, and their application is growing. They are generally more expensive than induction motor drives, but the advantage is that the efficiency is higher, which tends to lower the life cycle cost. The basic principles of synchronous machines and their characteristics were discussed in Chapter 2. As mentioned before, wound-field synchronous machines (WFSMs) are generally used in high-power (multi-megawatt) applications. On the other hand, permanent magnet synchronous machines (PMSMs) are used in low- to medium-power (up to several hundred horsepower) applications. A traditional line-start 60 Hz machine starts as an induction motor with a cage or damper winding, but locks into synchronous speed at steady state.

The general classifications of PM machines are radial flux (drum-type) and axial flux (or disk-type), and the former type is most commonly used. There are also classifications of sinusoidal and trapezoidal types, as discussed in Chapter 2. A sinusoidal machine can be a surface permanent magnet (SPM) type or an interior or buried permanent magnet (IPM) type. Synchronous reluctance machines (SyRMs), as the name indicates, do not have any separate field excitation. Variable-reluctance or double-reluctance (reluctance variation in both stator and rotor) machines can be stepper or switched reluctance types. The switched reluctance machine (SRM) does not strictly fall into the synchronous machine category.

Some typical applications of synchronous motor drives are:

- Fiber spinning mills
- Rolling mills
- Cement mills
- Ship propulsion

- Electric vehicles
- Servo and robotic drives
- MAGLEV – linear synchronous motor propulsion
- Variable-frequency starters for 60 Hz wound field synchronous motors
- Starters/generators for aircraft engines

In this chapter, we will study the control and estimation of wound-field and PM synchronous machine drives with voltage-fed inverters, current-fed inverters, and cycloconverters. Both scalar and vector control techniques will be covered. Both sensor and sensorless drives will be discussed. Generally, the machines should be considered as nonsalient pole, if not mentioned otherwise.

It should be mentioned here that most of the concepts in control and estimation which were developed in Chapter 8 for induction motor drives are also applicable to synchronous motor drives. Therefore, only the distinctive features of synchronous motor drives will be discussed in this chapter. Like single-phase (or split-phase), variable-speed induction motor drives, there are also single-phase PM synchronous motor drives in the fractional hp range for low-cost, low-performance applications, but these will not be covered. Similarly, linear and inductor-type synchronous machines are beyond the scope of this book. It is recommended that the reader first review the fundamentals of synchronous machines in Chapter 2 before studying this chapter.

9.2 SINUSOIDAL SPM MACHINE DRIVES

In a synchronous machine drive, as the name indicates, the speed of the machine is uniquely related to the frequency supplied by the inverter or cycloconverter. Unlike an induction machine, it will run at synchronous speed, or will not run at all. There are essentially two control modes for synchronous machine drives: one is the open-loop, true synchronous machine mode, where the motor speed is controlled by the independent frequency control of the converter; the other is the self-control mode, where the variable-frequency converter control pulses are derived from an absolute rotor position encoder mounted on the machine shaft.

9.2.1 Open Loop Volts/Hertz Control

An example of an independent frequency control is the open loop volts/Hz speed control shown in Figure 9.1. It is the simplest scalar control method of a synchronous machine, but it is achieved at the cost of inferior performance, unlike high-performance vector control, which will be described later. In topology and performance, the scheme is somewhat similar to the volts/Hz-controlled induction motor drive described in Chapter 8. This method of speed control is particularly popular in multiple synchronous reluctance or PM machine drives (as shown in Figure 9.1), where close speed tracking is essential among a number of machines for applications such as fiber spinning mills. Here, all the machines are connected in parallel to the same inverter so that they move in synchronism corresponding to the command frequency ω_e^* at the input. The phase voltage command V_s^* is generated through a function generator (FG), where

Sinusoidal SPM Machine Drives

the voltage is essentially maintained proportional to the frequency so that the stator flux ψ_s remains constant. Similar to the induction motor drive, a boost voltage is added near zero frequency to compensate the stator resistance drop. Maintaining a constant and rated stator flux permits nearly maximum available torque per ampere of stator current and fast transient response. The front end of the voltage-fed PWM inverter is supplied from the utility line through a diode rectifier and LC filter. The machine is normally built with a damper or cage winding to prevent oscillatory or underdamping behavior during the transient response.

The performance of the drive is explained in Figure 9.2, which indicates the motoring mode as well as the braking mode in the forward direction. The corresponding phasor diagram is shown in Figure 9.3. The phasor diagram is the same as Figure 2.31(a), except the stator resistance R_s is neglected, the field flux ψ_f is shown as the reference phasor, and the stator current I_s is shown with a lagging angle. An equivalent (but fictitious) constant field current I_f for a PM machine is indicated in the figure. Assume for simplicity that initially the load torque T_L on the machine shaft is zero. In Figure 9.2, the machine can be easily started from stand-still condition

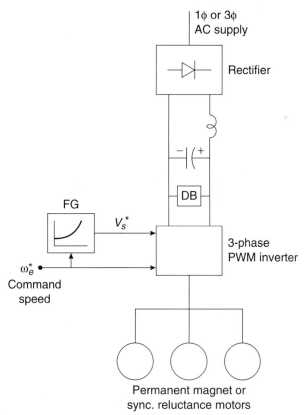

Figure 9.1 Open loop volts/Hz speed control of multiple PM synchronous motors

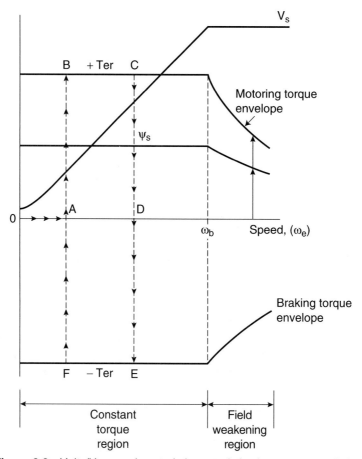

Figure 9.2 Volts/Hz speed control characteristics in torque-speed plane

at point O to point A by slowly increasing the frequency. At this point, the load torque T_L is gradually increased. At steady-state condition, $T_e = T_L$, the operating point will move vertically along AB in the first quadrant. The torque expression is given as

$$T_e = 3\left(\frac{P}{2}\right)\frac{\psi_s \psi_f}{L_s}\sin\delta \qquad (2.169)$$

$$= 3\left(\frac{P}{2}\right)\psi_s I_s \cos\varphi \qquad (2.170)$$

where δ = torque angle and $I_T = I_s \cos\phi$ is the in-phase component of the stator current. Therefore, with constant ψ_s, the δ angle and stator current I_s will increase gradually until the rated torque is reached at point B, where either the limit δ angle ($\pi/2$) or the rated stator current I_s is attained (whichever is earlier). Usually, the inverter current limit is reached before the machine

Sinusoidal SPM Machine Drives

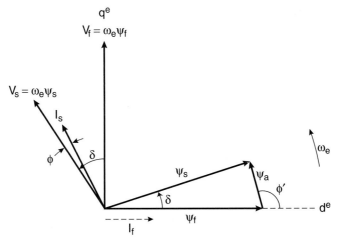

Figure 9.3 Phasor diagram of synchronous machine (motoring mode only)

stability limit. The operating point can be changed from B to C by slowly increasing the frequency command. It can then be brought back to point D by gradually decreasing T_L. At base speed ω_b, voltage V_s will saturate. Beyond this point, the machine will enter into field-weakening mode; therefore, the available torque will decrease due to reduced ψ_s, as shown. Like induction motor control, any sudden change in ω_e^* will make the system unstable because of the loss of synchronism. For variable-speed operation, the motor speed should be able to track the command frequency without losing synchronism. The rate of ω_e^* change or maximum acceleration/deceleration capability is dictated by the following equation:

$$J(\frac{2}{P})\frac{d\omega_e}{dt} = T_e - T_L \tag{2.105}$$

where J = moment of inertia, $\omega_e = P/2\, \omega_m$ is the synchronous electrical speed (r/s), P = number of poles, and ω_m = mechanical speed (r/s) Therefore, the maximum acceleration and deceleration capabilities are given, respectively, by the equations

$$\frac{d\omega_e^*}{dt} = +\frac{1}{J}(\frac{P}{2})(T_{er} - T_L) \tag{9.1}$$

$$\frac{d\omega_e^*}{dt} = -\frac{1}{J}(\frac{P}{2})(T_{er} + T_L) \tag{9.2}$$

where T_{er} = rated torque and T_L contributes to deceleration. At point A, if ω_e^* is ramped up, the developed T_e will jump to point B and the machine will accelerate along line BC until steady state is reached at point D. Similarly, the profile during deceleration will be D-E-F-A.

The recovered electrical energy in deceleration is dissipated in the dynamic brake (DB) installed in the dc link. Speed reversal is possible by reversing the phase sequence of the inverter.

The damper winding prevents hunting or oscillatory behavior, as mentioned before. However, it will induce the inverter harmonic frequency-related currents, which will tend to decrease the efficiency of the drive. Note that although the angular speed of parallel machines is identical, the angular positions will not be identical if the machine parameters are not matched or the load torques are different.

9.2.2 Self-Control Model

A self-controlled synchronous machine has close analogy with a dc machine. This requires clear explanation. In fact, a vector-controlled induction motor is also a self-controlled machine, and its analogy with a dc machine was discussed in Chapter 8. Figure 9.4(a) shows a PM dc motor, and Figure 9.4(b) illustrates a sinusoidal SPM machine with self-control. The7 stator winding of the machine is fed by an inverter that generates a variable-frequency variable voltage sinusoidal supply as in Figure 9.1. But, in this case, instead of controlling the inverter frequency indepen-

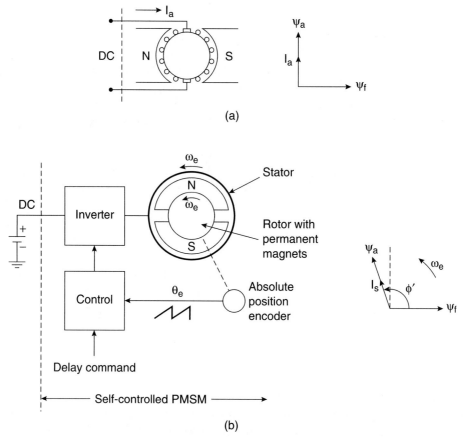

Figure 9.4 DC motor and self-controlled synchronous motor analogy, (a) Permanent magnet dc motor, (b) Self-controlled synchronous motor

Sinusoidal SPM Machine Drives

dently, the frequency and phase of the output wave are controlled by an absolute position sensor mounted on machine shaft, giving it self-control characteristics. Of course, the pulse train from the position sensor can be delayed by an external command, as shown in the figure.

To explain the dc machine analogy, consider the dc machine where the field flux ψ_f is supplied from the stator side as shown. The dc power from the external circuit is supplied to the rotor (armature) through brushes and commutators. The commutator segments, fixed on the rotor, are rotor position-sensitive. Basically, the commutators and brushes convert the input dc supply to ac, which is impressed in the armature winding. We can look upon the commutators and brushes as mechanical rotor (absolute) position-sensitive inverters. However, the field flux ψ_f and armature reaction flux ψ_a remain stationary in space at perpendicular positions, as shown on the right of Figure 9.4(a).

A self-controlled synchronous machine can be considered analogous to a dc motor except for the following differences:

- Unlike a dc machine, the field is rotating and the armature is stationary (often called an inside-out dc machine).
- Unlike a mechanical position-sensitive inverter, we have in this case an electronic inverter that is controlled by an absolute position encoder.
- Unlike stationary fluxes in space, the fluxes and phasor diagram are rotating at synchronous speed. The absolute position sensor gives the position of the field flux ψ_f, which is fixed on the rotor. If the inverter is current-controlled, the phase position of I_s can be controlled with respect to ψ_f at angle $\delta + \pi/2 - \varphi$ angle (see Figure 9.3).

This dc machine analogy gives the self-controlled synchronous machine various names, such as electronically commutated motor (ECM), brushless dc motor (BLDM or BLDC), or commutatorless-brushless motor, in the literature. However, the commercial name of BLDM (or BLDC) is restricted to the trapezoidal PM machine drive, which will be discussed later.

The self-controlled synchronous machine has several features which can be summarized as follows:

- An electronic commutator replaces the mechanical commutators and brushes, thus eliminating the disadvantages of the dc machine, such as maintenance and reliability problems, sparking, limitations in speed and power rating, difficulty to operate in corrosive and explosive environments, the limitation of altitude, and the EMI problem.
- Because of self-control, the machine does not show any stability or hunting problem of the traditional synchronous machine.
- The transient response can be similar to a dc machine.
- The phase angle between the current I_s and flux ψ_f can be controlled as necessary by delay control. The ψ_a is no longer fixed to the $\pi/2$ angle as in a dc machine.
- With a high-energy magnet, the rotor inertia can be made smaller, which is an advantage in a fast-response servo-type drive.

Because of so many favorable characteristics, synchronous machine drives almost exclusively use self-control.

9.2.3 Absolute Position Encoder

Synchronous machines have absolute location of rotor magnetic poles, which is unlike the location of slipping poles in an induction motor. Therefore, in a self-controlled synchronous machine, an absolute position encoder is mandatory. Sinusoidal PM machines require continuous information of rotor position with high accuracy. This is much more demanding than that of a trapezoidal machine or load-commutated inverter wound-field machine, which both require position information at discrete locations only. Position encoders can generally be classified into two types: optical-type and resolver-type.

9.2.3.1 Optical Encoder

Digital information about rotor position can be obtained directly from a coded disk that transmits or interrupts a light beam. The light beam may be generated by a light-emitting diode (LED), and the transmitted beam can be detected by photo-transistor. Figure 9.5 shows a binary-coded disk that consists of a number of concentric rings with binary coding and a light beam for each ring. The shaded area transmits light and indicates a digital count of 1. The simple four-ring disk gives a digital output of 0101, that is, a decimal count of 5 at the indicated position. The disk resets at a count of $2^4 = 16$, that is, its mechanical angle resolution is 360/16 = 22.5°, which corresponds to 45 electrical degrees for a four-pole machine. A practical disk with 14 rings gives a 14-bit position resolution, which corresponds to nearly 0.04 electrical degrees for a four-pole machine. In natural binary code, bit positions can change simultaneously, which may cause problems. For this reason, a grey-coded disk has been developed where only one bit change occurs per transition.

Another type of optical encoder in the form of a slotted (or coded) disk is shown in Figure 9.6(a), and (b) gives the encoder waves. The encoder has been specifically designed for a four-pole machine. The disk has a large number of slots in the outer perimeter, but two slots of π/2 angle at

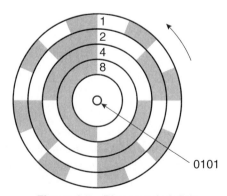

Figure 9.5 Binary-coded disk

Sinusoidal SPM Machine Drives

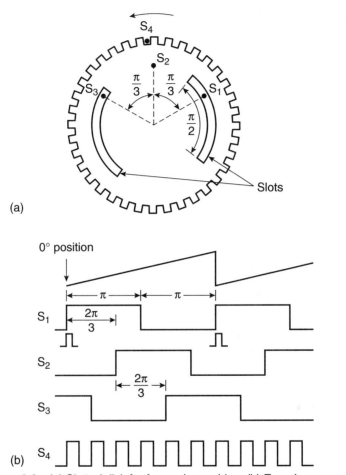

Figure 9.6 (a) Slotted disk for four-pole machine, (b) Encoder waves

an inner radius as shown. There are four stationary optical sensors (S_1–S_4), of which S_4 is mounted at the outer perimeter and S_1, S_2, and S_3 are at the inner radius with $\pi/3$ angle spacing as indicated. A sensor is defined here as an LED and photo-transistor pair, and a logic 1 is generated when the sensor is within the slot. Sensors S_1 to S_3 generate square waves at $2\pi/3$-angle phase shift, and S_4 generates a high-frequency pulse train. This type of absolute encoder can be used for the generation of firing angles for a three-phase, load-commutated, current-fed inverter for a WFSM drive, which will be described later. The Hall sensors that generate the logic waves (described later) used so commonly in trapezoidal machines work on a somewhat similar principle.

To determine absolute rotor position in a four-pole machine, assume for the present that only sensors S_1 and S_4 are present, where S_1 is aligned to the absolute zero position of the rotor. A pulse at the leading edge of the S_1 wave can reset and trigger an UP-counter, which counts the pulses generated by sensor S_4. The counter output, shown at the top of Figure 9.6(b), has a period of 2π electrical degrees, and it gives the absolute rotor angle position. If the number of slots on the disk perimeter is 720, the electrical angle resolution is 1.0°.

9.2.3.2 Analog Resolver with Decoder

Another type of position encoder which is mechanical and therefore more robust and environmental contamination-insensitive than the optical type is shown in Figure 9.7. It consists of two parts: an analog resolver and a resolver-to-digital converter (RDC). The analog resolver is basically a two-phase machine that is excited by a rotor-mounted field winding, which is excited by a carrier wave of several kHz frequency. The resolver is brushless because the rotor winding is excited by a revolving transformer whose primary is supplied from an oscillator, as shown. The stator windings of the resolver give the amplitude-modulated outputs given by

$$V_1 = AV_0 \sin \omega t \sin \theta \tag{9.3}$$

$$V_2 = AV_0 \sin \omega t \cos \theta \tag{9.4}$$

where ω = oscillator frequency, V_0 = oscillator voltage, A = effective transformation ratio between the transformer primary and output windings, and θ = electrical orientation angle of the rotor excitation winding as shown. If, for example, $\theta = 0$, the horizontal stator winding will be

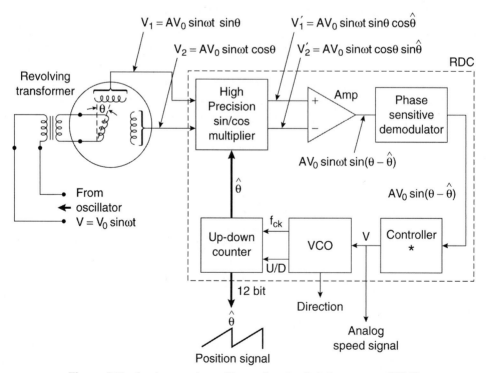

Figure 9.7 Analog resolver with resolver-to-digital converter (RDC)

decoupled, whereas the vertical winding will give maximum voltage output because of maximum coupling. The analog output signals can be used directly or they can be processed through an RDC to obtain the digital signal. The RDC can be considered as a close loop position tracking servo system. It consists of a number of components indicated in the figure. Additionally, it generates the carrier signal for the resolver (not shown). The high-precision sin/cos multiplier multiplies the input signals by $\cos\hat{\theta}$ and $\sin\hat{\theta}$, respectively, by the feedback estimated position signal $\hat{\theta}$ generated by the UP/DOWN-counter. Output signals V_1' and V_2' are subtracted from one another through an error amplifier to generate the $AV_0 \sin\omega t.\sin(\theta - \hat{\theta})$ signal at the output. A phase-sensitive demodulator converts this signal to $AV_0 \sin(\theta - \hat{\theta})$ at its output. This signal is processed by an integral-type controller, VCO (voltage-controlled oscillator), and UP/DOWN-counter to generate the estimated $\hat{\theta}$ signal at the output. The tracking error will be zero at steady state because of integration in the controller, and it will give the correct position signal at the counter output. Note that the VCO input (V) is an analog bipolar speed signal, which can be tapped for control and monitoring purposes. The polarity of V gives the direction signal. UP counting indicates a positive direction of rotation, whereas DOWN counting denotes reverse rotation The resolver system is definitely more expensive than the optical encoder.

9.2.4 Vector Control

The vector control principle of a sinusoidal SPM machine is somewhat simple. In Chapter 2, a sinusoidal SPM machine was described basically as a nonsalient pole machine with a large effective air gap. This makes the synchronous inductance L_s and the corresponding armature flux $\psi_a (= L_s I_s)$ very small, that is, $\psi_s \approx \psi_m \approx \psi_f$. For maximum torque sensitivity with the stator current (i.e., maximum efficiency), we can set $i_{ds} = 0$ and $\hat{I}_s = i_{qs}$, as shown in the phasor diagram of Figure 9.8, where the stator resistance R_s is neglected for simplicity. This condition also gives a minimal inverter power rating. The developed torque expression can be easily derived from Equation (2.170) in the form

$$T_e = \frac{3}{2}(\frac{P}{2})\hat{\psi}_f i_{qs} \qquad (9.5)$$

where $\hat{\psi}_f$ is the space vector magnitude ($\sqrt{2}\psi_f$) and $\psi_s \cos\varphi = \psi_s \cos\delta = \psi_f$. The equation indicates that the torque is proportional to i_{qs} and the power factor angle φ equals the torque angle δ. Figure 9.9 shows the vector control block diagram for the machine, where the stator command current i_{qs}^* is derived from the speed control loop. Its polarity is positive for motoring mode, but negative for regeneration mode. The rotating frame signals are converted to stator current commands with the help of unit vector signals ($\cos\theta_e$, $\sin\theta_e$) as shown. The position control loop can be added easily, if desired.

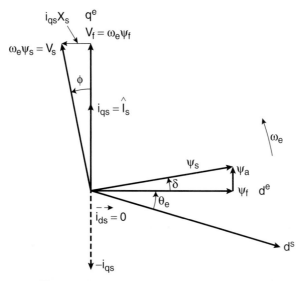

Figure 9.8 Vector control phasor diagram

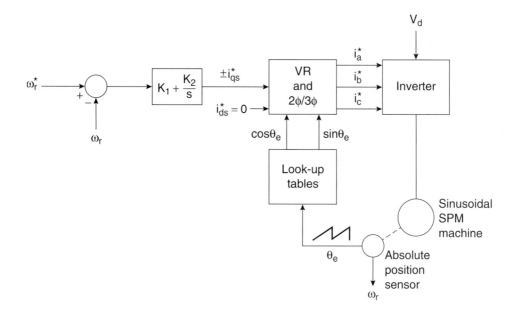

Figure 9.9 Vector control of sinusoidal SPM machine (constant torque region)

Sinusoidal SPM Machine Drives

This vector control strategy is somewhat similar to that of the induction motor vector control, which was shown in Figure 8.32, except for the following:

- The slip frequency $\omega_{sl} = 0$ because the machine always runs at synchronous speed ω_e.
- The magnetizing current $i_{ds}^* = 0$ because the rotor flux is supplied by the PM.
- The unit vector is generated from an absolute position sensor because, unlike the slipping poles of an induction motor, the poles are fixed on the rotor.

Note that the ψ_a and ψ_f phasors in Figure 9.8 are at quadrature like the dc machine shown in Figure 9.4(a), except these are rotating at synchronous speed. Such a drive has truly brushless dc motor characteristics. The machine operates at a small lagging power factor angle φ, which is shown in the figure. The vector control shown is valid only in the constant torque region. As the speed increases, voltages V_s and V_f increase proportionally to speed ω_e, and eventually, vector control is lost when the PWM controller saturates at the edge of the constant torque region.

9.2.4.1 Field-Weakening Mode

The speed of a sinusoidal SPM machine can be controlled beyond base speed ω_b by field-weakening control. However, the field-weakening speed range is small because of a weak armature reaction effect. This is explained with the help of the phasor diagram in Figure 9.10, and the corresponding torque-speed curve in Figure 9.11. As the stator voltage tends to saturate at the

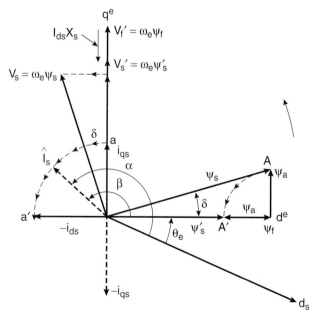

Figure 9.10 Phasor diagram showing field-weakening control

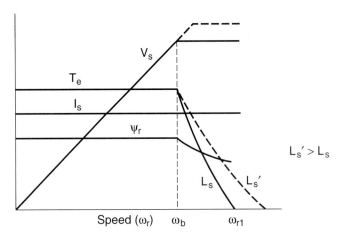

Figure 9.11 Torque-speed curve showing field-weakening control above base speed ω_b

edge of the constant torque region, and since $V_s = \omega_e \psi_s$, the stator flux ψ_s must be weakened beyond the base speed ω_b so that stator current control remains possible. This means that a demagnetizing current $-i_{ds}$ must be injected on the stator side. But, because of low armature reaction flux ψ_a, this demagnetization demands large i_{ds}. Within the specified machine stator current rating, therefore, it appears that the weakening of ψ_s is small, giving a small range of field-weakening speed control. It is also obvious that with constant ψ_f, V_f will increase proportionally with ω_e, and the overexcited machine will give a leading power factor at the machine terminal.

Now consider the phasor diagram in Figure 9.10, where the region of constant torque mode control remains the same as in Figure 9.8. With rated $\hat{I}_s = i_{qs}$, $\hat{I}_s$ can now be rotated anticlockwise in the a – a' locus so that $\hat{I}_s = \sqrt{i_{qs}^2 + i_{ds}^2}$; in other words, $-i_{ds}$ helps weaken the stator flux. With a constant magnitude of ψ_a, the ψ_s phasor is reduced and rotated in the locus A – A' as shown. At point A', $|\hat{I}_s| = |-i_{ds}|$, which corresponds to zero developed torque at speed ω_{r1}, as shown in Figure 9.11. The orientation of voltages V_s' and V_f' for this condition are shown in the phasor diagram and the corresponding machine power factor is zero leading. Obviously, within the rated stator current, the field-weakening range can be increased if the synchronous inductance L_s is increased to L_s', as indicated. Figure 9.12 shows the vector control block diagram for constant torque as well as field-weakening modes. A position servo is considered and an optional torque control loop has been added in the inner loop. In constant torque mode, $i_{ds}^* = 0$, as in Figure 9.9, but in field-weakening mode, flux $\hat{\psi}_s$ is controlled inversely with speed, with $-i_{ds}^*$ control generated by the flux loop. Within the torque loop, i_{qs} is controlled so that its maximum value is limited by

$$i_{qsm}^* = \sqrt{\hat{I}_s^2 - |i_{ds}|^2} \qquad (9.6)$$

Sinusoidal SPM Machine Drives

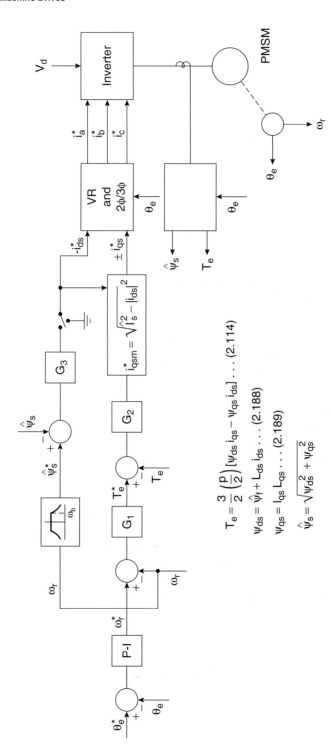

Figure 9.12 Vector control of sinusoidal SPM machine (including field-weakening region)

where $i_{qsm}{}^* = $ is the limit value and $\hat{I}_s = $ rated stator current (peak). The current model-based feedback signal processing equations discussed in Chapter 2 are shown in the lower part of the figure. Currents i_{ds} and i_{qs} can be derived from i_a and i_b with the help of the θ_e signal.

A somewhat simpler polar form of vector control implementation that operates in both the constant torque and field-weakening regions is shown in Figure 9.13. The scheme does not require a stator flux control loop and there is no switching in the i_{ds} loop as in Figure 9.12. Only the stator current phasor I_s is oriented at an appropriate angle, as shown in the phasor diagram of Figure 9.10. In Figure 9.13, the torque control loop generates the magnitude of stator current $|\hat{I}_s|$, which is oriented at angle α with respect to the d^s axis where $\alpha = \beta + \theta_e$, as shown in the figure. The angle θ_e is the absolute position angle, and β is the orientation angle of $\hat{I}_s$ with respect to the rotating d^e axis. Obviously, $\beta = \pi/2$ in the constant torque region, and in field-weakening mode, it is greater than $\pi/2$, as indicated. The β angle is generated from a function generator which has a value of $\pi/2$ in the constant torque region, but increases with speed beyond the base speed ω_b until the maximum value is π. The function generator is symmetrical for both positive and negative speed. The polarity of rotating frame signals is then converted to phase current commands by the following relations:

$$i_{ds}{}^s = |\hat{I}_s|\cos\alpha \qquad (9.7)$$

$$i_{qs}{}^s = |\hat{I}_s|\sin\alpha \qquad (9.8)$$

$$i_a{}^* = i_{qs}{}^s \qquad (9.9)$$

$$i_b{}^* = -\frac{1}{2}i_{qs}{}^s - \frac{\sqrt{3}}{2}i_{ds}{}^s \qquad (9.10)$$

$$i_c{}^* = -\frac{1}{2}i_{qs}{}^s + \frac{\sqrt{3}}{2}i_{ds}{}^s \qquad (9.11)$$

The polarity of developed torque depends on the sign of $\hat{I}_s$, which correspondingly determines the polarity of the β angle. The feedback signals can be computed as follows:

$$\hat{I}_s = \sqrt{i_{ds}{}^{s2} + i_{qs}{}^{s2}} \qquad (9.12)$$

$$\alpha = \pi - \tan^{-1}\frac{i_{qs}{}^s}{i_{ds}{}^s} \qquad (9.13)$$

$$T_e = \frac{3}{2}(\frac{P}{2})\hat{\psi}_f|\hat{I}_s|\sin\beta \qquad (9.14)$$

where $\beta = \alpha - \theta_e$.

Synchronous Reluctance Machine Drives

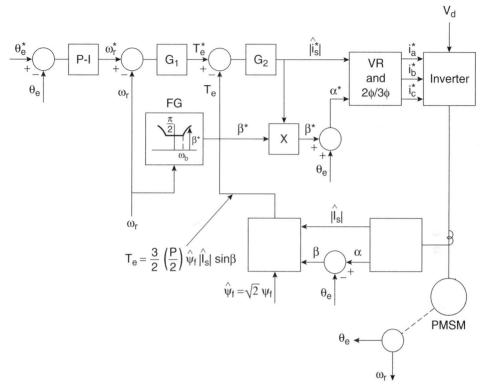

Figure 9.13 Vector control of sinusoidal SPM machine in polar form (including field-weakening region)

9.3 SYNCHRONOUS RELUCTANCE MACHINE DRIVES

As discussed in Chapter 2, synchronous reluctance machines (SyRMs) have salient poles, and they do not have any field winding or PM on the rotor. The machines are low-cost, rugged, have high-efficiency (ideally no rotor loss), and are capable of operating at very high speeds. The traditional SyRM has low saliency, that is, a low L_{dm}/L_{qm} ratio, which gives poor torque density, low power factor, and poor efficiency. For these reasons, most SyRM drives have been outdated by PM machine drives. However, the recent development of SyRMs by anisotropic axially laminated construction (see Figure 2.37) has made a much higher L_{dm}/L_{qm} ratio (8–10) possible, which has significantly improved torque density, power factor, and efficiency to the point where they are almost comparable to those of an induction motor. Their application has increased recently, although there are only a few manufacturers of this machine worldwide.

In many respects, the SyRM is similar to the sinusoidal PM machine, except that the field flux $\psi_f = 0$ and $L_{dm} \neq L_{qm}$. The d^e-q^e equivalent circuits of the machine are simple, and are shown in Figure 9.14, which can be derived directly from Figure 2.42. The machine may or may not have a cage or damper winding. Ideally, there is no core or copper loss in the rotor, but the inverter-fed harmonics will cause some copper loss in the damper winding, if present. The

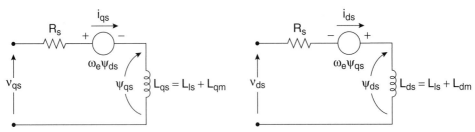

Figure 9.14 Rotating frame (d^e-q^e) equivalent circuits of SyRM

simplest operation of an SyRM is a line-start motor where the machine starts like an induction motor with the help of a cage winding, but pulls into synchronism at synchronous speed. One traditionally popular application of an SyRM is the multi-motor drive by open loop volts/Hz speed control shown in Figure 9.1. The performance curves in Figure 9.2 remain valid for the SyRM, except the developed torque is lower, the field-weakening range is larger, and the stability limit is reached at lower δ angle ($\pi/4$) (see Figure 2.34). With volts/Hz control, the motor requires damper winding to prevent hunting or oscillatory response. With self-control, the response is robust like a dc machine, and no damper winding is needed. Figure 9.15 shows the phasor diagram of the machine with standard symbols where $\psi_{ds} = L_{ds} i_{ds}$, $\psi_{qs} = L_{qs} i_{qs}$, and $V_s = \omega_e \psi_s$. The flux ψ_{ds} tends to saturate at higher i_{ds}. In fact, there is some cross-saturation effect of L_{ds} due to i_{qs} current. The stator resistance drop has been neglected for simplicity. Note that the ψ_f phasor and corresponding V_f phasor are absent. Since the stator supplies magnetizing current like an induction motor, the stator power factor angle φ is large.

Equation (2.183) in Chapter 2 shows the developed torque expression of a SyRM, which is rewritten as

$$T_e = 3(\frac{P}{2}) \psi_s^2 \frac{(L_{ds} - L_{qs})}{2 L_{ds} L_{qs}} \sin 2\delta \tag{2.183}$$

or

$$= \frac{3}{2}(\frac{P}{2}) \hat{\psi}_s^2 \frac{(L_{ds} - L_{qs})}{2 L_{ds} L_{qs}} \sin 2\delta \tag{9.15}$$

where space vector flux magnitude $\hat{\psi}_s = \sqrt{2} \psi_s$, P = number of poles, and δ = torque angle shown in Figure 9.15. Substituting $\sin 2\delta = 2 \sin \delta \cos \delta$, where $\sin \delta = \psi_{qs}/\hat{\psi}_s$ and $\cos \delta = \psi_{ds}/\hat{\psi}_s$, Equation (9.15) can be written in the form

$$T_e = \frac{3}{2}(\frac{P}{2}) \frac{(L_{ds} - L_{qs})}{L_{ds} L_{qs}} \psi_{qs} \psi_{ds} \tag{9.16}$$

or

$$T_e = \frac{3}{2}(\frac{P}{2})(L_{ds} - L_{qs}) i_{qs} i_{ds} \tag{9.17}$$

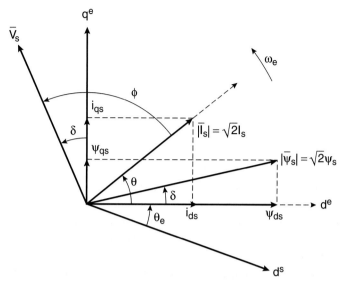

Figure 9.15 Phasor diagram of SyRM (shown with vectors)

This equation indicates that torque can be controlled by i_{ds}, i_{qs}, or both components. Equation (9.17) can also be written in the form

$$T_e = \frac{3}{2}(\frac{P}{2})(\psi_{ds}i_{qs} - \psi_{qs}i_{ds}) \qquad (9.18)$$

which is also the general torque equation given in Equation (2.114).

9.3.1 Current Vector Control of SyRM Drive

In this section, we will review several close loop self-control methods of an SyRM drive [3]. These can be classified as:

- Constant d^e-axis current (i_{ds}) control
- Fast torque response control
- Maximum torque/ampere control
- Maximum power factor control

Although these methods are defined as vector control because of the independent control of i_{ds} and i_{qs} components of stator current and vector transformation, they are not truly vector control because there is no orientation with machine flux.

9.3.1.1 Constant d^e-Axis Current (i_{ds}) Control

In this simple control method, the d^e-axis current i_{ds} is kept constant as shown in Figure 9.15, and the q-axis current i_{qs} is controlled to control the torque in the constant torque region. The torque equation, (9.17), can be written in the form

$$T_e = \frac{3}{2}\left(\frac{P}{2}\right)\left(1 - \frac{L_{qs}}{L_{ds}}\right)\psi_{ds} i_{qs} \tag{9.19}$$

This equation indicates that the torque is proportional to the product of ψ_{ds} and i_{qs}, and its polarity can be reversed by i_{qs} polarity. It looks similar to the torque equation of a vector-controlled induction motor or sinusoidal PMSM, but the coefficient $K = \frac{3}{2}\left(\frac{P}{2}\right)\left(1-\left(L_{qs}/L_{ds}\right)\right)$ is smaller. Figure 9.16 shows the block diagram of a speed control system with constant i_{ds} control. The absolute position sensor provides the θ_e signal for self-control and ω_r signal for close loop speed control. The speed loop error generates the i_{qs}^* current command that controls the torque. The polarity of this signal determines the torque polarity. The magnetizing current command i_{ds}^* is constant in the constant torque region ($\omega_r < \omega_b$), but it is reduced beyond the base speed ω_b for extended speed control operation. The i_{ds}^* and i_{qs}^* components of the stator current are vector-rotated (VR) and transformed to three-phase stator current commands in stationary frame with the help of the θ_e angle as shown. The actual inverter currents can be controlled by hysteresis-band PWM or synchronous current control, which are not shown in detail.

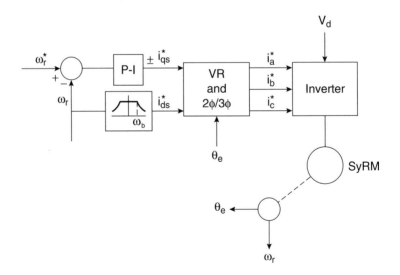

Figure 9.16 SyRM drive with constant i_{ds} control

Synchronous Reluctance Machine Drives

Note that the magnitude of stator flux $\hat{\psi}_s$ is given by

$$\hat{\psi}_s = \sqrt{\psi_{ds}^2 + \psi_{qs}^2} \tag{9.20}$$

or

$$\hat{\psi}_s = \sqrt{(L_{ds}i_{ds})^2 + (L_{qs}i_{qs})^2} \tag{9.21}$$

and its orientation angle with the d^e-axis is given as

$$\delta = \tan^{-1}\left(\frac{\psi_{qs}}{\psi_{ds}}\right) = \tan^{-1}\left(\frac{L_{qs}i_{qs}}{L_{ds}i_{ds}}\right) \tag{9.22}$$

These equations indicate the dependence of $\hat{\psi}_s$ and δ on both the i_{qs} and i_{ds} currents. Therefore, as i_{qs} is increased in Figure 9.16, both the $\hat{\psi}_s$ and δ angle increase, giving more of a coupling effect. In fact, controlling the in-phase and magnetizing current components at the stator terminal with stator flux-oriented vector control, which will be discussed later, can give better performance of the drive. Since $V_s = \omega_e \psi_s$, at higher speed, the PWM controller will tend to saturate. Current control in the extended speed range is possible by weakening the flux ψ_s, that is, ψ_{ds} by reducing i_{ds} inversely with speed ω_r in an open loop manner, as shown in the figure.

9.3.1.2 Fast Torque Response Control

In this control strategy, it is desired to have the fastest control response of the drive. The torque equation, (9.17), the can be modified by multiplying it with the square of the stator flux expression (9.21) in both the numerator and denominator as

$$T_e = \frac{3}{2}\left(\frac{P}{2}\right)\frac{(L_{ds} - L_{qs})i_{ds}i_{qs}\hat{\psi}_s^2}{(L_{ds}^2 i_{ds}^2 + L_{qs}^2 i_{qs}^2)} \tag{9.23}$$

which can be simplified in the form

$$T_e = \frac{3}{2}\left(\frac{P}{2}\right)\frac{(L_{ds} - L_{qs})\hat{\psi}_s^2 \tan\theta}{(L_{ds}^2 + L_{qs}^2 \tan^2\theta)} \tag{9.24}$$

where $\tan\theta = i_{qs}/i_{ds}$. This equation shows torque as a function of stator flux $\hat{\psi}_s$ and θ angle. The fastest response at maximum torque is possible by differentiating T_e with respect to $\tan\theta$ at a given ψ_s and equating to zero, that is, $\left.\frac{dT_e}{d(\tan\theta)}\right|_{\psi_s} = 0$. This operation on Equation (9.24) yields

$$\tan\theta = \frac{L_{ds}}{L_{qs}} \tag{9.25}$$

This optimum condition means

$$\frac{i_{qs}}{i_{ds}} = \frac{L_{ds}}{L_{qs}} \tag{9.26}$$

in other words,

$$\tan \delta = \frac{\psi_{qs}}{\psi_{ds}} = 1 \tag{9.27}$$

so that ψ_s is always oriented at angle $\pi/4$ with the d^e-axis. To get a clear perspective, let us express torque T_e, stator flux ψ_s, and stator current I_s in normalized form [3]. The $\hat{\psi}_s$ expression (9.21) can be written in the form

$$\hat{\psi}_s = \hat{I}_s \sqrt{\frac{L_{ds}^2 i_{ds}^2 + L_{qs}^2 i_{qs}^2}{i_{ds}^2 + i_{qs}^2}} \tag{9.28}$$

where $\hat{I}_s = \sqrt{i_{qs}^2 + i_{ds}^2}$ has been multiplied in both the numerator and denominator. Equation (9.28) can be derived in the form

$$\hat{\psi}_s = \hat{I}_s \sqrt{\frac{L_{ds}^2 + L_{qs}^2 \tan^2 \theta}{1 + \tan^2 \theta}} \tag{9.29}$$

Substituting Equation (9.25) in (9.29) and noting that the maximum value of ψ_s (ψ_{sm}) corresponds to the maximum I_s (I_{sm}), we can derive the relation

$$\hat{\psi}_{sm} = \frac{\sqrt{2} L_{ds} L_{qs}}{\sqrt{L_{ds}^2 + L_{qs}^2}} \hat{I}_{sm} \tag{9.30}$$

Substituting Equations (9.25) and (9.30) in (9.24), the maximum torque expression T_{em} can be derived as

$$T_{em} = \frac{3}{2} \left(\frac{P}{2}\right) \frac{L_{ds} L_{qs} (L_{ds} - L_{qs})}{(L_{ds}^2 + L_{qs}^2)} \hat{I}_{sm}^2 \tag{9.31}$$

Defining (pu) variables as

$$T_e(pu) = \frac{T_e}{T_{em}}, \; \psi_s(pu) = \frac{\hat{\psi}_s}{\hat{\psi}_{sm}}, \; \text{and } I_s(pu) = \frac{\hat{I}_s}{\hat{I}_{sm}} \tag{9.32}$$

we can combine Equations (9.24), (9.30), and (9.31) in the normalized form as

$$T_e(pu) = \frac{2(\frac{L_{qs}}{L_{ds}})\tan\theta}{\left[1+(\frac{L_{qs}}{L_{ds}})^2\tan^2\theta\right]}\psi_s(pu)^2 \qquad (9.33)$$

or

$$T_e(pu) = \frac{\left[1+(\frac{L_{qs}}{L_{ds}})^2\right]}{2(\frac{L_{qs}}{L_{ds}})}I_s(pu)^2\sin 2\theta \qquad (9.34)$$

Figure 9.17(a) shows a family of curves of $\psi_s(pu)$ and $T_e(pu)$ variables with several numerical values on the $I_s(pu) - \theta$ plane. Figure 9.17(b) shows a family of curves for $I_s(pu)$ and $\psi_s(pu)$ on the $T_e(pu) - \theta$ plane. All these curves are plotted with constant $L_{qs}/L_{ds} = 0.38$. The locus of operation at the optimum θ angle (Equation (9.25)) is shown on these figures. Note that operation always occurs at the peak of constant flux $\psi_s(pu)$ trajectory, giving fastest transient response. If the θ angle is negative, the drive goes into regenerative mode, as indicated on the left-hand side of Figure 9.17(a).

Figure 9.18 shows the control block diagram with this strategy. An additional torque control loop is introduced within the speed loop, which generates the $\pm i_{qs}^*$ current. The current $|i_{ds}^*|$ is generted from i_{qs}^* so that $\tan\theta = L_{ds}/L_{qs} = i_{qs}/i_{ds} =$ constant. The feedback torque can be calculated from Equation (9.17), the details of which are not shown. The saturation effect of L_{ds} can be compensated in the torque computation, if desired. The addition of torque loop with saturation compensation will give faster response than direct i_{qs} control. With absolute $|i_{ds}^*|$ excitation, the polarity of torque is determined by i_{qs}^* polarity.

The performance curves in Figure 9.17 also indicate operation at constant $\psi_s(pu) = 1$ and constant $i_{ds(pu)}$ trajectories. With constant rated flux linkage, the operating point is at A when unloaded, where the flux is generated solely by i_{ds} ($\theta = 0$) current. As the machine is loaded, the operating point travels along AB until it reaches point B at maximum torque. This means that the rated flux ψ_s is rotating anti-clockwise with increasing δ angle in Figure 9.15. If the machine is operated at constant $i_{ds} = i_{ds0}$, the operating point C corresponds to no-load operation at $\theta = 0$. With loading, the point travels along CB, increasing I_s and ψ_s until it reaches point B at maximum torque.

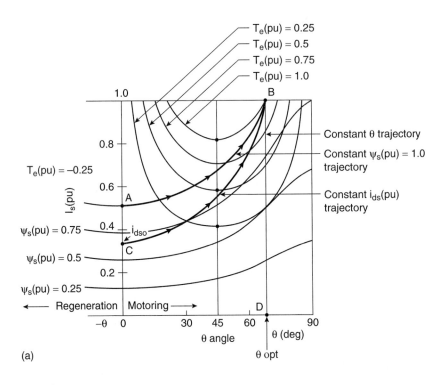

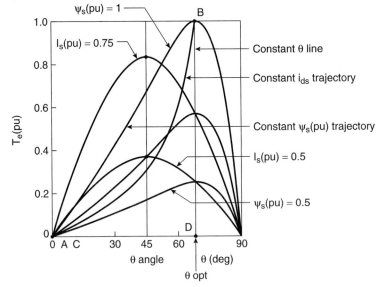

Figure 9.17 Performance characteristics of synchronous reluctance motor (L_{qs}/L_{ds} = 0.38) (a) Normalized plot of variables on θ angle – stator current $I_s(pu)$ plane, (b) Normalized plot of variables on θ angle – torque $T_e(pu)$ plane

Synchronous Reluctance Machine Drives

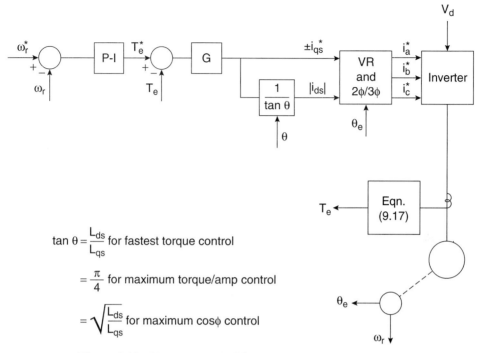

Figure 9.18 Vector control of SyRM drive at constant θ angle

9.3.1.3 Maximum Torque/Ampere Control

As mentioned before, this condition tends to give maximum drive efficiency. For this control strategy, it is necessary to develop a torque expression as a function of stator current I_s. Substituting $i_{ds} = \hat{I}_s \cos\theta$ and $i_{qs} = \hat{I}_s \sin\theta$ in Equation (9.17), we get

$$T_e = \frac{3}{2}\left(\frac{P}{2}\right)(L_{ds} - L_{qs})\hat{I}_s^2 \frac{\sin 2\theta}{2} \tag{9.35}$$

The maximum torque/ampere is obtained when $\theta = \pi/4$, that is, when $i_{ds} = i_{qs}$. The operating locus at this condition is shown on Figure 9.17. The control block diagram of Figure 9.18 is also valid here, except with this new θ angle input.

9.3.1.4 Maximum Power Factor Control

The machine terminal power factor expression $\cos\varphi$ can be found from the following expression:

$$T_e\omega_m = 3V_s I_s \cos\varphi = \frac{3}{2}\hat{V}_s\hat{I}_s\cos\varphi \tag{9.36}$$

In other words,

$$\cos\varphi = \frac{T_e \omega_m}{\frac{3}{2}\hat{V}_s \hat{I}_s} \quad (9.37)$$

where ω_m = speed in mechanical r/s, V_s = rms phase voltage ($\hat{V}_s/\sqrt{2}$), and I_s = rms phase current ($\hat{I}_s/\sqrt{2}$).

Substituting the following relations:

$$\omega_m = \omega_e\left(\frac{2}{P}\right) \quad (9.38)$$

$$\hat{V}_s = \omega_e \hat{\psi}_s \quad (9.39)$$

$$\hat{I}_s = \sqrt{i_{ds}^2 + i_{qs}^2} \quad (9.40)$$

in Equation (9.37) and with the help of Equations (9.17) and (9.21), we can derive a $\cos\varphi$ expression as

$$\cos\varphi = \frac{(\frac{L_{ds}}{L_{qs}} - 1)i_{ds}i_{qs}}{\left[i_{ds}^2(\frac{L_{ds}}{L_{qs}})^2 + i_{qs}^2\right]^{1/2}(i_{ds}^2 + i_{qs}^2)^{1/2}} \quad (9.41)$$

or

$$\cos\varphi = \frac{(\frac{L_{ds}}{L_{qs}} - 1)}{\left[(\frac{L_{ds}}{L_{qs}})^2 + \tan^2\theta\right]^{1/2}(\frac{1}{\tan^2\theta} + 1)^{1/2}} \quad (9.42)$$

where $\tan\theta = i_{qs}/i_{ds}$ has been substituted. The $\cos\varphi$ will be maximum when the denominator D of Equation (9.42) is minimum. Solving $d(D^2)/d(\tan\theta)$ and equating to zero, we get the relation

$$\tan\theta = \sqrt{\frac{L_{ds}}{L_{qs}}} \quad (9.43)$$

which means that to maintain the maximum power factor, ratio i_{qs}/i_{ds} should always be equal to $\sqrt{L_{ds}/L_{qs}}$. Figure 9.18, the control block diagram, is also valid in this case.

9.4 SINUSOIDAL IPM MACHINE DRIVES

The basic construction and characteristics of the interior or buried magnet machine were discussed in Chapter 2. As mentioned before, the torque developed in an IPM machine has two components: (1) the component due to field flux, and (2) the reluctance torque component. The general torque expression of a salient pole machine can be given by

$$T_e = \frac{3}{2}(\frac{P}{2})\left[\frac{\hat{\psi}_s \hat{\psi}_f}{L_{ds}}\sin\delta + \hat{\psi}_s^2 \frac{(L_{ds}-L_{qs})}{2L_{ds}L_{qs}}\sin 2\delta\right] \quad (9.44)$$

or

$$T_e = \frac{3}{2}(\frac{P}{2})(\psi_{ds}i_{qs} - \psi_{qs}i_{ds}) \quad (2.114)$$

where P = number of poles, $\psi_{ds} = \hat{\psi}_f + L_{ds}i_{ds}$, $\psi_{qs} = L_{qs}i_{qs}$, and $\hat{\psi}_s = \sqrt{\psi_{ds}^2 + \psi_{qs}^2}$. Note that Equation (9.44) is the same as Equation (2.177), except the rms phasors have been replaced by the corresponding peak values ($\hat{\psi}_f = \sqrt{2}\psi_f$ and $\hat{\psi}_s = \sqrt{2}\psi_s$). Substituting ψ_{ds} and ψ_{qs} expressions in Equation (2.114), we get

$$T_e = \frac{3}{2}(\frac{P}{2})\left[\hat{\psi}_f i_{qs} + (L_{ds}-L_{qs})i_{ds}i_{qs}\right] \quad (9.45)$$

This equation identifies the two torque components. In a sinusoidal SPM machine, the reluctance torque is zero, thus giving only the first component as shown in Equation (9.5), whereas in a reluctance machine, $\psi_f = 0$, giving the reluctance torque component in Equation (9.17). Note that in an IPM machine, $L_{qs} > L_{ds}$, unlike a wound-field, salient pole, synchronous machine.

Like a sinusoidal SPM machine, an IPM machine can be operated in line-start or open loop volts/Hz speed control mode, where a damper winding is essential. However, in self-control mode, the machine does not use a damper winding. In this section, we will discuss two control methods for self-controlled IPM machine drives.

9.4.1 Current Vector Control with Maximum Torque/Ampere

Like a synchronous reluctance motor, it is possible to control an IPM machine with the maximum torque/ampere principle [6] to minimize electrical losses, that is, to optimize the drive efficiency. This also means a minimum converter rating as well as its maximum efficiency. Of course, this does not mean the best transient response.

To derive maximum torque/ampere criteria, it is better to normalize the torque expression and express it as function of normalized stator current components. Defining the base torque as

$$T_{eB} = \frac{3}{2}\left(\frac{P}{2}\right)\hat{\psi}_f I_B \qquad (9.46)$$

where the base current I_B is defined as

$$I_B = \frac{\hat{\psi}_f}{L_{qs} - L_{ds}} = I_f \frac{L_{dm}}{L_{qs} - L_{ds}} \qquad (9.47)$$

The fictitious field current I_f in the machine can be treated as a constant. The defined T_{eB} and I_B expressions are for convenience only, and are not related to rated or maximum machine ratings. Substituting Equations (9.46) and (9.47) in (9.45), we get

$$T_e(pu) = \frac{T_e}{T_{eB}} = \frac{i_{qs}}{I_B} - \frac{i_{ds}}{I_B} \cdot \frac{i_{qs}}{I_B} \qquad (9.48)$$

or

$$T_e(pu) = i_{qs}(pu)\left[1 - i_{ds}(pu)\right] \qquad (9.49)$$

where $i_{qs}(pu) = i_{qs}/I_B$ and $i_{ds}(pu) = i_{ds}/I_B$.

Figure 9.19 shows the constant torque loci for $T_e(pu)$ as function of d- and q-axis stator current components $i_{ds}(pu)$ and $i_{qs}(pu)$.

Consider, for example, $T_e(pu) = 1.0$ locus in the second quadrant. Any radial distance on the locus from the origin represents the stator current magnitude $\hat{I}_s = \sqrt{i_{qs}^2 + i_{ds}^2}$. Point A on the locus represents minimum stator current; in other words, maximum torque/ampere criteria will be satisfied for $T_e(pu) = 1.0$ with stator current OA. For higher $T_e(pu)$, the corresponding optimum points are B, C, D, etc. Note that for positive torque, the polarity of i_{qs} is positive, whereas the polarity of i_{ds} is negative. Negative i_{ds} contributes an additive reluctance torque component in Equation (9.45). The polarity of torque can be reversed by reversing the i_{qs} current. The figure also shows a symmetric $\hat{I}_s$ trajectory for negative torque in the third quadrant. Figure 9.20 shows the plot of optimum $i_{ds}(pu)$ and $i_{qs}(pu)$ currents as functions of normalized torque $T_e(pu)$, which are derived from Figure 9.19. The information in Figure 9.20 can be used to develop the control strategy for maximum torque/ampere, which is shown in Figure 9.21. The speed control loop generates the torque command T_e^* as shown. From the T_e^* signal, command currents i_{ds}^* and i_{qs}^* are generated in feedforward manner with the help of function generators $FG1$ and $FG2$, respectively. The function generators use the curves in Figure 9.20 and convert them to actual signal values.

Sinusoidal IPM Machine Drives

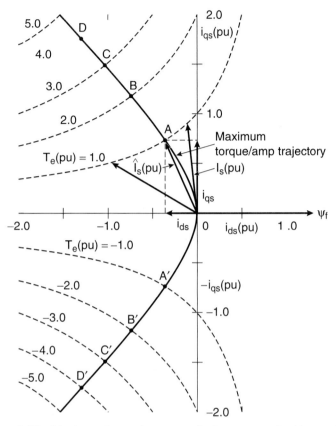

Figure 9.19 Maximum torque/ampere trajectory on constant torque loci

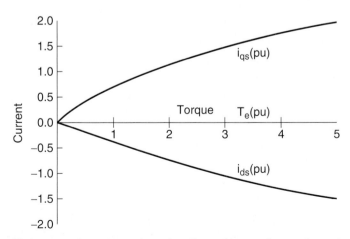

Figure 9.20 Stator current components as functions of torque for maximum torque/ampere

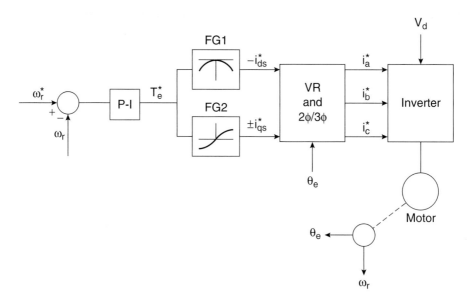

Figure 9.21 Speed control system of IPM machine with open loop maximum torque/ampere assignment (constant torque region only)

Note that the polarity of i_{ds}^* is always negative, irrespective of the torque polarity. However, the polarity of i_{qs}^* is dictated by the T_e^* polarity, as shown in the figure. The absolute position signal θ_e converts the rotating frame signals i_{ds}^* and i_{qs}^* into stationary frame phase current commands i_a^*, i_b^*, and i_c^*, as shown. The simple feedforward current control will work correctly if the machine parameter variation effect is neglected. In practice, the parameter variation (i.e., ψ_f, L_{ds}, and L_{qs}) effect in FG1 and FG2 will cause incorrect distribution of the i_{ds} and i_{qs} currents. Again, the control algorithm is only valid in the constant torque region, where PWM control of the inverter permits control of the desired currents.

9.4.2 Field-Weakening Control

It was mentioned before that an IPM machine (unlike a sinusoidal SPM machine) has a smaller effective air gap, which makes the magnetizing inductance larger; in other words, it has a stronger armature reaction effect. This means that the stator current has a stronger effect in weakening the flux that links the stator winding (ψ_s) (see Figure 9.10). This field-weakening capability permits the drive to operate in an extended speed range, beyond the base speed ω_b, for applications such as electric vehicle drives.

In the constant torque region with PWM mode current control, as the machine speed increases, the CEMF also increases proportionally, demanding more supply voltage. Eventually, the PWM controller saturates in square-wave mode when the current control is completely lost. In

Sinusoidal IPM Machine Drives

square-wave mode, the available peak value of fundamental stator voltage is $\hat{V}_s = 2V_d/\pi$, where V_d = dc link voltage of the inverter. This is related to steady-state v_{ds} and v_{qs} components of stator phase voltage as

$$\hat{V}_s = \sqrt{v_{ds}^2 + v_{qs}^2} \tag{9.50}$$

where

$$v_{qs} = \omega_e \psi_{ds}' + \omega_e \hat{\psi}_f \tag{9.51}$$

$$v_{ds} = -\omega_e \psi_{qs} \tag{9.52}$$

Equations (9.51) and (9.52) are derived from (2.188) and (2.189), respectively, at steady-state condition and neglect the stator resistance drop.

Substituting Equations (9.51) and (9.52) in (9.50) and expressing $\hat{V}_s = 2V_d/\pi$,

$$\frac{4V_d^2}{\pi^2} = (\omega_e \psi_{ds}' + \omega_e \hat{\psi}_f)^2 + \omega_e^2 \psi_{qs}^2 \tag{9.53}$$

Again, substituting ψ_{ds}' and ψ_{qs} in Equation (9.53) by $L_{ds} i_{ds}$ and $L_{qs} i_{qs}$, respectively, we get

$$(\omega_e L_{ds} i_{ds} + \omega_e \hat{\psi}_f)^2 + \omega_e^2 L_{qs}^2 i_{qs}^2 = \frac{4V_d^2}{\pi^2} \tag{9.54}$$

which can be modified to the form

$$\tag{9.55}$$

$$\frac{(i_{ds} + \frac{\hat{\psi}_f}{L_{ds}})^2}{(\frac{2V_d}{\pi \omega_e L_{ds}})^2} + \frac{i_{qs}^2}{(\frac{2V_d}{\pi \omega_e L_{qs}})^2} = 1$$

This is an equation of an ellipse in the form

$$\frac{(i_{ds} - C)^2}{A^2} + \frac{i_{qs}^2}{B^2} = 1 \tag{9.56}$$

where

$$A = \frac{2V_d}{\pi \omega_e L_{ds}} \quad \text{is the length of the semi-major axis} \tag{9.57}$$

$$B = \frac{2V_d}{\pi\omega_e L_{qs}} \quad \text{is the length of the semi-minor axis} \tag{9.58}$$

and

$$C = -\frac{\hat{\psi}_f}{L_{ds}} \quad \text{offset of the center on the } i_{ds} \text{ axis} \tag{9.59}$$

Equation (9.56) has been plotted in Figure 9.22, along with the optimum $\hat{I}_s$ trajectory.

The equation indicates that the center of the ellipse is fixed, but the size shrinks at increasing speed ω_e. For a certain torque demand at point 1 on the $\hat{I}_s$ trajectory, as the speed increases, the square-wave voltage limit ellipse shrinks until it crosses point 1 at speed ω_{e3}, when the current control is totally lost. To activate the current control, operating point 1 or any point outside the ellipse must be pushed inside the ellipse.

Consider, for example, the current demand corresponding to operating point 2. As the speed increases and the voltage limit ellipse shrinks below point 2, the i_{qs}^* command can be lowered to point 3 for the current control to remain valid. Basically, this means weakening the stator flux ψ_s in the constant power region (i.e., weakening the CEMF) so that the current control is restored beyond the base speed. Note that the available torque at point 3 is higher than that at point 1. Figure 9.23 shows a control block diagram that permits field-weakening as well as constant torque region control, and it is basically an extension of Figure 9.21. The feedback current i_{ds} can be easily calculated from the stator currents by inverse vector rotation (VR^{-1}) with

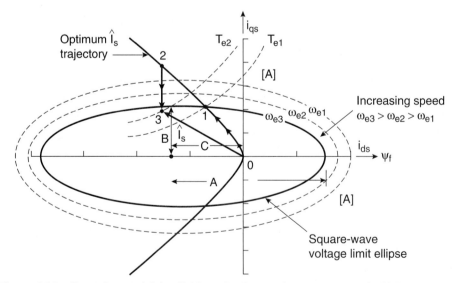

Figure 9.22 Operation explaining field-weakening mode current control with square-wave voltage limit ellipse and optimum $\hat{I}_s$ trajectory

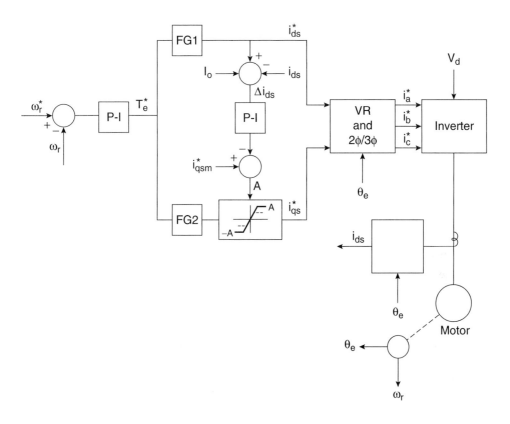

Figure 9.23 Speed control system for constant torque and flux-weakening regions

the help of position signal θ_e (unit vector). In the constant torque region, currents i_{ds}^* and i_{ds} track well, giving negligible Δi_{ds} error as shown. However, the loop error tends to build up as the current controller saturates at and beyond point 1. The error Δi_{ds} through a P-I controller lowers the limit curve of i_{qs}^* until Δi_{ds} approaches zero, that is, the i_{ds} control is restored. This corresponds to point 3 in Figure 9.22 for speed ω_{e3}. As the speed increases further, point 3 is pushed lower, giving lower torque at higher speed. For negative torque, current i_{qs} is negative and the operating region is symmetrically on the lower side of the ellipse.

9.4.3 Vector Control with Stator Flux Orientation

The control algorithm described above is somewhat simple, but it is truly not a vector control because there is no orientation of stator currents with the flux. Besides, there is no direct control of machine stator flux. Therefore, the transient response of the drive may not be optimal. We will now describe a vector-controlled IPM machine drive with stator flux ψ_s orientation in some detail. The stator flux will be controlled directly by a feedback loop. In an IPM machine, the magnet flux ψ_f is constant (or nearly constant), and since there is a strong armature reaction effect, it is natural to control the stator flux magnitude and have stator current orientation with ψ_s for optimum transient response. The control will be described with particular emphasis on the electrical vehicle (EV) drive [7].

Figure 9.24 shows the torque-speed characteristics of the machine in forward motoring mode, which also shows the phase voltage V_s and stator flux ψ_s curves. The machine has two regions of operation: the PWM or constant torque region, where current control permits vector control, and the square-wave constant power or field-weakening region. The boundary between the two regions depends on the dc input (or battery) voltage V_d, which may vary widely for EV. With higher V_d ($V_{d2} > V_{d1}$), the limit of V_s increases proportionally, thus increasing the maximum power output proportionally. The stator flux of the machine can be programmed as a function of torque for efficiency improvement at light load, which is shown in Figure 9.25. The light load efficiency optimization principle was discussed for the induction motor drive in Chapter 8. Since core loss is a function of flux as well as frequency, a family of these curves with torque and frequency will give improved efficiency. The boundary between the PWM and SW modes in Figure 9.24 is slanting because of the flux programming effect. The flux-speed relation on this curve can be given as

$$\psi_s = \frac{V_s}{\omega_r} = \omega_b \psi_{sr} = \omega_e \psi_f \qquad (9.60)$$

where ψ_s = stator flux, ψ_{sr} = stator flux at rated torque T_{er}, ψ_f = magnet flux, ω_b = base speed, ω_r = speed, and ω_c = crossover speed.

The phasor diagram of the machine, which is valid in the steady-state condition, is shown in Figure 9.26. For forward rotation of the motor, the phasor diagram can be considered as rotat-

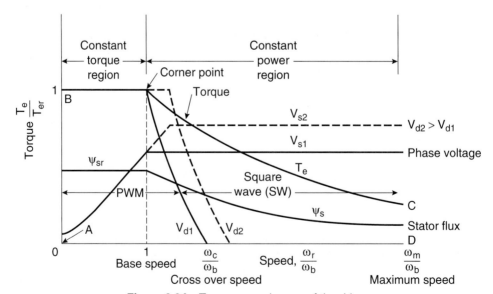

Figure 9.24 Torque-speed curve of the drive

Sinusoidal IPM Machine Drives

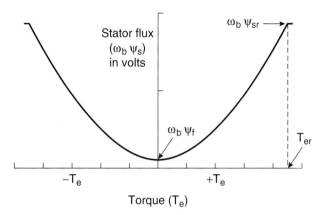

Figure 9.25 Stator flux ($\omega_b \psi_s$) program with torque for efficiency improvement

ing in the anti-clockwise direction at synchronous speed ω_e with respect to the stationary d^s axis, and at any instant, the angle between the magnet axis d^e and the d^s axis is $\theta_e = \omega_e t$. The armature reaction flux phasor ψ_a is given by the following magnitude

$$\psi_a = \frac{\hat{\psi}_a}{\sqrt{2}} = \frac{1}{\sqrt{2}}\sqrt{\psi_{ds}'^2 + \psi_{qs}'^2} \qquad (9.61)$$

or

$$\psi_a = \frac{1}{\sqrt{2}}\sqrt{(L_{ds}i_{ds})^2 + (L_{qs}i_{qs})^2} \qquad (9.62)$$

This adds with the magnet flux ψ_f to constitute the stator flux ψ_s, which is aligned at an angle δ with the d^e axis as shown. The stator phase voltage V_s and speed EMF V_f, given by $\omega_e \psi_s$ and $\omega_e \psi_f$, respectively, are perpendicular to the respective flux components. A new set of $d^{e'}$-$q^{e'}$ axes, which align with ψ_s and V_s, respectively, are shown in the figure. The line current I_s is shown at an arbitrary lagging power factor angle φ. The current I_s can be resolved as the i_{ds}-i_{qs} pair on the d^e-q^e axes or the I_M-I_T pair on the $d^{e'}$-$q^{e'}$ axes. The in-phase or torque component of current I_T contributes to the active input power; therefore, it controls the torque. The reactive or magnetizing component of current I_M, which is aligned with the stator flux, can control ψ_s. However, ψ_s is a function of both I_T and I_M currents. The torque angle δ is shown as positive in the motoring mode. The developed torque expression (9.18) can be expressed with these new axes as

$$T_e = \frac{3}{2}(\frac{P}{2})(\psi_{ds}^{e'} I_T + \psi_{qs}^{e'} I_M) \qquad (9.63)$$

or

$$T_e = \frac{3}{2}(\frac{P}{2})\hat{\psi}_s I_T \qquad (9.64)$$

because $\psi_{qs}^{e'} = 0$ and $\psi_{ds}^{e'} = \hat{\psi}_s$. For negative torque, the angle δ and the active current I_T become negative and the flux triangle is shifted in the fourth quadrant, as shown in the figure.

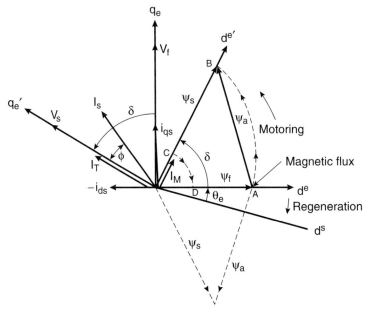

Figure 9.26 Phasor diagram of IPM machine

In the constant torque region, the torque can be controlled by current I_T, whereas the flux ψ_s can be maintained at a value determined by the ψ_s program in Figure 9.25. The dotted curve AB in Figure 9.26 is the locus of phasor ψ_s when the torque is increased from zero to the rated value in the constant torque region (see Figure 9.24). At point A, the torque $T_e = 0$, $I_T = 0$, and therefore, $\psi_s = \psi_f$. As T_e increases, ψ_s increases, demanding an increase of I_T and I_M until the rated torque is reached.

Figure 9.27 shows the current program, which gives the functional relation between I_M and I_T. It is interesting to note that in a wound-field synchronous machine, the flux ψ_s is controlled by field current, and therefore, the machine can always be operated at unity power factor. In an IPM machine, on the other hand, the flux is controlled by the stator-fed lagging current I_M; therefore, the machine operates at a lagging power factor. At the edge of the constant torque region, the control of I_T and I_M, and correspondingly the vector control, is lost because of saturation of the current controller. Then, the machine enters into the SW constant power region. Of course, ψ_s can be further weakened as a function of speed in the constant power region to permit current control, which will restore vector control in this region.

Figure 9.28 shows a simplified control block diagram with a stator flux-oriented vector control. Basically, it is a close loop torque control system for traction-type applications, as mentioned before. The position and speed control loops can be used, if desired. The figure includes some switches for transitioning to square-wave (SW) mode, which will be explained later. The feedback signal processing block receives the machine stator currents i_a and i_b, the

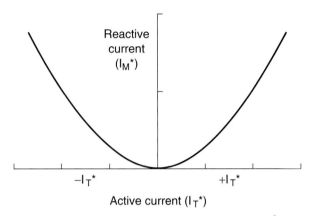

Figure 9.27 Current program showing relation between reactive (I_M^*) and active (I_T^*) current

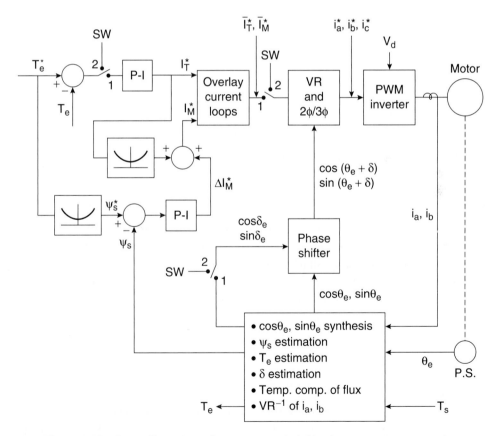

Figure 9.28 Stator flux-oriented vector control of drive in constant torque region

rotor position angle θ_e in the form of a ramp (0 – 360° electrical) and the stator temperature T_s. It then does the following computations:

- $\sin \theta_e$ and $\cos \theta_e$ synthesis
- Stator flux ψ_s estimation
- Inverse vector rotation of stationary frame currents i_a and i_b
- Temperature compensation of *NdFeB* magnet flux ψ_f
- Torque Te estimation
- Torque angle δ estimation

This block will be explained later. In Figure 9.28, the command torque T_e^* is compared with the feedback torque, and the resulting error generates the active current command I_T^* through a P-I compensator. The reactive or magnetizing current command I_M^* is then generated from I_T^* through the feedforward current program shown in Figure 9.27. The control system has a flux control loop, where the command flux ψ_s^* is generated from the command torque through the flux program shown in Figure 9.25. The flux loop generates an incremental magnetizing current ΔI_M^* to supplement the current program output, and helps maintain the desired ψ_s irrespective of the machine's parameter variation effect. The flux program can be replaced by constant ψ_{sr}^* if constant rated flux operation is desired.

The command I_T^* and I_M^* signals are then processed through the high-gain overlay current control loops, VR, and $2\varphi/3\varphi$ transformation to generate the three-phase stator current commands as shown in the figure. The overlay current control and vector rotation are shown in detail in Figure 9.29. The redundant current control loops, in addition to the vector control, permit the following two functions: (1) simple hysteresis-band (HB) current control of the inverter (although it is not harmonically optimum), and (2) partial saturation of the HB current controller (overmodulation), which helps smooth the transition to the SW mode, which will be described later.

Assume for the present that all the feedback signals shown are available from the feedback signal processing block of Figure 9.28. The d^e-q^e axes currents i_{ds} and i_{qs} can be converted to I_M and I_T currents by the following current coordinate shifter equations:

$$I_T = i_{qs} \cos\delta - i_{ds} \sin\delta \tag{9.65}$$

$$I_M = i_{qs} \sin\delta + i_{ds} \cos\delta \tag{9.66}$$

P-I control of the I_T and I_M loops assures tracking of command and actual currents as long as the current controller remains effective. In normal PWM (undermodulation) mode, the operation of the overlay loops is redundant and the loop outputs $I_T^{*'}$ and $I_M^{*'}$ remain equal to the respective command signals. These current signals are then vector rotated so that $I_T^{*'}$ and $I_M^{*'}$

Sinusoidal IPM Machine Drives

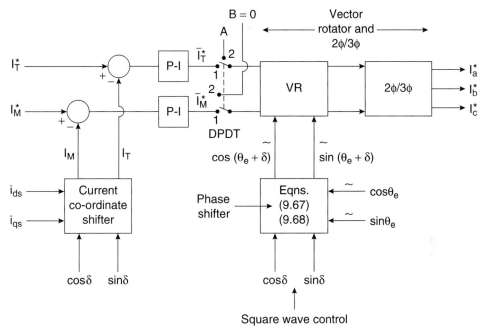

Figure 9.29 Overlay current control strategy with vector rotation

are aligned to the V_s and ψ_s phasors, respectively, as shown in Figure 9.26. The unit vector signals for this phase alignment can be computed in the phase shifter as

$$\cos(\theta_e + \delta) = \cos\theta_e \cos\delta - \sin\theta_e \sin\delta \qquad (9.67)$$

$$\sin(\theta_e + \delta) = \sin\theta_e \cos\delta + \cos\theta_e \sin\delta \qquad (9.68)$$

As the speed increases in the constant torque region, the machine CEMF V_f increases and the current controller enters into the quasi-PWM or overmodulation region. At this condition, the output of the overlay loops becomes higher than the respective command input, while the command and feedback signals match because of P-I control. As the speed increases further, the number of chops per cycle in the current controller decreases, and eventually in SW mode, the current controller saturates completely and the overlay loops become ineffective. Note that instead of overlay current control loops, the standard synchronous current control with voltage PWM can also be used.

9.4.3.1 Feedback Signal Processing

$\cos\theta_e - \sin\theta_e$—The synthesis of unit vector signals $\cos\theta_e$ and $\sin\theta_e$, with the help of a microcomputer lookup table, is simple and will not be discussed further.

Stator Flux Estimation—The accurate estimation of ψ_s is difficult because of inductance saturation with rotor saliency, which creates complex cross-coupling effects between the direct and quadrature flux axes. Extensive modeling and simulation substantiated by laboratory experimentation suggest [7] the following types of flux equations:

$$V_q' = \omega_b \psi_{qs} = \omega_b i_{qs} L_{qs} \tag{9.69}$$

where

$$L_{qs} = L_{qs0} - \frac{i_{ds}}{B} - \frac{i_{qs}}{C} - \frac{i_{ds}^2}{D} - \frac{i_{qs}^2}{E} - \frac{i_{ds} i_{qs}}{F} \tag{9.70}$$

and

$$V_d' = \omega_b \psi_f + \omega_b L_{ds} i_{ds} \tag{9.71}$$

where $\omega_b L_{ds}$ = constant and

$$\omega_b \psi_f = \omega_b \psi_{f0} + \frac{i_{ds}}{H} + \frac{i_{qs}}{I} - \frac{i_{ds}^2}{J} + \frac{i_{qs}^2}{K} + \frac{i_{ds} i_{qs}}{L} \tag{9.72}$$

All the constants in the above equations depend on the particular machine used. The voltages V_q' and V_d' represent the corresponding flux components at base speed ω_b. The above equations indicate that parameters L_{qs} and ψ_f are functions of i_{ds} and i_{qs}, but L_{ds} essentially remains constant.

VR^{-1} of i_a and i_b—The currents i_{ds} and i_{qs} can be derived from the phase currents by $3\varphi/2\varphi$ conversion and inverse vector rotation, as discussed before. The expressions are

$$i_{ds} = i_a \sin\theta_e - \frac{1}{\sqrt{3}}(i_a + 2i_b)\cos\theta_e \tag{9.73}$$

$$i_{qs} = i_a \cos\theta_e + \frac{1}{\sqrt{3}}(i_a + 2i_b)\sin\theta_e \tag{9.74}$$

Temperature Compensation of Flux—Both the V_q' and V_d' flux equations require compensation for machine rotor temperature variation because the high-energy *NdFeB* magnet has approximately 0.1%/°C negative temperature coefficient. With the rise in rotor temperature, the magnet flux ψ_f decreases, which correspondingly reduces saturation of the q^s axis flux, thereby increasing its magnitude. The compensated flux equations can be given in the form

$$V_d = V_d' + K_d(75° - T_R) \tag{9.75}$$

$$V_q = V_q'\left[1 - K_q(75° - T_R)\right] \tag{9.76}$$

where T_R = rotor temperature (°C), K_d, K_q = constant coefficients, and 75 °C is taken as the reference temperature. This means that if T_R = 75 °C, the flux equations, (9.69) and (9.71), do not require any compensation. For higher T_R, V_d decreases, but V_q increases, and vise versa.

Unfortunately, Equations (9.75) and (9.76) require rotor temperature information, which is difficult to measure or estimate. Since rotor dissipation is negligible, temperature T_R can be estimated roughly from stator temperature information with the help of a dynamic thermal model of the machine, which is quite involved. A simplified first-order thermal model can be defined in the following difference equation form:

$$T_R(n) = T_R(n-1) + \frac{1}{\tau}[T_s(n) - T_R(n-1)] \qquad (9.77)$$

where $T_R(n)$, $T_R(n-1)$ = rotor temperature at nth and $(n-1)$ second, respectively, and τ = approximate thermal time constant. The stator flux with rotor temperature compensation can then be given as

$$\hat{V}_b = \omega_b \hat{\psi}_s = \sqrt{V_d^2 + V_q^2} \qquad (9.78)$$

Torque and δ angle Estimation—Once the d^e and q^e axes fluxes and currents have been estimated, the torque can be computed by Equation (2.182) as

$$T_e = \frac{3}{2}(\frac{P}{2})(\psi_{ds} i_{qs} - \psi_{qs} i_{ds}) \qquad (2.114)$$

or

$$T_e = \frac{3}{2}(\frac{P}{2})\frac{1}{\omega_b}(V_d i_{qs} - V_q i_{ds}) \qquad (9.79)$$

The $\sin \delta$ and $\cos \delta$ signals can then be calculated as

$$\sin \delta = \frac{V_q}{\hat{V}_b} \qquad (9.80)$$

$$\cos \delta = \frac{V_d}{\hat{V}_b} \qquad (9.81)$$

9.4.3.2 Square-Wave (SW) Mode Field-Weakening Control

An SW mode field-weakening control can have the advantage of control simplicity and some efficiency improvement because of the reduced switching loss of the inverter. However, the disadvantages are non-optimum transient response and complexity of transitioning between the vector control PWM mode and SW mode. The basic idea for SW torque control is to orient the axis of voltage V_s (q_e' axis) (see Figure 9.26) at the desired δ angle with respect to the V_f (q^e) axis in feedforward manner. Figure 9.30 shows the simplified control block diagram in SW mode.

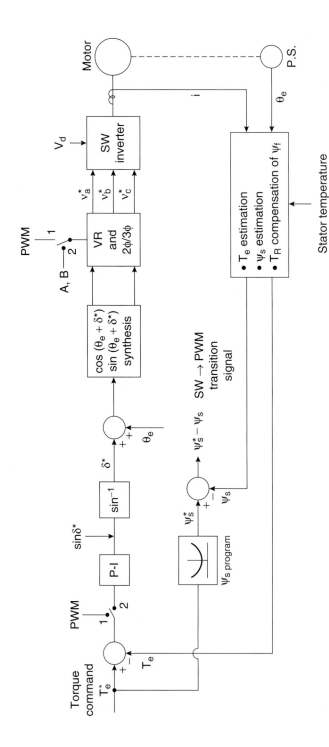

Figure 9.30 SW mode control block diagram in constant power (field-weakening) region

Sinusoidal IPM Machine Drives

The switches shown in the figure indicate the control structure change from PWM mode (1 – PWM mode, 2 – SW mode). The torque control loop generates the sin δ^* command through the P-I compensator because of the following torque relation:

$$T_e = 3\left(\frac{P}{2}\right)\left[\frac{\psi_s \psi_f}{L_{ds}}\sin\delta + \psi_s^2 \frac{(L_{ds} - L_{qs})}{2L_{ds}L_{qs}}\sin 2\delta\right] \tag{2.177}$$

The sin δ^* signal is then converted to angle δ^* using the $\sin^{-1}$ lookup table. The δ^* angle is then added with the rotor angle θ_e, as shown. The resulting angle $(\theta_e + \delta^*)$ is then converted to $\cos(\theta_e + \delta^*)$ and $\sin(\theta_e + \delta^*)$ unit vector components. The δ^* angle control is implemented with the help of a VR in Figure 9.29, after switching the position of the DPDT (double-pole double-throw) switch to position 2 and considering $\cos(\theta_e + \delta^*)$ and $\sin(\theta_e + \delta^*)$ as the control signals. Figure 9.31 explains in detail the phase shift angle control principle for phase a with the help of waveforms. The θ_e angle ramp from the position sensor is phase-advanced by adding the δ^* angle, and correspondingly, $\cos(\theta_e + \delta^*)$ and $\sin(\theta_e + \delta^*)$ waves are generated by the lookup table. Using VR and $2\varphi/3\varphi$ transformation of input constants A and B and substituting B = 0, we can derive the phase voltage signals as

$$i_{qs}^s = v_{qs}^{s*} = A\cos(\theta_e + \delta^*) \tag{9.82}$$

$$i_{ds}^s = v_{ds}^{s*} = -A\sin(\theta_e + \delta^*) \tag{9.83}$$

$$i_a^* = v_{a0}^* = A\cos(\theta_e + \delta^*) \tag{9.84}$$

$$i_b^* = v_{b0}^* = A\cos(\theta_e + \delta^* - \frac{2\pi}{3}) \tag{9.85}$$

$$i_c^* = v_{c0}^* = A\cos(\theta_e + \delta^* + \frac{2\pi}{3}) \tag{9.86}$$

Note that constant A is very large, and therefore, the theoretical phase current commands i_a^*, i_b^*, and i_c^* should be replaced by the phase voltage commands v_{a0}^*, v_{b0}^*, and v_{c0}^*, respectively, because the steep sides of these control waves will force the HB current controller to switch at the edge of each half-cycle and thus generate SW phase voltages. In Figure 9.31, it is evident that the fundamental component of the v_{a0} square wave is leading at angle $(\pi/2 + \delta)$ with respect to the ψ_f wave (see Figure 9.26). Other phase voltages are generated at the $2\pi/3$ phase shift angle with respect to the v_{a0} wave. Evidently, the polarity of torque reverses by the reversal of the δ angle. As speed increases in the constant power region with the limit torque (Figure 9.24) to point C, the stator flux ψ_s decreases correspondingly with constant δ angle. Finally, if the torque decreases to zero (point D) at the highest speed, the ψ_s phasor becomes oriented to the d^e axis. It can easily be seen from the phasor diagram of Figure 9.26 that the machine operates at

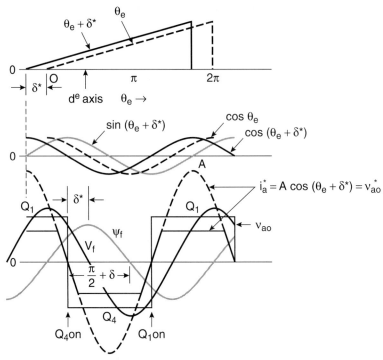

Figure 9.31 Implementation of square-wave control through VR and HB current controller

low leading power factor at high-speed, light load condition, thus reducing the converter-machine efficiency. If the torque is zero in the constant power region, ideally the machine operates at zero leading power factor.

9.4.3.3 PWM — Square-Wave Sequencing

The PWM and SW modes of operation, as described above, can be combined through a fast and smooth transitioning algorithm. Figure 9.32 shows the transition sequence diagram, indicating all the conditionals and actions, and Figure 9.33 indicates its operation on the torque-speed curve. The transition conditional from PWM to SW mode is dictated by the IGBT gate drive pulse count N over two fundamental frequency cycles. In the PWM controller, count N decreases as the converter approaches the SW mode through the quasi-PWM (overmodulation) region. For a specified minimum count of N, transition occurs at SW mode. The torque-speed curve in Figure 9.33 indicates two boundary lines where PWM $\rightarrow$ SW and SW $\rightarrow$ PWM transitions occur. These lines are separated by a hysteresis band and lie below the nominal boundary line. The voltage jump (DE) for the PWM-to-SW transition and the corresponding flux increase (de) are shown in the figure. The SW torque control loop is activated after initializing the P-I output with estimated $\sin \delta$ from the feedback signals and transferring the switch to point 2 $I_T^* = A$, $I_M^* = 0$ in Figure 9.29.

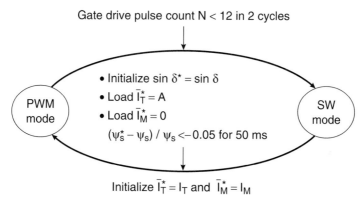

Figure 9.32 Sequence diagram between PWM vector control and SW modes

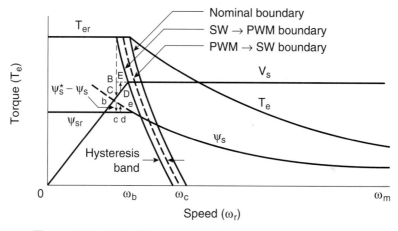

Figure 9.33 PWM/SW mode transitions on torque-speed curve

The criterion for the SW-to-PWM mode transition is dictated by the flux loop error $\psi_s^* - \psi_s$ as indicated in Figure 9.30. As speed decreases in the SW mode and crosses the SW → PWM boundary line in Figure 9.33, transition occurs in the PWM mode, causing the voltage jump (BC) and corresponding stator flux jump (bc), respectively, in Figure 9.33. Before activating the PWM mode, the overlay current loop P-I compensators are initialized by the respective estimated I_T and I_M values.

9.5 TRAPEZOIDAL SPM MACHINE DRIVES

9.5.1 Drive Operation with Inverter

It was mentioned before that a trapezoidal PM machine is basically a surface magnet nonsalient pole machine that induces three-phase trapezoidal voltage waves at the machine terminal due to concentrated full-pitch windings in the stator. Between sinusoidal and trapezoidal PM machines with self-control, the latter gives performance closer to that of a dc motor. For this reason, it is widely known as a brushless dc motor (BLDM or BLDC). Basically, it is an electronic motor and requires a three-phase inverter in the front end, as shown in Figure 9.34. The machine is represented in the figure by a three-phase equivalent circuit, where each phase consists of stator resistance R_s, equivalent self inductance L_s, and a trapezoidal CEMF wave in series. In self-control mode, the inverter acts like an electronic commutator that receives switching logical pulses from the absolute position sensor. The drive is also commonly known as an electronically commutated motor (ECM). Basically, the inverter can operate in the following two modes:

- $2\pi/3$ angle switch-on mode
- Voltage and current control PWM mode

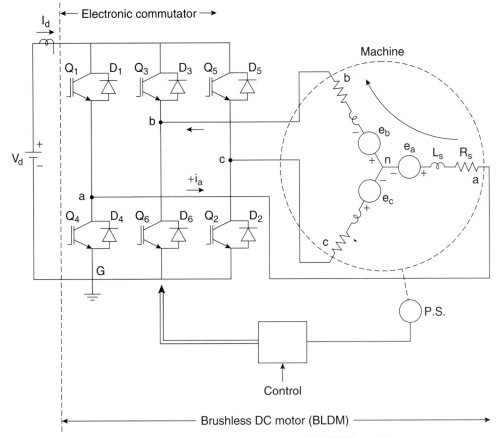

Figure 9.34 Converter circuit with trapezoidal SPM machine

9.5.1.1 2π/3 Angle Switch-on Mode

Inverter operation in this mode is explained with the help of the waveforms shown in Figure 9.35. The six switches of the inverter ($Q_1 - Q_6$) operate in such a way so as to place the input dc current I_d symmetrically for the $2\pi/3$ angle at the center of each phase voltage wave. The angle α shown is the advance angle of current wave with respect to voltage wave. In this case, α is zero. It can be seen that at any instant, two switches are on, one in the upper group and another in the lower group. For example, from instant t_1, Q_1, and Q_6 are on when the supply voltage V_d and line current I_d are placed across line ab (phase a and phase b in series) so that I_d is positive in phase a, but negative in phase b. Then, after $\pi/3$ interval (the middle of phase a), Q_6 is turned off and Q_2 is turned on, but Q_1 continues conduction for the full $2\pi/3$ angle. This switching commutates $-I_d$ from phase b to phase c while phase a continues to carry $+I_d$ as shown. The conduction pattern changes every $\pi/3$ angle, indicating six switching modes in a full cycle. The absolute position sensor dictates the switching or commutation of the devices at the precise instants of the waves. It can easily be seen from Figure 9.34 that at any instant, two phase CEMFs ($2V_c$) appear in series across the inverter input (neglecting R_s and L_s drops). The power flow to the machine at any

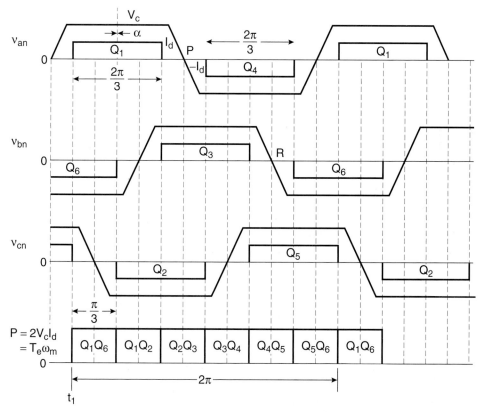

Figure 9.35 Stator phase voltages and current waves indicating the converter conducting devices

instant is ideally constant and is given by $P = 2V_c I_d$, which is indicated at the bottom of Figure 9.35. The inverter is basically operating as a rotor position-sensitive electronic commutator (similar to a mechanical commutator in a dc machine). Therefore, although it is an ac drive with an inverter and synchronous motor, dc machine-like behavior (without brushes and commutators) makes it popularly known as a brushless dc motor (BLDM). Note that instead of a constant dc voltage source V_d, the supply can be a variable voltage or current source.

9.5.1.2 PWM Voltage and Current Control Mode

In the previous discussion, the inverter switches were controlled to give commutator function only when the devices were sequentially on for $2\pi/3$-angle duration. In addition to the commutator function, it is possible to control the switches in PWM chopping mode (buck converter) for controlling voltage and current continuously at the machine terminal. Figure 9.36 shows the waveforms with chopping mode, current-controlled operation of the inverter. There are essentially two chopping modes: feedback (FB) mode and freewheeling (FW) mode. In both these modes, the devices are turned on and off on a duty cycle basis to control the machine average current I_{av} and the corresponding average voltage V_{av}. Consider, for example, the FB mode for the $\pi/3$ interval from instant t_1 when Q_1 and Q_6 are chopping together. When $Q_1 Q_6$ are on, phase a and phase b currents will increase ($V_d > 2V_c$). However, when the devices are turned off, the current will decrease because of feedback through diodes $D_3 D_4$, and the average machine terminal voltage V_{av} will be determined by the duty ratio. The average phase voltage is given as $V_{av} = V_c + I_{av} R_s$. A single dc link current sensor will sense the average machine current because machine current is always flowing in the line.

In FW mode, chopping control is done with only one device. For example, in Figure 9.34, all the upper devices (Q_1, Q_3, and Q_5) are kept on sequentially in the middle of the respective positive voltage half cycles. In the $\pi/3$ interval following t_1, device Q_6 is chopping. When Q_6 is on, V_d is applied across ab and the current increases. When Q_6 is turned off, decaying freewheeling current (due to CEMF) flows through Q_1 and D_3, short-circuiting the motor terminals. In Figure 9.36, six-step current waves are shown with rise and fall times at the leading and trailing edges, respectively, because in the practical case, commutation takes a finite amount of time. Note that the motoring mode (buck converter) shown in Figure 9.36 can be converted to the regenerative mode (boost converter) by shifting the current waves by π angle (i.e., $\alpha = \pi$).

The PWM mode is useful for starting a BLDM for the $2\pi/3$ angle switch-on mode with initial current limit control. In this mode, the machine is gradually brought to full speed by current-limited starting, and then transferred to the $2\pi/3$ switch-on mode. This is like the starter operation of a conventional dc motor. This mode also permits continuous speed control of a BLDM, which will be described next.

9.5.2 Torque-Speed Curve

Steady-state dc machine-like torque-speed characteristics of a BLDM can easily be derived by first considering the machine's operation in the $2\pi/3$ angle switch-on mode. Neglecting the

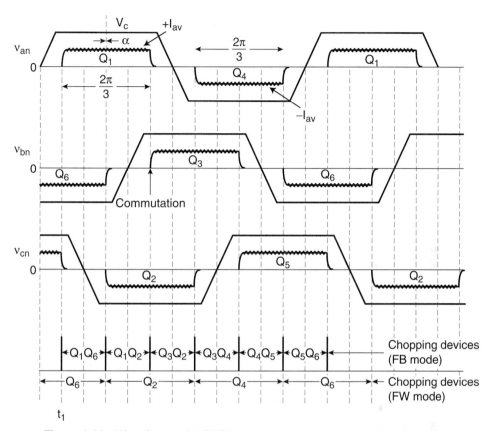

Figure 9.36 Waveforms with PWM current control mode operation of inverter

losses, the power input and developed torque are related by the following expression (see Figure 9.35):

$$P = e_a i_a + e_b i_b + e_c i_c = 2I_d V_C = T_e \omega_m \qquad (9.87)$$

or

$$T_e = \frac{2I_d V_C}{\omega_m} = \frac{P I_d V_C}{\omega_r} \qquad (9.88)$$

where V_c = phase CEMF, I_d = dc line current, $\omega_m = \omega_r(2/P)$ speed in mechanical r/s, and P = number of poles. The CEMF is proportional to speed, that is,

$$V_C = K\omega_r \qquad (9.89)$$

The dc circuit equation is

$$V_d = 2R_s I_d + 2V_C \qquad (9.90)$$

Substituting Equation (9.89) in (9.88), we get

$$T_e = K_1 I_d \qquad (9.91)$$

where $K_1 = K \cdot P$.

Define the base torque T_{eb} as

$$T_{eb} = K_1 I_d \big|_{I_d = I_{sc}} = \frac{K_1 V_d}{2R_s} \qquad (9.92)$$

where $I_{sc} = V_d/2R_s$ is the short-circuit current derived from Equation (9.90). Similarly, define the base speed as

$$\omega_{rb} = \omega_r \big|_{I_d = 0} = \frac{V_d}{2K} \qquad (9.93)$$

which is derived from Equations (9.89) and (9.90). Therefore, the torque-speed relation in terms of ω_{rb} and T_{eb} can be derived from Equations (9.90), (9.92), and (9.93) as

$$\omega_r = \frac{V_d - 2R_s I_d}{2K} = \frac{V_d}{2K} - \left(\frac{V_d}{2K}\right) \frac{K_1 I_d}{\left(\frac{K_1 V_d}{2R_s}\right)}$$

$$= \omega_{rb}\left[1 - \frac{T_e}{T_{eb}}\right] \qquad (9.94)$$

or

$$T_e(pu) = 1 - \omega_r(pu) \qquad (9.95)$$

where $T_e(pu) = T_e/T_{eb}$ and $\omega_r(pu) = \omega_r/\omega_{rb}$.

This normalized dc machine-like torque-speed relation has been plotted in Figure 9.37 for variable dc voltage V_d. The no-load speed can be controlled by varying V_{av} by chopper action of the inverter. The no-load speed droops linearly with load due to the stator resistance R_s drop.

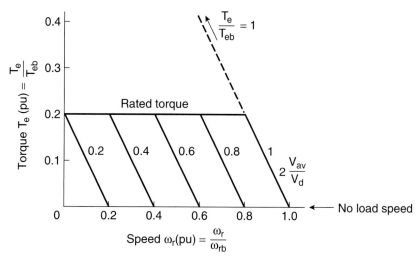

Figure 9.37 Normalized torque-speed curves showing drooping speed with load

9.5.3 Machine Dynamic Model

The dynamic machine model is very simple compared to that of a sinusoidal machine. It can be given in the following general form [11]:

$$\begin{bmatrix} v_{an} \\ v_{bn} \\ v_{cn} \end{bmatrix} = \begin{bmatrix} R_s & 0 & 0 \\ 0 & R_s & 0 \\ 0 & 0 & R_s \end{bmatrix} \begin{bmatrix} i_a \\ i_b \\ i_c \end{bmatrix} + S \begin{bmatrix} L & M & M \\ M & L & M \\ M & M & L \end{bmatrix} \begin{bmatrix} i_a \\ i_b \\ i_c \end{bmatrix} + \begin{bmatrix} e_a \\ e_b \\ e_c \end{bmatrix} \quad (9.96)$$

where L = self inductance, M = mutual inductance between phases, e = phase CEMF, v = applied phase voltage, and S = Laplace operator. Since

$$i_a + i_b + i_c = 0 \quad (9.97)$$

that is,

$$Mi_b + Mi_c = -Mi_a \quad (9.98)$$

Substituting Equation (9.98) in (9.96), the state variable form of the equation is

$$L_s \begin{bmatrix} \dfrac{di_a}{dt} \\ \dfrac{di_b}{dt} \\ \dfrac{di_c}{dt} \end{bmatrix} = \begin{bmatrix} 1 & 0 & 0 \\ 0 & 1 & 0 \\ 0 & 0 & 1 \end{bmatrix} \begin{bmatrix} v_{an} \\ v_{bn} \\ v_{cn} \end{bmatrix} - \begin{bmatrix} R_s & 0 & 0 \\ 0 & R_s & 0 \\ 0 & 0 & R_s \end{bmatrix} \begin{bmatrix} i_a \\ i_b \\ i_c \end{bmatrix} - \begin{bmatrix} e_a \\ e_b \\ e_c \end{bmatrix} \quad (9.99)$$

where $L_s = L - M$ is the equivalent self-inductance per phase. The torque and mechanical equations are

$$T_e = \left(\frac{P}{2}\right) \frac{e_a i_a + e_b i_b + e_c i_c}{\omega_r} \quad (9.100)$$

$$J\left(\frac{2}{P}\right) \frac{d\omega_r}{dt} = T_e - T_L \quad (9.101)$$

Equations (9.99), (9.100), and (9.101) describe the complete dynamic model of the machine and can be used for computer simulation study.

9.5.4 Drive Control

9.5.4.1 Close Loop Speed Control in Feedback Mode

Figure 9.38 shows a close loop speed control system of a BLDM drive with the inverter operating in PWM feedback mode. An absolute position sensor, usually a set of three low-cost Hall sensors (see Figure 9.39), is placed on the stator side at the edge of the rotor poles so as to generate three $2\pi/3$-angle phase-shifted square waves (see Figure 9.6). These encoder waves are co-phasal to the respective phase voltage waves. A decoder circuit then converts these waves into the six-step waves shown in Figure 9.38. Of course, any encoder described in Figures 9.5–9.7 with suitable signal processing can be used. In the figure, the speed control loop generates the phase current magnitude I_a^* command, which is positive for motoring but negative for regeneration. This signal is then enabled with appropriate polarity to the respective phases with the help of decoder output. The actual phase currents track the command currents by hysteresis-band current control. At any instant, two phase currents are enabled, one with positive polarity and another with negative polarity. Consider, for example, the motoring mode time duration when phase a positive current $+i_a^*$ and phase b negative current $-i_b^*$ commands are enabled by the decoder. Devices Q_1 in phase a and Q_6 in phase b are turned on simultaneously to increase $+i_a$ and $-i_b$, respectively. When the currents (equal in magnitude) tend to exceed the hysteresis band, both the devices are turned off at the same instant to initiate current feedback through the diodes. In isolated neutral load, only two current sensors are sufficient. Again, in feedback mode, since phase currents always flow in the input, only one dc current sensor should be enough. The inverter uses power MOSFETs with 12-volt input. With higher voltage and higher power, IGBTs can be used.

Trapezoidal SPM Machine Drives

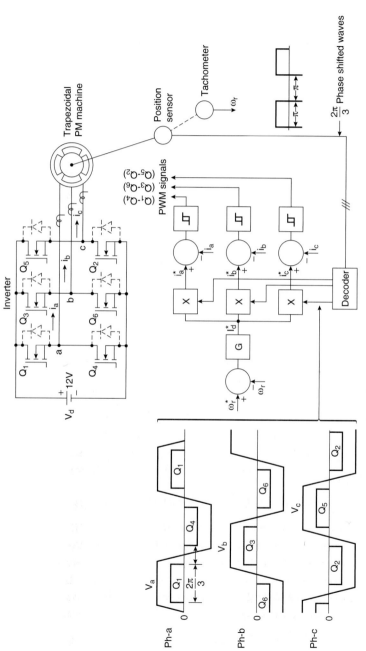

Figure 9.38 Close loop BLDM speed control in PWM feedback mode

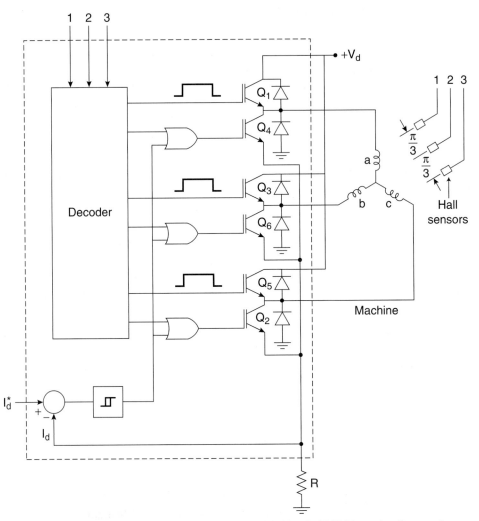

Figure 9.39 Close loop current control of BLDM drive in PWM freewheeling mode

9.5.4.2 Close Loop Current Control in Freewheeling Mode

Figure 9.39 shows a low-cost power-integrated circuit (PIC) for a BLDM, where the phase currents are controlled by the PWM freewheeling mode. Hall sensing and the decoding logic principle were discussed previously. In this case, three upper devices (Q_1, Q_3, and Q_5) of the inverter are turned on sequentially in the middle of the respective positive voltage half-cycles, whereas the lower devices (Q_4, Q_6, and Q_2) are chopped in sequence for $2\pi/3$ angles in the respective negative voltage half-cycles with the help of a decoder for controlling the current I_d^*. If, for example, Q_6 is turned on with Q_1 on, the phase a and phase b currents $+i_a$ and $-i_b$ will build up. When Q_6 is turned off, the decaying freewheeling current will flow through D_3 and Q_1. Only one dc current sensor to the ground, as shown, senses all the phase currents.

9.5.5 Torque Pulsation

The BLDM drive has the advantages of being a simple machine with somewhat higher power density, simple discrete position sensors, and simple control compared to a sinusoidal machine. However, its disadvantage is the pulsating torque problem. A typical developed torque wave is shown in Figure 9.40, which contains a dc or average component and pulsating components [8]. There are essentially three sources of pulsating torque: (1) high frequency component, (2) rounding effect with a period of $\pi/3$ angle, and (3) a commutation transient every $\pi/3$ angle.

The high-frequency pulsating torque component is induced due to PWM control of the inverter that places a ripple current in the phases. This pulsating torque effect is negligible due to inertia filtering of the motor. The rounding effect occurs because the induced phase voltages are somewhat rounded (quasi-sinusoidal because of flux leakage paths to the adjacent magnet poles), which causes rounded power curves (see ideal power curve in Figure 9.35). This is a major source of sixth harmonic torque pulsation. The commutation effect, that is, the current transfer from the outgoing to the incoming phase, occurs every $\pi/3$ angle and takes a finite time, as shown in Figure 9.36. During commutation, a transient torque is developed because the sum of the two commutating currents is rarely constant. Consider the instantaneous torque expression (9.100) during commutation of current from phase b to phase c, which is given as

$$T_e = \frac{e_a i_a + e_b i_b + e_c i_c}{\omega_r}\left(\frac{P}{2}\right) \quad (9.102)$$

Assuming e_a, e_b, and e_c are constant and equal in magnitude and speed ω_r is constant, the pulsating torque will be zero if

$$|i_b + i_c| = i_a = I_{av} \quad (9.103)$$

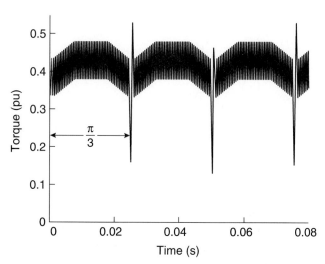

Figure 9.40 Developed torque indicating pulsating torque components

Unfortunately, due to finite phase inductance, the sum of the commutating currents is never constant and this is the reason for the generation of pulsating torque. The increment of current in phase a as a result of phase b-to-phase c commutation can sometimes be fatal. The harmful effect of sixth harmonic torque will be less with a greater number of poles since the pulsation period is $\pi/3$ electrical angle.

9.5.6 Extended Speed Operation

It is possible to control the speed of a BLDM beyond the base speed ω_b like a sinusoidal SPM machine. The six-step PWM current control mode, as discussed above, is possible only when the supply voltage $V_d > 2V_c$, where V_c = phase CEMF. Since V_c is proportional to speed, the PWM current control will be lost at higher speeds, when the controller is fully saturated at $V_d = 2V_c$. At this condition, the phase current amplitude and torque fall off sharply and the inverter goes into voltage source mode. Figure 9.41 shows the typical normalized torque-speed curves of a BLDM in the extended speed region ($\omega_r/\omega_b > 1.0$) [8]. The solid curves indicate the regular six-step current control mode for leading α angle (see Figure 9.36), where $\alpha = 0$ means that the six-step current wave is aligned at the center of the phase voltage wave. The torque curve at $\alpha = 0$ indicates the quick decay of torque at $V_d = 2V_c$ until torque is zero at $\omega_r(pu) = 1.3$. The torque can be extended to a higher speed if the α angle is advanced. This means that the voltage V_d can be switched on for the phase current to build up when V_c has not yet developed sufficiently. However, this initial current falls fast as full CEMF builds up. In this mode of operation, the peak phase current and pulsating torque will be higher at higher speeds. The speed control range can be extended further if $2\pi/3$ conduction mode is gradually reverted to regular π angle conduction mode (shown by dotted curves). In this mode, the inverter operates as a normal square-wave inverter for a sinusoidal machine and three devices conduct at any instant. The

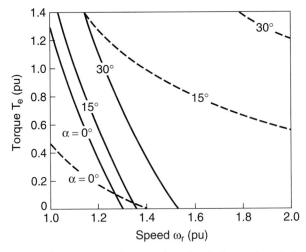

Figure 9.41 Torque-speed curves in extended speed control region

additional advantage in this mode is the reduction of pulsating torque compared to $2\pi/3$ mode of operation. In the absence of any damper winding, this mode of operation can have the problem of hunting at a transient condition.

9.6 WOUND-FIELD SYNCHRONOUS MACHINE DRIVES

9.6.1 Brush and Brushless dc Excitation

Wound-field synchronous machines require dc current excitation in the rotor winding. The field flux ψ_f can be controlled by the field current I_f, which permits machine operation at any desired power factor, leading, lagging, or unity. The traditional method of field excitation consists of a phase-controlled thyristor rectifier that supplies to the field winding through slip rings and brushes. The disadvantages of slip rings and brushes can be avoided by brushless excitation as shown in Figure 9.42. The machine under consideration is shown on the right. A three-phase rotory transformer with a secondary winding on the rotor is mounted on the same shaft. The rotary transformer is nothing but a wound rotor induction motor (WRIM). The primary or stator winding is supplied from 60 Hz supply bus through a thyristor ac phase controller where the voltage V_L' (at same frequency) is controlled by firing angle α. The stator magnetic field of the WRIM rotates at fixed speed ω_e' corresponds to 60 Hz frequency. However, the speed of rotor magnetic field ω_e, which corresponds to the synchronous motor speed, is dictated by the inverter frequency. The slip voltage V in the rotor winding is rectified to dc by the rotor-mounted diode rectifier and supplies dc current I_f in the field winding. Note that the WRIM does not have any slip rings or brushes. The field current I_f is controlled by the firing angle α. As slip S increases at lower machine speeds, the slip voltage also increases, tending to increase I_f, but α angle is retarded to counteract this increase. The brushless excitation adds cost and complexity to the machine, and the transient response of field current is somewhat slow in comparison with brush excitation.

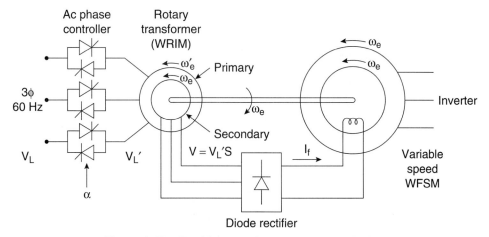

Figure 9.42 Brushless synchronous motor excitation

9.6.2 Load-Commutated Inverter (LCI) Drive

As discussed in Chapter 6, a current-fed thyristor inverter can be load-commutated with a wound-field synchronous machine load by operating the machine at leading power factor, which makes the converter very simple and economical. Load-commutated inverter drives are very popular in multi-MW drives for compressors, pumps, blowers, ship propulsion, etc.

Self-control principle of PM synchronous motor drives with voltage-fed inverters is also valid for load-commutated inverter drives. Figure 9.43 shows a self-controlled thyristor inverter where the inverter firing pulses are derived from the absolute position sensor through a delay control circuit. Figure 9.44 shows the fundamental frequency phasor diagram of the machine with load commutation of the inverter, where the stator current I_s leads the stator voltage V_s by angle φ in motoring mode. For simplicity, the saliency and stator resistance of the machine are neglected. The field flux ψ_f location, which is related to the rotor position, is established by the field current. The magnitude of the stator current I_s, controlled by the front-end rectifier, determines the magnitude of the armature reaction flux $\psi_a = I_s L_s$, and the phasor can be positioned at

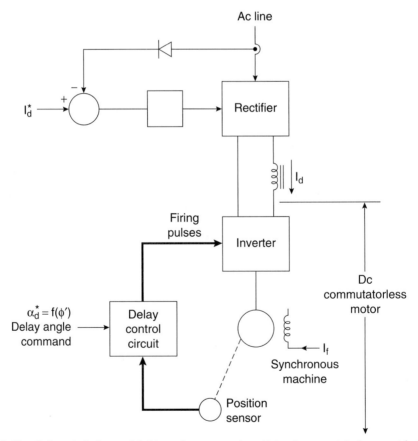

Figure 9.43 Self-control of wound-field synchronous motor with load-commutated, current-fed inverter

Wound-Field Synchronous Machine Drives

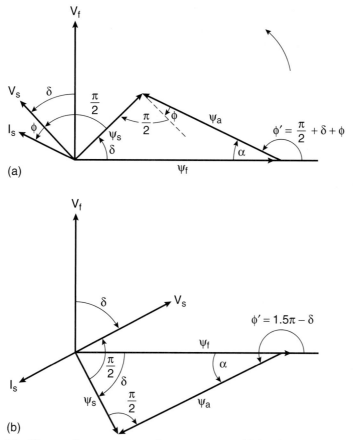

Figure 9.44 Phasor diagrams of synchronous motor with load-commutated inverter, (a) Motoring, (b) Regeneration

a desirable angle by the delay angle command α_d^* shown in Figure 9.43. In Figure 9.44(a), the phasor ψ_a leads ψ_f by the angle φ' so that the resultant stator flux ψ_s induces the stator voltage V_s, which lags at angle φ with respect to the stator current I_s. The angle φ' can be given as

$$\varphi' = \pi - \alpha = \delta + (\frac{\pi}{2} + \varphi) \tag{9.104}$$

where δ = torque angle. Figure 9.44(b) shows the phasor diagram in regenerative mode, where the motor is working as a generator, and the rectifier and inverter reverse their roles. In this mode, the machine-side converter operates as a diode rectifier ($\alpha = \pi$) with V_s and I_s phasors in opposite directions, as shown. The commutation in diode rectifier mode also occurs by CEMFs of the machine.

9.6.2.1 Control of LCI Drive with Constant γ Angle

For reliable operation of a load-commutated inverter and minimum reactive current loading of a machine-converter system, it is desirable to have load commutation at constant turn-off angle γ (see Figure 6.11). A complete speed control system of a synchronous machine with constant γ angle control is shown in Figure 9.45 [13], where the power circuit is the same as in Figure 9.43. The drive is shown to operate in the constant torque region in motoring mode, where the stator flux ψ_s is maintained constant in an open loop manner. The control consists of the following elements:

- Speed and dc link current control
- Delay angle or φ' angle control
- Field flux or field current control
- Generation of ψ_f^*, α_d^*, δ^*, and μ^* command signals

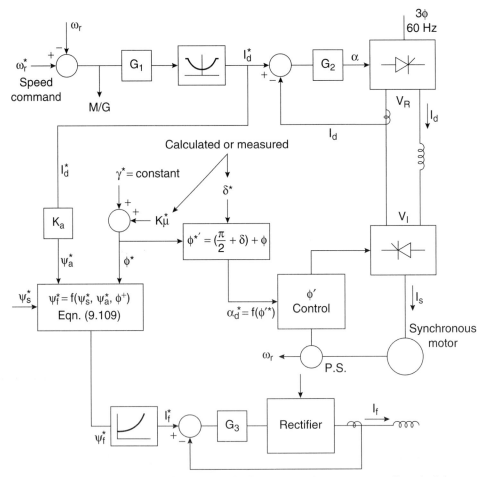

Figure 9.45 Control block diagram of LCI drive with constant turn-off angle (γ)

The speed and current controls are shown in the upper part of the figure, and these are essentially identical to a dc motor drive system with a phase-controlled rectifier. The command speed ω_r^* is compared with the actual speed ω_r and the resulting error through a P-I controller and absolute value circuit generates the current command I_d^*. The I_d control loop controls the rectifier firing angle α. Ideally, in full-speed motoring mode, $\alpha = 0°$, whereas in full-speed regeneration mode, $\alpha = \pi$. It can be shown that the developed torque of the machine is proportional to current I_d. Neglecting losses in the machine, we can equate the mechanical power output with the input power as

$$P_m = T_e \omega_m = T_e \omega_r \left(\frac{2}{P}\right) = 3 V_s I_s \cos\varphi \qquad (9.105)$$

where all are in standard symbols. Therefore, the torque can be expressed as

$$T_e = \frac{P_m}{\omega_m} = 3(\frac{P}{2}) \psi_s I_s \cos\varphi = K I_s \qquad (9.106)$$

where $\psi_s = V_s/\omega_e$, $K = 3(P/2) \psi_s \cos\varphi$, and the φ angle is treated as constant approximately. As indicated in Figure 9.45, $\varphi^* = \gamma^* + k\mu^*$, where μ = commutation overlap angle and k is a fraction. This definition of leading power factor angle φ should be evident from Figure 6.11. For an ideal six-step current wave,

$$I_s = \frac{\sqrt{6}}{\pi} I_d \qquad (9.107)$$

which can be substituted in Equation (9.106) to determine the torque as a function of current I_d as

$$T_e = \frac{3\sqrt{6}}{\pi} (\frac{P}{2}) \psi_s I_d \cos\varphi = K' I_d \qquad (9.108)$$

The two additional control variables in the system are field current I_f and phase angle φ'. These are to be controlled so that the phasor diagrams shown in Figure 9.44 are satisfied at all load and speed conditions. The command current I_f^* is generated from the flux command ψ_f^* through a function generator that corrects the saturation effect. In a brush excitation system, as indicated in the figure, a phase-controlled rectifier can be controlled in a close loop manner to control the current I_f. Considering first the motoring mode, the command ψ_f^* can be calculated from the geometry of the flux triangle by the following relation:

$$\psi_f^* = \sqrt{\psi_s^{*2} + \psi_a^{*2} - 2\psi_s^* \psi_a^* \cos(\frac{\pi}{2} + \varphi^*)} \qquad (9.109)$$

where ψ_s^* = constant, $\psi_a^* = L_s I_s = K_a I_d^*$, and $\varphi^* = \gamma^* + k\mu^*$. The signal ψ_a^* can be easily generated from I_d^* as shown. The signal φ^* requires information about μ, which can be calculated or measured.

The μ angle can be estimated from the equation

$$\mu = \cos^{-1}\left[\cos\gamma^* - \frac{2L_c I_d^*}{\sqrt{6}\psi_s^*}\right] - \gamma^* \qquad (9.110)$$

which is easily derived from the equation

$$\cos\gamma - \cos(\mu + \gamma) = \frac{2\omega_e L_c I_d}{\sqrt{6} V_s} \qquad (3.49)$$

substituting $\psi_s = V_s/\omega_e$.

In the running condition of the drive, the μ angle can be measured more accurately by the scheme shown in Figure 9.46. The voltage across each inverter thyristor (Q_1 to Q_6) is sensed through a proper isolation amplifier and comparator to generate the γ angle signal, as explained on the left for device Q_2. The signals for all the devices flowing in sequence at $\pi/3$ angle are logically ORed and fed to an R-S flip-flop as shown. For the Q_2 device, the flip-flop is set by the Q_4 firing pulse, but is reset at the leading edge of the Q_2 γ angle, giving its μ angle at the output as shown. All six μ angles are measured and the average value is computed for the digital control in Figure 9.45.

The phase angle φ'^*, given by Equation (9.104), can be determined from the δ and φ angles. The δ angle can also be estimated from Equations (2.169) and (9.108) as

$$\delta^* = \sin^{-1}\left[\frac{\sqrt{6}L_s \cos\varphi}{\pi\psi_f} I_d^*\right] \qquad (9.111)$$

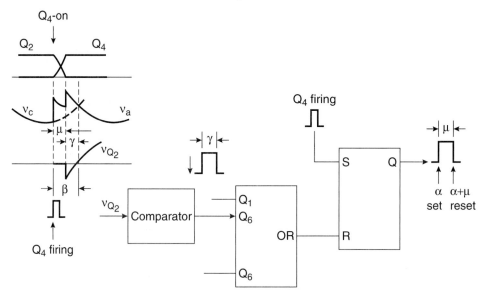

Figure 9.46 Measurement of commutating overlap angle μ

Wound-Field Synchronous Machine Drives

In all these control signal estimations, an extra margin of safety against commutation failure is desirable wherever there is doubt about an estimation's accuracy. Note that with load commutation, the control response is inherently sluggish. Any transient loading of the machine is likely to induce commutation failure.

9.6.2.2 Delay angle α_d or φ' Angle Control

At this point, all the control elements have been discussed except the φ' or delay angle control shown in Figure 9.45. Since a six-step inverter requires six discrete firing pulses at $\pi/3$ intervals within a cycle, the position sensor described in Figure 9.6 is satisfactory. Let us first consider the general delay angle α_d control block diagram shown in Figure 9.47; Figure 9.48 shows the corresponding waves for phase a only in motoring mode. Each channel of the respective phase delay control circuit consists of a programmable DOWN-counter followed by a D flip-flop. The outputs of the flip-flops are combined in a pulse distributor, which generates a $2\pi/3$-angle-wide firing pulse train for each phase group in the correct phase position. Consider, for example, firing pulse generation for phase a only, where the reference signal input is S_1 and

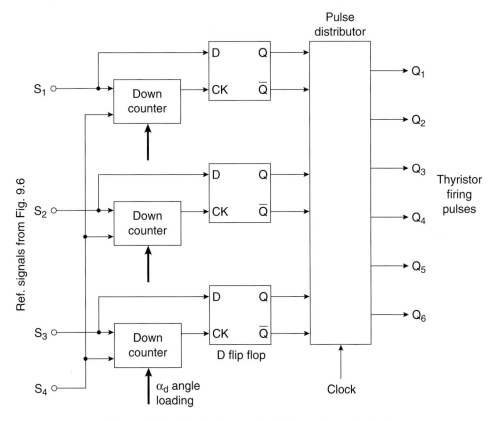

Figure 9.47 Block diagram for delay angle α_d control

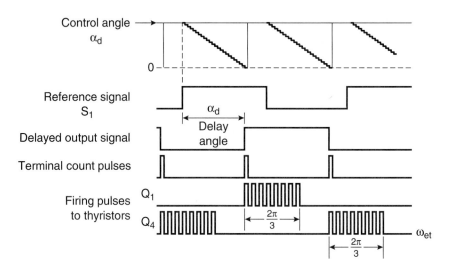

Figure 9.48 Waveforms for delay control in motoring mode (phase a only)

the corresponding waves are shown in Figure 9.48. At the leading edge of the reference signal, the delay angle α_d is loaded in the counter and is decremented with the sensor S_4 pulse train. At the terminal count, a clock pulse is applied to the D flip-flop. The outputs of the D flip-flops are square waves, which are like reference signals but phase-delayed by the α_d angle. These are then logically combined and ANDed with a clock to generate $2\pi/3$-wide thyristor firing pulse trains. Figure 9.48 shows the firing pulses for Q_1 and Q_4, which constitute the phase a group.

The position of the reference signal from S_1 is not necessarily aligned in the same phase as the ψ_f wave. The delay angle α_d is generated with respect to the leading edge of the reference wave to generate firing pulses in both the motoring and regeneration modes as so as to satisfy the φ' angles shown in Figure 9.44. The reference signal from S_1 is shown aligned at lagging angle A with respect to ψ_f in Figure 9.49; Figure 9.50 shows the corresponding waveforms. The α_d, φ', and φ angles are marked in the figure. Note that the fundamental component of current i_{af} lags the S_1 leading edge by angle $\alpha_d - \pi/6$. In regeneration mode, the α_d angle is such that the Q_1 current pulse is opposite in phase with respect to the phase voltage v_a wave as shown in Figure 9.50. The phase b and phase c waves are displaced at $2\pi/3$-angle lagging and leading, respectively.

From Figure 9.50, we can write

$$A + (\alpha_d - \frac{\pi}{6}) = 2\pi - \varphi' \qquad (9.112)$$

or

$$\alpha_d = K - \varphi' \qquad (9.113)$$

where $K = 2\pi + \pi/6 - A$. This equation indicates a linear relation between the α_d and φ' angles. If the φ^* signal is updated in Figure 9.45, the α_d angle can be calculated from Equation (9.113)

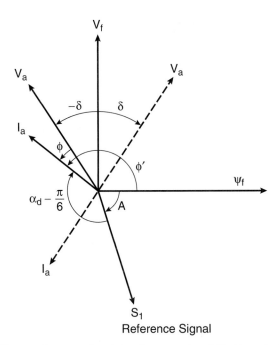

Figure 9.49 Phasor diagram showing reference signal S_1 at angle A (phase a only)

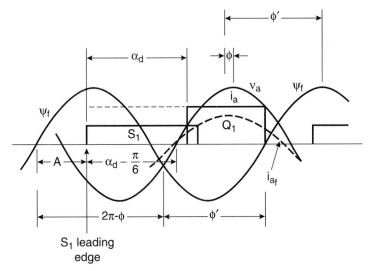

Figure 9.50 Waveforms showing relation between α_d and φ' angles (phase a only) in motoring mode

and inverter firing signals can be generated for both the motoring and regeneration modes with the help of Figure 9.47. In regeneration mode, Equation (9.113) is also valid, except $\varphi^* = 1.5\pi - \delta$, as shown in Figure 9.44(b). The motoring/regeneration logic command can be obtained from the error polarity of the speed loop. If, for example, $A = 80°$, $\varphi = 10°$, and $\delta = 40°$, then in motoring mode; $\varphi' = 140°$ and $\alpha_d = 170°$ (see Figure 9.50), and in regenerating mode; $\varphi' = 230°$ and $\alpha_d = 80°$.

So far, drive operation has been discussed in the constant torque region. As the speed builds up, the dc link voltage V_R increases, ultimately saturating the rectifier and thus losing the current control. Higher range speed control is possible by weakening the flux ψ_s inversely proportional to speed so as to make the current control effective.

9.6.2.3   Control with Machine Terminal Voltage Signals

The operation of a load-commutated inverter drive with a shaft position sensor can be extended down to zero speed, where the initial start-up range of speed (0 to 5%–10% of base speed) can be controlled by the dc link current interruption method discussed in Chapter 6. The position sensor can be eliminated, and instead, machine terminal voltage signals can be sensed for self-control if the speed does not fall below a minimum value. The method of generating machine-side converter firing pulses and load commutation by terminal voltage sensing is analogous to firing pulse generation and line commutation of the line-side converter by line voltage sensing. The machine terminal voltage waves contain large commutation spikes that are difficult to process in analog form. Attempting to filter these transients causes a frequency-sensitive phase delay, which is difficult to compensate. A zero-crossing method of terminal voltage sensing is shown in Figure 9.51.

Machine line voltages v_{ab}, v_{bc}, and v_{ca} are processed through zero-crossing detectors to generate the three square reference waves with $2\pi/3$-angle phase shift as shown. The details of a zero-crossing detector with an optical isolation scheme are shown in Figure 9.51(b). The opto-transistor flip-flop is insensitive to commutation spikes and gives a square-wave output, which is in phase with the input line voltage. Any phase error due to the threshold voltage of the opto-coupler can be compensated as a function of machine speed. The voltage-sensing circuit also provides input to a frequency multiplier whose operating principle is explained in Figure 9.52. As shown in the block diagram, the machine fundamental frequency is multiplied by a factor of 360. A conventional phase-locked loop (PLL) with a limited frequency locking range may not be satisfactory for a wide range of machine frequency variation. The output square waves of the voltage-sensing circuit are passed through the edge detectors to generate a pulse train at $6f_e$ frequency (f_e = machine frequency), which is used to load an 8-bit latch from an UP-counter and then reset the counter with a delay as shown. The counter is clocked by the frequency $f_c/360$, obtained from a crystal clock through a frequency divider. The latch loads the digital word $W = (1/6 f_e) \cdot (f_c/360)$ to a programmable DOWN-counter, which is clocked by the frequency $f_c/6$. This results in an output pulse train at frequency $f_0 = 360 f_e$. All the square reference waves V_1, V_2, and V_3 in Figure 9.51 are fed to S_1, S_2, and S_3 channels, respectively, and the f_0 signal is fed to S_4 in the delay angle control block diagram, Figure 9.47, to generate the inverter firing pulses. The explanatory waveforms

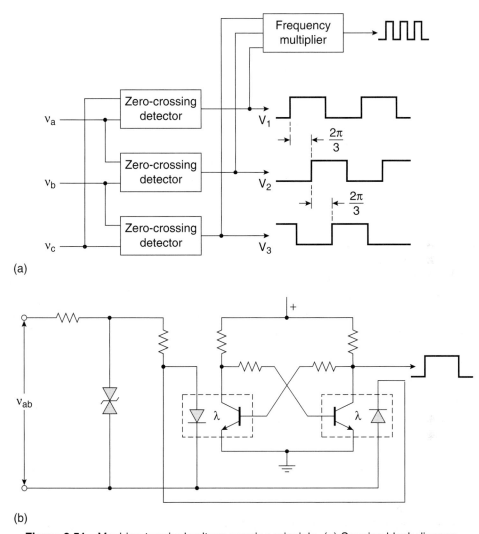

Figure 9.51 Machine terminal voltage sensing principle, (a) Sensing block diagram, (b) Zero-crossing detector with optical isolation

of one channel (S_1) for firing at α angle are shown in Figure 9.53, which indicates that the incoming thyristors are fired at advance angle β with reference to the respective line voltage wave. The angle α^* command can be generated from the relation

$$\alpha^* = \pi - \beta = \pi - (\gamma^* + \mu) \tag{9.114}$$

where γ = constant turn-off angle and μ = overlap angle, which is measured or calculated. The α angle can be easily controlled for the drive to operate in both the motoring and regeneration modes.

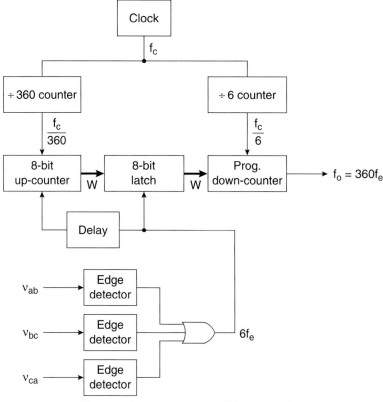

Figure 9.52 Frequency multiplier principle

9.6.2.4 Phase-Locked Loop (PLL) γ Angle Control

The PLL method of rectifier control discussed in Chapter 3 can be extended to a load-commutated inverter drive. Figure 9.54 shows the control block diagram for constant γ angle control by PLL. The desired γ^* angle for safe commutation is compared with the measured γ angle as shown, and the error through a P-I controller generates the input dc voltage for the voltage-controlled oscillator (VCO). The VCO supplies the pulse train to a ring counter, which through a pulse distributor, generates the firing pulses for the inverter. The γ angle is available from the six thyristors at $\pi/3$ intervals. If the actual γ angle tends to decrease by loading the machine, the VCO input voltage will tend to increase, which will advance firing angle α until the desired γ angle is restored. The speed and flux loops are independent, as shown. As more speed is demanded, the torque is increased by increasing dc current I_d. The stator flux ψ_s is controlled by a feedback loop by controlling the field current I_f. All three control loops operating concurrently satisfy the phasor diagrams in Figure 9.44. In regeneration mode detected by the error polarity of speed loop, the γ angle is advanced to near π angle. In field-weakening mode, the flux ψ_s^* is reduced inversely with speed so that the γ angle can be maintained constant. The

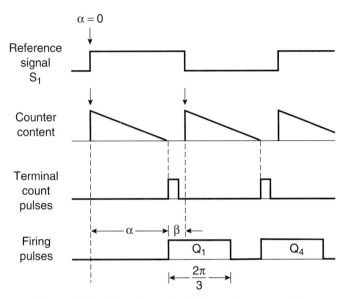

Figure 9.53 Waveforms for firing angle α generation

PLL method of control is simpler, but the disadvantage is that the transient response is further slowed because of the sluggish response of the PLL.

With a constant γ angle, the available turn-off time $t_{off} = \gamma/\omega_e$ for thyristors increases as the frequency decreases, giving an unnecessarily large leading power factor angle. Instead of the turn-off angle, the turn-off time t_{off} can be maintained constant by the PLL method.

9.6.3 Scalar Control of Cycloconverter Drive

High-power, wound-field synchronous motors can be operated at unity power factor when excited by phase-controlled thyristor cycloconverters. In this case, the thyristors are commutated by the line (line commutation). Cycloconverter drives are very popular for cement mills, mine hoists, rolling mills, etc. Cycloconverter drive control is essentially similar to that of a voltage-fed inverter drive. For high performance, vector control can be used. Otherwise, scalar control, such as open loop volts/Hz control, may be satisfactory.

Figure 9.55 shows a simple scalar control method with position sensor-based self-control, and Figure 9.56 shows the corresponding phasor diagram for unity power factor operation. The diagram is identical to Figure 9.44(a), except the phase angle φ between the voltage and current phasors is reduced to zero. The magnitude of the ψ_s phasor remains invariant with the speed. However, its phase position varies with the stator current I_s so that it remains orthogonal to the ψ_a phasor and the locus of point A describes a circle. From Figure 9.56(a), the angle φ' and field current I_f can be described graphically as functions of stator current I_s, as shown in Figure 9.56(b). Any saturation effect in field flux ψ_f can be included in the I_f function generator. The error in the speed control loop generates the stator current command I_s^*, which is the principal control vari-

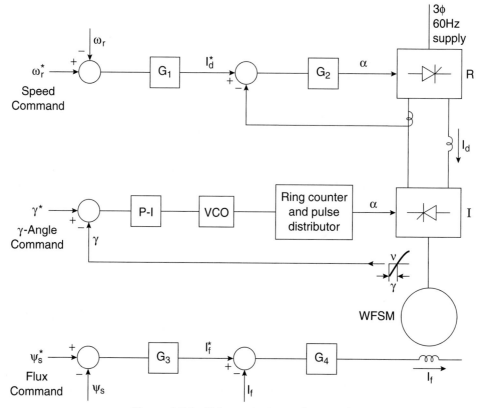

Figure 9.54 PLL constant γ angle control

able in the system. This in-phase stator current is proportional to the developed torque. The other control variables φ^* and I_f^* are generated from I_s^* through the respective function generators. The position sensor and encoder provide the sinusoidally varying unit signals $\cos\theta_e$ and $\sin\theta_e$ and the speed signal ω_r. The $\cos\theta_e$ wave is aligned to be in phase with the ψ_f phasor. The two-phase unit signals $\cos\theta_e$, $\sin\theta_e$ are converted to three-phase unit signals by the relations

$$U_a = \cos\theta_e \tag{9.115}$$

$$U_b = \cos(\theta_e - \frac{2\pi}{3}) = -\frac{1}{2}\cos\theta_e + \frac{\sqrt{3}}{2}\sin\theta_e \tag{9.116}$$

$$U_c = \cos(\theta_e + \frac{2\pi}{3}) = -\frac{1}{2}\cos\theta_e - \frac{\sqrt{3}}{2}\sin\theta_e \tag{9.117}$$

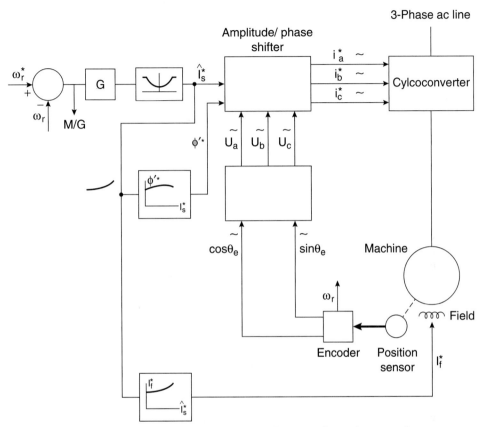

Figure 9.55 Cycloconverter drive with scalar control at unity power factor

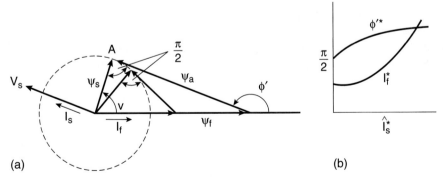

Figure 9.56 (a) Phasor diagram for motoring at unity power factor, (b) Phase angle and field current as functions of stator current

Each of the three-phase unit signals is then multiplied by the magnitude I_s^* and phase-shifted by the angle φ'^* to generate the phase current command signals as follows:

$$i_a^* = \hat{I}_s^* U_a \angle \varphi'^* \qquad (9.118)$$

$$i_b^* = \hat{I}_s^* U_b \angle \varphi'^* \qquad (9.119)$$

$$i_c^* = \hat{I}_s^* U_c \angle \varphi'^* \qquad (9.120)$$

The phasor diagram in the regeneration mode of the system is shown in Figure 9.57, where V_s and I_s phasors are opposite in phase. The control remains identical with the motoring mode, except that the angle φ'^* is negative. The polarity of φ' can be controlled by the polarity of the speed loop error. Note that the field current I_f has sluggish response compared to those of I_s and φ', thus giving sluggish response of the drive. The phasor diagrams with decoupling between ψ_s and I_s phasors are only valid in steady-state condition.

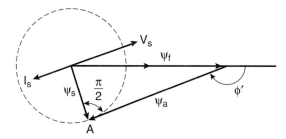

Figure 9.57 Phasor diagram in regeneration mode at unity power factor

9.6.4 Vector Control of Cycloconverter Drive

The transient response of a cycloconverter-fed synchronous motor drive can be enhanced considerably by vector control instead of scalar control, as previously discussed. In the constant torque region, if the developed torque is increased at the constant rated stator flux, the field current I_f is required to be boosted to satisfy the phasor diagram of Figure 9.56. The response of the field current is slow because of the large time constant, and as a result, the machine response is slow. The response can be considerably improved by vector control, where the magnetizing current in the direction of the stator flux can be transiently injected from the machine stator side to compensate the sluggish field flux rise. However, in steady state, this current falls to zero so that the unity machine terminal power factor condition is satisfied. It should be mentioned here that in induction motor vector control, the rotor flux, which is solely established by stator-fed magnetizing current i_{ds}, can be easily maintained constant, and torque can be boosted almost instantaneously by the decoupling torque component i_{qs} of the stator current. The synchronous machine torque response with vector control is therefore somewhat sluggish compared to that of the induction machine.

Wound-Field Synchronous Machine Drives

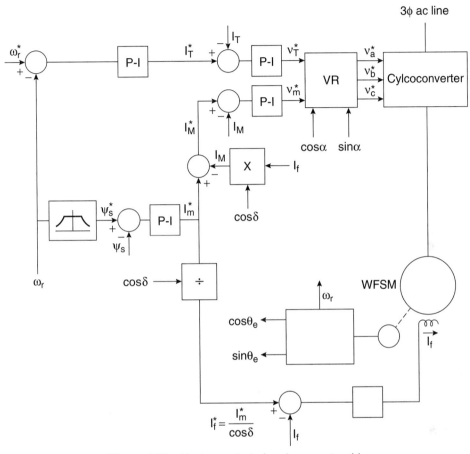

Figure 9.58 Vector control of cycloconverter drive

Figure 9.58 shows the vector control of a cycloconverter-fed drive, and Figure 9.59 shows the corresponding phasor diagram. The vector control is applicable in the constant torque region as well as in the field-weakening regions. In the latter case, the stator flux command ψ_s^* is reduced inversely with speed as indicated to prevent current controller saturation. In the figure, the speed control loop error generates the torque component of current I_T^* through a P-I controller. The stator flux ψ_s is controlled by a feedback loop. The flux loop error generates the magnetizing current command I_m^* required to establish the desired ψ_s. In the phasor diagram, the current I_T is aligned in the direction of the $q^{e'}$ axis which is also the direction of phase voltage V_s, and the current I_m is aligned on the $d^{e'}$ axis, which is, the direction of ψ_s. At steady state, the phasor diagram indicates that phasors ψ_s and ψ_a are at quadrature. This means that at steady state, $I_s = I_T$ and is co-phasal with the stator voltage V_s, which leads the ψ_s phasor by $\pi/2$ angle. This assures the unity power factor condition of the machine at steady-state operating condition. The current phasors I_m, I_f, and

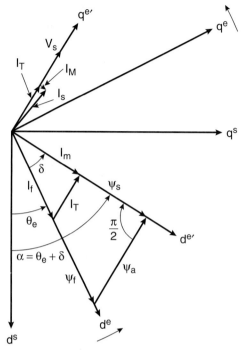

Figure 9.59 Phasor diagram for vector control indicating transient boosting of magnetizing current I_M

I_T form a right-angled triangle, which is proportional to the corresponding flux-linkage triangle. At steady state, the field current I_f can be related to the magnetizing current I_m by the relation

$$I_f = \frac{I_m}{\cos \delta} \qquad (9.121)$$

where δ is the angle between the I_m and I_f phasors. This control equation establishes the field current command I_f^*, which is then controlled by the feedback loop. The flux component of stator current I_M^* is given by the relation

$$I_M^* = I_m^* - I_f \cos \delta \qquad (9.122)$$

This equation gives a finite value of I_M^* when I_f has not been fully established. At the steady-state condition, $I_M^* = 0$ so that Equation (9.121) is satisfied. As shown in the phasor diagram, current I_M is aligned in the direction of ψ_s (perpendicular to I_T). Currents I_T^* and I_M^* are then controlled by the synchronous current control method. Basically, the command currents are compared with the respective feedback currents and the errors generate the voltage commands v_T^* and v_M^* through the P-I compensators. These voltage signals are then vector-rotated by the unit vector signals $\cos \alpha$ and $\sin \alpha$ to generate the phase voltage commands of the cycloconverter. Consider, for example, a sudden increase in speed command ω_r^* in the constant torque region. For the constant flux ψ_s, the current I_m^* will tend to remain constant, but the angle δ will increase and will therefore demand higher field current I_f^*. But, the response delay of I_f will cause a finite I_M^* by Equation (9.122), which will assist to maintain a constant ψ_s. As the current I_f builds up gradually, I_M^* will be diminished until it vanishes completely at steady state when equation (9.121) is satisfied. During the transient condition, the stator current I_s is given by

$$\hat{I}_s = \sqrt{I_T^2 + I_M^2} = \sqrt{2}I_s \tag{9.123}$$

and the machine power factor temporarily deviates from unity.

The complete feedback signal processing for vector control is shown in Figure 9.60, and the corresponding equations for the estimation blocks are given below the figure. In the current model estimation, the machine phase currents are sensed and first converted to the d^s-q^s frame and then to the d^e-q^e frame by the unit vector signals $\sin\theta_e$ and $\cos\theta_e$, which are available from the position encoder. The rotating frame currents i_{ds} and i_{qs}, along with the field current I_f, solve the flux expressions of ψ_{ds} and ψ_{qs} from the d^e-q^e equivalent circuits of the synchronous machine shown in Figure 2.36 as

$$\begin{aligned}\psi_{ds} &= i_{ds}L_{ls} + i_{dm}'L_{dm} \\ &= i_{ds}L_{ls} + L_{dm}(I_f + i_{ds})\frac{(R_{dr} + SL_{ldr})}{R_{dr} + S(L_{ldr} + L_{dm})}\end{aligned} \tag{9.124}$$

$$\begin{aligned}\psi_{qs} &= i_{qs}L_{ls} + i_{qm}L_{qm} \\ &= i_{qs}L_{ls} + L_{qm}i_{qs}\frac{(R_{qr} + SL_{lqr})}{R_{qr} + S(L_{lqr} + L_{qm})}\end{aligned} \tag{9.125}$$

where S is the Laplace operator. Note that the damper winding helps to reduce the delay in the transient response as well as in the commutation overlap angle μ. The current model-based stator flux vector (ψ_{ds}^s, ψ_{qs}^s) synthesis is essential if the machine is required to operate at very low speeds. However, the machine parameter variation remains a problem in the current model. If the machine speed exceeds typically above 5 percent of the base speed, there is an option to switch to the voltage model-based flux synthesis (stator resistance neglected), as shown in the figure. The voltage model is essentially independent of machine parameters and integration is no problem at higher frequencies. However, such a transition can introduce jerk, unless the variables are properly initialized for the incoming model.

9.6.5 Vector Control with Voltage-Fed Inverter

The vector control principles discussed above for the cycloconverter drive are equally valid for voltage-fed inverter drives. Figure 9.61 shows the vector control of a double-sided GTO, three-level PWM converter-based rolling mill drive with synchronous machine. Currently,

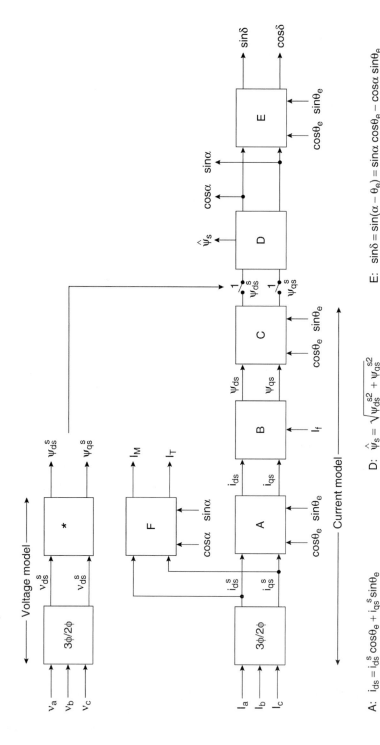

Figure 9.60 Feedback signal processing for vector control

such high-capacity voltage-fed inverter drives are being used to replace cycloconverter drives. For the inverter-machine section, there are close loop speed and stator flux controls as described before. The speed loop generates the torque component of current I_T^* and the flux loop generates the flux component of current I_m^*. The rest of the control lumped under vector control is similar to Figure 9.58. The feedback signal processing is similar to Figure 9.60 and is not shown in the figure. The line-side converter permits a sinusoidal line current at a programmable power factor, which can be leading, lagging, or unity. It is also responsible for the regulation of dc link voltage V_d. The vector control of a line-side converter was described in Figure 8.39. As shown in the figure, the V_d control loop generates the in-phase line current I_P^* and the reactive current I_Q^* is set to zero for unity power factor operation. Both I_P^* and I_Q^* are controlled by synchronous current control with unit vector signals $\cos\theta_e$ and $\sin\theta_e$ derived from the line. Both Figures 9.58 and 9.61 are valid for four-quadrant speed control.

9.7 SENSORLESS CONTROL

In this section, we will briefly review position sensorless control of PM synchronous motor drives, including trapezoidal and sinusoidal machines. The justification of sensorless control is the same as that of induction motor drives discussed in Chapter 8, that is, cost savings and reliability improvement. Again, like sensorless control of induction motor drives, the estimation of speed and position (absolute) is not difficult if the machine is running above a certain minimum speed, but becomes extremely difficult at low speeds, particularly from the standstill condition. In fact, we have already discussed sensorless control of LCI wound-field synchronous motor drives based on terminal voltage sensing when the machine operates typically above 5 percent of base speed.

9.7.1 Trapezoidal SPM Machine

The sensorless control of trapezoidal SPM machine drive (or BLDM) is somewhat simple because only two devices conduct at any time and device commutation occurs at the discrete interval of $\pi/3$ angle. Among a number of suggested methods, we will discuss only two methods: a method based on CEMF or terminal voltage sensing, and another method based on stator third-harmonic voltage sensing.

9.7.1.1 Terminal Voltage Sensing

In this method, as the name indicates, machine terminal voltages are sensed and processed to derive inverter drive signals directly, as shown in Figure 9.62; Figure 9.63 gives the explanatory waveforms [21]. The terminal voltage waves v_{aG}, v_{bG}, and v_{cG}, with respect to ground (see Figure 9.34), can be measured directly. Note that the phase voltage waves v_{an}, v_{bn}, and v_{cn}, shown in Figure 9.35, can be obtained by subtracting the dc bias voltage $0.5V_d$ from v_{aG}, v_{bG}, and v_{cG} waves, respectively. The mid-point P in the v_{aG} wave corresponds to a positive-to-negative zero crossing point in the v_{an} wave. Note that in Figure 9.35, Q_3-to-Q_5 commutation occurs at $\pi/2$-angle delay with respect to the P point. Similarly, Q_5-to-Q_1 and Q_1-to-Q_3 commutations

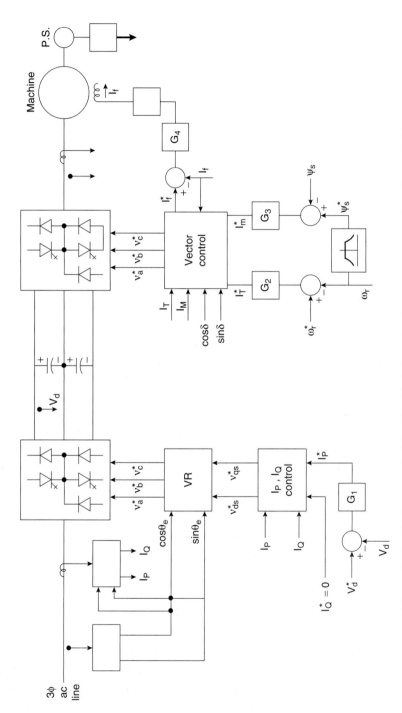

Figure 9.61 Vector control of double-sided, three-level converter system for rolling mill drive

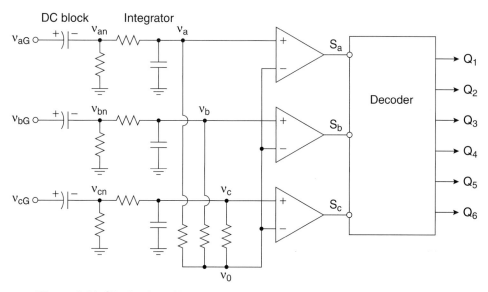

Figure 9.62 Derivation of inverter drive signals from machine terminal voltages

occur from the corresponding points of the v_{bn} and v_{cn} phase voltages. Looking into the voltage waves carefully, it can be seen that the commutation of lower devices of the inverter occurs sequentially at a delay angle of $\pi/2$ with respect to the negative-to-positive zero crossing point of these waves. This pattern of inverter drive signals can be derived from Figure 9.62. The machine terminal voltages v_{aG}, v_{bG}, and v_{cG} are sensed and passed through dc blocking capacitors to derive the phase voltages v_{an}, v_{bn}, and v_{cn}, respectively. These trapezoidal wave signals are then integrated to derive the near triangular waves v_a, v_b, and v_c, as shown in Figure 9.63. Note that the peak of the triangular wave v_a corresponds to point P. A three-phase, balanced, wye-connected resistor bank is connected at v_a, v_b, and v_c, and the voltage wave v_0 at the neutral point is indicated in the figure. The v_0 wave essentially contains a third harmonic (will be derived later), which crosses zero at the instants of zero crossing the v_a, v_b, and v_c waves. The three triangular waves are compared with the v_0 wave to derive the square-wave logic signals S_a, S_b, and S_c at $2\pi/3$-phase displacement as shown. These signals are then processed through a decoder to generate the inverter drive signals. This decoder is basically the same as in the pulse distributor discussed in Figure 9.47. The phase relationship of the inverter drive signals with terminal voltages can easily be verified to be the same as in Figure 9.35. The machine terminal voltages will contain some harmonic ripple because of PWM control of the phase currents, but integration will essentially filter the ripple.

With the terminal voltage sensing method discussed above, it is necessary that the machine run at a finite speed to develop sufficient CEMF, which means that the sensorless control is not valid at very low speeds, including the start-up condition. Initial speed can be developed by starting the machine in open loop synchronous machine mode and then transitioning to

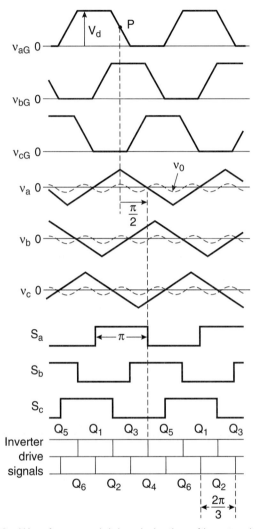

Figure 9.63 Waveforms explaining derivation of inverter drive signals

brushless mode with terminal voltage signals, as explained in Figure 9.64. The open loop control signals for the devices shown on the left are obtained from a frequency signal, which is gradually accelerated. Then, at a critical speed, the control is transitioned to the voltage sensing signals shown on the right. However, the machine may fall out of step at the transition if not properly syn-chronized. The criteria for smooth transitioning are explained in the figure. The phase deviation δ/T at the trailing edges of the voltage sensing square-wave signal S_a and the open loop control signal for Q_3, as shown in the figure, are measured and transition occurs when the deviation falls below a critical value α. The correct synchronization between the S_a wave and Q_3 drive signal is evident from the waves in Figure 9.63.

Sensorless Control

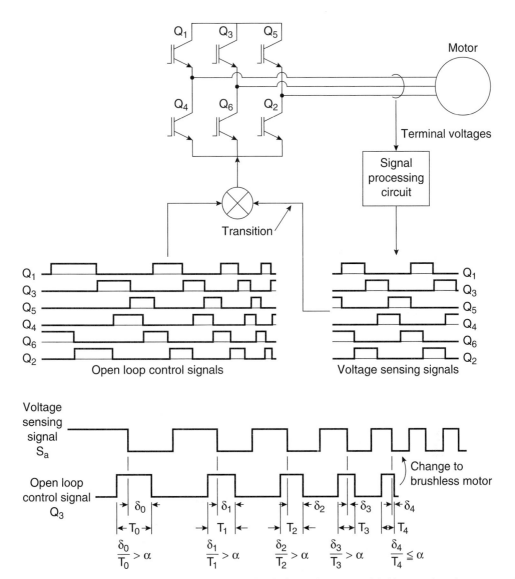

Figure 9.64 Power circuit showing the method of sensing stator third harmonic voltage

9.7.1.2 Stator Third Harmonic Voltage Detection

This method [22] depends on the detection of a third harmonic voltage wave at the stator neutral point of the machine. Consider the simplified power circuit shown in Figure 9.65, where a balanced, wye-connected resistor bank R is connected to create the neutral point N and two series dc link resistors R_d are connected to create the mid-point O of the dc supply. The third harmonic voltage wave under consideration appears between the machine neutral point n and the

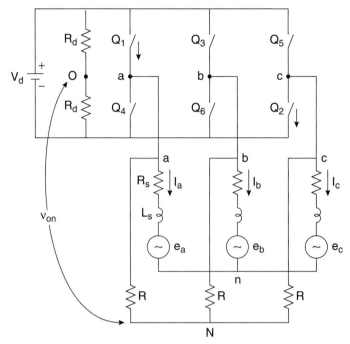

Figure 9.65 Waveforms explaining derivation of inverter drive signals from third harmonic voltage wave

resistor bank neutral N. The derivation of this wave can be made as follows: From Figure 9.65, the phase a voltage equation can be written as

$$v_{an} = R_s i_a + L_s \frac{di_a}{dt} + e_a \tag{9.126}$$

where R_s and L_s are the phase resistance and equivalent inductance, respectively. The Fourier series for the trapezoidal-shaped phase a CEMF e_a will contain only odd harmonics and can be given in the form

$$e_a = E[\cos\omega_e t + K_3 \cos 3\omega_e t + K_5 \cos 5\omega_e t + ...] \tag{9.127}$$

where E = peak value of the fundamental component. Equations (9.126) and (9.127) can also be written for the other phases. The sum of the phase voltages can be expressed in the form

$$v_{an} + v_{bn} + v_{cn} = R_s(i_a + i_b + i_c) + L_s\left(\frac{di_a}{dt} + \frac{di_b}{dt} + \frac{di_c}{dt}\right) + (e_a + e_b + e_c) \tag{9.128}$$
$$= 0 + 0 + 3EK_3 \cos 3\omega_e t + v_{hf}$$

where $i_a + i_b + i_c = 0$, and all the odd triplen harmonics (which are co-phasal) will appear in this expression. The third harmonic term and all the other high-frequency harmonic components

Sensorless Control

together (v_{hf}) are shown separately. We can write the following loop equations for the three phases:

$$v_{an} + v_{nN} + v_{Na} = 0 \tag{9.129}$$

$$v_{bn} + v_{nN} + v_{Nb} = 0 \tag{9.130}$$

$$v_{cn} + v_{nN} + v_{Nc} = 0 \tag{9.131}$$

Adding these equations, we get

$$v_{Nn} = \frac{1}{3}(v_{an} + v_{bn} + v_{cn}) + \frac{1}{3}(v_{Na} + v_{Nb} + v_{Nc}) \tag{9.132}$$

Substituting Equation (9.128) in (9.132) and noting that $v_{Na} + v_{Nb} + v_{Nc} = 0$, we get

$$v_{Nn} = EK_3 \cos 3\omega_e t + \frac{1}{3} v_{hf} \tag{9.133}$$

This third harmonic voltage wave v_{Nn} is nearly triangular-shaped (as indicated in Figure 9.66). If the machine neutral point is not available, this voltage wave can be derived indirectly, as explained below. We can write the following phase voltage equations with respect to the dc center-point O:

$$v_{a0} + v_{ON} + v_{Nn} + v_{na} = 0 \tag{9.134}$$

$$v_{b0} + v_{ON} + v_{Nn} + v_{nb} = 0 \tag{9.135}$$

$$v_{c0} + v_{ON} + v_{Nn} + v_{nc} = 0 \tag{9.136}$$

At any instant, two switches are conducting in the bridge. Assume, for example, that switches Q_1 and Q_2 are conducting. At this condition, we can write the following relations:

$$v_{a0} = +0.5V_d, \ v_{c0} = -0.5V_d, \ v_{nc} = +0.5V_d, \text{ and } v_{na} = -0.5V_d \tag{9.137}$$

Substituting Equations (9.137) and (9.133) in (9.136), we get

$$v_{ON} = -v_{Nn} = -EK_3 \cos 3\omega_e t + \frac{1}{3} v_{hf} \tag{9.138}$$

The waveform of v_{ON} in correct phase relation with the phase voltage v_{an} is shown in Figure 9.66. It can be shown that for other combinations of on switches (see Figure 9.36), such as Q_3Q_2, Q_3Q_4, Q_4Q_5, Q_5Q_6, and Q_1Q_6, Equation (9.138) is also valid. Therefore, the voltage v_{ON} can be used to represent the stator third harmonic voltage in the absence of machine neutral.

Figure 9.66 explains how this third harmonic voltage wave is utilized to derive the commutation instants of the inverter switches. The v_{ON} wave is integrated to derive a relatively harmonic-free, near-sinusoidal waveform with a $\pi/2$ phase-lag angle, as shown in the figure. It

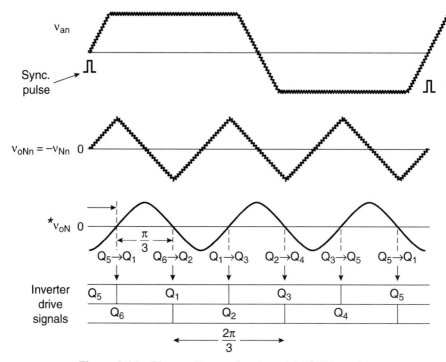

Figure 9.66 Phasor diagram for sinusoidal SPM machine

can be seen that negative-to-positive zero crossing of the $\int v_{oN}$ wave gives commutation instants of the upper devices, whereas positive-to-negative zero crossing points give commutation instants of the lower devices. The zero crossing point of the v_{an} wave, as shown, is also essential for synchronization purposes. Noise filtering of this wave for synchronization signal detection is not affected by phase delay because the control algorithm uses this information only to keep track of zero crossing of the $\int v_{oN}$ wave. Finally, note that to use this algorithm, the machine requires separate start-up, as discussed before, so that sufficient CEMF is developed.

9.7.2 Sinusoidal PM Machine (PMSM)

Unlike trapezoidal SPM machine, the sensorless control of a sinusoidal SPM machine is generally more difficult because three devices in the inverter conduct at any instant and a continuous position signal of the rotor is required for the control. Again, among a number of suggested techniques, we will discuss briefly only three methods: terminal voltage and current sensing, inductance variation or saliency effect, and state estimation based on the extended Kalman filter (EKF) method. In fact, all these methods depend on sensing machine terminal voltages and currents.

9.7.2.1 Terminal Voltage and Current Sensing

This is the simplest method, where unit vectors are derived directly from machine terminal voltages and currents to implement stator flux-oriented vector control, which was explained in Figure 9.26.

Sensorless Control

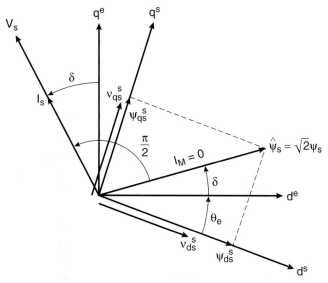

Figure 9.67 Phasor diagram for sinusoidal SPM machine

Figure 9.67 shows the phasor diagram for the machine, and Figure 9.68 shows a diagram of a vector-controlled drive based on estimation from machine terminal voltages. The estimation equations can be given as

$$\psi_{ds}^{s} = \int (v_{ds}^{s} - R_s i_{ds}^{s}) dt \tag{8.32}$$

$$\psi_{qs}^{s} = \int (v_{qs}^{s} - R_s i_{qs}^{s}) dt \tag{8.33}$$

$$\hat{\psi}_s = \sqrt{\psi_{ds}^{s^2} + \psi_{qs}^{s^2}} \tag{8.34}$$

$$\cos(\theta_e + \delta) = \frac{\psi_{ds}^{s}}{\hat{\psi}_s} \tag{9.139}$$

$$\sin(\theta_e + \delta) = \frac{\psi_{qs}^{s}}{\hat{\psi}_s} \tag{9.140}$$

$$\omega_e = \frac{d\theta_e}{dt} = \frac{(v_{qs}^{s} - i_{qs}^{s} R_s)\psi_{ds}^{s} - (v_{ds}^{s} - i_{ds}^{s} R_s)\psi_{qs}^{s}}{\hat{\psi}_s^{2}} \tag{8.70}$$

where $\cos(\theta_e + \delta)$ and $\sin(\theta_e + \delta)$ are the unit vector signals and all other symbols are in standard nomenclature. The d^s and q^s axes currents and voltages can be derived from the respective

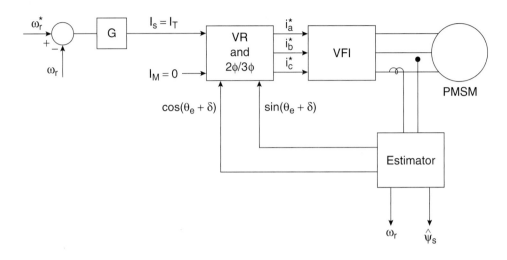

Figure 9.68 Sensorless vector control of sinusoidal SPM drive based on terminal voltage and currents

phase variables by the usual transformation equations. The estimated speed in sensorless control can be calculated from Equation (8.68), which was derived in Chapter 8 for an induction motor. Although unity power factor control ($I_M = 0$) is indicated in the figures, any other power factor control is possible. Note that the control strategy is valid for both SPM and IPM drives.

The above estimation equations based on the machine voltage model are valid only above a minimum speed (typically above 5 percent of base speed) when sufficient CEMF is developed. This means that start-up operation must be implemented by some other method. In Figure 9.68, the drive can be started by open loop current or the voltage control method (see Figure 9.1), with independent frequency control and then ramping up the frequency command. Beyond the critical frequency, the control is transitioned to sensorless self-control. Of course, the usual disadvantages of open loop scalar control and transition jerk must be tolerated.

9.7.2.2 Inductance Variation (saliency) Effect

It was mentioned before that a sinusoidal IPM machine is basically a salient pole machine ($L_{ds} < L_{qs}$), which means that the self-inductance seen by a phase winding will vary with rotor position. Therefore, information on the position of the rotor can be obtained by knowing the phase inductances. The stator self inductances of phase a, phase b, and phase c for a two-pole machine can be given as

$$L_{aa} = L_{al} + L_{aa0} + L_{g2} \cos(2\theta_e) = L_{a0} + L_{g2} \cos(2\theta_e) \tag{9.141}$$

$$L_{bb} = L_{al} + L_{aa0} + L_{g2} \cos(2\theta_e + \frac{2\pi}{3}) = L_{a0} + L_{g2} \cos(2\theta_e + \frac{2\pi}{3}) \tag{9.142}$$

Sensorless Control

$$L_{cc} = L_{al} + L_{aa0} + L_{g2}\cos(2\theta_e - \frac{2\pi}{3}) = L_{a0} + L_{g2}\cos(2\theta_e - \frac{2\pi}{3}) \qquad (9.143)$$

where L_{al} = leakage inductance of the stator winding, L_{aa0} = magnetizing inductance due to the fundamental air gap flux, L_{g2} = self-inductance component due to the rotor's position-dependent flux, and θ_e = rotor electrical angle. The mutual inductance between the stator phases will also vary with the θ_e angle and can be given as

$$L_{ab} = L_{ba} = -0.5 L_{aa0} + L_{g2}\cos(2\theta_e - \frac{2\pi}{3}) \qquad (9.144)$$

$$L_{bc} = L_{cb} = -0.5 L_{aa0} + L_{g2}\cos(2\theta_e) \qquad (9.145)$$

$$L_{ca} = L_{ac} = -0.5 L_{aa0} + L_{g2}\cos(2\theta_e + \frac{2\pi}{3}) \qquad (9.146)$$

The direct and quadrature axes synchronous inductances, which are not functions of the θ_e angle, can be given as

$$L_{ds} = 1.5(L_{aa0} - L_{g2}) + L_{al} \qquad (9.147)$$

$$L_{qs} = 1.5(L_{aa0} + L_{g2}) + L_{al} \qquad (9.148)$$

From these equations, we can write

$$L_{al} + 1.5 L_{aa0} = \frac{L_{ds} + L_{qs}}{2} \qquad (9.149)$$

$$L_{g2} = \frac{L_{qs} - L_{ds}}{3} \qquad (9.150)$$

Consider a high switching frequency ($f_{sw} > 10$ kHz) of the inverter so that the inductance variation within a switching period can be neglected. With this assumption, the instantaneous voltage equation for phase a can be given as

$$v_a = R_s i_a + L_{sa}\frac{di_a}{dt} + e_a \qquad (9.151)$$

where v_a = applied voltage, R_s = phase a resistance, $L_{sa} = L_{aa} - L_{ab}$ is the synchronous inductance of phase a, and e_a = phase a CEMF. Combining Equations (9.141)–(9.146) and (9.149)–(9.150), we can write

$$L_{sa} = L_{aa} - L_{ab} = A + B\left[\cos(2\theta_e) - \cos(2\theta_e - \frac{2\pi}{3})\right] \qquad (9.152)$$

$$L_{sb} = L_{bb} - L_{bc} = A + B\left[\cos(2\theta_e + \frac{2\pi}{3}) - \cos(2\theta_e)\right] \quad (9.153)$$

$$L_{sc} = L_{cc} - L_{ca} = A + B\left[\cos(2\theta_e - \frac{2\pi}{3}) - \cos(2\theta_e + \frac{2\pi}{3})\right] \quad (9.154)$$

where $A = 0.5(L_{ds} + L_{qs})$ and $B = 1/3(L_{qs} - L_{ds})$. These phase synchronous inductances are plotted in Figure 9.69 as functions of angle θ_e, indicating their variation at double the fundamental frequency. It appears that if we know the values of L_{sa}, L_{sb}, and L_{sc} at any instant, we can determine the rotor angle θ_e from this figure.

The phase synchronous inductances of an IPM machine can be calculated analytically from the instantaneous voltage and current information [23]. They can also be evaluated directly from the following expression:

$$L_{sa} = \frac{v_a - e_a - R_s i_a}{\frac{di_a}{dt}} \quad (9.155)$$

where

$$\frac{di_a}{dt} \simeq \frac{\Delta i_a}{\Delta t} = \frac{i_a(t_2) - i_a(t_1)}{\Delta t} \quad (9.156)$$

and

$$e_a = K \cdot \omega_e = K \frac{d\theta}{dt} = K\left[\frac{\theta_e(t_2) - \theta_e(t_1)}{\Delta t}\right] \quad (9.157)$$

Times t_2 and t_1 are the previous subsequent sampling instants. The CEMF e_a as a function of speed can be considered constant between the interval Δt.

Similar equations can be used to calculate L_{sb} and L_{sc} by measuring the phase b and phase c terminal signals of the machine, respectively. The estimated inductance signals can be used to determine the absolute position angle θ_e from Figure 9.69 with the help of a lookup table. Precision estimation (which may be difficult) is mandatory for correct rotor angle information.

9.7.2.3 Extended Kalman Filter (EKF)

Among the few observer-based position and speed estimation methods of PM synchronous machines suggested in the literature, the EKF method offers a powerful computation-intensive alternative. The EKF method for speed estimation of an induction motor was discussed in Chapter 8. It is basically a full-order observer (in the presence of noise) for recursive state estimation of nonlinear dynamic systems in real time by a powerful DSP. The observer senses the machine terminal voltages v_{ds}^s and v_{qs}^s and currents i_{qs}^s and i_{ds}^s and

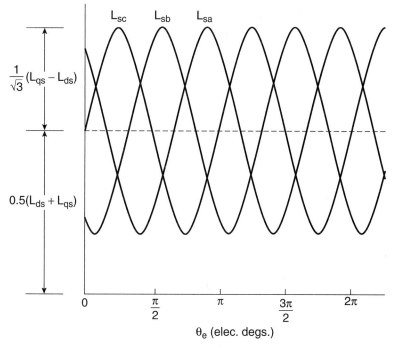

Figure 9.69 Variation of phase a, phase b and phase c synchronous inductances at different rotor position angle θ_e

estimates the position θ_e and speed signal ω_e. The machine model used in the EKF algorithm can be based on either a stationary frame or synchronously rotating frame. However, the former simplifies the computation and will be considered here. We will consider a surface magnet-type machine only ($L_{ds} = L_{qs}$) and discuss the augmented stationary frame model to be used in the EKF algorithm. The stationary frame voltage equations in terms of stator flux linkages can be given as

$$v_{ds}^{s} = R_s i_{ds}^{s} + \frac{d}{dt}(\psi_{ds}^{s}) \tag{9.158}$$

$$v_{qs}^{s} = R_s i_{qs}^{s} + \frac{d}{dt}(\psi_{qs}^{s}) \tag{9.159}$$

From the phasor diagram of Figure 9.10, we can write the following expression of stator flux in complex form:

$$\overline{\psi}_s = L_s \overline{I}_s + \overline{\psi}_f \tag{9.160}$$

where L_s = synchronous inductance. Equation (9.160) can be resolved into d^s-q^s components as follows:

$$\psi_{ds}^{\,s} = L_s i_{ds}^{\,s} + \hat{\psi}_f \cos\theta_e \tag{9.161}$$

$$\psi_{qs}^{\,s} = L_s i_{qs}^{\,s} + \hat{\psi}_f \sin\theta_e \tag{9.162}$$

Differentiating Equations (9.161) and (9.162) and substituting in (9.158) and (9.159), respectively, we get

$$v_{ds}^{\,s} = R_s i_{ds}^{\,s} + L_s \frac{d}{dt}(i_{ds}^{\,s}) - \hat{\psi}_f \omega_e \sin\theta_e \tag{9.163}$$

$$v_{qs}^{\,s} = R_s i_{qs}^{\,s} + L_s \frac{d}{dt}(i_{qs}^{\,s}) + \hat{\psi}_f \omega_e \cos\theta_e \tag{9.164}$$

These equations contain the states $i_{ds}^{\,s}$, $i_{qs}^{\,s}$, ω_e, and θ_e, where the latter two variables are estimated. The dynamic model of the machine in state-variable form can be expressed as

$$\frac{dX}{dt} = f(X) + BU \tag{9.165}$$

$$Y = CX \tag{9.166}$$

where $X = [i_{ds}^{\,s}\ i_{qs}^{\,s}\ \omega_e\ \theta_e]^T$ is the state vector, $U = [v_{ds}^{\,s}\ v_{qs}^{\,s}]^T$ is the input vector, $Y = [i_{ds}^{\,s}\ i_{qs}^{\,s}]^T$ is the output vector, and the matrices $f(X)$, B and C are given as

$$f(X) = \begin{bmatrix} -\dfrac{R_s}{L_s} i_{ds}^{\,s} + \dfrac{\hat{\psi}_f}{L_s} \omega_e \sin\theta_e \\ -\dfrac{R_s}{L_s} i_{qs}^{\,s} - \dfrac{\hat{\psi}_f}{L_s} \omega_e \cos\theta_e \\ 0 \\ \omega_e \end{bmatrix} \tag{9.167}$$

$$B = \begin{bmatrix} \dfrac{1}{L_s} & 0 \\ 0 & \dfrac{1}{L_s} \\ 0 & 0 \\ 0 & 0 \end{bmatrix} \qquad (9.168)$$

$$C = \begin{bmatrix} 1 & 0 & 0 & 0 \\ 0 & 1 & 0 & 0 \end{bmatrix} \qquad (9.169)$$

Note that the machine model described by Equations (9.165) and (9.166) is nonlinear and the state vector has been augmented with the variables to be estimated, ω_e and θ_e. The two extra equations in (9.165) are $d\omega_e/dt = 0$ and $d\theta_e/dt = \omega_e$, with the assumption that infinite inertia are not true in reality. The required correction is made by EKF. The standard model contains a speed relation with torque in the form $J(2/P)d\omega_e/dt = T_e - T_L$. For the digital EKF implementation, the machine model must be expressed in the discrete-time form as

$$X(k+1) = f\left[X(k), k\right] + B(k)U(k) + V(k) \qquad (9.170)$$

$$Y(k) = C(k)X(k) + W(k) \qquad (9.171)$$

where $V(k)$ and $W(k)$ are the system noise vector and measurement noise vector, respectively. Detailed steps for the digital implementation of EKF can be given with a flowchart similar to Figure 8.50. The estimation accuracy in the EKF algorithm becomes poor at low speeds because of machine parameter variation and the inaccuracy involved with low-voltage signal measurement.

9.8 SWITCHED RELUCTANCE MOTOR (SRM) DRIVES

Switched reluctance motors (SRMs) and their performance characteristics were discussed briefly in Chapter 2. Although it is not a synchronous machine, a brief description of the SRM drive will be included in this chapter to complete the discussion on ac drives.

Figure 2.45 in Chapter 2 described an SRM as a device with has eight stator poles and six rotor poles. A pair of opposite stator poles is supplied by a converter phase winding, as shown, which carries unidirectional current. The current pulses in the phases are synchronized with the rotor position; therefore, an absolute position encoder is mandatory. Figure 9.70 shows an SRM drive system where a four-phase, voltage-fed IGBT converter excites the respective machine phases. The machine is shown with a position encoder that also generates the speed signal. Phase a, for example, is excited by turning on IGBTs Q_a and Q_a'. When the devices are turned off, the energy stored in the inductance flows to the source through the feedback diodes. All four machine phases are excited sequentially in synchronism with the rotor position to get unidirectional torque. The converter is unipolar because the machine current is unipolar. In the

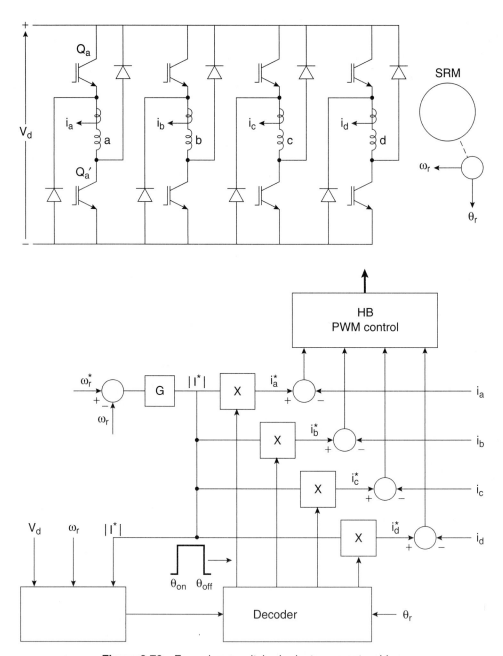

Figure 9.70 Four-phase switched reluctance motor drive

Switched Reluctance Motor (SRM) Drives

speed control system shown in the figure, the speed loop generates the absolute current command $|I^*|$, which is related with the torque given by Equation (2.194). In the constant torque region, the phase current magnitude is controlled by HB PWM technique. A particular phase is enabled by the commutation angles θ_{on} and θ_{off} acting at the position decoder output shown in the figure. At high speeds, as explained before, the current control is lost due to high CEMF, which leaves only the single pulse angle control mode. Figure 9.71 explains the generation of commutation angles for phase a only. For simplicity, the numerical values of angles are assigned for the inductance profile, where the beginning of the rise of the inductance profile is considered the reference point (0°). For the six-pole rotor, the period of the inductance profile is $\pi/3$. The drive can be controlled in all four quadrants, but the figure only shows waveforms in phase a for forward motoring and forward regeneration in the constant torque region. In the

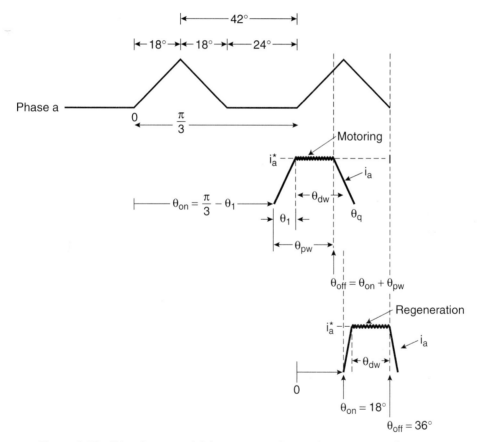

Figure 9.71 Waveforms explaining commutation angles generation for phase a

motoring mode, the current i_a is turned on at advance angle θ_1, and it rises linearly to the magnitude I at the reference point ($0°$), which is given by the relation

$$\theta_1 = I^* \frac{L_m \omega_r}{V_d} \qquad (9.172)$$

where L_m = minimum inductance, ω_r = motor speed, and V_d = dc link voltage. This makes the turn-on angle $\theta_{on} = \pi/3 - \theta_1$, as shown in the figure. The current I is maintained constant by HB current control, and then turned off at angle θ_{off} given by

$$\theta_{off} = \theta_{on} + \theta_1 + \theta_{dw} = \theta_{on} + \theta_{pw} \qquad (9.173)$$

where θ_{dw} = dwell angle (a fixed value) and θ_{pw} = pulse width angle. The current falls to zero at angle θ_q. The dwell angle is restricted so that θ_q does not extend much in the negative inductance slope. For the regenerative braking condition, which is indicated by the polarity of the speed loop error, the conduction pulse slides to a negative inductance slope with $\theta_{on} = 18°$ and $\theta_{off} = 36°$, as shown in the figure. A torque control loop can be added within the speed loop to enhance the response, but the feedback torque computation becomes somewhat involved. A position control loop can also be added over the speed loop, if desired.

9.9 SUMMARY

This chapter gave a broad review of control and feedback signal estimation of different types of synchronous motor drives. The machine types covered include sinusoidal SPMs (surface permanent magnets), sinusoidal IPMs (interior permanent magnets), synchronous reluctance machines, trapezoidal SPMs, and wound-field synchronous machines. In the beginning, we briefly discussed open loop volts/Hz control for a single machine or multiple machines in parallel. Most of the chapter, however, dealt with self-controlled drives with the help of absolute position sensors.

Sinusoidal PM machine drives with vector control and trapezoidal machine drives give dc machine-like performance; therefore, they can be defined as brushless dc motors (BLDMs). However, trapezoidal machine drives are popularly known as BLDMs or ECMs (electronically commutated motors) because of their closer analogy with dc motor drives. Synchronous motors are generally more expensive than induction motors, particularly with high-energy magnets, but the advantages are that control is somewhat simple and the drive efficiency is higher, which tends to make the life-cycle cost lower. The machines (except synchronous reluctance motors) can be operated near unity power factor, which reduces the machine and converter loss and lowers the converter's KVA rating. Of course, the need for an absolute position sensor is a disadvantage.

Extended speed operation in field-weakening mode, which is essential in many applications such as electric vehicles and spindle drives, is very convenient with induction motors, but is generally difficult with PM machines. To achieve the highest level of power in variable-speed drives, wound-field synchronous machines with load-commutated thyristor inverters (LCIs) are very commonly used. High-power cycloconverter drives with vector control are recently being replaced by double-sided, voltage-fed PWM converter drives using wound-field synchronous machines. A powerful microcomputer or DSP is normally used for the implementation of control and feedback estimation of a drive. It is desirable to have a modeling and simulation study on a computer before developing a prototype. For very complex systems, multiple DSPs or a DSP with ASIC chips may be used.

Sensorless control with the parameter variation problem has recently received wide attention in the literature. Speed and position sensorless control is not difficult if the machine speed is above a critical value, but zero-speed start-up or very low-speed sensorless control remains a challenge. The chapter covered a number of sensorless control techniques, including a brief discussion of EKF, which was also covered in Chapter 8. A detailed discussion of EKF, which is quite involved, is beyond the scope of this book.

Finally, a brief discussion on the control of switched reluctance motor (SRM) drives was included for completeness. This type of drive, particularly its sensorless control, is receiving attention in the literature. Many control techniques for induction motor drives, such as DTC control, adaptive and optimal control, etc. can be extended for synchronous machine drives, but these were not discussed separately. They are left as exercises to knowledgeable readers.

REFERENCES

1. T. M. Jahns, "Motion control with permanent magnet ac machines", *Proc. of the IEEE*, vol. 82, pp. 1241–1252, Aug. 1994.
2. P. Vas, *Sensorless Vector and Direct Torque Control*, Oxford, London, 1998.
3. A. Chiba and T. Fukao, "A closed loop control of super high speed reluctance motor for quick torque response", *IEEE IAS Annu. Meet. Conf. Rec.*, pp. 289–294, 1987.
4. R. E. Betz, "Control of synchronous reluctance machines", *IEEE IAS Annu. Meet. Conf. Rec.*, pp. 456–462, 1991.
5. P. Pilley (Ed.), Performance and Design of Permanent Magnet AC Motor Drives, Tutorial Course, *IEEE IAS Industrial Drives Committee*, Oct. 1989.
6. T. M. Jahns, G. R. Kliman, and T. W. Neumann, "Interior permanent magnet synchronous motors for adjustable speed drives", *IEEE Trans. on Ind. Appl.*, vol. 22, pp. 738–747, July/Aug. 1986.
7. B. K. Bose, "A high performance inverter-fed drive system of an interior permanent magnet synchronous machine", *IEEE Trans. on Ind. Appl.*, vol. 24, pp. 989–997, Nov./Dec. 1988.
8. T. M. Jahns, "Torque production in permanent magnet synchronous motor drives with rectangular current excitation", *IEEE Trans. on Ind. Appl.*, vol. 20, pp. 803–813, July/Aug. 1984.
9. B. K. Bose and P. M. Szczesny, "A microcomputer-based control and simulation of an advanced IPM synchronous machine drive system for electric vehicle propulsion", *IEEE Trans. on Ind. Elec.*, vol. 35, pp. 547–559, Nov. 1988.
10. T. M. Jahns, "Flux-weakening regime operation of an interior permanent magnet synchronous motor drive", *IEEE Trans. on Ind. Appl.*, vol. 23, pp. 681–689, July/Aug. 1987.
11. P. Pillay and R. Krishnan, "Modeling, simulation, and analysis of permanent-magnet motor drives, Part II: the brushless dc motor drive", *IEEE Trans. on Ind. Appl.*, vol. 25, pp. 274–279, Mar./Apr., 1989.
12. H. Stemmler, "High power industrial drives", *Proc. of the IEEE*, vol. 82, pp. 1266–1286, Aug. 1994.
13. J. Leimgruder, "Stationary and dynamic brhaviour of a speed controlled synchronous motor with constant cos φ or commutation limit line control", *Conf. Rec. IFAC Symp. On Control in Power Elec. and Electrical Drives*, pp. 463–473, 1977.

14. H. Le-Huy, R. Perret, and D. Roye, "Microprocessor control of a current-fed synchronous motor drive", *Conf. Rec. IEEE/IAS Annu. Meet.*, pp. 873–880, 1979.
15. H. Le-Huy, A. Jakubowicz, and P. Perret, "A self-controlled synchronous motor drive using terminal voltage sensing", *Conf. Rec., IEEE/IAS Annu. Meet.*, pp. 562–569, 1980.
16. H. Stemmler, "Drive system and electrical control equipment of the gearless tube mill", *Brown Boveri Rev.*, pp. 120–128, Mar. 1970.
17. K. H. Bayer, H. Waldmann, and M. Weibelzahl, "Field-oriented close loop control of a synchronous machine with the new transvector control system", *Siemens Review*, vol. 39, pp. 220–223, 1972.
18. T. Nakano, H. Ohsawa, and K. Endoh, "A high performance cycloconverter fed synchronous machine drive system", *IEEE Int'l. Semicon. Power Conv. Conf. Rec.*, pp. 334–341, 1982.
19. H. Okayama et al., "Large capacity high performance 3-level GTO inverter system for steel main rolling mill drives", *IEEE/IAS Annu. Meet. Conf. Rec.*, pp. 174–179, 1996.
20. K. Rajashekara, A. Kawamura, and K. Matsuse (Ed.), *Sensorless Control of AC Motor Drives*, IEEE Press, Piscataway, 1996.
21. K. Iizuka, H. Uzuhashi, M. Kano, T. Endo, and K. Mohri, "Microcomputer control for sensorless brushless motor", *IEEE Trans. on Ind. Appl.*, vol. 21, pp. 595–601, May/June 1985.
22. J. C. Moreira, "Indirect sensing for rotor flux position of permanent magnet ac motors operating in a wide speed range", *IEEE IAS Annu. Meet. Conf. Rec.*, pp. 401–407, 1994.
23. A. B. Kulkarni and M. Ehsani, "A novel position sensor elimination technique for the interior permanent magnet synchronous motor drive", *IEEE Trans. on Ind. Appl.*, vol. 28, pp. 144–150, Jan./Feb. 1992.
24. R. Dhaouadi, N. Mohan, and L. Norum, "Design and implementation of an extended Kalman filter for the state estimation of a permanent magnet synchronous motor", *IEEE Trans. on Pow. Elec.*, vol. 6, pp. 491–497, July 1991.
25. B. K. Bose, T. J. E. Miller, P. M. Szczesny, and W. H. Bicknell, "Microcomputer control of switched reluctance motor", *IEEE Trans. on Ind. Appl.*, vol. 22, pp. 708–715, July/Aug. 1985.

CHAPTER 10

Expert System Principles and Applications

10.1 INTRODUCTION

The terms expert system (ES), fuzzy logic (FL), artificial neural network (ANN), and genetic algorithm (GA) belong to an area called artificial intelligence (AI), which is an important branch of computer science or computer engineering. Recently, the area of AI has penetrated deeply into electrical engineering, and their applications in power electronics and motion control appears very promising. Lotfi Zadeh, the inventor of fuzzy logic, has classified computing into "hard computing," or precise computation, and "soft computing," or approximate computation. The areas of expert systems and traditional data processing can be categorized as hard computing, whereas soft computing encompasses fuzzy logic, neural network, and probabilistic reasoning techniques, such as genetic algorithms, chaos theory, belief networks, and parts of learning theory.

What is AI? The goal of AI is planting human intelligence in a computer so that the computer can think intelligently like a human being. A system with embedded computational intelligence is often defined as an "intelligent" system, which has a "learning," "self-organizing" or "self-adapting" capability. AI was originally defined as "computer processes that attempt to emulate the human thought processes that are associated with activities that require the use of intelligence." Can a computer really think and make intelligent decisions? Or, is it as smart as the program developed and implanted in it by a human programmer? Computer intelligence has been debated since its inception and will possibly continue forever. However, there is no denying the fact that the human brain, which is the source of intelligent thinking, is the most complex machine on earth, and we hardly understand it and its behavior in thinking, learning, and reasoning for complex problems. Neuro-biologists and behavioral scientists have investigated the structure of the brain and its functioning for a long time. However, our present knowledge about the brain is extremely inadequate, and will possibly remain so for a long time in spite of painstaking research.

However complex the human thought process, there is no denying the fact that computers can have adequate intelligence to help solve problems that are difficult to solve in traditional ways. Therefore, it is true that today, AI techniques are being applied extensively in solving our problems in areas such as industrial process control, medicine, geology, agriculture, information management, military science, and space technology, just to name a few. However limited computational intelligence, it has some superiority over human intelligence in several aspects. The computer can process a problem very fast compared to that of a human being, it can work relentlessly without becoming tired and fatigued, and its problem-solving capability is not impaired by anger, emotion, boredom, and the other frailties of a human being.

The advent of AI technology has brought a new challenge to power electronics engineers who are already struggling with complex, fast-advancing, and interdisciplinary areas of the technology. In this chapter, we will study the principles of ES and a number of application examples will be given from the literature. FL and ANN will be discussed in Chapters 11 and 12, respectively. Hopefully, the readers will be able to formulate their own problems and solutions from these example applications.

10.2 EXPERT SYSTEM PRINCIPLES

ES, sometimes called classical AI, is the forerunner of all the AI techniques, and has been traditionally recognized as the most important branch of AI. Since the 1960s and in the early 1970s, it was felt that computers had severe limitations for solving only algorithmic-type problems. It was during this period and in the 1980s that ES theory and applications proliferated.

Historically, it was perceived in early times that the human brain made decisions on the basis of logical yes/no or true/false reasoning. In 1984, George Boole published his historical article, "Investigations on the Laws of Thought," which gave birth to Boolean algebra and set theory. Gradually, the advent of electronic circuits and solid-state integrated circuits ushered in the modern era of digital computers using Von Neumann-type sequential computation. Digital computers were known as "intelligent" machines because of their capability to process problems with human thought-like yes (or logical 1)/no (or logical 0) reasoning. Of course, using the same binary logic, computers can also solve complex scientific, engineering, and other data processing problems.

An ES is basically an "intelligent" computer program that is designed to implant the expertise of a human being in a certain domain. A human being may have multiple areas of expertise. Consider a power electronics technician who has a special or domain expertise for fault diagnosis of a drive system. He or she can apply power to the drive and measure voltages and currents at appropriate points with the help of an oscilloscope and other measuring instruments. Then, based on his or her knowledge or expertise, he or she can conclude which power semiconductor devices are not healthy, or which elements in the control electronics are giving faulty performance. He or she has acquired this domain knowledge by education and experience over a prolonged period of time. The question is: Is it possible to embed the same knowledge in

Expert System Principles

a computer program so that it can automate similar measurements and make similar decisions replacing the human expert? The answer is a qualified "yes" because we need to understand that no computer program, however sophisticated, can ever replace human thinking.

How does an ES program work and how it is different from a conventional CAD program? A CAD program is normally characterized by heavy numeric computation and light logical computation, and its algorithm can usually be described by a flowchart. The program can have knowledge embedded in it, and the computation flow path can be altered by the logic signals. If the embedded knowledge requires alteration or updating, a time-consuming program change will be required. An ES program's structure and features are entirely different. Figure 10.1 shows the basic elements of an ES. These are:

- Knowledge base
- Inference engine
- User interface

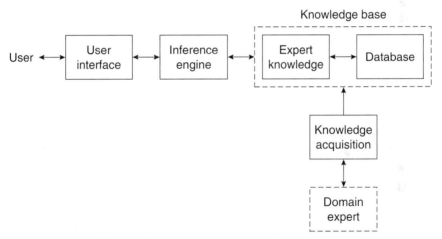

Figure 10.1 Basic elements of expert system

10.2.1 Knowledge Base

The core of an ES is the representation of knowledge or expertise in the knowledge base, and for this reason, an ES is often called a knowledge-based system. The domain knowledge is acquired from the domain expert for implanting in the computer program. The domain expert, say the power electronics technician, may not have the software expertise to write the ES program. Knowledge engineering is defined as an area that deals with the acquisition of knowledge and the structuring or representation of this knowledge in efficient ES software. Of course, a domain expert and a knowledge engineer can be the same person. The knowledge base, as shown, consists of two parts: expert knowledge and a database with data, facts, and statements that support the expert knowledge. A limited amount of data and computational capability may be contained in the expert knowledge. If the data is large, a separate database may be required. The database is equivalent to catalog information for the human knowledge interface. Similarly, separate computational routines may be linked with the expert knowledge.

The knowledge is basically organized in the form of a set of IF.... THEN production rules, and for this reason, an ES is often called a rule-based system. The segments of IF and THEN statements are normally connected by logical AND, OR, and NOT operations for drawing the inference or conclusion. Figure 10.2 shows the rule base of an ES in matrix form where X, Y, and Z are the variables, defined as parameters, and these parameter values are shown in the matrix. The matrix contains 16 rules, which can be read as follows:

Rule 1: IF $X = A$ AND $Y = A'$, THEN $Z = a$
Rule 2: IF $X = B$ AND $Y = A'$, THEN $Z = b$
Rule 3: IF $X = C$ AND $Y = A'$, THEN $Z = c$
Rule 4: IF $X = D$ AND $Y = A'$, THEN $Z = d$
·
·
·
Rule 16: IF $X = D$ AND $Y = D'$, THEN $Z = p$

Note that similar IF ... THEN rules are also used in finite-state machine computer programs. In an ES, the rules are normally defined in abbreviated rule language (ARL) form, and the parameter values may be logical, numeric, a mathematical expression, statement, or the execution of a command. Rule 1 in the diagnostic system may actually read as follows:

IF phase a output voltage is zero AND Q_4 collector voltage is high
THEN Q_1 gate is open-circuited

where X = phase a output voltage, $Y = Q_4$ collector voltage, and $Z = Q_1$ gate. The respective parameter values are "zero," "high," and "open-circuited." A rule is "fired" or executed if the antecedent (also called premise) or conditional part is true, and then the consequent THEN statement or action part is executed.

Y \ X	A	B	C	D
A'	a	b	c	d
B'	e	f	g	h
C'	i	j	k	l
D'	m	n	o	p

Figure 10.2 Matrix of rule base

Complex rules can be designed to handle probability through certainty factors and probability-based models, such as the Bayesian approach. Rules can also be hybrid by mixing in fuzzy logic (FL), which will be discussed in Chapter 11.

Unlike a CAD program, one outstanding feature of the knowledge base is that it is structured in such a way that the knowledge and data can be easily altered or updated. This easy alteration may be important if know-how of the problem changes or technology advancement mandates the alteration. Sometimes, on-line alteration of the knowledge base is possible based on "machine learning." Knowledge can sometimes be defined as "shallow" or "deep." Shallow knowledge is that which is directly obtained from the domain expert (see Figure 10.1), whereas deep knowledge-based rules can be derived from a system model, simulation response, etc. that is based on the designer's or researcher's knowledge. The knowledge in a knowledge base can also be categorized as declarative or fact-like knowledge and procedural or method-like knowledge. Declarative knowledge is basically what to do (i.e., kernel of knowledge base), whereas procedural knowledge is concerned with how to do (i.e., knowledge base organization with frames, rules, and parameters).

10.2.1.1 Frame Structure

ES knowledge is usually structured in the form of a tree that consists of a root frame and a number of subframes, as shown in Figure 10.3. A simple knowledge base can have only one frame, which is the root frame, whereas a large and complex knowledge base may be structured on the basis of multiple frames, as indicated in the figure. The root frame and subframes are organized such that each has its respective rule cluster and parameters. The frame-based structure gives modular hierarchical organization to a large and complex knowledge base. The root frame is the primary core and may have child subframes (A, B, and C) and grandchild subframes (D, E, and F), as indicated in the figure. Generally, a subframe has access to the parameters of its ancestors, but not to those of its descendants. However, a frame can have access to the rules of its descendants, but not to those of its ancestors. These are for the convenience of efficient knowledge organization.

Each frame and subframe can be considered a subdomain of expert knowledge. Consider, for example, the problem of ac drive product selection from a large company for a certain application. The root frame may correspond to the expertise of a general sales engineer, and subframes A, B, and C may correspond to the application engineer's expertise in induction motor drives, PM synchronous motor drives, and specialty motor drives, respectively. The user may not have much knowledge of ac drives. He or she interfaces the root frame in the beginning and holds a dialog with its rule base regarding his or her application needs. The root frame makes some calculations, consults the company's drive catalog, and determines that an induction motor is the best selection. It then directs the user to subframe A for further consultation, where calculations are made and a catalog is consulted with the help of its rule base. Finally, a drive product is selected. Subframe D may correspond to delivery information, and subframe E may contain price and installation information. These subframes are consulted by the user, and finally, the user gets the results of the consultation. A subframe can be activated or "instantiated" for con-

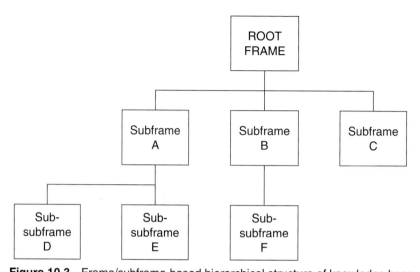

Figure 10.3 Frame/subframe-based hierarchical structure of knowledge base

sultation by the parent frame by a simple rule. A frame or subframe can access external programs for calculation or a database (such as a catalog) for consultation.

10.2.1.2 Meta-Knowledge

In a large and complex knowledge base, meta-rules and other forms of meta-knowledge can increase the efficiency with which the ES reaches a conclusion. Unlike conventional programs, an ES is said to have learning capability, as mentioned before, because of its meta-knowledge. Meta-knowledge is knowledge about the operation of a knowledge base, and meta-rules are rules about the operation of the rules. A knowledge base usually has two levels of operation: the usual domain level and the meta-level. Meta-level knowledge determines the most efficient strategy of operation that the domain level can take. The knowledge base can be made to learn from experience which rules are most useful, that is, most likely to be fired. Then, the ES can test these rules during consultation. This means that all the rules in Figure 10.2 may not be tested for a problem and not in the same sequence. Avoidance of test for the unlikely rules will enhance the speed of execution, that is, improve the efficiency of a knowledge base search. This is particularly true for ES-based real-time control of a power electronic system. An example of a meta-rule to guide the order of a rule search (see Figure 10.2) within a frame is

> Meta-Rule 1: IF rules 2, 5, and 9 are skipped 15 times consecutively AND rule 1 is never tested
> THEN do not test rule 1 AND test rules 2, 5, and 9 at the end.

10.2.1.3 ES Language

What software language is most appropriate for an ES? In principle, any language can be used, but some languages are more efficient than others. Since the major part of an ES program's processing is usually in symbolic logic, a symbolic processing language, such as PROLOG or LISP, has traditionally been popular. Of course, these languages have limited numeric computation capabilities. LISP and its dialects are particularly strong candidates because of their power and flexibility. Numeric computation-intensive, high-level languages such as BASIC, FORTRAN, PASCAL, MATLAB, and the like have also been used, but they have limited symbolic processing capabilities. The inherent drawbacks of high-level languages are that they are slow, but user-friendly, and can be used mainly for off-line computation. For real-time control applications of power electronic systems, a fast, lower level language, such as C or even assembly language, may be essential.

10.2.2 Inference Engine

The inference engine, as shown in Figure 10.1, interfaces to the knowledge base and the user through the user interface. Basically, it is the control software, and as the name indicates, it draws an inference or conclusion by testing the knowledge base. This means that the inference engine tests the rules in systematic order and draws a conclusion of the problem. It is similar to the human mind which tests its domain knowledge (possibly in the form of a rule-set) to draw a conclusion. The rules whose premise or conditional part is true will be "fired," or validated, and the control action determined by the THEN statement will be activated. The rules are tested by the inference engine by using forward chaining, backward chaining, or a combination of both, called mixed chaining. A forward chaining rule (defined as an antecedent rule) works on the principle of deductive logic; that is, the premise part is tested first, and if it is true, then the rule is activated. A backward chaining rule (defined as a consequent rule), on the other hand, works on the principle of inductive logic; that is, the inference engine hypothesizes the inference or consequent part of the rule and then tests backward for the premise part to be true for the rule's validity. This is like a medical doctor assuming that a patient has a certain disease and then trying to match the symptoms with it to validate a diagnosis. It is interesting to note that one of the most successful ES programs is MYCIN, a medical diagnostic program of infectious diseases, which was developed in the early 1970s by Stanford University. The inference engine normally defaults to the backward chaining rule unless forward chaining is specifically requested by an antecedent property of the rule. For example, a subframe instantiation must use a forward chaining rule. In an ES, both forward and backward chaining rules may be mixed strategically for best performance.

10.2.3 User Interface

The user or client interface of ES is very important because often, the user is an unskilled or semi-skilled person who may not have much knowledge of the problem he or she is trying to solve. The user must communicate with the computer in natural language (such as English) in a very user-friendly manner because he or she is usually not familiar with the intricacies of ES software and its language. The ES normally seeks values of the parameters that are used for evaluating rules in the rule base. These parameter values are provided through the user interface. The information may be requested via a menu, or direct input. An example questionnaire and the corresponding inputs for ac drive product selection may be as follows:

What is the drive application? ...."pump"

What is the speed range? ...."10 to 1400 rpm"

What is the supply voltage? ...."220"

Do you need a reversing drive? ...."yes"

Based on the user's inputs to such a questionnaire, the knowledge base is searched and relevant rules are fired. The solutions then appear on the monitor. If the ES is designed for real-time control, the sensors supply signal values and the resulting control action may be something like boosting the command current, sounding an alarm, tripping a circuit breaker, or printing a message.

The ES user dialog has essentially two parts: one gets data or parameter values from the user, as indicated above, and the other provides consultation results and information to the user. Consultation results may be in the form of text, numerical data, or a picture (such as a block diagram of the recommended drive system). When questions are posed to the user, they may appear meaningless because the technical expertise of the user in the particular domain may be low. Therefore, **HELP** messages can be provided on the screen (at the user's command) for dialog questions that do not appear straightforward. These messages give technical explanation as background to these questions. For example, suppose that during the consultation for an automated test on a drive system, the user does not know how the test will be performed. The **HELP** command can clearly explain the steps of the test, giving a technical description for each step. One of the key objectives in writing an ES program is operator or user training in a particular technical area with these **HELP** messages. This is like explaining MICROSOFT WORD™ to a person with a **HELP** command. These messages can be very extensive, incorporating data, explanation, mathematical analysis, circuit diagrams, simulation, and experimental results in graphical form. In addition to **HELP** messages, an ES has the capability of explaining the questions asked by the program using **WHY** and **HOW** user commands. The **WHY** command explains why the value of a particular parameter is needed when the user is prompted for it. For example, when an ES asks for line voltage, the **WHY** message will explain that the converter devices and machine voltage rating depend on this parameter. This clarifies any confusion or ambiguity on the part of the user, in spite of his or her technical knowledge, and helps to answer the question that can be deciphered by the ES. The **HOW** command explains to the user how the ES arrived at a certain conclusion. This explanation convinces the user that a sensible and valid conclusion has been reached. In all such messages, pictures can be drawn extensively to supplement the texts with the help of graphic programs.

10.3 EXPERT SYSTEM SHELL

Basically, an ES shell provides a software environment (i.e., a development system) for designing an ES program. It provides an efficient and user-friendly software platform to the knowledge engineer for building an ES program. The developed program for a user or client can be consulted within the shell or exported to a compatible computer for consultation. A number of PC-based commercial ES shells are available [3] (such as GURU, PC PLUS, ART, EXSYS, etc.), and hardly anybody builds an ES program from ground zero unless it is very specialized in nature.

10.3.1 Shell Features

A good shell should have the features as discussed above, which can be summarized as follows:

- Highly interactive environment for program development and testing
- Good debugging and value verification aids
- Easy access to external databases and programs
- Ability to incorporate graphic files
- Mechanism to handle uncertainty or fuzziness from both the developer and from the end-user
- Ability to explain in English (or other natural languages) why the system is asking for information and how it reached a conclusion
- Windows-oriented interface with extensive on-line help
- Comprehensive rule entry language that is similar to English or an other natural language
- Full screen editor
- Efficient meta-rules for sophisticated rule control
- Frame capability that allows the knowledge base structure to be divided into logically different but related segments
- Ability to trigger actions, such as reading instruments and updating displays
- Additional means (other than inferencing and prompting) to evaluate and set parameter values
- Ability to extend knowledge base through the use of user-defined LISP functions

The reader should first be familiar with these features of the shell before starting development of a program. A detailed discussion of the shell is beyond the scope of this chapter.

10.3.2 External Interface

The interface of the shell with external programs is shown in Figure 10.4. A limited amount of data, logical, and arithmetic capabilities can be directly embedded in the ES program, but for a larger amount of data, such as product catalog information, database files should be constructed. With the help of dBASE functions, the shell can obtain, change, add, remove, or print information from the database. The spreadsheets can also be consulted from the shell. For complex calculations

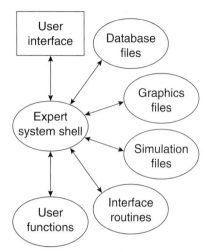

Figure 10.4 Shell interface with external files for expert system program development

such as solving differential equations, the ES program can access an external program, as shown in Figure 10.5. The input data can be loaded, the program can be executed by command, and then the output data can be read as shown. Similarly, the shell can control and transfer data to and from a simulation program, which will be discussed later. ES-aided simulation of a power electronic system with auto-tuning of parameters is very important before building a breadboard or prototype. A powerful feature of the shell is that it can integrate pictures in the knowledge base, as mentioned before. Additionally, user-defined or written functions can be embedded in the shell.

10.3.3 Program Development Steps

The flowchart for application program development is shown in Figure 10.6.

The shell permits program rule-based development in a template setting with English-like ARL, as mentioned before. Once the program is designed by systematically defining the frames,

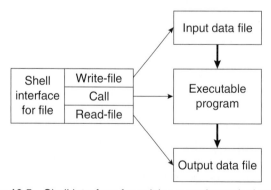

Figure 10.5 Shell interface for solving a mathematical program

Expert System Shell

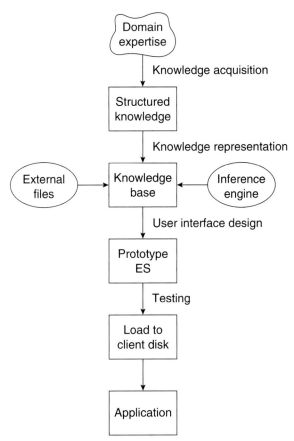

Figure 10.6 ES program development flowchart

rules, parameters, and their properties, the program is compiled and ready for execution with the built-in inference engine and user-supplied parameter values and external files as shown in Figure 10.4. After compiling, every rule is translated into English by the shell and checked for correctness. After program development and testing, a run-time version or client program is created. This program is then loaded to a disk to operate on a stand-alone basis in a compatible operating system environment. The inference engine in the shell is automatically loaded with the knowledge base and user interface to the client program. The user dialog appears in pure English, as mentioned before. The client program can be generated in LISP for non-time-critical applications or in C for less memory- and time-critical real-time applications. When the program is resident in the shell, the developer can easily alter or update it, but no program modification is possible in the client environment.

10.4 DESIGN METHODOLOGY

At the beginning, a problem must be analyzed in sufficient detail to justify whether it deserves to be solved by an ES program. A conventional algorithmic program may be sufficient. Also, FL or ANN, which are both described in later chapters, can also be considered for solving the problem, if necessary.

A systematic procedure for building an ES program is usually given in the manual of the shell being used. The discussions so far on the ES principles and the shell can be applied to the design of an ES. A number of application examples given later will clarify the design methodology. In the case of a complex project, it is advisable to write a simple demo program first to get confidence. A summary of the design procedure is as follows:

- Collect all the information or knowledge related to the problem and organize it in the form of matrices, charts, etc. It is assumed that the domain expert and knowledge engineer are the same person.
- Select a suitable software language for the program. Most off-line programs can be based on LISP or its dialect. Real-time DSP-based controls should be based on the C language. Often, a LISP-based shell can directly generate a C-based program.
- Select a suitable PC-based shell from those available on the market.
- Analyze the knowledge part and partition it into a number of hierarchical frames. For a simple problem, only one frame is adequate. Define the properties of the frames as demanded by the shell.
- Determine the parameters of each frame and define their properties as guided by the shell.
- Write the rules in each frame in ARL and define their properties.
- Plan all the pictures or artwork related to the project and prepare graphic files.
- Organize all the data, statements, and catalog information in the form of database files.
- Design the user dialog to get consultation data.
- Design HELP or explanatory information for the user.
- Develop other external programs and their interface, if necessary, and integrate with the knowledge base.
- Debug and test the complete program (which may have to be done in several stages).
- Finally, generate the client program on CD-ROM and deliver. The inference engine is automatically loaded in the client program.

The users' manual of a shell usually gives enough guidance for development of program. For a newcomer to this area, it is desirable first to develop a simple demo program to gain familiarity with the process.

10.5 APPLICATIONS

ES technology has now practically reached the mature state. Unfortunately, its application in the power electronics and drives areas is very limited, although vast potential exists for more

Applications

extensive applications. One possible reason may be that power electronics engineers are not familiar with this technology. ES applications in power electronics may include system model generation from static and dynamic test data, optimum selection of a commercial product based on its application, automated design of a converter circuit or complete system design, auto-tuning a simulated or experimental close loop system, on-line and off-line fault diagnostics, fault-tolerant control, automated system performance tests, plotting of efficiency and power factor curves, etc. In fact, ES technology can be used in the computer-integrated manufacturing (CIM) of a power electronic product, which may include system analysis, design, simulation study, debugging, tests, performance optimization, etc. In most applications, it is better to use an ES program as an aid rather than solely depending on the results dictated by the ES. In this section, we will briefly discuss a few example applications.

10.5.1 P-I Control Tuning of a Drive

In a P-I-controlled drive, the tuning of the proportional and integral gains of a simulated or experimental system can be done with the help of an ES. In the past, several commercial auto-tuned P-I-D controllers for general-purpose, higher order linear control systems were available. Figure 10.7 shows a vector-controlled drive system where the proportional and integral gains K_p and K_i are being tuned in the speed control loop. The expert controller contains the knowledge base for tuning the controller. It is assumed that initially, the P-I parameters will be loaded such that the system will remain within the stability limit. The initial parameters can be derived from the knowledge of the plant parameters. A square-wave auxiliary test signal, as shown in the figure, is injected as the speed command ω_r^* and the pattern of the error response e are retrieved. From the knowledge base, the controller can look into the error response and determine how the K_p and K_i parameters are to be modified to get the correct tuning. Since $K_p + K_i/S = K_i/S(1+S/\alpha)$ (where $\alpha = K_i/K_p$), it is evident that in the second-order drive system, reducing K_i will reduce the loop gain constant as well as reduce the crossover frequency, whereas reducing K_p will only decrease the crossover frequency with a constant gain. A typical rule may read as follows:

IF the damping constant $\xi < 0.4$
  THEN reduce K_i by 20% and increase K_p by 10%

10.5.2 Fault Diagnostics

Fault diagnostics in industrial plants are possibly the most popular and earliest applications of ES programs. The diagnostics may be "off-line" or "on-line." In off-line diagnostics (defined as troubleshooting), the plant is shut down and the ES is used to identify the fault. In this case, the expertise of the diagnostic technician is embedded in the knowledge base. The procedure can be static or dynamic. A static ES dictates the troubleshooting procedure step-by-step to the operator, and the observed symptoms are fed at the input in the form of a dialog. The ES then searches the knowledge base and gives the conclusion on the screen. The dynamic diagnostic system is more

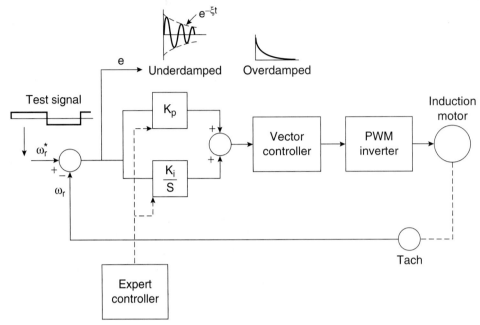

Figure 10.7 ES-based P-I tuning of vector-controlled drive system

sophisticated. Here, signals are applied to the system automatically, the responses are retrieved through sensors, and finally, conclusions are drawn without user intervention.

On-line diagnostics based on an ES can also be very complex. Here, the main objective is to maintain reliability and safety of the operating plant and avoid unnecessary shut-down. For example, in a voltage-fed inverter drive, the ac line voltages and currents, dc link voltage and current, IGBT gate drive logic signals, machine stator voltages and currents, and stator winding temperature signals can be sensed and fed to a DSP that embeds the diagnostic program. The monitoring, alarm processing, and protection functions can easily be integrated with the diagnostic program. For a large and complex drive plant, the ES can be designed to guide start-up, monitor the general health of the drive in its operating conditions, avoid preventable shut-down, and give fault-tolerant control of the system. A typical rule in a fault-tolerant control can read as

    IF dc link voltage < 200 V AND
      ac line voltage = 0 AND
      machine speed > 1500 rpm
    THEN reduce machine speed by 20% by regenerative braking

In this program, the drive shut-down is prevented for temporary loss of line power by pumping up the dc link voltage with regenerative braking.

10.5.3 Selection of Commercial ac Drive Product

With the help of an ES, a semi-skilled user can select a drive product best suited for his or her application. In a general case, the user determines his or her application needs and then consults a company application or sales engineer extensively to decide the product to be purchased. The application engineer, with his or her knowledge of drive technology and company products, makes some calculations, consults the catalog, and then recommends the appropriate drive. For a large company, there may be a general applications engineer, who after preliminary consultation, may direct the user to the respective application engineer with expertise in induction motor drives or synchronous motor drives. In an ES, the application engineer's expertise and product catalogs may be implanted in the knowledge base. The knowledge can be partitioned into frames, as discussed previously. The user holds a conversation with the ES and supplies all the relevant application information. Based on the dialog, the knowledge and database are searched, and then the appropriate product is recommended. The user may have power electronics expertise and indicate a choice of an induction or synchronous machine, or a voltage-fed or current-fed inverter. This user may also select a power device and have a preference for a particular company's product. The program may give a default recommendation on all these aspects, explaining the basis of its recommendations. A typical rule in ARL may be

```
IF :: MOTOR = CAGE-ROTOR-INDUCTION-MOTOR AND
     APPLICATION = PUMP AND
     LINE VOLTAGE = 230-V AND
     PHASE = THREE AND
     POWER = 20-HP AND
     SPEED REVERSAL = NO
THEN :: PRODUCT = COMPANY X- TYPE-531-TFY
```

Figure 10.8 shows a simplified flow diagram for this program's operation stages.

The motor short-time oversizing need, if any, is dictated by the acceleration-deceleration requirement to be sure that developed torque remains below the breakdown torque and temperature remains within the safe limit. The corresponding short-time inverter rating need is determined so that the device current and temperature rise remain within the safe limit.

10.5.4 Configuration Selection, Design, and Simulation of a Drive System

The design of a complete ac drive for a certain application is a complex task due to a number of constraints and trade-offs the designer has to make in the various stages of the design. In the traditional approach, the power electronics engineer with the expertise of drive technology designs the system using paper, pencil, calculator, CAD programs, and component catalogs. He or she then simulates the system and optimizes performance with the iteration of circuit and control parameters. A laboratory breadboard is then built to verify performance before finally building the prototype. In this application of an ES, the expertise of the drive design is embedded in the knowledge base to help a semi-skilled designer automate the drive configuration selection,

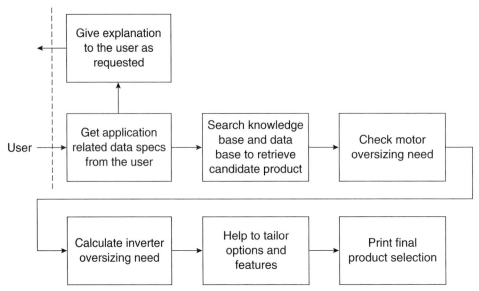

Figure 10.8 Simplified flow diagram for drive product selection

motor rating design, converter design, control design, system simulation design with real-time control software, and finally, to download the software to an experimental drive system. Figure 10.9 shows the flowchart for the design. A considerable amount of cost and time savings is possible with this approach. Figure 10.10 shows the frame structure of the knowledge base.

10.5.4.1 Configuration Selection

The selection of an ac drive configuration involves the selection of a suitable type of drive motor, the type of the power converter topology, and the type of control from the qualitative application specifications supplied by the user. The user is asked to furnish the type of the application, such as pump, fan, elevator, electric vehicle, etc. The other inputs from the user could be his or her preference for the type of machine, converter, and control, the application needs for speed reversal, regenerative braking, line DPF, line harmonic current distortion, acoustic noise, control loops (position, speed, torque), acceleration/deceleration needs, etc. Figure 10.11 shows a simplified selection tree for drive configuration where converters are assumed to be of the voltage-fed type and the bold enclosures indicate that indirect vector-controlled induction motor drive is under consideration here. Each drive configuration has special characteristics, and the recommendation will depend on the matching of these characteristics with the user's needs. If the user inputs are contradictory, the knowledge base will identify them and flag the user. Within the constraints, a knowledgeable designer can also inject his or her own preferences.

10.5.4.2 Motor Ratings Design

Considering that a standard NEMA-classified, cage-type induction motor with a suitable number of poles, efficiency, power factor, and service factor is being used in the present application, this step will determine the machine ratings from the loading and supply considerations,

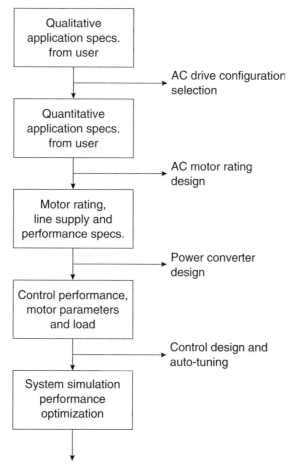

Figure 10.9 AC drive design flowchart

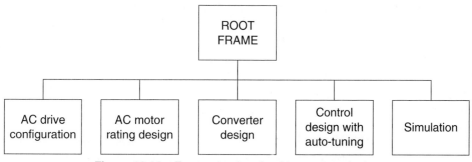

Figure 10.10 Frame structure for drive system design

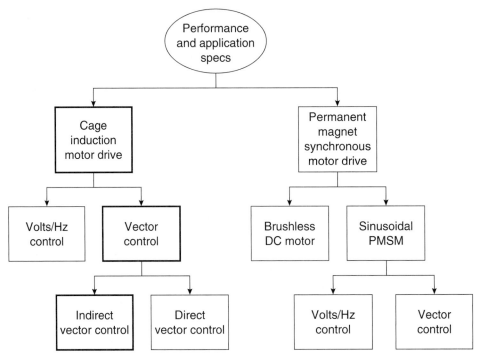

Figure 10.11 AC drive configuration selection tree

that is, power, base speed, field-weakening speed range, steady-state torque, acceleration/deceleration torque, voltage and frequency range, stator current, flux, etc. These ratings will help to select the machine with the help of the appropriate machine database.

10.5.4.3 Converter Design

Once the machine ratings are known and the line supply conditions are given, the converter system design can be automated with the help of a knowledge base. Figure 10.12 shows the converter system with an indirect vector-controlled drive considered in this case. The design of the converter system can be resolved into the design of a diode rectifier in the front end, a PWM IGBT-based inverter associated with polarized RC snubbers, and a dc link electrolytic filter capacitor. A dynamic brake can be used in the dc link, but is not considered in this case. Once the devices are selected, the conduction and switching losses of the power devices are calculated, and based on the thermal model, the heat sink sizes are also determined. Finally, the circuit of the designed converter system with a table of designated components is displayed on the screen. Again, the characteristics and design equations for the rectifier, inverter, filter, and snubber are embedded in the knowledge base with the help of rules and parameters. The catalog information regarding power semiconductors, capacitors,

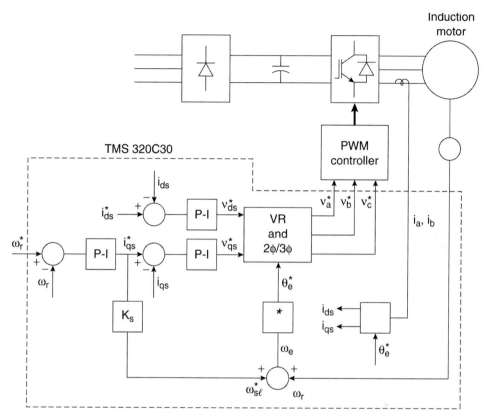

Figure 10.12 Converter system with indirect vector-controlled drive

and snubber components is stored in the database. All the graphical information should be converted into equations or a lookup table. Two sample rules can be given as follows:

Rule 1: IF :: (SUPPLY-INPUT-VOLT IS KNOWN)
    THEN :: (POWER_IGBT_BVCEO) = (FIX(2.15*SUPPLY-INPUT VOLT))
    AND SNUB_D_PIV = (FIX(2.15*SUPPLY-INPUT-VOLT)) AND
    SNUB-C-VOLT = (FIX(2.15*SUPPLY-INPUT-VOLT))

Rule 2: IF :: (MOTOR-MAX-POWER-RATING IS KNOWN AND SUPPLY-INPUT-VOLT
    IS KNOWN)
    THEN :: (ID-AV = (MOTOR-MAX-POWER-RATING/((0.85*0.85)
    * 0.95)*1.414235)*SUPPLY-INPUT-VOLT)

Rule 1 determines the voltage ratings of the IGBT, snubber diode, filter capacitor, and snubber capacitor from the line voltage rating and Rule 2 calculates maximum average dc link current from the motor power and line voltage rating. It is assumed that motor efficiency = 85 percent, inverter efficiency = 95 percent, and minimum line voltage = 85 percent.

10.5.4.4 Control Design and Simulation Study

Once the converter and machine designs are complete, the remaining tasks are the control design and system simulation study with the help of an ES. The objectives of the simulation study are:

- Verify that all voltages and currents are within the safe limit of the converter and machine.
- Iterate the R and C components of the snubber so that turn-on peak current and turn-off peak voltage are within the desired limits.
- Design the real-time control software in the C language so that after compiling, it can be directly downloaded to the experimental drive system.
- Tune the P-I gains for the feedback loops.

The indirect vector-controlled drive with a synchronous current control, shown in Figure 10.12, was considered here. From our knowledge of machine parameters, the slip gain K_s and initial values of the P-I gains of the i_{ds} loop, i_{qs} loop, and ω_r loop can be set. This initial gain setting must be conservative so that the drive remains well within the stability limit. The i_{ds} loop, i_{qs} loop, and speed loop are tuned in sequence, observing the error signal pattern, as discussed previously. The controller software is developed in the C language, and the corresponding object code must be loaded in a Texas Instruments DSP-type TMS320C30.

After designing the machine, converter, and controller, the ES passes the parameter values of the design to the system simulation, as shown in Figure 10.13. The simulation of the designed system, controlled by the ES according to the user's choices, then passes the results back to the ES. The knowledge base contains the expertise for guiding simulation, making intelligent observations, and optimizing system performance. The simulation results can be further processed, if desired. For example, the current wave can be processed by the FFT routine to determine harmonic distribution and THD.

The simulation block, as shown in the figure, is hybrid in nature; that is, the converter and machine (plant) are simulated in slow, non-real-time language, such as MATLAB/SIMULINK, whereas the controller is simulated in real time with the C language. The objective of real-time controller simulation is that the control software, after proper tuning, can be directly downloaded to the experimental system. The advantage of this procedure is obvious. The plant and controller simulations occur continuously in a sequential manner within a certain sampling time with the help of a batch file. The plant simulation is very slow compared to the controller simulation and, at the end of each simulation period, it halts and passes the parameter values to the controller. Similarly, the controller halts after its simulation interval and passes the parameter values to the plant simulation. Note that the format of the commands and data may not be compatible between the ES and simulation programs. Similarly, the plant simulation and controller simulation interfaces may not be compatible. A conversion routine (indicated by C) may be required in each interface. At any time, if the user requests a display of the response, all the simulation programs go into halt mode.

Glossary

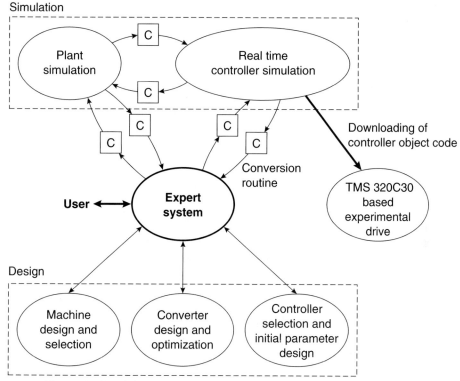

Figure 10.13 ES-based design and simulation of vector-controlled drive

10.6 GLOSSARY

Abbreviated rule language (ARL) – A language that abbreviates rules in English language for implementation by ES shell.

Antecedent rule – A rule that is fired or activated by the forward chaining principle.

Artificial intelligence (AI) – A branch of computer science where computer programs are used to model or simulate some portion of human reasoning or brain activity.

Backward chaining – An inference method where the system starts from a desired conclusion and then finds rules that could have caused that conclusion.

Boolean logic – A logic or reasoning method using Boolean variables (0 and 1).

Database – An organized set of data, facts, and conclusions used by the rules in an expert system.

Declarative knowledge – Fact-like, what-to-do knowledge that is the kernel of the knowledge base.

Domain expert – A person who has extreme proficiency at problem-solving in a particular domain or area.

Forward chaining – An inference method where the IF portion of a rule is tested first to arrive at a conclusion.

Frame – A tree-like structuring of a knowledge base.

Inference – A process of reaching a conclusion using logical operations and facts.

Inference engine – The part of an expert system that tests rules and tends to reach a conclusion.

Knowledge acquisition – The process of transferring knowledge from a domain expert to a knowledge engineer.

Knowledge base – The portion of an expert system that contains rules and data or facts.

Knowledge engineering – The process of translating domain knowledge from a human expert into the knowledge base software.

Knowledge representation – The process of structuring knowledge about a problem into appropriate software.

Meta-rule – A rule that describes how other rules should be used.

Parameter – The variables contained in a rule.

Procedural knowledge – Software structuring for the representation of knowledge.

Rule – A statement in an expert system, usually in an IF-THEN format, which is used to arrive at a conclusion.

Rule base – A cluster of rules that constitute a knowledge base.

Shell – A software development system that is used to develop an expert system program.

Symbolic logic – Solving a logical problem by manipulating symbols.

10.7 SUMMARY

This chapter gives a brief overview of expert system principles and several applications in power electronics area. For a newcomer to this area, it is advisable to first select a shell, understand its features with the help of examples, and then develop and test a simple demo program before applying an ES to a complex application. ES techniques (often defined as classic AI) and their applications flourished during the 1970s and 1980s in many different areas, but unfortunately, their application in power electronics area has been limited. Mainly, the reason is unfamiliarity of the technology in the power electronics community. A glossary was added at the end of this chapter for the convenience of the reader.

In the author's opinion, the vast potential of the ES has yet to be tapped for power electronics applications. Recently, fuzzy logic and neural network technologies, which will be described in the remaining chapters, have practically camouflaged the importance of the ES technology.

REFERENCES

1. B. K. Bose, "Expert system, fuzzy logic, and neural network applications in power electronics and motion control", *Proc. of the IEEE*, vol. 82, pp. 1303–1323, Aug. 1994.
2. M. W. Firebaugh, *Artificial Intelligence*, Boyd and Fraser, Boston, 1988.
3. R. G. Vedder, "PC based expert system shells: Some desirable and less desirable characteristics", *Expert Syst.*, vol. 6, pp. 28–42, Feb. 1989.
4. K. L. Anderson et al., "A rule based adaptive PID controller", MS Thesis, University of Maryland, May 1989.
5. C. Daoshen and B. K. Bose, "Expert system based automated selection of industrial AC drives", *IEEE IAS Annu. Meet. Conf. Rec.*, pp. 387–392, 1992.
6. K. Debebe, V. Rajagopalan, and T. S. Sanker, "Expert systems for fault diagnosis of VSI-fed ac drives", *IEEE IAS Annu. Meet. Conf. Rec.*, pp. 368–373, 1991.
7. K. Debebe, V. Rajagopalan, and T. S. Sanker, "Diagnostics and monitoring for ac drives", *IEEE IAS Annu. Meet. Conf. Rec.*, pp. 370–377, 1992.
8. S. M. Chhaya and B. K. Bose, "Expert system based automated design technique of a voltage-fed inverter for induction motor drive", *IEEE IAS Annu. Meet. Conf. Rec.*, pp. 770–778, 1992.
9. S. M. Chhaya and B. K. Bose, "Expert system aided automated design, simulation and controller tuning of ac drive system", *IEEE IECON Conf. Rec.*, pp. 712–718, 1995.
10. S. M. Chhaya, "Expert system based automated design of AC drive systems", Ph.D. Dissertation, University of Tennessee, Knoxville, Aug. 1995.

CHAPTER 11

Fuzzy Logic Principles and Applications

11.1 INTRODUCTION

What is fuzzy logic (FL)? FL is another class of AI, but its history and applications are more recent than those of the expert system (ES). In Chapter 10, it was mentioned that according to George Boole, human thinking and decisions are based on "yes"/"no" reasoning, or "1"/"0" logic. Accordingly, Boolean logic was developed, and ES principles were formulated based on Boolean logic. It has been argued that human thinking does not always follow crisp "yes"/"no" logic, but is often vague, qualitative, uncertain, imprecise, or fuzzy in nature. For example, in terms of "yes"/"no" logic, a thinking rule may be

"IF it is not raining AND outside temperature is less than 80°F
THEN take a sightseeing trip for more than 100 miles"

In actual thinking, it might be

"IF weather is good AND outside temperature is mild
THEN take a long sightseeing trip"

Based on the nature of fuzzy human thinking, Lotfi Zadeh, a computer scientist at the University of California, Berkeley, originated the "fuzzy logic," or fuzzy set theory, in 1965. In the beginning, he was highly criticized by the professional community, but gradually, FL captured the imagination of the professional community and eventually emerged as an entirely new discipline of AI. The general methodology of reasoning in FL and ES by "IF... THEN..."

statements or rules is the same; therefore, it is often called "fuzzy expert system." For example, an ES rule for speed control in a variable-speed drive may be

> IF speed of the motor is greater than 1500 rpm AND the machine stator temperature is between 60°F and 100°F
> THEN set the stator current i_{qs} less than 10 amps

The same rule in FL may read as

> IF speed of the motor is high and stator temperature is medium
> THEN set the stator current i_{qs} low

FL can help to supplement an ES, and it is sometimes hybrided with the latter to solve complex problems. FL has been successfully applied in process control, modeling, estimation, identification, diagnostics, military science, stock market prediction, etc.

In this chapter, we will discuss the principles of FL and some of its applications in power electronic systems. The Fuzzy Logic Toolbox in the MATLAB environment, which includes an example program development, will then be introduced. Neuro-fuzzy systems will be discussed in Chapter 12.

11.2 FUZZY SETS

FL, unlike Boolean logic, deals with problems that have fuzziness or vagueness, as mentioned before. The classical set theory is based on Boolean logic, where a particular object or variable is either a member of a given set (logic 1), or it is not (logic 0). On the other hand, in fuzzy set theory based on FL, a particular object has a degree of membership in a given set that may be anywhere in the range of 0 (completely not in the set) to 1 (completely in the set). For this reason, FL is often defined as multi-valued logic (0 to 1), compared to bi-valued Boolean logic. It may be mentioned that although FL deals with imprecise information, the information is processed in sound mathematical theory, which has been advanced in recent years.

Before discussing the FL theory, it should be emphasized here that basically, a FL problem can be defined as an input/output, static, nonlinear mapping problem through a "black box," as shown in Figure 11.1. All the input information is defined in the input space, it is processed in the black box, and the solution appears in the output space. In general, mapping can be static or dynamic, and the mapping characteristics are determined by the black box's characteristics. The black box cannot only be a fuzzy system, but also an ES, neural network, general mathematical system, such as differential equations, algebraic equations, etc., or anything else.

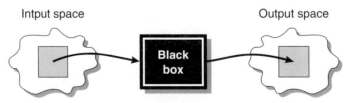

Figure 11.1 Input/output mapping problem

11.2.1 Membership Functions

A fuzzy variable has values that are expressed by the natural English language. For example, as shown in Figure 11.2(a), the stator temperature of a motor as a fuzzy variable can be defined by the qualifying linguistic variables Cold, Mild, or Hot, where each is represented by a triangular or straight-line segment membership function (MF). These linguistic variables are defined as fuzzy sets or fuzzy subsets. An MF is a curve that defines how the values of a fuzzy variable in a certain region are mapped to a membership value μ (or degree of membersip) between 0 and 1. The fuzzy sets can have more subdivisions such as Zero, Very Cold, Medium Cold, Medium Hot, Very Hot, etc. for a more precise description of the fuzzy variable. In Figure 11.2(a), if the temperature is below 40° F, it belongs completely to the set Cold, that is, the MF value is 1; whereas for 55° F, it is in the set Cold by 30 percent (MF = 0.3) and to the set Mild by 50 percent (MF = 0.5). At temperature 60° F, it belongs completely to the set Mild (MF = 1) and not in the set Cold and Hot (MF = 0). If the temperature is above 80° F, it belongs completely to the set Hot (MF = 1), where MF = 0 for Cold and Mild. In Figure 11.2(b), the corresponding crisp or Boolean classification of the variable is given for comparison. For the temperature range below 55° F, it belongs to the set Cold (MF = 1); between 55° F to 65° F, it belongs to the set Mild (MF = 1); and above 65° F, it belongs to the set Hot only (MF = 1). The sets are not members (MF = 0) beyond the defined ranges. The numerical interval (20° F to 90° F) that is relevant for the description of a fuzzy variable is defined as the universe of discourse in Figure 11.2(a).

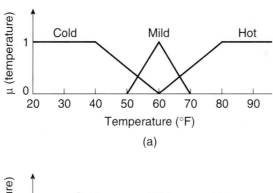

(a)

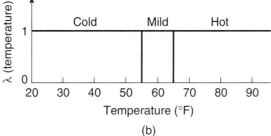

(b)

Figure 11.2 Representation of temperature using (a) Fuzzy sets, (b) Crisp set

An MF can have different shapes, as shown in Figure 11.3. The simplest and most commonly used MF is the triangular-type, which can be symmetrical or asymmetrical in shape. A trapezoidal MF (symmetrical or unsymmetrical) has the shape of a truncated triangle. Two MFs are built on the Gaussian distribution curve: a simple Gaussian curve and a two-sided composite of two different Gaussian curves. The bell MF with a flat top is somewhat different from a Gaussian function. Both the Gaussian and bell functions are smooth and non-zero at all points. A sigmoidal-type MF can be open to the right or left. Asymmetrical and closed (not open to the right or left) MFs can be synthesized using two sigmoidal functions, such as difference sigmoidal (difference between two sigmoidal functions) and product sigmoidal (product of two sigmoids). Polynomial-based curves can have several functions, including asymmetrical polynomial curve open to the left (Polynomial-Z) and its mirror image, open to the right (Polynomial-S), and one that is zero at both ends but has a rise in the middle (Polynomial-Pi).

In addition to these types, any arbitrary MF can be generated by the user. In practice, one or two types of MFs (such as triangular and Gaussian) are more than enough to solve most problems. A singleton is a special type of MF that has a value of 1 at one point on the universe of discourse and zero elsewhere (a vertical spike). MFs can be represented by mathematical functions, segmented straight lines (for triangular and trapezoidal shapes), and look-up tables.

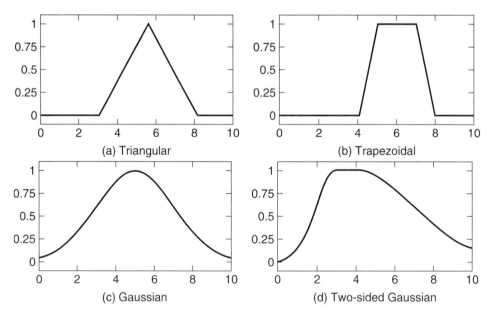

Figure 11.3 Different types of membership functions; (a) Triangular, (b) Trapezoidal, (c) Gaussian, (d) Two-sided Gaussian

(Continued)

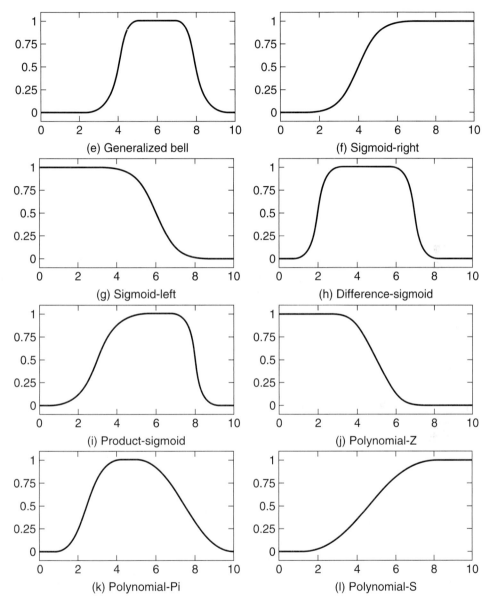

Figure 11.3 (Cont.) Different types of membership functions; (e) Generalized bell, (f) Sigmoid-right, (g) Sigmoid-left, (h) Difference-Sigmoid, (i) Product-Sigmoid, (j) Polynomial-Z, (k) Polynomial-Pi, (l) Polynomial-S

11.2.2 Operations on Fuzzy Sets

The basic properties of Boolean logic are also valid for FL. Figure 11.4 shows the logical operations of OR, AND, and NOT on fuzzy sets A and B using triangular MFs and compares them with the corresponding Boolean operations on the right. Let $\mu_A(x)$, $\mu_B(x)$ denote the degree of membership of a given element x in the universe of discourse X (denoted by $x \in X$).

Union: Given two fuzzy sets A and B, defined in the universe of discourse X, the union ($A \cup B$) is also a fuzzy set of X, with the membership function given as

$$\mu_{A \cup B}(x) \equiv \max\left[\mu_A(x), \mu_B(x)\right]$$
$$\equiv \mu_A(x) \vee \mu_B(x) \tag{11.1}$$

where the symbol "$\vee$" is a maximum operator. This is equivalent to Boolean OR logic.

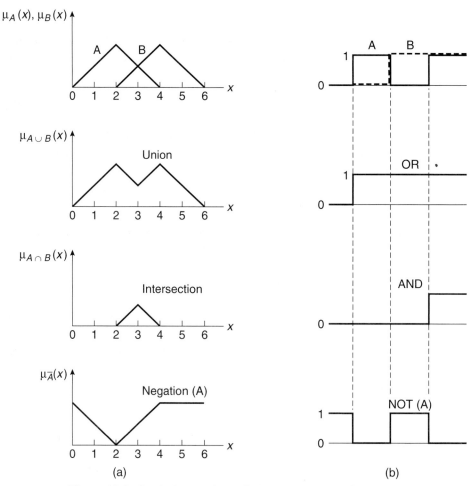

Figure 11.4 Logical operations of (a) Fuzzy sets, (b) Crisp sets

Intersection: The intersection of two fuzzy sets A and B in the universe of discourse X, denoted by $A \cap B$, has the membership function given by

$$\mu_{A \cap B}(x) \equiv \min[\mu_A(x), \mu_B(x)] \\ \equiv \mu_A(x) \wedge \mu_B(x) \quad (11.2)$$

where "$\wedge$" is a minimum operator. This is equivalent to Boolean AND logic.

Complement or Negation: The complement of a given set A in the universe of discourse X is denoted by $\overline{A}$ and has the membership function

$$\mu_{\overline{A}}(x) \equiv 1 - \mu_A(x) \quad (11.3)$$

This is equivalent to the NOT operation in Boolean logic.

In FL, we can also define the following operations:

Product of two fuzzy sets: The product of two fuzzy sets A and B defined in the same universe of discourse X is a new fuzzy set, $A.B$, with an MF that equals the algebraic product of the MFs of A and B,

$$\mu_{A.B}(x) \equiv \mu_A(x) . \mu_B(x) \quad (11.4)$$

which can be generalized to any number of fuzzy sets in the same universe of discourse.

Multiplying Fuzzy Set by a Crisp Number: The MF of fuzzy set A can be multiplied by a crisp number k to obtain a new fuzzy set called product $k.A$. Its MF is

$$\mu_{kA(x)} \equiv k.\mu_A(x) \quad (11.5)$$

Power of a Fuzzy Set: We can raise fuzzy set A to a power m (positive real number) by raising its MF to m. The m power of A is a new fuzzy set, A^m, with MF

$$\mu_{A^m}(x) \equiv [\mu_A(x)]^m \quad (11.6)$$

Fuzzy set properties, as discussed above, are useful in performing additional operations using fuzzy sets. Consider the fuzzy sets A, B, and C defined over a common universe of discourse X. The following properties are valid for crisp and fuzzy sets, but some are more specific to fuzzy sets.

Double Negation:

$$\overline{(\overline{A})} = A \quad (11.7)$$

Idempotency:

$$A \cup A = A \\ A \cap A = A \quad (11.8)$$

Commutativity:

$$A \cap B = B \cap A$$
$$A \cup B = B \cup A \quad (11.9)$$

Associative Property:

$$(A \cup B) \cup C = A \cup (B \cup C)$$
$$(A \cap B) \cap C = A \cap (B \cap C) \quad (11.10)$$

Distributive Property:

$$A \cup (B \cap C) = (A \cup B) \cap (A \cup)C$$
$$A \cap (B \cup C) = (A \cap B) \cup (A \cap C) \quad (11.11)$$

Absorption:

$$A \cap (A \cap B) = A$$
$$A \cup (A \cap B) = A \quad (11.12)$$

De Morgan's Theorems:

$$\overline{A \cup B} = \overline{A} \cap \overline{B}$$
$$\overline{A \cap B} = \overline{A} \cup \overline{B} \quad (11.13)$$

In fuzzy sets, all these properties can be expressed using the MF of the sets involved and the basic definition of union, intersection, and complement. For example, the distributive property in Equation (11.11) can be written in the

$$\mu_A(x) \vee (\mu_B(x) \wedge \mu_C(x)) = (\mu_A(x) \vee \mu_B(x)) \wedge (\mu_A(x) \vee \mu_C(x))$$
$$\mu_A(x) \wedge (\mu_B(x) \vee \mu_C(x)) = (\mu_A(x) \wedge \mu_B(x)) \vee (\mu_A(x) \wedge \mu_C(x)) \quad (11.14)$$

Similarly, the De Morgan's theorems in (11.13) can be written in the form

$$\overline{\mu_A(x) \vee \mu_B(x)} = \mu_{\overline{A}}(x) \wedge \mu_{\overline{B}}(x)$$
$$\overline{\mu_A(x) \wedge \mu_B(x)} = \mu_{\overline{A}}(x) \vee \mu_{\overline{B}}(x) \quad (11.15)$$

11.3 FUZZY SYSTEM

A fuzzy inference system (or fuzzy system) basically consists of a formulation of the mapping from a given input set to an output set using FL, as indicated in Figure 11.1. This mapping process provides the basis from which the inference or conclusion can be made. A fuzzy inference process consists of the following five steps:

- Step 1: Fuzzification of input variables
- Step 2: Application of fuzzy operator (AND, OR, NOT) in the IF (antecedent) part of the rule
- Step 3: Implication from the antecedent to the consequent (THEN part of the rule)
- Step 4: Aggregation of the consequents across the rules
- Step 5: Defuzzification

Let us first take a simple non-technical example of FL application and illustrate all the above five steps. Figure 11.5 shows a typical fuzzy inference system for restaurant tipping, where Food and Service are the input fuzzy variables (0–10 range) and Tip is the output fuzzy variable (0–25% range). The output is the aggregation of the evaluation of the three rules shown in the system. Normally, the tipping rules are evaluated in our mind to come to the decision of how much tip we should give. But, fuzzy systems theory can be applied to compute the precise output, which is explained in Figure 11.6.

In this example, the input variable Service is represented by three fuzzy sets (see Figure 11.2) Poor, Good, and Excellent, which correspond to curved MFs. The variable Food is represented by two fuzzy sets, Bad and Delicious, which correspond to straight-line MFs. The output variable Tip is represented by three sets Cheap, Average, and Generous, which correspond to triangular MFs. The universe of discourse for the input variables is 0–10, whereas for the output variable is 0%–25%. The processing of the three rules 1, 2, and 3 in the horizontal direction is shown in the figure. Consider, for example, that the score of the quality of service is 3. This crisp input, when referred to MF Poor, gives the output $\mu = 0.3$, which is the result of fuzzification (Step 1). If the score for Food is 8 and is referred to MF Bad, the result of fuzzification is $\mu = 0$, as shown in the figure. Once the

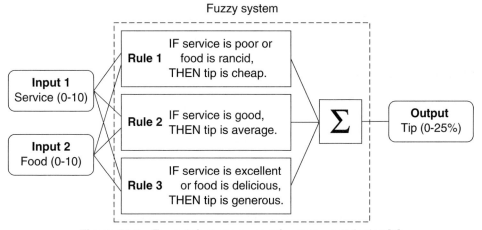

Figure 11.5 Fuzzy inference system for restaurant tipping [3]

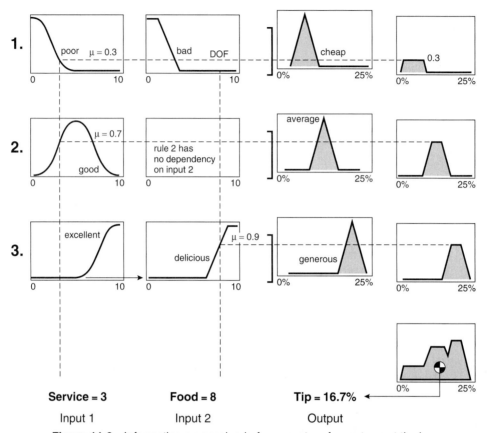

Figure 11.6 Information processing in fuzzy system for restaurant tipping

inputs have been fuzzified, we know the degree to which each part of the antecedent of a rule has been satisfied. In the rule, the OR or max operator is specified, and therefore, between the two values, 0.3 and 0, the result of the fuzzy operator is 0.3, that is, the 0.3 value is selected (Step 2). This is also defined as the degree of fulfillment (DOF) of a rule. If, on the other hand, the rule contains an AND or minimum operator, the value 0 will be selected. The implication step helps to evaluate the consequent part of a rule. In this rule, the output MF Cheap is truncated at the value $\mu = 0.3$ to give the fuzzy output (Step 3) shown. All three rules are evaluated in the same manner and their contributions are shown on the right. These outputs are combined or aggregated in a cumulative manner to result in the final fuzzy output (Step 4) shown at the bottom right of the figure. Finally, the fuzzy output (area) is converted to crisp output (Tip = 16.7%); a single number, which is defined as defuzzification (Step 5). Typically, it is the centroid or center of gravity of the area. Note that in Figure 11.5, the information is processed in the forward direction only in a parallel manner and the input/output mapping property is evident.

So far, we have given a simple example of a fuzzy restaurant tipping system to clarify some of the concepts in a fuzzy system. In general, there will be a matrix of rules similar to the

Fuzzy System

ES rule matrix shown in Figure 10.1. If, for example, there are 7 MFs for input variable X and 5 MFs for input variable Y, then there will be all together $7 \times 5 = 35$ rules (see Table 11.1).

11.3.1 Implication Methods

The implication step (Step 3) was introduced in the above example for the evaluation of individual rules. There are a number of implication methods in the literature, so we will study a few of the types that are frequently used.

11.3.1.1 Mamdani Type

Mamdani, one of the pioneers in the application of FL in control systems, proposed this implication method which, in fact, has been applied in the above tipping example. This is the most commonly used implication method. Let us again consider three rules in a fuzzy system, which are given in general form given as

Rule 1: IF X is negative small (NS) AND Y is zero (ZE)
THEN Z is positive small (PS)

Rule 2: IF X is zero (ZE) AND Y is zero (ZE)
THEN Z is zero (ZE)

Rule 3: IF X is zero (ZE) AND Y is positive small (PS)
THEN Z is negative small (NS)

where X and Y are the input variables, Z is the output variable, and NS, ZE, and PS are the fuzzy sets. Figure 11.7 explains the fuzzy inference system with the Mamdani method for inputs $X = -3$ and $Y = 1.5$. Note that all the rules have an AND operator. From the figure, the DOF of Rule 1 can be given as

$$DOF_1 = \mu_{NS}(X) \wedge \mu_{ZE}(Y) = 0.8 \wedge 0.6 = 0.6 \tag{11.16}$$

where $\wedge$ = minimum operator and $\mu_{NS}(X)$ and $\mu_{ZE}(Y)$ are the MFs of X and Y, respectively. The rule output is given by the truncated MF PS', as shown in the figure. Similarly, for Rules 2 and 3, we can write

$$DOF_2 = \mu_{ZE}(X) \wedge \mu_{ZE}(Y) = 0.4 \wedge 0.6 = 0.4 \tag{11.17}$$

$$DOF_3 = \mu_{ZE}(X) \wedge \mu_{PS}(Y) = 0.4 \wedge 1.0 = 0.4 \tag{11.18}$$

The corresponding fuzzy output MFs are ZE' and NS', respectively, as indicated in the figure. The total fuzzy output is the union (OR) of all the component MFs,

$$\mu_{OUT}(Z) = \mu_{PS'}(Z) \vee \mu_{ZE'}(Z) \vee \mu_{NS'}(Z) \tag{11.19}$$

which is shown in the lower right part of the figure. The defuzzification to convert the fuzzy output to crisp output will be discussed later.

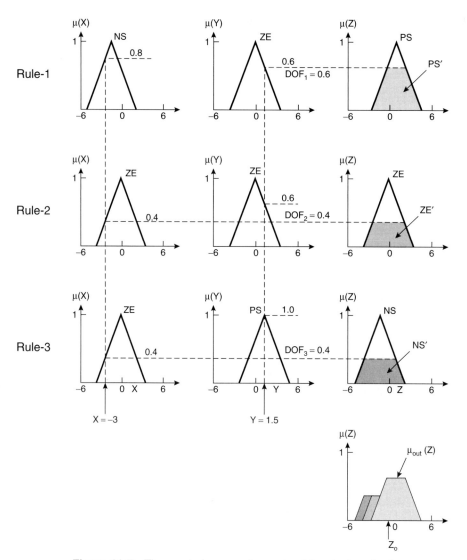

Figure 11.7 Three-rule fuzzy system using Mamdani method

11.3.1.2 Lusing Larson Type

In this method, the output MF is scaled instead of being truncated, as shown in Figure 11.8. In this case, the same three rules are considered as well as the same inputs, of $X = -3$ and $Y = 1.5$ giving DOFs of $DOF_1 = 0.6$, $DOF_2 = 0.4$, and $DOF_3 = 0.4$. The output MF PS of Rule 1 is scaled so that the output is PS' with a peak value of 0.6 as shown. Similarly, Rules 2 and 3 give output MFs ZE' and NS', each with a peak value of 0.4 as indicated. The total output MF is given by Equation (11.19). The output area is somewhat different from that of the Mamdani method, and the corresponding crisp output will differ slightly.

Fuzzy System

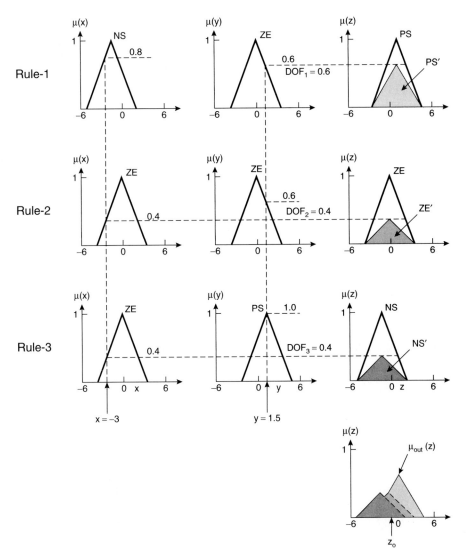

Figure 11.8 Three-rule fuzzy system using Lusing Larson method

11.3.1.3   Sugeno Type

The Sugeno, or Takagi-Sugeno-Kang method of implication was first introduced in 1985. The difference here is that unlike the Mamdani and Lusing Larson methods, the output MFs are only constants or have linear relations with the inputs. With a constant output MF (singleton), it is defined as the zero-order Sugeno method, whereas with a linear relation, it is

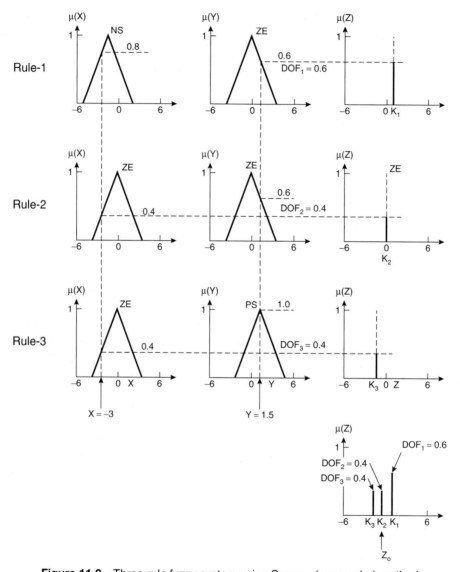

Figure 11.9 Three-rule fuzzy system using Sugeno (zero-order) method

known as the first-order Sugeno method. Figure 11.9 shows a three-rule fuzzy system using the Sugeno zero-order method where the rules read as

Rule 1: IF X is NS AND Y is ZE
THEN $Z = K_1$
Rule 2: IF X is ZE AND Y is ZE
THEN $Z = K_2$
Rule 3: IF X is ZE AND Y is PS
THEN $Z = K_3$

The constants K_1, K_2, and K_3 are crisply defined constants, as shown in the figure, in the consequent part of each rule. The output MF in each rule is a singleton spike, which is multiplied by the respective DOF to contribute to the fuzzy output of each rule. These MFs (truncated vertical segments) are then aggregated to constitute the total fuzzy output as shown in the figure. Fortunately, it can be shown that if the Mamdani, Lusing Larson, and Sugeno methods are applied to the same problem, the output will be approximately the same.

The more general first-order Sugeno method, as shown in Figure 11.10, has the rules in the form

Rule 1: IF X is NS AND Y is ZE
THEN $Z = Z_1 = A_{01} + A_{11}X + A_{21}Y$

Rule 2: IF X is ZE AND Y is ZE
THEN $Z = Z_2 = A_{02} + A_{12}X + A_{22}Y$

Rule 3: IF X is ZE AND Y is PS
THEN $Z = Z_3 = A_{03} + A_{13}X + A_{23}Y$

where all the A's are constants. An easy way to visualize a first-order system is to think of each rule as defining the location of a moving singleton, that is, the singleton output spikes can move around in a linear fashion in the output space depending on the input signal values. Higher order Sugeno methods are also possible, but are not of much practical use. In power electronics applications, we will later give one example with a Sugeno first-order method and the remaining examples with the Mamdani method. The Sugeno method is widely used in adaptive neuro-fuzzy inference systems (ANFISs) which are discussed in Chapter 12.

11.3.2 Defuzzification Methods

So far, we have discussed the different steps of a fuzzy inference system, except the final defuzzification method. The result of the implication and aggregation steps is the fuzzy output, which is the union of all the outputs of individual rules that are validated or "fired." Conversion of this fuzzy output to crisp output is defined as defuzzification. We will now discuss a few important methods of defuzzification.

11.3.2.1 Center of Area (COA) Method

In the COA method (often called the center of gravity method) of defuzzification, the crisp output Z_0 of the Z variable is taken to be the geometric center of the output fuzzy value $\mu_{out}(Z)$ area, where $\mu_{out}(Z)$ is formed by taking the union of all the contributions of rules whose DOF > 0 (see, for example, the lower right part of Figure 11.7). The general expression for COA defuzzification is

$$Z_0 = \frac{\int Z \cdot \mu_{out}(Z)}{\int \mu_{out}(Z) dZ} \tag{11.20}$$

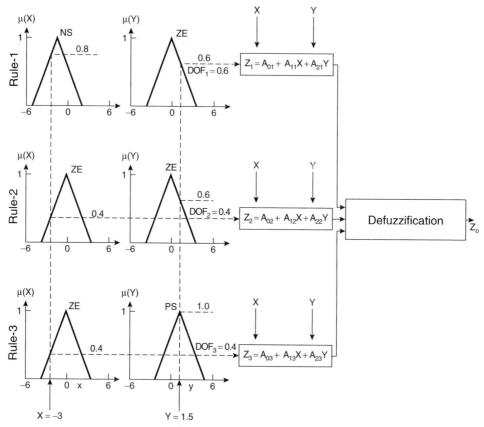

Figure 11.10 Three-rule fuzzy system using Sugeno (first-order) method

With a discretized universe of discourse, the expression is

$$Z_0 = \frac{\sum_{i=1}^{n} Z_i \mu_{out}(Z_i)}{\sum_{i=1}^{n} \mu_{out}(Z_i)} \qquad (11.21)$$

For example, Figure 11.11 shows a simple fuzzy output for a two-rule system where the COA formula gives crisp output as

$$Z_0 = \frac{1.\frac{1}{3}+2.\frac{2}{3}+3.\frac{2}{3}+4.\frac{2}{3}+5.\frac{1}{3}+6.\frac{1}{3}+7.\frac{1}{3}}{\frac{1}{3}+\frac{2}{3}+\frac{2}{3}+\frac{2}{3}+\frac{1}{3}+\frac{1}{3}+\frac{1}{3}} = 3.7 \qquad (11.22)$$

Fuzzy System

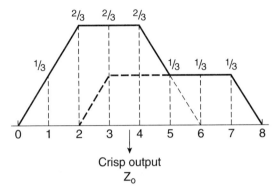

Figure 11.11 Defuzzification of output for a two-rule system

COA defuzzification is a well-known method and it is often used in spite of some amount of complexity in the calculation. Note that if the areas of two or more contributing rules overlap, the overlapping area is counted only once, which should be evident from Figure 11.11.

11.3.2.2 Height Method

In the height method of defuzzification, the COA method is simplified to consider only the height of each contributing MF at the mid-point of the base. For example, in Figure 11.11,

$$Z_0 = \frac{3.\frac{2}{3} + 5.\frac{1}{3}}{\frac{2}{3} + \frac{1}{3}} = 3.67 \tag{11.23}$$

which is slightly less than the 3.7 obtained by COA method.

11.3.2.3 Mean of Maxima (MOM) Method

The height method of defuzzification is further simplified in the MOM method, where only the highest membership function component in the output is considered. According to this method, the output $Z_0 = 3$ in Figure 11.11. If M such maxima are present, then the formula is

$$Z_0 = \sum_{m=1}^{M} \frac{Z_m}{M} \tag{11.24}$$

where $Z_m = m^{th}$ element in the universe of discourse, where the output MF is at the maximum value, and M = number of such elements.

11.3.2.4 Sugeno Method

In the Sugeno method, defuzzification is very simple. For example, in the zero-order method shown in Figure 11.9, the defuzzification formula is

$$Z_0 = \frac{K_1.DOF_1 + K_2 DOF_2 + K_3.DOF_3}{DOF_1 + DOF_2 + DOF_3} \quad (11.25)$$

whereas in the first-order method shown in Figure 11.10, the formula is

$$Z_0 = \frac{Z_1.DOF_1 + Z_2 DOF_2 + Z_3.DOF_3}{DOF_1 + DOF_2 + DOF_3} \quad (11.26)$$

11.4 FUZZY CONTROL

11.4.1 Why Fuzzy Control?

The control algorithm of a process that is based on FL or a fuzzy inference system, as discussed above, is defined as a fuzzy control. In general, a control system based on AI is defined as intelligent control. A fuzzy control system essentially embeds the experience and intuition of a human plant operator, and sometimes those of a designer and/or researcher of a plant. The design of a conventional control system is normally based on the mathematical model of a plant. If an accurate mathematical model is available with known parameters, it can be analyzed, for example, by a Bode or Nyquist plot, and a controller can be designed for the specified performance. Such a procedure is tedious and time-consuming, although CAD programs are available for such design. Unfortunately, for complex processes, such as cement plants, nuclear reactors, and the like, a reasonably good mathematical model is difficult to find. On the other hand, the plant operator may have good experience for controlling the process.

Power electronics system models are often ill-defined. Even if a plant model is well-known, there may be parameter variation problems. Sometimes, the model is multivariable, complex, and nonlinear, such as the dynamic *d-q* model of an ac machine. Vector or field-oriented control of a drive can overcome this problem, but accurate vector control is nearly impossible, and there may be a wide parameter variation problem in the system. To combat such problems, various adaptive control techniques were discussed in Chapter 8. Fuzzy control, on the other hand, does not strictly need any mathematical model of the plant. It is based on plant operator experience and heuristics, as mentioned previously, and it is very easy to apply. Fuzzy control is basically an adaptive and nonlinear control, which gives robust performance for a linear or nonlinear plant with parameter variation. In fact, fuzzy control is possibly the best adaptive control among the techniques discussed so far.

11.4.2 Historical Perspective

The history of fuzzy control applications is very interesting. Since the development of FL theory by Zadeh in 1965, its first application to control a dynamic process was reported by

Fuzzy Control

Mamdani in 1974, and by Mamdani and Assilian in 1975. These were extremely significant contributions because they stirred widespread interest by later workers in the field. Mamdani and Assilian were concerned with the control of a small laboratory steam engine. The control problem was to regulate the engine speed and boiler steam pressure by means of the heat applied to the boiler and the throttle setting of the engine. The process was difficult because it was nonlinear, noisy, and strongly coupled, and no mathematical model was available. The fuzzy control designed purely from the operator's experience by a set of IF...THEN rules was found to perform well and was better than manual control.

In 1976, Kickert and Lemke examined the fuzzy control performance of an experimental warm water plant, where the problem was to regulate the temperature of water leaving a tank at a constant flow rate by altering the flow of hot water in a heat exchanger contained in the tank. The success of Mamdani and Assilian's work led King and Mamdani (1977) to attempt to control the temperature in a pilot-scale batch chemical reactor by a fuzzy algorithm. Tong (1976) also applied FL to the control of a pressurized tank containing liquid. These results indicated that fuzzy control was very useful for complex processes and gave superior performance over conventional P-I-D control.

FL applications in power electronics and motor drives are somewhat recent. Li and Lau (1989) applied FL to a microprocessor-based servo motor controller, assuming a linear power amplifier. They compared the fuzzy-controlled system performance with that of P-I-D control and MRAC and demonstrated the superiority of the fuzzy system. Da Silva et al. (1987) developed a fuzzy adaptive controller and applied it to a four-quadrant power converter for the first time. Gradually, fuzzy control gathered momentum to other applications in the power electronics and drives areas.

11.4.3 Control Principle

Consider the fuzzy speed controller block in a vector-controlled drive system shown in Figure 11.12. The controller observes the pattern of the speed loop error signal and correspondingly updates the output DU so that the actual speed ω_r matches the command speed ω_r^*. There

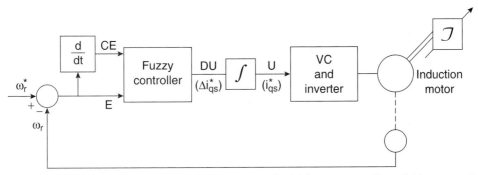

Figure 11.12 Fuzzy speed controller in vector-controlled drive system with variable moment of inertia (J)

are two input signals to the fuzzy controller, the error $E = \omega_r^* - \omega_r$ and the change in error, CE, which is related to the derivative dE/dt of error. In a discrete system, $dE/dt = \Delta E/\Delta t = CE/T_s$, where $CE = \Delta E$ in the sampling time T_s. With constant T_s, CE is proportional to dE/dt. The controller output DU in a vector-controlled drive is Δi_{qs}^* current. This signal is summed or integrated to generate the actual control signal U or current i_{qs}^*. From the physical operation principle of the system, we can write a simple control rule in FL as

> IF E is near zero (ZE) AND CE is slightly positive (PS)
> THEN the controller output DU is small negative

where E and CE are the input fuzzy variables, DU is the output fuzzy variable, and ZE, PS, and NS are the corresponding fuzzy set MFs. The implementation of this fuzzy control is illustrated in Figure 11.13 with the help of triangular MFs. With the values of $E = -1$ and $CE = 1.8$, as shown, the Mamdani method will give the output DU or $\Delta i_{qs}^* = -2$ amps. As discussed in the previous section, generally more than one fuzzy rule is fired and the contribution of the individual rules is combined at the output. Figure 11.14 illustrates the principle of two-rule control with the Mamdani method where the rules are

> Rule 1: IF E = ZE AND CE = NS THEN DU = NS
>
> Rule 2: IF E = PS AND CE = NS THEN DU = ZE

where DU represents the output. For the given rule base of the control system, the fuzzy controller computes a meaningful control action for a specific input condition of the variables. The term "composition" is often used for the inference to generate the output fuzzy control signal. There are a number of composition rules in the literature, but the most common one is the MAX-MIN (or SUP-MIN) composition, which is illustrated in Figure 11.14 for the two stated rules. Note that the output MF of each rule is given by a MIN (minimum) operator, whereas the combined fuzzy output is given by a MAX (maximum) operator.

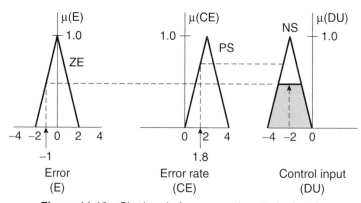

Figure 11.13 Single-rule fuzzy speed control principle

Fuzzy Control

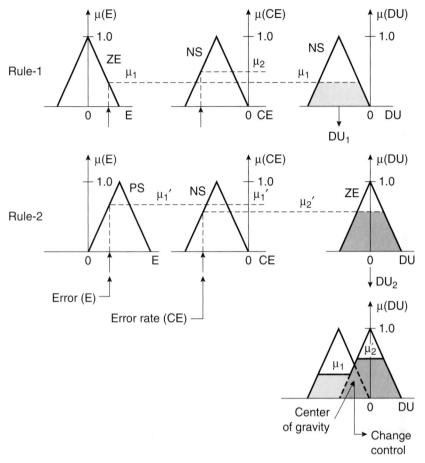

Figure 11.14 Two-rule fuzzy speed control principle

Since the fuzzy controller is basically an input/output static nonlinear mapping, we can write the controller action in the form

$$K_1 E + K_2 CE = DU \qquad (11.27)$$

where K_1 and K_2 are nonlinear coefficients or gain factors. Including the summation process shown in Figure 11.12, we can write

$$\int DU = \int K_1 E dt + \int K_2 CE dt \qquad (11.28)$$

or

$$U = K_1 \int E dt + K_2 E \qquad (11.29)$$

which is nothing but a fuzzy P-I controller with nonlinear gain factors. Extending the same principle, we can write a fuzzy control algorithm for P and P-I-D control as follows:

Fuzzy P control example rule: IF E = positive small (PS) THEN U = positive big (PB)

In othe words, $KE = U$, where K is nonlinear gain

Fuzzy P-I-D control example rule: IF E = PS AND CE = NS AND C^2E = PS THEN DU = ZE,

where C^2E is the derivative of CE. The control can be written in the form

$$K_1 E + K_2 CE + K_3 C^2 E = DU \qquad (11.30)$$

and including the output integration

$$\int DU = \int K_1 E dt + \int K_2 CE dt + \int K_3 C^2 E dt \qquad (11.31)$$

or

$$U = K_1 \int E dt + K_2 E + K_3 \frac{d}{dt}(E) \qquad (11.32)$$

which is nothing but a *P-I-D* controller. The nonlinear adaptive gains in the fuzzy controller that are varied on-line give the power to the fuzzy controller to make the system response robust in the presence of parameter variation and load disturbance.

The general structure of a fuzzy feedback control system is shown in Figure 11.15. The loop error E and change in error CE signals are converted to the respective per unit signals e and ce by dividing by the respective scale factors, that is, $e = E/GE$ and $ce = CE/GC$. Similarly, the output plant control signal U is derived by multiplying the per unit output by the scale factor GU, that is, $DU = du.GU$, and then summed to generate the U signal.

The advantage of fuzzy control in terms of per unit variables is that the same control algorithm can be applied to all the plants of the same family. Besides, it becomes convenient to design the fuzzy controller. The scale factors can be constant or programmable; programmable scale factors can control the sensitivity of operation in different regions of control or the same

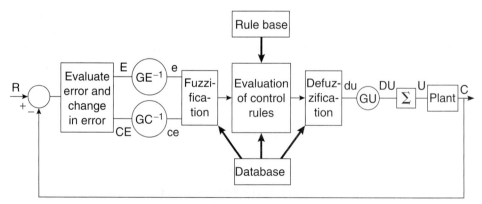

Figure 11.15 Structure of fuzzy control in feedback system

control strategy can be applied in similar response loops. The processes of fuzzification, evaluation of control rules from the rule base and database (defined as the knowledge base), and defuzzification have already been discussed.

11.4.4 Control Implementation

There are essentially two methods for implementation of fuzzy control. The first involves rigorous mathematical computation for fuzzification, evaluation of control rules, and defuzzification in real time. This is the generally accepted method and will be illustrated in our example applications described later. An efficient C program is normally developed with the help of a FL tool, such as the Fuzzy Logic Toolbox (described later) in the MATLAB (Math Works, Inc.) environment. The program is compiled and the object program is loaded in a DSP (digital signal processor) for execution. Commercial (digital or analog) ASIC chips are also available for implementation. The second method is the look-up table method, where all the input/output static mapping computation (fuzzification, evaluation of control rules, and defuzzification) is done ahead of time and stored in the form of a large look-up table for real-time implementation. Instead of one look-up table, there may be hierarchical tables (coarse, medium, and fine) [6]. Look-up tables require large amounts of memory for precision control, but their execution may be fast. A neural network (described in Chapter 12) can also be trained to emulate a fuzzy controller.

11.5 GENERAL DESIGN METHODOLOGY

The discussions given above will give guidance to readers for the formulation of an FL application for a certain problem and its implementation. A number of example applications in power electronics, which are given later, will make the concepts more clear. In summary, the general design procedure for fuzzy control can be given as follows:

1. First, analyze whether the problem has sufficient elements to warrant a FL application; otherwise, apply a conventional method. For example, in a linear feedback system where the mathematical model is known and there is no parameter variation or load disturbance problem, FL has little advantage. An ES or neural network can also be used, if necessary.
2. Get all the information from the operator of the plant to be controlled. Get information about the design and operational characteristics of the plant from the plant designer, if possible.
3. If a plant model is available, develop a simulation program with conventional control and study the performance characteristics.
4. Identify the functional elements where FL can be applied.
5. Identify the input and output variables of each fuzzy system.
6. Define the universe of discourse of the variables and convert to corresponding per unit variables as necessary.
7. Formulate the fuzzy sets and select the corresponding MF shape of each. For a sensitive variable, more fuzzy sets are needed. If a variable requires more precision near steady state, use more crowding of MFs near the origin.

8. Formulate the rule table. (This step and the previous one are the main design steps, which require intuition and experience about the process.)

9. If the mathematical plant model is available, simulate the system with the fuzzy controller. Iterate the fuzzy sets and rule table until the performance is optimized. For a plant without a model, the fuzzy system must be designed conservatively and then fine-tuned by the test results on the operating plant.

10. Implement the control in real time and further iterate to improve performance.

11.6 APPLICATIONS

Fuzzy logic has been widely applied in power electronic systems. Applications include speed control of dc and ac drives, feedback control of converter, off-line P-I and P-I-D tuning, nonlinearity compensation, on-line and off-line diagnostics, modeling, parameter estimation, performance optimization of drive systems based on on-line search, estimation for distorted waves, and so on. In this section, a few example applications from the literature will be reviewed.

11.6.1 Induction Motor Speed Control

Consider the fuzzy speed control system shown in Figure 11.12 [5][14], where the input signals are E and CE and the output signal is DU, as explained before. Figure 11.16 shows the fuzzy sets and corresponding triangular MF description of each signal.

The fuzzy sets are defined (the linguistic definition is immaterial) as follows:

Z = Zero	PS = Positive Small,	PM = Positive Medium
PB = Positive Big	NS = Negative Small	NM = Negative Medium
NB = Negative Big	PVS = Positive Very Small	NVS = Negative Very Small

The universe of discourse of all the variables, covering the whole region, is expressed in per unit values. All the MFs are asymmetrical because near the origin (steady state), the signals require more precision. There are seven MFs for $e(pu)$ and $ce(pu)$ signals, whereas there are nine MFs for the output. All the MFs are symmetrical for positive and negative values of the variables. Table 11.1 shows the corresponding rule table for the speed controller. The top row and left column of the matrix indicate the fuzzy sets of the variables e and ce, respectively, and the MFs of the output variable $du(pu)$ are shown in the body of the matrix. There may be $7 \times 7 = 49$ possible rules in the matrix, where a typical rule reads as

IF $e(pu)$ = PS AND $ce(pu)$ = NM THEN $du(pu)$ = NS

Some blocks in the rule table may remain vacant, giving less number of rules.

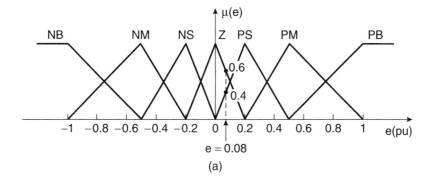

(a)

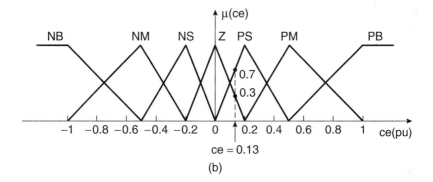

(b)

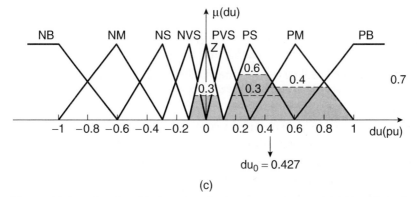

(c)

Figure 11.16 Membership functions for fuzzy speed control (a) Error ($e(pu)$), (b) Change in error ($ce(pu)$), (c) Change in output control ($du(pu)$)

Table 11.1 Rule Matrix for Fuzzy Speed Control

du(pu) ce(pu) \ e(pu)	NB	NM	NS	Z	PS	PM	PB
NB	NVB	NVB	NVB	NB	NM	NS	Z
NM	NVB	NVB	NB	NM	NS	Z	PS
NS	NVB	NB	NM	NS	Z	PS	PM
Z	NB	NM	NS	Z	PS	PM	PB
PS	NM	NS	Z	PS	PM	PB	PVB
PM	NS	Z	PS	PM	PB	PVB	PVB
PB	Z	PS	PM	PB	PVB	PVB	PVB

The general considerations in the design of the controller are:

1. If both $e(pu)$ and $ce(pu)$ are zero, then maintain the present control setting $du(pu) = 0$.
2. If $e(pu)$ is not zero but is approaching this value at a satisfactory rate, then maintain the present control setting.
3. If $e(pu)$ is growing, then change the control signal $du(pu)$ depending on the magnitude and sign of $e(pu)$ and $ce(pu)$ to force $e(pu)$ towards zero.

As mentioned above, the rule matrix and MF description of the variables are based on the knowledge of the system and their fine-tuning may be time-consuming for optimal performance. For a simulation-based system design, controller tuning with the help of the MATLAB Fuzzy Logic Toolbox (for example) may be reasonably fast. Recently, neural network and genetic algorithm techniques have been proposed for tuning MFs. Adaptive Neuro-Fuzzy Inference System (ANFIS) will be discussed in Chapter 12.

The algorithm for fuzzy speed control in detail can be summarized as follows. A numerical example is included in each step for clarity (refer to Figures 11.15 and 11.16)

1. Sample speeds ω_r^* and ω_r.
2. Compute error E, change in error CE, and their per unit values as follows:
$E(k) = \omega_r^* - \omega_r$
$CE(k) = E(k) - E(k-1)$
$e(pu) = E(k)/GE$

$ce(pu) = CE(k)/GC$
$[E = 0.8, CE = 1.3, GE = 10, GC = 10, e(pu) = 0.8/10 = 0.08, ce(pu) = 1.3/10 = 0.13]$

3. Identify the interval index I and J for $e(pu)$ and $ce(pu)$, respectively
   $[I = 1, J = 1]$
4. Calculate the degree of membership of $e(pu)$ and $ce(pu)$ for the relevant fuzzy sets
   $[\mu_Z(e) = 0.6, \mu_{PS}(e) = 0.4, \mu_Z(ce) = 0.3, \mu_{PS}(ce) = 0.7)]$
5. Identify the four valid rules from Table 11.1 (stored as a look-up table) for Z and PS values of $e(pu)$ and $ce(pu.)$ These are:

R1: IF $e(pu) = Z$ AND $ce(pu) = Z$ THEN $du(pu) = Z$
R2: IF $e(pu) = Z$ AND $ce(pu) = PS$ THEN $du(pu) = PS$
R3: IF $e(pu) = PS$ AND $ce(pu) = Z$ THEN $du(pu) = PS$
R4: IF $e(pu) = PS$ AND $ce(pu) = PS$ THEN $du(pu) = PM$

Calculate the DOF of each rule using the AND or min operator

$[DOF_1 = \min \{\mu_Z(e), \mu_Z(ce)\} = \min \{0.6, 0.3\} = 0.3$
$DOF_2 = \min \{\mu_Z(e), \mu_{PS}(ce)\} = \min \{0.6, 0.7\} = 0.6$
$DOF_3 = \min \{\mu_{PS}(e), \mu_Z(ce)\} = \min \{0.4, 0.3\} = 0.3$
$DOF_4 = \min \{\mu_{PS}(e), \mu_{PS}(ce)\} = \min \{0.4, 0.7\} = 0.4]$

6. Retrieve the amount of correction $du(pu)_i$ ($i = 1, 2, 3, 4$) corresponding to each rule in Table 11.1

$[du(pu)_1 = 0$ for Z corresponding to $DOF_1 = 0.3$
$du(pu)_2 = 0.35$ for PS corresponding to $DOF_2 = 0.6$
$du(pu)_3 = 0.35$ for PS corresponding to $DOF_3 = 0.3$
$du(pu)_4 = 0.6$ for PM corresponding to $DOF_4 = 0.4]$

7. Calculate the crisp output $du(pu)_0$ by the height defuzzification method

$$[du(pu)_0 = \frac{0.3 \times 0 + 0.6 \times 0.35 + 0.3 \times 0.35 + 0.4 \times 0.6}{0 + 0.35 + 0.35 + 0.6} = 0.427]$$

Figure 11.17 shows the typical speed loop response with stepped load torque T_L, and Figure 11.18 shows a similar response with four times the rotor inertia. The performances are superior to conventional P-I control. The robustness in the response is evident from the results.

11.6.2 Flux Programming Efficiency Improvement of Induction Motor Drive

It was discussed in Chapter 8 that in variable-frequency drives, the machines are normally operated at the rated flux to give maximum developed torque per ampere and optimum transient response. However, at light loads, this causes excessive core loss, impairing the efficiency of the drive. To improve the efficiency, flux programming at light loads by the open loop or close loop method was discussed. The efficiency optimization control based on an on-line search of optimum flux is very attractive. Figure 11.19 explains the on-line search method for an indirect vector-controlled induction motor drive (see Figure 11.20). Assume that the machine operates

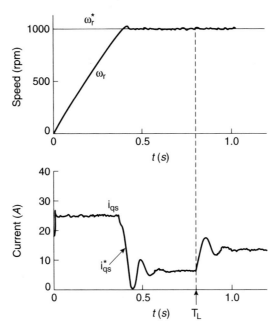

Figure 11.17 Fuzzy control response with steps of speed command and load torque with nominal inertia (J)

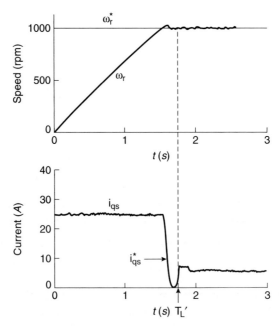

Figure 11.18 Fuzzy control response with steps of speed command and load torque with four times the nominal inertia ($4J$)

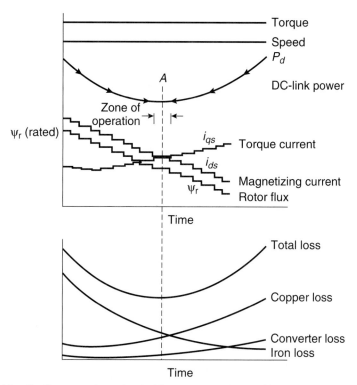

Figure 11.19 On-line search method of flux programming efficiency improvement control

initially at the rated flux in steady state with low load torque at a certain speed, as indicated. The rotor flux ψ_r is decremented in steps by reducing the magnetizing component i_{ds} of the stator current. This results in an increase of the torque component of current i_{qs} (due to speed control loop) so that the developed torque remains the same for steady-state operation. As the core loss decreases with a decrease of flux, the copper loss increases, but the system loss (machine and converter loss) decreases, improving the overall efficiency. This is reflected in the decrease of the dc link power P_d, as shown for the same output power. The search is continued until the system settles at the minimum input power (i.e., maximum efficiency) point, A. Any search attempt beyond point A adversely affects efficiency and forces the search direction such that operation always settles at point A. This type of algorithm has the advantages of the control not requiring knowledge of machine parameters, it has complete insensitivity to parameter variation, and it is universally applicable to any arbitrary machine.

Figure 11.20 shows the block diagram of vector control incorporating the FL-based efficiency optimizer [7]. The total control implemented by a DSP is indicated by the dashed outline. Although other methods of control can be used, the fuzzy control has the advantages of being able to handle noisy and inaccurate input signals, and the step size of the i_{ds} decrement is adaptive (initially large–then small), so that fast convergence in the control is attained. In Figure 11.20, the speed control loop generates the torque component of current i_{qs}^*, as shown. The vector rotator receives

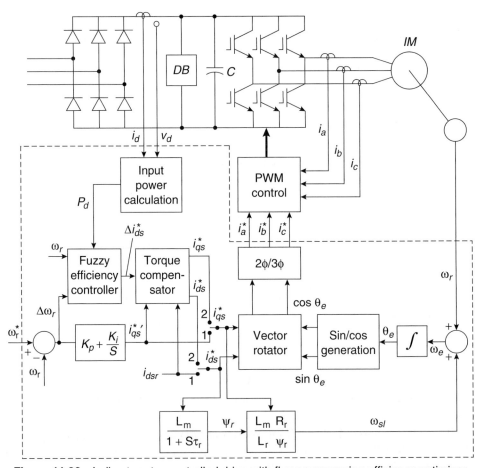

Figure 11.20 Indirect vector-controlled drive with flux programming efficiency optimizer

the torque and magnetizing current commands i_{qs}^* and i_{ds}^*, respectively, from two positions of a switch: (1) the transient position, where the magnetizing current is established to the rated value i_{dsr} and the torque current $i_{qs}^{*\prime}$; and (2) the steady-state position, where the magnetizing and torque currents i_{ds}^*, i_{qs}^* are generated by the fuzzy efficiency controller and feedforward torque compensator, which will be explained later. The fuzzy controller becomes effective only at steady-state condition, that is, when the speed loop error $\Delta\omega_r$ approaches zero. Figure 11.21 shows the fuzzy efficiency controller block diagram. The dc link power $P_d(k)$ is sampled and compared with the previous value to determine the decrement (or increment) $\Delta P_d(k)$. In addition, the last magnetization current segment's $(\Delta i_{ds}(pu)^*(k-1))$ polarity is reviewed. Based on these input signals, the decrement step of $\Delta i_{ds}(pu)$ is generated from the fuzzy inference system. The normalizing scale factors P_b and I_b are programmable and are given by the expressions

$$P_b = A\omega_r + B \tag{11.33}$$

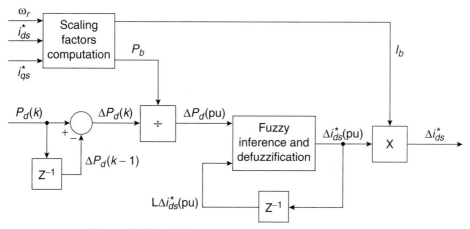

Figure 11.21 Efficiency optimizer control block diagram

and

$$I_b = C_1\omega_r - C_2\hat{T}_e + C_3 \tag{11.34}$$

where A, B, C_1, C_2, and C_3 are constants (determined by simulation study) and $\hat{T}_e$ is the estimated torque given by

$$\hat{T}_e = K'_t i_{ds}^* i_{qs}^* \tag{11.35}$$

The programmable scale factors make these variables insensitive to the operating point on the torque-speed plane.

Variables $\Delta P_d(pu)$ and $\Delta i_{ds}^*(pu)$ are each described by seven asymmetric triangular MFs, whereas $L\Delta i_{ds}^*(pu)$ has only two (positive and negative) MFs, as shown in Figure 11.22. Table 11.2 shows the corresponding rule table where a typical rule reads as

IF $\Delta P_d(pu)$ = Negative Small (NS) AND $L\Delta i_{ds}(pu)$ = Negative (N)
THEN $\Delta i_{ds}(pu)$ = Negative Small (NS)

The basic idea is that if the last control action indicates a decrease of dc link power, then proceed searching in the same direction, and the control magnitude should be somewhat proportional to the measured dc link power change. When the control action results in an increase of $P_d (\Delta P_d > 0)$, the search direction is reversed. At steady state, the operation oscillates around point A (Figure 11.19) with very small step size.

11.6.2.1 Pulsating Torque Compensation

Figure 11.23 explains the principle of feedforward pulsating torque compensation. This compensator functions to compensate the loss of torque due to the decrement of i_{ds} by injecting an equivalent Δi_{qs}^* so that the developed torque remains the same. Otherwise, the slow compensation of i_{qs} by the speed loop creates a large pulsating torque at a low sampling frequency

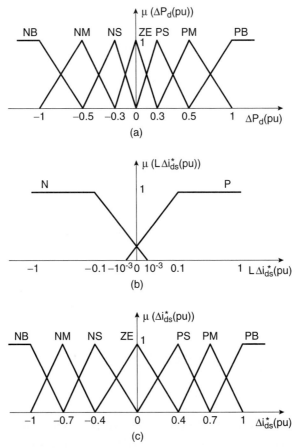

Figure 11.22 Membersip functions of fuzzy variables

Table 11.2 Rule Matrix for Efficiency Improvement

$L\Delta i_{ds}(pu)$ / $\Delta P_d(pu)$	N	P
PB	PM	NM
PM	PS	NS
PS	PS	NS
ZE	ZE	ZE
NS	NS	PS
NM	NM	PM
NB	NB	PB

Applications

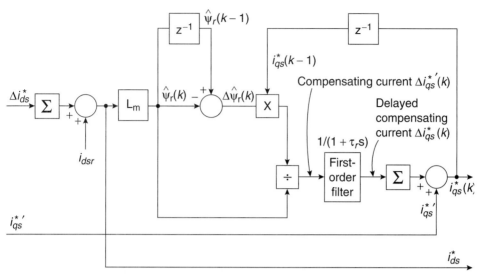

Figure 11.23 Feedforward pulsating torque compensation block diagram

(sometimes causing resonance effect), which may be harmful for the drive. The compensating current $\Delta i_{qs}^{*\prime}(k)$ can be calculated as

$$\Delta i_{qs}^{*\prime}(k) = \frac{\hat{\psi}_r(k-1) - \hat{\psi}_r(k)}{\hat{\psi}_r(k)} \cdot i_{qs}^{*}(k-1) \tag{11.36}$$

where $\hat{\psi}_r(k)$ is the rotor flux vector magnitude.

Since the response of $\Delta i_{qs}^{*\prime}(k)$ is practically instantaneous, whereas Δi_{ds}^{*} responds with the delay of rotor time constant τ_r for correct torque matching, the $\Delta i_{qs}^{*\prime}(k)$ signal must be delayed by the same τ_r in the first-order filter shown in the figure. Figure 11.24(a) shows the efficiency optimizer performance, and (b) indicates the response of the torque compensator. If the $\Delta \omega_r$ signal appears in Figure 11.20 due to a change of speed command or a load torque change, the fuzzy control is abandoned and full flux is established by transferring both switches to position 1. The control thus transitions from efficiency optimization mode to transient response optimization mode.

11.6.3 Wind Generation System

Wind electrical power generation systems have recently attracted a lot of attention because they are cost-effective, environmentally clean, and safe renewable power sources compared to fossil fuels and nuclear power generation. Compared to traditional variable-pitch, fixed-speed wind turbines (horizontal or vertical axis), variable-speed wind turbine (VSWT) systems with the help of power electronics have found more acceptance recently. Although power electronics are expensive, the larger energy capture of VSWT systems makes the life-cycle cost of these systems lower. In this section, we will describe a FL-based efficiency optimization and perfor-

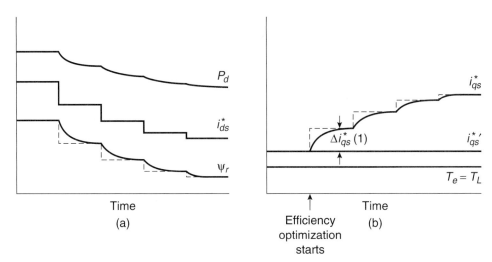

Figure 11.24 (a) Performance of fuzzy efficiency controller, (b) Response of feedforward torque compensator

mance enhancement control of a wind generation system using a cage-type induction generator. A description of the power system is essential before explaining the fuzzy control.

11.6.3.1 Wind Turbine Characteristics

Either horizontal or vertical axis wind turbines can be used in wind generation systems. The vertical Darrieus (egg beater) type, which is under consideration here, has the advantages that it can be installed on the ground, accepting wind from any direction. However, the disadvantages are that the turbine is not self-starting and there is a large pulsating torque, which depends on wind velocity and turbine speed. Figure 11.25 shows the torque-speed curves of a wind turbine at different wind velocities. At a particular wind velocity (V_{W1}), if the turbine speed ω_r decreases from ω_{r1}, the developed torque increases, reaches the maximum value at point B, and then decreases at lower ω_r. Superimposed on the family of curves is a set of constant power lines (dotted), indicating the points of maximum power output for each wind speed. This means that at a particular wind velocity, the turbine speed is to be varied to get maximum power delivery (highest aerodynamic efficiency), and this point (for example, C) deviates from the maximum torque point, as indicated in the figure. Since the torque-speed characteristics of a wind generation system are analogous to those of a motor-blower system (except the turbine runs in the reverse direction), the torque follows the square law characteristics $T_e = K \omega_r^2$, and the output power follows the cube law $P_0 = K \omega_r^3$, as indicated in the figure.

11.6.3.2 System Description

Figure 11.26 shows the block diagram of the wind generation system incorporating all the control elements [8]. The turbine at the left (a vertical type) is coupled to the cage-type induction generator through a speed-up gear ratio (not shown). The variable-frequency, variable-voltage power generated by the machine is rectified to dc by a PWM voltage-fed rectifier that also

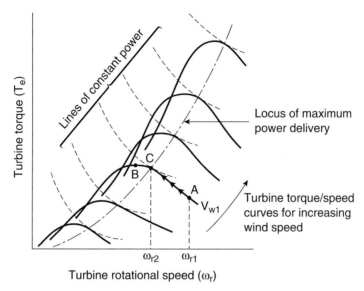

Figure 11.25 Torque-speed curves of fixed-pitch wind turbine at different wind velocities

supplies the excitation current (lagging) to the machine. The dc link power is inverted to 220 V, 60 Hz ac through a PWM inverter and fed to the utility grid. The line current is sinusoidal at unity power factor, as indicated. The generator speed ω_r is controlled by an indirect vector control with a torque control for stiffness and a synchronous current control in the inner loops. The machine flux ψ_r is controlled by open loop control of the excitation current i_{ds}, but in normal condition, the flux is set to the rated value for fast transient response. The line-side converter is also vector-controlled, using direct vector control and synchronous current control in the inner loops. The output power P_0 is controlled to control the dc link voltage V_d as shown in the figure. Because an increase in P_0 causes a decrease in V_d, the voltage loop error polarity has been inverted. The tight regulation of V_d within a small tolerance band requires feedforward power injection in the power control loop, as shown. The insertion of line filter inductance L_s creates some coupling effect, which is eliminated by a decoupler in the synchronous current control loops (not shown). The power can be controlled to flow easily in either direction. The vertical turbine is started with a motoring torque. As the speed develops, the machine goes into generating mode. The machine is shut down by regenerative braking.

11.6.3.3 Fuzzy Control

The system uses three fuzzy controllers whose functions can be summarized as follows:

FLC-1: On-line search of generator speed to maximize output power
FLC-2: On-line search of machine excitation current to optimize machine efficiency at light load
FLC-3: Robust control of generator speed

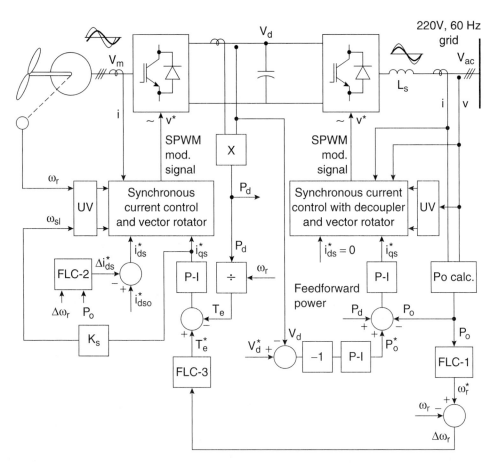

Figure 11.26 FL-based block diagram of wind generation system

Controllers FLC-2 and FLC-3 are essentially the same as described before, and will not be discussed further. With the cubic law power output, as mentioned before, the machine is at light load at reduced speed; therefore, flux programming control by an on-line search of the excitation current will improve its efficiency. The vertical turbine generates oscillatory torque which tends to make the machine speed jittery. Besides, the vortex in the wind velocity tends to disturb the speed. The fuzzy P-I control by FLC-3 makes the speed control robust against these disturbances.

Generator Speed Tracking Control (FLC-1) – The fuzzy controller FLC-1 maximizes the aerodynamic efficiency of the turbine at any wind velocity to extract maximum power output from the wind turbine. Since power is the product of torque and speed and turbine power equals line output power (neglecting losses in the system), the torque-speed curves of Figure 11.25 can be translated into line power (P_0) – generator speed (ω_r) curves, as shown approximately in Figure 11.27. For a particular wind velocity, the function of FLC-1 is to search the generator speed until the system settles down at the maximum output power condition. The figure also shows the

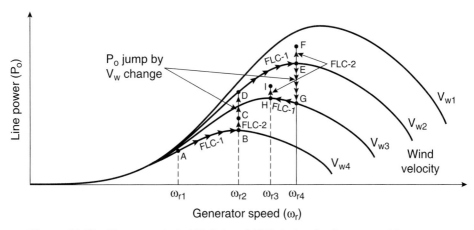

Figure 11.27 Fuzzy control of FLC-1 and FLC-2 showing increase of line power

effect of FLC-2 control. For example, at a wind velocity of V_{W4} in Figure 11.27, the power output will be at A if the generator speed is ω_{r1}. FLC-1 will alter the speed in steps on the basis of an on-line search until it reaches speed ω_{r2}, where the output power is maximum at B. At the end of FLC-1 control, FLC-2 control becomes effective and the improvement in generator efficiency increases the power output and brings the operating point to C. If the wind velocity increases to V_{W2}, the output power will jump to point D, and FLC-1 will bring the operating point to E by searching when the speed to ω_{r4} is attained. Then, FLC-2 control will bring the operating point to F. The profile for a decrease of wind velocity to V_{W3} is also indicated in the figure.

Figure 11.28 shows the block diagram of FLC-1. By incrementing (or decrementating) speed ω_r^*, the corresponding increment (or decrement) of output power P_0 is estimated. If ΔP_0 is positive with the last positive $\Delta \omega_r^* (L\Delta \omega_r^*)$, the search is continued in the same direction. If, on the other hand, $+\Delta \omega_r^*$ causes $-\Delta P_0$, the direction of search is reversed. The speed oscillates by a small increment when it reaches the optimum condition. The normalized form of variables $\Delta P_0(pu)$, $\Delta \omega_r^*(pu)$, and $L\Delta \omega_r^*(pu)$ are described by triangular MFs, as shown in Figure 11.29, and the control rules are given by the matrix in Table 11.3. In Figure 11.28, the output $\Delta \omega_r$ is added with some amount of the $L\Delta \omega_r^*$ signal to avoid local minima due to wind vortex and torque ripple. Again, the controller operates on a per unit basis so that the response is insensitive to system variables and the algorithm is universal to any similar system. The scale factors *KPO* and *KWR*, shown in Figure 11.28, are a function of generator speed so that the control becomes somewhat insensitive to speed variation. Although scale factors are normally generated by mathematical expression, in this case, they are generated by fuzzy computation indicating that *KPO* and *KWR* increase with speed. Figure 11.30 gives the MFs and rule matrix for scale factor computation. The whole fuzzy controller is designed from heuristic knowledge of the system.

Again, the advantages of fuzzy control are obvious. As mentioned previously, it will accept noisy and inaccurate signals and it provides an adaptively decreasing step size in the search that leads to fast convergence. Also, note that wind velocity information is not needed and the system parameter variation does not affect the search.

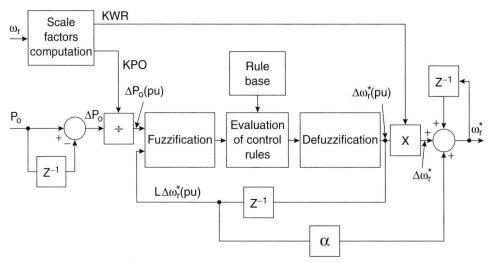

Figure 11.28 Block diagram of fuzzy controller FLC-1

Table 11.3 Rule Matrix for FLC-1

$L\Delta\omega_r^*(pu)$ / $\Delta P_o(pu)$	P	ZE	N
PVB	PVB	PVB	NVB
PB	PB	PVB	NB
PM	PM	PB	NM
PS	PS	PM	NS
ZE	ZE	ZE	ZE
NS	NS	NM	PS
NM	NM	NB	PM
NB	NB	NVB	PB
NVB	NVB	NVB	PVB

Figure 11.31 shows the system efficiency improvement by FLC-1 and FLC-2 with variable wind velocity considering the generator operation at constant 940 rpm. As the wind velocity increases from a low value, the efficiency gain by FLC-2 decreases because of higher generator loading. The gain due to FLC-1 by tuning the generator speed to optimal value can be very high at low wind velocity although P_0 may be small. It decreases to zero near 0.7(pu) wind velocity

Applications

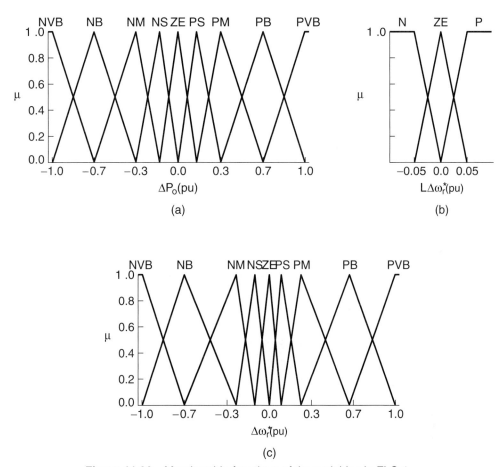

Figure 11.29 Membership functions of the variables in FLC-1

(where the generator speed of 940 rpm is optimal) and then increases again. This efficiency variation can be explained with the help of Figure 11.27.

11.6.4 Slip Gain Tuning of Indirect Vector Control

On-line slip gain (K_s) tuning of an indirect vector controlled induction motor drive with the help of MRAC was discussed in Chapter 8 (see Figure 8.38). FL principles can be applied in solving similar problems. Figure 11.32 shows the block diagram of an indirect vector-controlled drive where a fuzzy tuning controller has been incorporated. Figure 11.33 shows the details of fuzzy MRAC-based tuning controller [10]. The scheme depends on reference model computation of reactive power Q^* and the d-axis voltage v_{ds}^* at the machine terminal for the ideally tuned condition.

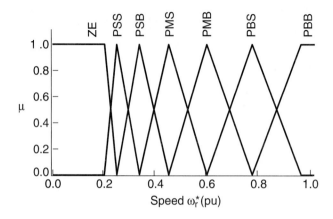

Figure 11.30 Membership functions and rule table for scale factor computation

11.6.4.1 Derivation of Q^* and v_{ds}^*

From the d^e-q^e equivalent circuits of Figure 2.23, the stator equations are

$$v_{qs} = R_s i_{qs} + \frac{d}{dt}(\psi_{qs}) + \omega_e \psi_{ds} \tag{2.92}$$

$$v_{ds} = R_s i_{ds} + \frac{d}{dt}(\psi_{ds}) - \omega_e \psi_{qs} \tag{2.93}$$

At steady-state condition under vector control (see Figure 8.36), we can write

$$\frac{d}{dt}(\psi_{qs}) = 0 \tag{11.37}$$

$$\frac{d}{dt}(\psi_{ds}) = 0 \tag{11.38}$$

$$\psi_{ds} = L_s i_{ds} \tag{11.39}$$

$$\psi_{qs} = L_s i_{qs} - \frac{L_m}{L_r} i_{qs} L_m = (L_s - \frac{L_m^2}{L_r}) i_{qs} \tag{11.40}$$

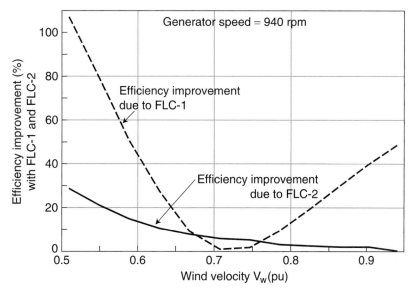

Figure 11.31 Efficiency improvement by controllers FLC-1 and FLC-2 at different wind velocities (1.0 pu = 31.5 mph)

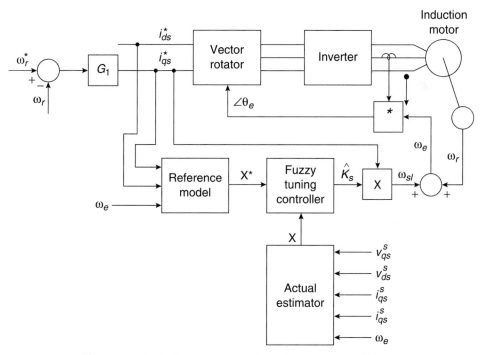

Figure 11.32 Indirect vector control with fuzzy slip gain tuner

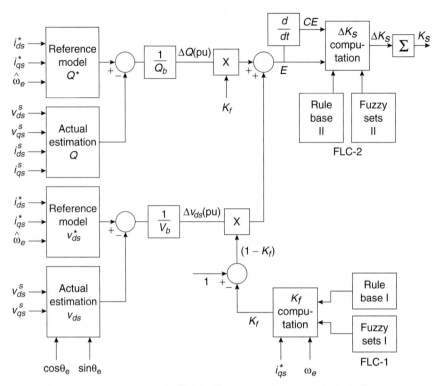

Figure 11.33 FL-based MRAC slip gain tuning control block diagram

Substituting Equations (11.37)–(11.40) in (2.92) and (2.93) and neglecting resistance R_s, we get

$$v_{qs} = \omega_e L_s i_{ds} \tag{11.41}$$

$$v_{ds} = -\omega_e (L_s - \frac{L_m^2}{L_r}) i_{qs} \tag{11.42}$$

The steady-state reactive power expression in a machine is given as [11]

$$Q = v_{qs} i_{ds} - v_{ds} i_{qs} \tag{11.43}$$

Substituting Equations (11.41) and (11.42) in (11.43), the reference model Q^* at the ideally tuned condition is

$$Q^* = \hat{\omega}_e (L_s i_{ds}^{*2} + L_\sigma i_{qs}^{*2}) \tag{11.44}$$

where $L_\sigma = L_s - L_m^2/L_r$. By knowing i_{ds}^*, i_{qs}^* and the estimated value of ω_e (see Equation (8.70)), the reference model Q^*, as shown in Figure 11.33, can be estimated.

Applications

The reference model v_{ds}^{*} is given from Equations (2.93) and (11.40) as

$$v_{ds}^{*} = R_s i_{ds}^{*} - \hat{\omega}_e L_\sigma i_{qs}^{*} \qquad (11.45)$$

Note that the reference models are functions of machine parameters which might vary and thus contribute inaccuracy.

The foregoing reference models are then compared with the actual estimation of the respective quantities given by

$$Q = v_{qs}^{s} i_{ds}^{s} - v_{ds}^{s} i_{qs}^{s} \qquad (11.46)$$

$$v_{ds} = v_{qs}^{s} \sin\theta_e + v_{ds}^{s} \cos\theta_e \qquad (2.75)$$

where $\cos\theta_e$ and $\sin\theta_e$ are the unit vector components. The loop errors are divided by the respective scale factors to derive the per unit variables $\Delta Q(pu)$ and $\Delta v_{ds}(pu)$ for manipulation in the fuzzy controller. There are, in fact, two fuzzy controllers in Figure 11.33, and each is designed with the respective rule base and fuzzy sets. Controller FLC-1, as shown, generates a weighting factor K_f that permits the appropriate distribution of the Q control and v_{ds} control on the torque-speed (i.e., $i_{qs} - \omega_e$) plane. The objective is to assign a high sensitivity to the tuning control by the dominant use of the Q control in the low-speed, high-torque region and the v_{ds} control in the high-speed, low-torque region. The FLC-1 control is simple and is given by an 2×2 rule matrix. A typical rule can be given as

IF speed (ω_e) is high (H) AND torque (i_{qs}) is low (L)
THEN weighting factor (K_f) is low (L)

The combined error signal for both loops in Figure 11.33 is given as

$$E = K_f \Delta Q(pu) + (1 - K_f)\Delta v_{ds}(pu) \qquad (11.47)$$

Fuzzy controller FLC-2 generates the corrective incremental slip gain ΔK_s based on the combined detuning error E and its derivative CE, as shown. This is basically the fuzzy P-I controller as discussed before for the speed control of the motor. The objective is to provide an adaptive feedback control for fast convergence at any operating point, irrespective of the strength of the E and CE signals. Note that in the ideally tuned condition of the system, both the reference model and actual estimation signals will match, and correspondingly, the E and CE signals will be zero and the slip gain K_s will be set to the correct value K_{s0}. If the system becomes detuned (for example, with rotor resistance variation), the actual Q and v_{ds} signals will deviate from their respective reference values and the resulting error will alter K_s until the system becomes tuned, that is, $E = 0$. The effects of detuning on torque and rotor flux transients are

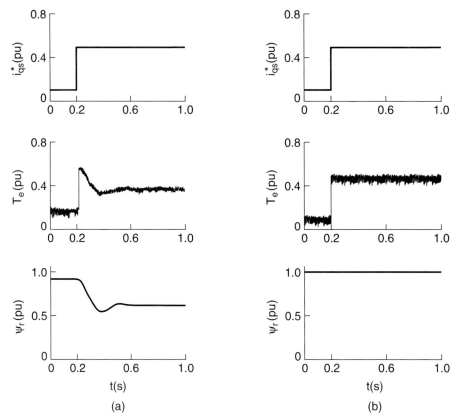

Figure 11.34 Fuzzy tuner performance: (a) Detuned slip gain, (b) Tuned slip gain

illustrated in Figure 11.34. Figure 11.34(a) shows a detuned system where the slip gain was set to twice the correct value ($K_s/K_{s0} = 2$), resulting in higher order dynamics for torque and flux transients as well as reduced torque and flux at steady-state condition. In Figure 11.34(b), on the other hand, the tuned slip gain condition gives ideal transient response.

11.6.5 Stator Resistance R_s Estimation

It was mentioned in Chapter 8 that compensation for the stator resistance variation is important for the correct estimation of flux (ψ_s or ψ_r), torque (T_e), speed (ω_r), and frequency (ω_e), particularly at low speed. Since the stator resistance variation is primarily a function of the stator winding temperature T_s, information about T_s is necessary for the compensation. The average value of T_s can be determined by mounting thermistor probes at several points in the stator, but some type of "sensorless" estimation is desirable.

FL principles can be applied for the approximate estimation of stator resistance by estimating the stator temperature [12]. Consider that a small 5 hp standard (NEMA Class B) induction motor with a shaft-mounted cooling fan and a set of five distributed thermistors mounted on the stator (for measurement of the stator temperature) is under test. The motor is operated with vector control at the rated flux condition. The machine is mounted on a dynamometer, which is operated at speed control mode. At each speed (i.e., frequency) setting, the drive torque (i.e., the stator current) is varied in steps and the stator temperature rise $(\Delta T_{ss} = T_s - T_A,$ where T_A = ambient temperature) is recorded at the steady-state condition. Figure 11.35 shows the experimentally determined ΔT_{ss} curves as functions of stator current I_s and frequency ω_e with ambient temperature $T_A = 25°$ C. Note that at higher frequencies (i.e., speed), the iron loss increases, which tends to a higher temperature rise, but the dominant cooling effect of the shaft-mounted fan essentially decreases the temperature. The curve below the minimum stator current, which corresponds to the magnetizing current for the rated flux, was extrapolated to the vertical axis. The temperature rise is small at low stator currents for the range of frequency variation.

The experimental curves in Figure 11.35 were used to formulate the fuzzy MFs in Figure 11.36 and the corresponding rule matrix in Table 11.4. Basically, the fuzzy estimator algorithm interpolates the $\Delta T_{ss}(pu)$ signal as a function of stator current and frequency. Numerous MFs with crowding at low frequency indicates more accurate estimation of $\Delta T_{ss}(pu)$ near zero speed.

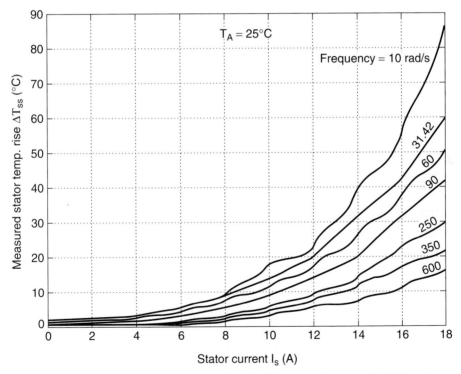

Figure 11.35 Measured stator temperature rise vs. stator current at different frequencies (at steady state)

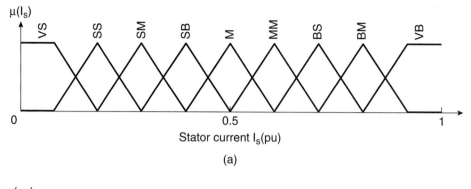

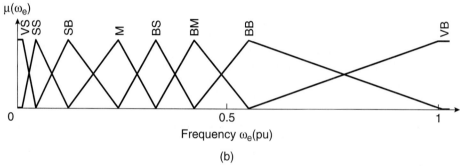

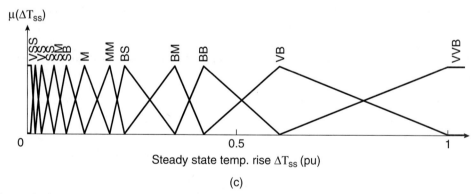

Figure 11.36 Fuzzy estimation MFs, (a) Stator current $I_s(pu)$, (b) Frequency $\omega_e(pu)$, (c) Steady state temperature rise $\Delta T_{ss}(pu)$

Figure 11.37 shows the complete estimation block diagram of stator resistance, which includes a thermal time constant curve and the thermistor network (shown dotted). The thermistor network was used for the generation of Figure 11.35 (as mentioned before), the calibration of T_s, and the estimation of the machine thermal time constant, which will be discussed next. In the present machine, cooling or heat transfer occurs by natural convection as well as by forced cooling by a shaft-mounted fan. The fan's cooling effect is related to the machine's speed.

Table 11.4 Rule Base for $\Delta T_{ss}(pu)$ Estimation

$I_s(pu)$ \ $\omega_e(pu)$	VS	SS	SB	M	BS	BM	BB	VB
VS	VS	VS	VVS	VVS	VVS	VVS	VVS	VVS
SS	SS	SS	VS	VS	VVS	VVS	VVS	VVS
SM	SM	SM	SS	SS	VS	VS	VVS	VVS
SB	SB	SB	SM	SM	SS	SS	VS	VS
M	MM	M	SB	SB	SM	SM	SS	SS
MM	BS	MM	M	M	SB	SB	SM	SM
BS	BB	BM	MM	MM	M	M	SB	SM
BM	VB	BB	BM	BS	BS	MM	M	SB
VB	VVB	VB	BB	BM	BM	BS	MM	M

The dynamic thermal model of the machine can be approximately represented by a first-order low-pass filter $1/(1 + \tau S)$, as indicated in the figure, where τ = approximate thermal time constant, which is a nonlinear function of speed (or frequency). The relation between τ and ω_e, as shown in Figure 11.37, was determined experimentally as follows: The machine, operating at rated flux, was mounted on the dynamometer and steps of torque (or stator current) were applied at different speed settings. The thermal time constant τ in each case was determined from the average transient temperature rise data on the thermistor network, as mentioned before.

In Figure 11.37, the steady-state ΔT_{ss} can be estimated from the measured (or estimated) values of stator current I_s and frequency ω_e through the fuzzy algorithm described by Figure 11.36 and Table 11.4. Basically, it is fuzzy interpolation of ΔT_{ss} in Figure 11.35 as mentioned before. Generally, four rules will be valid at any instant, and a typical rule may be

   IF stator current ($I_s(pu)$) is small-medium (SM)
     AND the frequency ($\omega_e(pu)$) is medium (M)
     THEN the temperature rise ($\Delta T_{ss}(pu)$) is small-small (SS)

Once the steady-state ΔT_{ss} is estimated by the fuzzy estimator, it is converted to a dynamic temperature rise through the low-pass filter and added to the ambient temperature T_A to derive the actual stator temperature T_s. Then, the derivation of actual resistance R_s by the linear expression shown in the figure becomes straightforward. The measured T_s by the thermistor network helps to calibrate the estimated T_s and iterate the estimation algorithm. Finally, the thermistor network is removed, and the estimation algorithm is applied to all machines of the same family. Figure

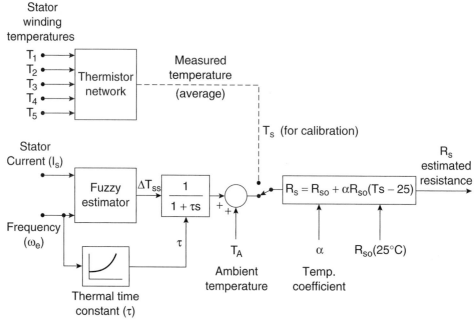

Figure 11.37 Fuzzy estimation block diagram of stator resistance R_s (shown with calibrating thermistor network)

11.38(a) shows the typical T_s estimation accuracy as a function of time at different stator currents, but constant speed, and Figure 11.38(b) shows the corresponding R_s estimation tracking accuracy.

11.6.6 Estimation of Distorted Waves

Power converters are characterized for generation of complex voltage and current waves, and it is often necessary to determine their parameters, such as total rms current I_s, fundamental rms current I_f, active power P, reactive power Q, displacement factor (DPF), distortion factor (DF), and power factor (PF). These parameters can be measured by electronic instrumentation (hardware and software) or estimated by mathematical model, FFT analysis, etc. FL principles can be applied for fast and reasonably accurate estimation of these parameters due to input/output nonlinear mapping (or pattern recognition) property [13]. Of course, fuzzy algorithm development is laborious and requires a large number of iterations for good accuracy. In this section, we will illustrate the estimation of a diode rectifier line current wave by two fuzzy algorithms: Mamdani method and Sugeno (first-order) method.

A typical line current wave for a three-phase diode bridge rectifier with balanced line voltage supply is shown in the lower part of Figure 11.39. The rectifier supplies a voltage-fed inverter-machine load. The parameters of the wave are characterized by the width W and height H of the pulse, which vary with loading for any given value of line voltage and line inductance.

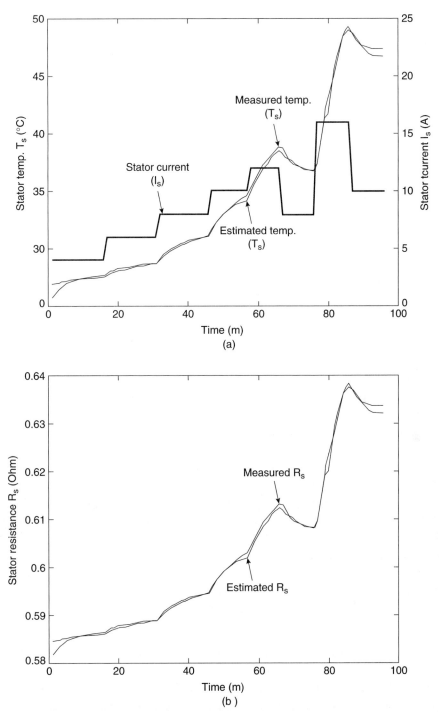

Figure 11.38 (a) Stator temperature estimator performance with dynamically varying stator current but at constant speed [357 rpm], (b) Corresponding resistance estimator performance

11.6.6.1 Mamdani Method

The fuzzy estimation for the rectifier input current wave by the Mamdani method is explained in Figure 11.39. The wave parameters W and H are defined by 6 and 11 MFs, respectively, giving all together 66 rules in the rule matrix(not shown). The number of MFs for rms current I_s and fundamental rms current I_f is 16, but the DPF has only 6 (same as W) MFs. In Figure 11.39, generally four rules are fired, as indicated, and an example rule is

IF H = PMS AND W = PSB
THEN I_s = PMM, I_f = PSB and DPF = PMS

Since the PF is given directly by the relation

$$PF = DPF \cdot \frac{I_f}{I_s} \qquad (11.48)$$

it is determined directly from the estimated I_s, I_f, and DPF. The MFs and rule matrix are iterated on the basis of simulation results until the desired accuracy is obtained. Note that one rule gives multiple outputs, and the asymmetrical MF sets are nonidentical because each output has a different degree of nonlinearity.

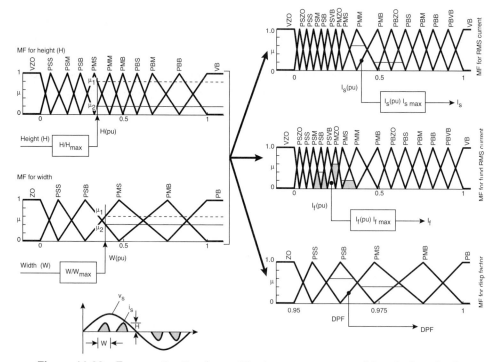

Figure 11.39 Fuzzy estimation for rectifier input current wave (Mamdani method)

11.6.6.2 Sugeno Method

The estimation, as discussed above, is now repeated using the Sugeno first-order method to assess its validity. Figure 11.40 explains the algorithm, where basically fuzzy and analytical methods are combined. Here, the parameter W is considered as the input variable and is described by 8 MFs. The actual W and H parameters of the wave are fed as input to the output linear equations, as shown. This method needs example data to fine-tune the coefficients of the linear equations for correct estimation. Multi-regression linear analysis was used on the data to determine the coefficients, which were then fine-tuned by the simulation results. There are only 8 rules in this case, compared to 66 rules in the Mamdani method.

The estimations in both methods are then compared with the actual values in Figure 11.41 with a gradually increasing inverter machine load current in the dc link. Both methods give very good accuracy in the estimation except the PF, which shows a large error due to the cumulative error contribution of I_s, I_f, and the DPF.

11.7 FUZZY LOGIC TOOLBOX

In this section, we will introduce a commonly used fuzzy system development tool and then give an example application for a drive system. The Fuzzy Logic Tool box (Math Works, Inc.) [3] is a user-friendly fuzzy program development tool in the MATLAB environment. Development can be done using either graphical user interface (GUI) or command-line functions. The toolbox also includes fuzzy clustering and an adaptive neuro-fuzzy inference system (ANFIS) technique. The ANFIS will be discussed in Chapter 12. Once a fuzzy program is developed, its performance can be tested by embedding it in the SIMULINK simulation of the system for fine-tuning, and then, the final stand-alone C program (with the fuzzy inference engine) can be generated, compiled, and downloaded to a DSP for real-time implementation. Or else, the program can be embedded in other external applications.

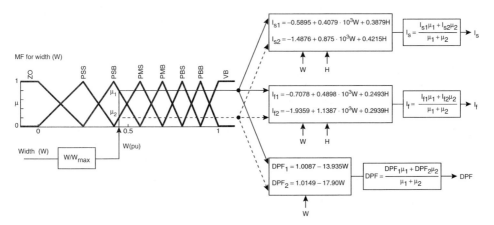

Figure 11.40 Fuzzy estimation for rectifier input current wave (Sugeno method)

610   Chapter 11 • Fuzzy Logic Principles and Applications

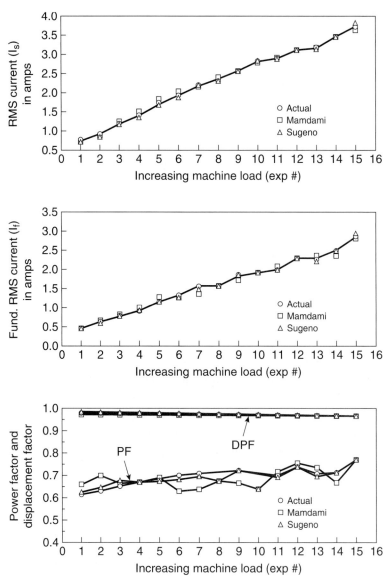

Figure 11.41   Fuzzy estimation accuracy comparison

There are five primary graphical tools for building, editing, and observing fuzzy inference systems in the Fuzzy Logic Toolbox. They are

- Fuzzy Inference System (FIS) Editor
- Membership Function (MF) Editor
- Rule Editor

- Rule Viewer
- Surface Viewer

11.7.1 FIS Editor

The FIS Editor, as shown in Figure 11.42, displays general information about a fuzzy system. At the top left, the names of the defined input fuzzy variables are indicated, and at the right, the output variables are shown. The MFs shown in the boxes are simple icons and do not indicate the actual MFs. Below this, the system name and inference method (either Mamdani or Sugeno) are indicated. At the lower left, the various steps of the inference process, which are user-selectable, are shown. At the lower right, the name of the input or output variable, its associated MF type, and its range are displayed. All the Editor and Viewer boxes depict the development of the restaurant tipping system, which was mentioned previously.

11.7.2 Membership Function Editor

The MF Editor displays and permits editing of all the MFs associated with the input and output variables. Figure 11.43 shows the user interface of the MF Editor. At the upper left, the FIS variables whose MFs can be set are shown. Each setting includes a selection of the MF type and the number of MFs of each variable. At the lower right, there are controls that permit you to

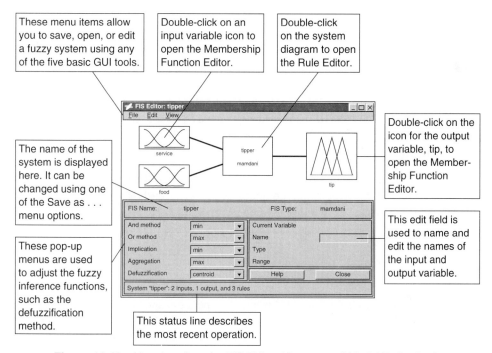

Figure 11.42 User interface for FIS Editor (Courtesy of Math Works, Inc.)

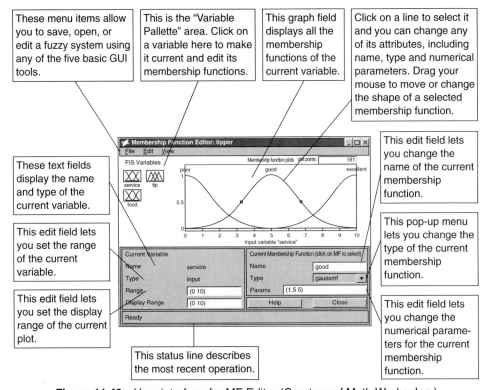

Figure 11.43 User interface for MF Editor (Courtesy of Math Works, Inc.)

change the name, type, and parameters (shape) of each MF, once you have selected it, the MFs of the current variable, which are being edited, are displayed in the graph. At lower left, information about the current variable is given. In the text field, the range (universe of discourse) and display range of the current plot of the variable under consideration can be changed.

11.7.3 Rule Editor

Once the rule matrix is designed on paper and the fuzzy variables are defined in the FIS Editor, construction of the actual rules by the Rule Editor is fairly easy, as shown in Figure 11.44. The logical connectives of rules, AND, OR, and NOT can be selected by buttons. The rules can be changed, deleted, or added, as desired.

11.7.4 Rule Viewer

Once the fuzzy algorithm has been developed, the Rule Viewer, as shown in Figure 11.45, essentially gives a micro view of the FIS, where the operation and contribution of each rule is explained in detail. Each rule is a row of plots, and each column is a variable. Three rules in the restaurant tipping system (connected by OR logic) are depicted in the figure. For a certain setting of the input variables (in this case: 5, 5), the output contribution of each rule, the total

Fuzzy Logic Toolbox

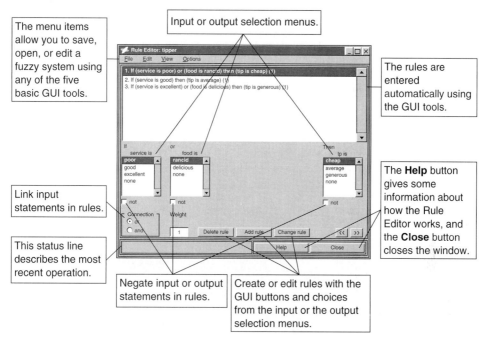

Figure 11.44 User interface for Rule Editor (Courtesy of Math Works, Inc.)

fuzzy output, and the corresponding defuzzified output are shown. The Rule Viewer, with its roadmap of operation of the whole fuzzy algorithm, permits fine-tuning of the MFs and rules.

11.7.5 Surface Viewer

After the fuzzy algorithm has been developed, the Surface Viewer permits you to view the mapping relations between the input variables and output variables, as shown in Figure 11.46. The plot may be three-dimensional, as shown, or two-dimensional. For a larger number of input/output variables, the variables for the Surface Viewer can be selected. Again, by closely examining the Surface Viewer, the algorithm can be iterated.

11.7.6 Demo Program for Synchronous Current Control

So far, we have discussed the basic elements of the Fuzzy Logic Toolbox that work in the MATLAB environment. Now, we will consider an example application for a drive. Figure 11.47 shows an indirect vector-controlled induction motor drive that incorporates fuzzy synchronous current control loops. The command i_{ds}^*, which corresponds to the rated rotor flux, is constant. Command i_{qs}^* is generated from the outer speed control loop, which has P-I control. In this example, a fuzzy P-I control will be developed for the i_{qs} and i_{ds} loops with the help of the Toolbox, and their response will be compared with the respective traditional P-I control in a simulated system.

Figure 11.48 shows the MFs of the fuzzy i_{ds} control loop. All the fuzzy variables, error e, change in error ce, and change in output cu (same as du in Figure 11.16), are described in normal-

614 Chapter 11 • Fuzzy Logic Principles and Applications

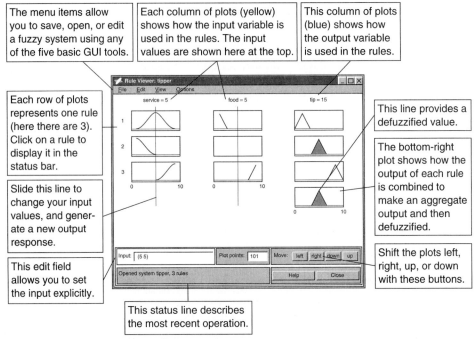

Figure 11.45 User interface for Rule Viewer (Courtesy of Math Works, Inc.)

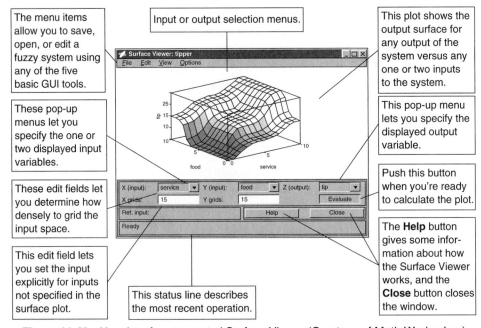

Figure 11.46 User interface to control Surface Viewer (Courtesy of Math Works, Inc.)

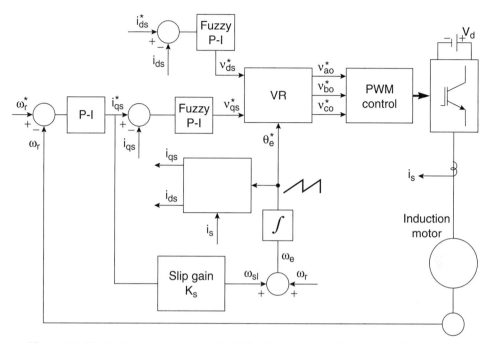

Figure 11.47 Indirect vector-controlled drive incorporating fuzzy i_{qs} and i_{ds} controls

ized form, as shown. Variables e and ce have 7 asymmetrical triangular MFs each, whereas cu has 11 MFs of similar shape. There are $7 \times 7 = 49$ rules all together, which are generated with the help of the Toolbox, as shown in Table 11.5. The MFS and rule matrix are the same for the i_{qs} loop.

As mentioned previously, the MFs and rule table are generated from the experience of the system operation. Once the preliminary control algorithms for both loops are developed, they are incorporated into the SIMULINK simulation of the drive system shown in Figure 11.47.

Figure 11.49 shows the simplified SIMULINK simulation block diagram of Figure 11.47. A discussion of SIMULINK was given in Chapter 5. The details of the "PWM inverter" and "induction motor d-q model" were given in Figures 5.67 and 5.68, respectively. The block "vector controller," which incorporates the fuzzy controllers, is shown in detail in Figure 11.50, and it is self-explanatory. In this figure, the block "command voltage generator" is essentially the same as the "VR" block in Figure 11.47. The synchronous currents i_{qs} and i_{ds} are generated from the block "abc-syn," which simulates the following equations:

$$i_{qs}{}^s = \frac{2}{3}i_a - \frac{2}{3}i_b - \frac{1}{3}i_c \tag{2.72}$$

$$i_{ds}{}^s = -\frac{1}{\sqrt{3}}i_b + \frac{1}{\sqrt{3}}i \tag{2.73}$$

$$i_{qs} = i_{qs}{}^s \cos\omega_e t - i_{ds}{}^s \sin\omega_e t \tag{2.74}$$

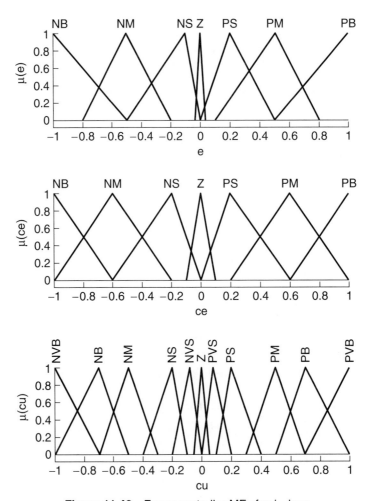

Figure 11.48 Fuzzy controller MFs for i_{ds} loop

$$i_{ds} = i_{qs}^{s} \sin\omega_e t + i_{ds}^{s} \cos\omega_e t \tag{2.75}$$

Once the fuzzy controllers were developed and incorporated into the simulated drive system, the simulation performance helped in the iteration of the controllers. The fuzzy controllers were fine-tuned in several stages of that iteration.

Figure 11.51 shows the controller surface, where the horizontal axes are e and ce, and the vertical axis is the output signal cu. The surface indicates that cu is positive high when both e and ce are positive high. On the other hand, when e and ce are negative high, the output signal cu is also negative high.

Table 11.5 Fuzzy controller rule table for i_{ds} loop

1. If (e is NB) and (ce is NB) then (cu is NVB)
2. If (e is NB) and (ce is NM) then (cu is NVB)
3. If (e is NB) and (ce is NS) then (cu is NB)
4. If (e is NB) and (ce is Z) then (cu is NM)
5. If (e is NB) and (ce is PS) then (cu is NS)
6. If (e is NB) and (ce is PM) then (cu is NVS)
7. If (e is NB) and (ce is PB) then (cu is Z)
8. If (e is NM) and (ce is NB) then (cu is NVB)
9. If (e is NM) and (ce is NM) then (cu is NB)
10. If (e is NM) and (ce is NS) then (cu is NM)
11. If (e is NM) and (ce is Z) then (cu is NS)
12. If (e is NM) and (ce is PS) then (cu is NVS)
13. If (e is NM) and (ce is PM) then (cu is Z)
14. If (e is NM) and (ce is PB) then (cu is PVS)
15. If (e is NS) and (ce is NB) then (cu is NB)
16. If (e is NS) and (ce is NM) then (cu is NM)
17. If (e is NS) and (ce is NS) then (cu is NS)
18. If (e is NS) and (ce is Z) then (cu is NVS)
19. If (e is NS) and (ce is PS) then (cu is Z)
20. If (e is NS) and (ce is PM) then (cu is PVS)
21. If (e is NS) and (ce is PB) then (cu is PS)
22. If (e is Z) and (ce is NB) then (cu is NM)
23. If (e is Z) and (ce is NM) then (cu is NS)
24. If (e is Z) and (ce is NS) then (cu is NVS)
25. If (e is Z) and (ce is Z) then (cu is Z)
26. If (e is Z) and (ce is PS) then (cu is PVS)
27. If (e is Z) and (ce is PM) then (cu is PS)
28. If (e is Z) and (ce is PB) then (cu is PM)
29. If (e is PS) and (ce is NB) then (cu is NS)
30. If (e is PS) and (ce is NM) then (cu is NVS)
31. If (e is PS) and (ce is NS) then (cu is Z)
32. If (e is PS) and (ce is Z) then (cu is PVS)
33. If (e is PS) and (ce is PS) then (cu is PS)
34. If (e is PS) and (ce is PM) then (cu is PM)
35. If (e is PS) and (ce is PB) then (cu is PB)
36. If (e is PM) and (ce is NB) then (cu is NVS)
37. If (e is PM) and (ce is NM) then (cu is Z)
38. If (e is PM) and (ce is NS) then (cu is PVS)
39. If (e is PM) and (ce is Z) then (cu is PS)
40. If (e is PM) and (ce is PS) then (cu is PM)
41. If (e is PM) and (ce is PM) then (cu is PB)
42. If (e is PM) and (ce is PB) then (cu is PVB)
43. If (e is PB) and (ce is NB) then (cu is Z)
44. If (e is PB) and (ce is NM) then (cu is PVS)
45. If (e is PB) and (ce is NS) then (cu is PS)
46. If (e is PB) and (ce is Z) then (cu is PM)
47. If (e is PB) and (ce is PS) then (cu is PB)
48. If (e is PB) and (ce is PM) then (cu is PVB)
49. If (e is PB) and (ce is PB) then (cu is PVB)

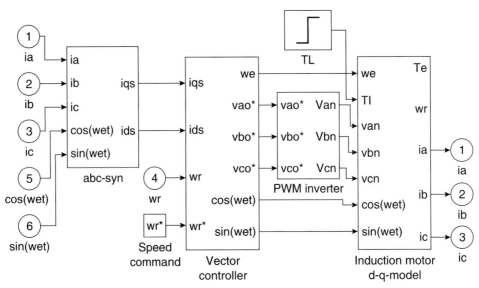

Figure 11.49 SIMULINK simulation block diagram of the drive system in Figure 11.47

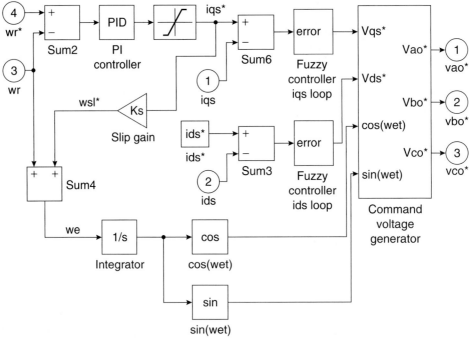

Figure 11.50 SIMULINK simulation block diagram of "vector control" showing fuzzy i_{qs} and i_{ds} controls

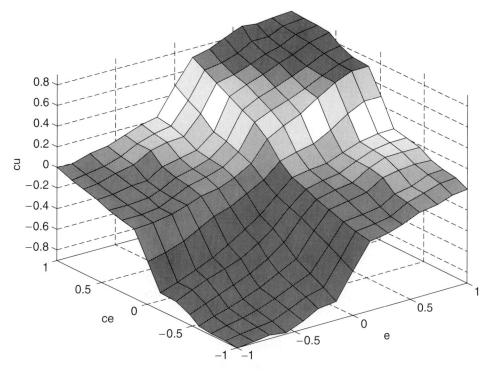

Figure 11.51 Control surface of fuzzy controller

The drive system was simulated with both fuzzy controllers and ordinary P-I controllers and their performances were compared. Figure 11.52 shows the response of the i_{qs} loop with the fuzzy control as well as the P-I control with a stepped speed command, as indicated. The responses are found to be essentially identical. The corresponding response for the i_{ds} loop is shown in Figure 11.53. In this case, the fuzzy control indicates robustness, eliminating transients in the P-I control.

11.8 GLOSSARY

Aggregation – Combination of consequents of each rule.

Antecedent – The initial (or "IF") part of a fuzzy rule.

COA (center of area) defuzzification – A method of calculating crisp output from the center of gravity of the output membership function (also called centroid defuzzification).

Consequent – The final (or "THEN") part of a rule.

Degree of membership – A number between 0 and 1 that represents the output value of a membership function.

Defuzzification – The process of transforming a fuzzy output of a fuzzy system into a crisp output.

DOF (degree of fulfillment) – The degree to which the antecedent part of a fuzzy rule is satisfied.

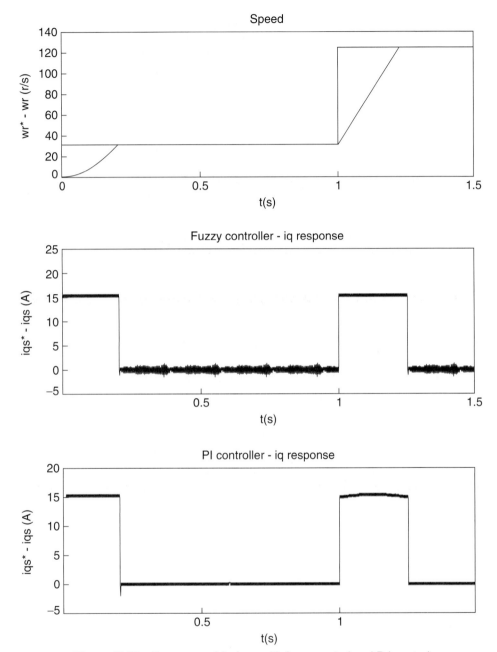

Figure 11.52 Response of i_{qs} loop with fuzzy control and P-I control

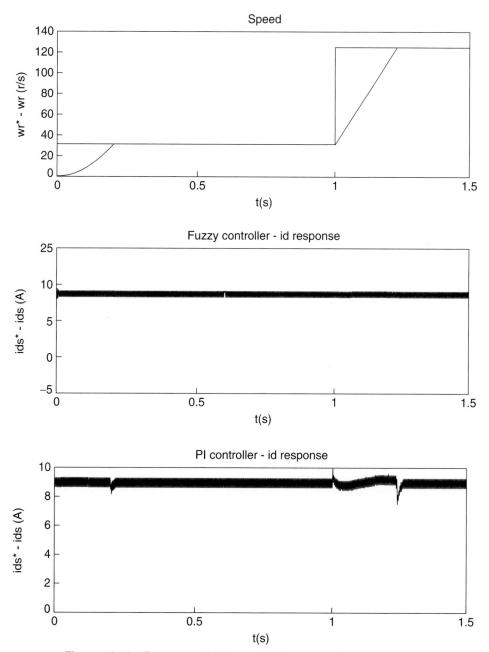

Figure 11.53 Response of i_{ds} loop with fuzzy control and P-I control

Fuzzification – The process of generating a membersip value for a fuzzy variable using a membership function.

Fuzzy composition – A method of deriving fuzzy control output from given fuzzy control inputs.

Fuzzy control – A process control that is based on fuzzy logic and is normally characterized by "IF … THEN…" rules.

Fuzzy system – A system that uses fuzzy reasoning to map an input space to an output space.

Fuzzy implication – The process of shaping the fuzzy set in the consequent part based on the antecedent part of a rule.

Fuzzy rule – IF/THEN rule relating input (conditions) fuzzy variables to output (actions) fuzzy variables.

Fuzzy set (or fuzzy subset) – A set consisting of elements having degrees of membership varying between 0 (nonmember) to 1 (full member). It is usually characterized by a membership function and associated with linguistic values, such as SMALL, MEDIUM, etc.

Fuzzy set theory – A set theory based on fuzzy logic.

Fuzzy variable – A variable that can be defined by fuzzy sets.

Height defuzzification – A method of calculating a crisp output from a composed fuzzy value by performing a weighted average of individual fuzzy sets. The heights of each fuzzy set are used as weighting factors in the procedure.

Linguistic variable – A variable (such as temperature, speed, etc.) whose values are defined by language, such as LARGE, SMALL, etc.

Membership function (MF) – A function that defines a fuzzy set by associating every element in the set with a number between 0 and 1.

MOM (mean of maxima) defuzzification – A method of calculating crisp output from an output membership function where the highest MF component is considered only.

Singleton – A special fuzzy set that has the membership value of 1 at a particular point and 0 elsewhere.

SUP-MIN composition – A composition (or inference) method for constructing the output membership function by using the maximum and minimum principle.

Universe of discourse – The range of values associated with a fuzzy variable.

11.9 SUMMARY

This chapter broadly reviewed FL principles and their application in power electronic systems. FL, as a part of AI, and its difference from an ES, was explained. The analogy and the differences between FL and Boolean logic were explained, particularly relating to operations on fuzzy sets. The principal steps of a fuzzy inference system were explained and illustrated with the restaurant tipping example for clarity of understanding. Several implication and defuzzification methods were explained. The chapter particularly emphasized the control applications of FL. After explaining the basic control principles, a number of applications in power electronic systems including a vector-controlled induction motor drive, an efficiency improvement control by an on-line search of flux, a wind electric generation system, slip gain tuning by a fuzzy MRAC, a stator resistance estimation, and an estimation of a distorted wave, were reviewed from the literature. The fuzzy speed control algorithm was described in detail with a numerical example for clarity.

Numerous software and hardware tools are available for the development and implementation of a fuzzy system. The salient features of a popular tool, MATLAB/Fuzzy Logic Toolbox, were reviewed, and finally, an example application using the Toolbox was discussed. A Glossary was added to the end of this chapter. Hopefully, the knowledge gained in this chapter will help the reader use FL in new applications.

REFERENCES

1. B. K. Bose, "Expert system, fuzzy logic and neural network applications in power electronics and motion control", *Proc. IEEE*, vol. 82, pp. 1303-1323, Aug. 1994.

2. L. H. Tsoukalas and R. E. Uhrig, *Fuzzy and Neural Approaches in Engineering*, Wiley, NY , 1997.

3. Math Works, *Fuzzy Logic Toolbox User's Guide*, Jan., 1998.

4. T. Takagi and M. Sugeno, "Fuzzy identification of a system and its applications to modeling and control", *IEEE Trans. Syst. Man and Cybern.*, vol. 15, pp. 116-132, Jan./Feb. 1985.

5. G. C. D. Sousa and B. K. Bose, "A fuzzy set theory based control of a phase controlled converter dc drive", *IEEE Trans. of Ind. Appl.*, vol. 30, pp. 34-44, Jan./Feb. 1994.

6. Y. F. Li and C. C. Lau, "Development of fuzzy algorithm for servo systems", *IEEE Control Syst. Magazine*, Apr. 1989.

7. G. C. D. Sousa, B. K. Bose, and J. G. Cleland, "Fuzzy logic based on-line efficiency optimization control of an indirect vector controlled induction motor drive", *IEEE Trans. of Ind. Elec.*, vol. 42, pp. 192-198, Apr. 1995.

8. M. G. Simoes, B. K. Bose, and R. J. Spiegel, "Design and performance evaluation of a fuzzy logic based variable speed wind generation system", *IEEE Trans. of Ind. Appl.*, vol. 33, pp. 956-965, July/Aug. 1997.

9. T. M. Rowan, R. J. Kerkman, and D. Leggate, "A simple on-line adaptation for indirect field orientation of an induction machine", *IEEE Trans. on Ind. Appl.*, vol. 42, pp. 129-132, Apr. 1995.

10. G. C. D.Sousa, B. K. Bose, and K. S. Kim, "Fuzzy logic based on-line tuning of slip gain for an indirect vector controlled induction motor drive", *IEEE IECON Conf. Rec.*, pp. 1003-1008, 1993.

11. H. Akagi, Y. Kanazawa, and A. Nabae, "Instantaneous reactive power compensators comprising switching devices without energy storage components", *IEEE Trans. on Ind. Appl.*, vol. 20, pp. 625-630, May/June 1984.

12. B. K. Bose and N. R. Patel, "Quasi-fuzzy estimation of stator resistance of induction motor", *IEEE Trans. of Pow. Elec.*, vol. 13, pp. 401-409, May 1998.

13. M. G. Simoes and B. K. Bose, "Application of fuzzy logic in the estimation of power electronic waveforms", *IEEE IAS Annu. Meet. Conf. Rec.*, pp. 853-861, 1993.

14. I. Miki, N. Nagai, S. Nishigama, and T. Yamada, "Vector control of induction motor with fuzzy PI controller", *IEEE IAS Annu. Meet. Conf. Rec.*, pp. 342-346, 1991.

CHAPTER 12

Neural Network Principles and Applications

12.1 INTRODUCTION

The artificial neural network (ANN), often called the neural network, is the most generic form of AI for emulating the human thinking process compared to the rule-based ES and FL, which were discussed in the previous two chapters. The cerebral cortex of a human brain is said to contain around 100 billion nerve cells or biological neurons, which are interconnected to form a biological neural network. The memory and intelligence of the human brain and the corresponding thinking process are generated by the action of this neural network. Although the structure of a biological neuron is known, the way they are interconnected is not well-known. An ANN tends to emulate the biological nervous system of the human brain in a very limited way by an electronic circuit or computer program.

However inferior the ANN model of the biological nervous system, it tends to solve many important problems. An ANN (also defined as a neurocomputer or connectionist system in the literature) is particularly suitable for solving pattern recognition and image processing-type problems that are difficult to solve by conventional digital computer. On the other hand, a digital computer is very efficient in solving ES problems and somewhat less efficient in solving FL problems.

Pattern recognition or input/output mapping (see Figure 11.1) constitutes the core of neurocomputation, which will be explained later in this chapter. Basically, this mapping is possible by the associative memory property of the human brain. This property helps us to remember or associate the name of a person when we see his or her face, or recognize an apple by its color and shape. Just like a human brain remembers and learns, an ANN is normally trained (not programmed) to learn by example input/output, associating patterns. This is like teaching alphabet characters to a child, where the characters are shown and their names are pronounced repeatedly.

Broadly, the ANN technology has been applied in process control, identification, forecasting, diagnostics, robot vision, and financial problems, just to name a few.

The history of the ANN technology is old and fascinating, and its brief review may be of interest to the reader. Its history predates the ES and FL technologies, and even the advent of modern digital computers. In 1943, McCulloch and Pitts first proposed a network composed of binary-valued artificial neurons that were capable of performing simple threshold logic computations. In 1949, Hebb proposed a network training rule that was called Hebb's rule. Most modern ANN training rules have their origin in Hebb's rule, or a variation of it. In the 1950s, the dominant figure in neural network research was Rosenblatt at Cornell Aeronautical Laboratory, who invented the Perceptron. The Perceptron is a representation of a biological sensory model, such as an eye. In the 1960s, Widrow and Hoff proposed Adaline and Madaline (many Adalines), and trained the network by their delta rule, which was the forerunner of the modern back propagation training method.

The lack of expected performance of these networks, coupled with the glamour of the von Neumann digital computer in the late 1960s and 1970s, practically camouflaged the evolution of the neural network. The modern era of the neural network with rejuvenated research practically started in 1982, when Hopfield, a professor at Cal Tech, presented his invention at the National Academy of Science. Since then, many network models and learning rules have been introduced. Since the beginning of the 1990s, the neural network as AI has captivated the attention of a large segment in the scientific community. The technology is now advancing rapidly, and it is expected to have a significant impact on our society in this century.

In this chapter, we will introduce the artificial neuron and describe several neural network structures. The method of training a neural network with example data will be discussed. Hybrid neuro-fuzzy techniques will be briefly reviewed. After reviewing neural network principles, a number of applications in power electronics and the drives area will be described from the literature. Finally, the MATLAB-based Neural Network Toolbox will be introduced, and a demo program will be included.

12.2 THE STRUCTURE OF A NEURON

12.2.1 The Concept of a Biological Neuron

The structure of an artificial neuron in a neural network is inspired by the concept of a biological neuron, which is shown in Figure 12.1. Basically, it is a processing element (PE) in a brain's nervous system that receives and combines signals from other similar neurons through thousands of input paths called dendrites. If the combined signal is strong enough, the neuron "fires," producing an output signal along the axon, which connects to dendrites of thousands of other neurons. Each input signal coming along a dendrite passes through a synapse or synaptic junction, as shown. This junction is an infinitesimal gap in the dendrite which is filled with neurotransmitter fluid that either accelerates or retards the flow of electrical charges. These electrical signals flow to the nucleus or soma of the junction. The adjustment of the impedance or conductance of the synaptic gap by the neurotransmitter fluid contributes to the "memory"

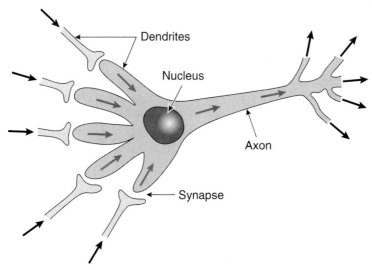

Figure 12.1 Structure of biological neuron

or "learning process" of the brain. According to this theory of the neuron, we are led to believe that the brain has the distributed memory or intelligence characteristics giving it the property of associative memory, but not like a digital computer's central storage memory which is addressed by the CPU. Otherwise, a surgical patient, when recovering from anesthesia, would forget everything that happened in the past.

12.2.2 Artificial Neuron

An artificial neuron is a concept whose components have a direct analogy with the biological neuron. Figure 12.2 shows the structure of an artificial neuron, reminding us of an analog summer-like computation. It is also called a neuron, PE (processing element), neurode, node, or cell. The input signals $X_1, X_2, X_3, \ldots X_N$ are normally continuous variables but can also be discrete pulses. Each input signal flows through a gain or weight, called a synaptic weight or connection strength, whose function is analogous to that of the synaptic junction in a biological neuron. The weights can be positive (excitory) or negative (inhibitory), corresponding to "acceleration" or "inhibition," respectively, of the flow of electrical signals in a biological cell. The summing node accumulates all the input weighted signals, adds the bias signal b, and then passes to the output through the activation function, which is usually nonlinear in nature, as shown in the figure. Mathematically, the output expression can be given as

$$Y = F(S) = F\left[\sum_{K=1}^{N} X_K W_K + b\right] \tag{12.1}$$

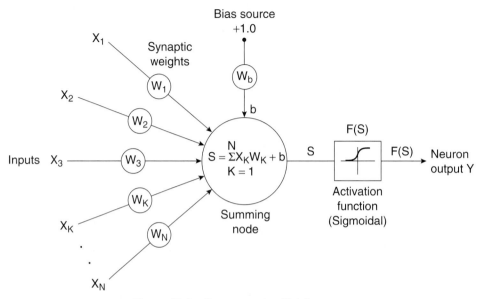

Figure 12.2 Structure of artificial neuron

12.2.2.1 Activation Functions of a Neuron

Figure 12.3 shows a number of possible activation functions (also called transfer function) in a neuron. The simplest of all is the linear activation function, where the output varies linearly with the input but saturates at ±1 as shown with a large magnitude of the input. It can also be a unidirectional-type function. The activation function can be a step or threshold type that passes logical 1 if $S > 0$, or logical 0 if $S < 0$. A positive or negative bias can be introduced in S to alter the threshold value. In the signum activation function shown in (c) with zero bias, the output is +1 if $S > 0$, or −1 if $S < 0$. A single neuron with a threshold activation function is known as a single-layer Perceptron, whereas the same neuron with a signum activation function is known as an Adaline.

The most commonly used activation functions are nonlinear, continuously varying types between two asymptotic values 0 and 1 or −1 and +1. These are, respectively, the sigmoidal function (also called the log-sigmoid or logistic function) shown in (d) and the hyperbolic tan function (also called tan-sigmoid) shown in (e) of the figure. The mathematical expressions of these functions are shown with the figure, where α is the gain or coefficient that adjusts the slope or sensitivity. The hyperbolic tan function is shown with $\alpha = 2$, 1, and 0.5. These functions are differentiable, and the derivative $(dF(S))/dS$ is maximum at $S = 0$, and gradually decreases with an increasing value of S in either direction. With a very high gain, the linear (unidirectional) and sigmoidal functions approach a threshold function, and the bi-directional linear and hyperbolic tan functions approach the signum function. The activation function can also be a Gaussian type, as shown in Figure 11.3(c). All these functions are characterized as squashing functions because they squash or limit the dynamic operation of a neuron between the asymptotic values. Note that

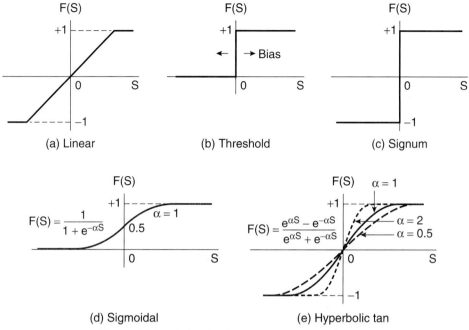

Figure 12.3 Activation functions of artificial neuron

a nonlinear activation function contributes to the nonlinear transfer characteristics of a neuron, which permits nonlinear input/output mapping in a neural network. This nonlinear property is lost if a linear activation function is selected.

12.3 ARTIFICIAL NEURAL NETWORK

The interconnection of artificial neurons results in an ANN, and its objective is to emulate the function of a human brain to solve scientific, engineering, and many other real-life problems. As mentioned before, the interconnection of biological neurons is not well-understood, but scientists have come up with numerous neural network models and many more are yet to come. These networks can generally be classified as feedforward and feedback (or recurrent) types. In a feedforward network, signals from neuron to neuron flow only in the forward direction, whereas in a recurrent network, the signals can flow in a forward as well as a backward or lateral direction. A few network models can be listed from the literature [4] as follows:

Feedforward:

- Perceptron
- Adaline and Madaline
- Back Propagation Network
- Radial Basis Function Network (RBFN)

- General Regression Network
- Modular Neural Network (MNN)
- Learning Vector Quantization (LVQ) Network
- Probabilistic Neural Network (PNN)
- Fuzzy Neural Network (FNN)

Recurrent:
- Hopfield Network
- Boltzmann Machine
- Kohonen's Self-Organizing Feature Map (SOFM)
- Recirculation Network
- Brain-State-in-a-Box (BSB)
- Adaptive Resonance Theory (ART) Network
- Bi-directional Associative Memory (BAM)

A network can be defined as static or dynamic, depending on whether it is a simulating static or dynamic system. It has been claimed [8] that any problem that can be solved by a recurrent network can also be solved by a feedforward network with the proper external connections. In this chapter, we will discuss in detail only a few topologies which are commonly used, particularly for power electronics applications. At present, almost 90 percent of applications are covered by the feedforward architecture, particularly the back propagation network mentioned previously.

Figure 12.4 shows the structure of most commonly used feedforward, multilayer, back propagation-type network. The name "back propagation" comes from its training method, which will be discussed later. For the present, neglect the training block diagram shown in the lower right of the figure. Often, it is called a multi-layer Perceptron (MLP)-type network The network is shown with three input signals (X_1, X_2, and X_3) and two output signals (Y_1 and Y_2). In general, these signals may be logical, discrete bi-directional, or continuous signals. The circles represent the neurons, and the interconnection between the neurons is shown by links. The links always carry signals in the forward direction. In fact, the circle contains the summing node of the neuron with the activation function and the synaptic weights are shown by dots in the links (often the dots are omitted). As mentioned previously, for the unipolar output (continuous or logic) sigmoidal activation function, and for bipolar output (continuous or discrete), the hyperbolic tan activation function is normally used. Sigmoidal outputs can be clamped to generate logical outputs, whereas hyperbolic tan outputs can be clamped to generate ± 1 output in each channel.

The network shown has three layers of neurons: input layer, hidden layer, and output layer. With the number of neurons indicated in each layer, it is defined as a 3-5-2 network. The hidden layer functions to associate the input and output layers. The input and output layers (defined as buffers) have neurons equal to the respective number of signals. The input layer neurons have linear activation functions with unity slope (or no activation function), but there is a scale factor with each input to convert them to per unit (normalization) signals. Similarly, the output signals are converted from per unit signals to actual signals by denormalization, as indicated. Since the input layer simply acts as a distributor of signals to the hidden layer, it is often defined as a two-layer network. A constant bias source supplies the bias signal to the hidden layer and output layer (not shown) neurons through a weight, as indicated. The bias circuits are often omitted for simplicity. There may be more than one hidden layer. The number of hidden layers and the number of neurons in each hidden layer depend on the complexity of the problem to be solved.

Note that there is no self, lateral, or feedback connection of neurons. The network is "fully connected" when each of the neurons in a given layer is connected with each of the neurons in the next layer (as indicated in Figure 12.4), or "partially connected" when some of these connections are missing.

The architecture of the neural network makes it evident that basically, it is a fast and massive parallel input-parallel output multidimensional computing system, where computation is done in a distributed manner, compared to a slow, sequential computation in a conventional digital computer that takes the help of centralized storage memory and CPU. Note that the network does not contain

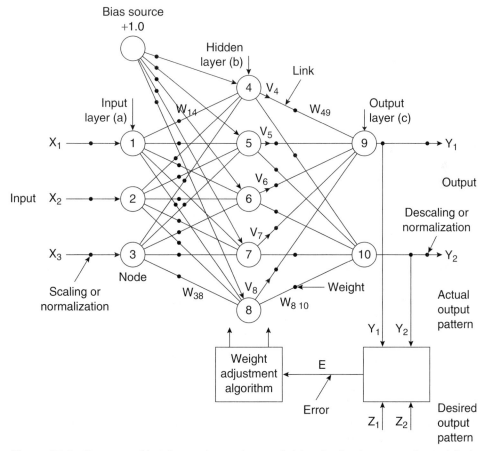

Figure 12.4 Structure of feedforward neural network (showing back propagation training)

any dynamic element. An analog or digital ASIC chip is suitable for ANN computation. However, computation can also be done serially with a high-speed DSP.

12.3.1 Example Application: $Y = A\sin X$

To make the idea clear, Figure 12.5 gives a simple application example of a feedforward three-layer network to synthesize the function $Y = A\sin X$, where X is the single input and Y is the single output. All of the actual weights of the network are shown in the figure. The input layer neuron symbols are omitted and shown as distributors of signals (without per unit conversion). Although the ANN looks like a look-up table generator, it actually is not. However, the network has signal interpolation capability (due to its learning or intelligence), which will be explained next.

12.3.2 Training of Feedforward Neural Network

How can a neural network perform useful functions? Basically, it performs the function of input/output nonlinear mapping or pattern recognition, like the fuzzy inference system indicated in Figure 11.1. This is possible due to its property of associative memory. With a set of input data patterns, the network can be "trained" (not programmed like a digital computer) to give corresponding desired patterns at the output. This is like teaching the alphabet to a child, as mentioned before. When the alphabet characters are shown and their names are pronounced repeatedly, the

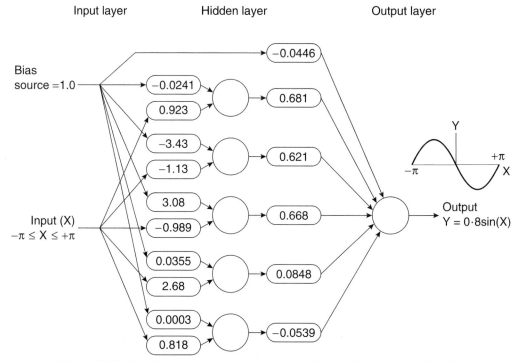

Figure 12.5 Feedforward neural network for synthesis of function $Y = A\sin X$

child learns the alphabet (i.e., his or her biological network is trained), and then he or she can identify them by sight. This is essentially pattern recognition by a supervised learning process.

A trained neural network, like a human brain, can associate a large number of output patterns where each corresponds to an input pattern (called input/output pattern matching). The network has the capability to "learn" because of the distributed intelligence contributed by the weights. This input/output pattern matching is possible if the appropriate weights are selected. In Figure 12.4, there are all together 32 weights (including the bias circuit), and by altering these weights, we can get 32 degrees of freedom at the output for a fixed input signal pattern. We can write output signals as function of inputs as

$$Y_1 = f(X_1, X_2, X_3) \tag{12.2}$$

$$Y_2 = f(X_1, X_2, X_3) \tag{12.3}$$

where $f(.)$ indicates a general functional relation. The signals X_1, X_2, and X_3 may be static (see Figure 12.6) or time functions (Figure 12.35). For each input pattern of signals, there will be a corresponding desired output pattern. It has been proved that such a three-layer ANN can convert any continuous function at the input to any desired function at the output. For this reason, it is often defined as a universal function approximator. The term "approximator" comes from the fact that there will always be some finite small error between the desired output pattern and the actual output pattern for a given input pattern. This is the definition of "soft computing" compared to "hard computing" where computation is precise. The vector and matrix notation is often convenient when dealing with inputs, outputs, and weights. In Figure 12.4, assume that the hidden layer neuron outputs are V_4, V_5, V_6, V_7, and V_8, as indicated. Assuming a hyperbolic tan activation function and neglecting the bias circuit for simplicity, the output of the hidden layer in matrix form can be given as

$$\begin{bmatrix} V_4 \\ V_5 \\ V_6 \\ V_7 \\ V_8 \end{bmatrix} = \tanh \left\{ \begin{bmatrix} W_{14} & W_{24} & W_{34} \\ W_{15} & W_{25} & W_{35} \\ W_{16} & W_{26} & W_{36} \\ W_{17} & W_{27} & W_{37} \\ W_{18} & W_{28} & W_{38} \end{bmatrix} \cdot \begin{bmatrix} X_1 \\ X_2 \\ X_3 \end{bmatrix} \right\} \tag{12.4}$$

or

$$V_b = \tanh(W_{ab} \cdot X_a) \tag{12.5}$$

where V_b is the output vector of layer b, which is given as the dot product of the weight or connectivity matrix W_{ab} and the input layer signal vector X_a. Similarly, the network output signals can be given in matrix form as

$$\begin{bmatrix} Y_1 \\ Y_2 \end{bmatrix} = \tanh \left\{ \begin{bmatrix} W_{49} & W_{59} & W_{69} & W_{79} & W_{89} \\ W_{4\,10} & W_{5\,10} & W_{6\,10} & W_{7\,10} & W_{8\,10} \end{bmatrix} \cdot \begin{bmatrix} V_4 \\ V_5 \\ V_6 \\ V_7 \\ V_8 \end{bmatrix} \right\} \qquad (12.6)$$

or

$$Y_c = \tanh(W_{bc}.V_b) \qquad (12.7)$$

Combining Equations (12.5) and (12.7), we get

$$Y_c = \tanh\left[W_{bc}.\tanh(W_{ab}.X_a)\right] \qquad (12.8)$$

These equations indicate that if the input signals and weights are known, the output signals can be calculated. The matrix and vector representation eases computation by the MATLAB program. For a given input pattern, the actual output pattern is computed and compared with the desired output pattern, and weights can be adjusted by a training algorithm until pattern matching occurs; that is, the error between the desired pattern and actual pattern becomes acceptably small. Such training should be continued with a large number of input/output example data patterns. At the end of training, the network should be capable not only of recalling all the example input-output patterns, but also of interpolating and extrapolating (a limited amount) them. This tests the learning capability of the network instead of the simple look-up table function mentioned for Figure 12.5.

12.3.2.1 Learning Methods

In general, an ANN can have three methods of learning: (1) supervised learning by a teacher, that is, taking the help of a training algorithm; (2) unsupervised or self-learning; and (3) reinforced learning. Supervised learning was discussed above. In unsupervised learning, the network is simply exposed to a number of inputs, and it organizes itself in such a way so as to come up with its own classifications for inputs (stimulation-reaction mechanisms, similar to the way a child learns language initially). With reinforced learning, the learning performance is verified by a critic. Neural network training is a complex subject and it is an exciting topic for current research. Needless to say, a network topology can be proposed easily, but without a satisfactory training method, it becomes useless. Supervised learning is used most widely, and in this chapter, this method will be emphasized.

12.3.2.2 Alphabet Character Recognition by an ANN

Before discussing any training algorithm, the pattern recognition characteristics of a feedforward neural network will be illustrated by a popular optical character recognition (OCR) problem to make the idea clear. This is shown in Figure 12.6. The problem in this case is to convert the English alphabet characters into a 5-bit binary code (considered data compression) so that all together, $2^5 = 32$ different characters can be coded theoretically. Here, the letter "A" is

Artificial Neural Network

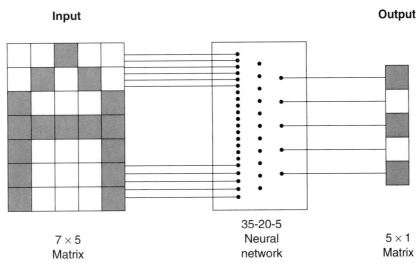

Figure 12.6 Mapping of letter "A" by a five-bit code

represented by a 5 × 7 matrix of inputs consisting of logical 0's and 1's, where the shaded area corresponds to 1 and the unshaded area corresponds to 0. The input vector of 35 logical signals is connected to the respective 35 neurons at the input layer. The three-layer network has 20 neurons in the hidden layer and 5 neurons in the output layer, corresponding to the 5 bits 10101 for the letter A, as indicated. The network uses a sigmoidal activation function, which is clamped to logical 0 and 1 at the output. The input/output pattern mapping is performed by supervised learning, that is, altering the network weights to the desired values. Assuming the network is fully connected and there are no bias signals, the network has all together (35 × 20 + 20 × 5) = 800 weights. The network thus has 800 degrees of freedom for input/output pattern mapping due to the 800 variable weights.

If the letter "B" is impressed at the input and the corresponding desired output map is 10001, the actual output will be totally distorted with the previous training weights. The network undergoes another round of training until the desired output pattern is satisfied. These new weights will deviate the output if "A" is impressed again. This back-and-forth training will satisfy output patterns for both "A" and "B". In this way, a large number of training exercises will eventually train the network to satisfy input/output mapping for all 32 characters. Evidently, the nonlinearity of the network with logical clamping at the output makes such pattern recognition possible. This property of a neural network is also called the pattern classification property.

It is also possible to train the network for inverse mapping; that is, with the input vector 10101, the output vector maps the letter "A", as shown in Figure 12.7. The procedure for training is the same as above. This case is like data expansion instead of data compression. It is possible to cascade Figures 12.6 and 12.7 so that the same letter is reproduced. This arrangement has the advantage of character transmission through a narrowband channel. Instead of cascading, a single network (35–20–35), as shown in Figure 12.8, can also make this conversion. This type of network

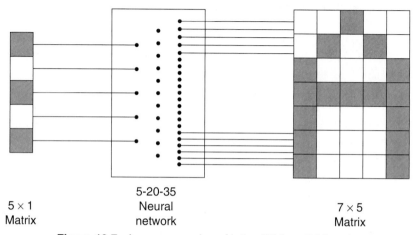

Figure 12.7 Inverse mapping of letter "A" from 5-bit code

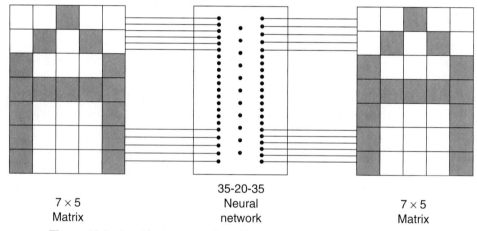

Figure 12.8 Input/output mapping of letter "A" by autoassociative network

is called an autoassociative network, as compared to the heteroassociative network discussed before. The benefit for autoassociative mapping is that if the input pattern is distorted, the output mapping will be clean and crisp because the network is trained to reproduce the nearest crisp output. This inherent noise or distortion-filtering (image recovery) property of a neural network is very important in many applications.

A neural network is often characterized as fault-tolerant. This means that if a few weights become erroneous or several connections are destroyed, the network remains virtually unaffected. This is because the knowledge or intelligence is distributed throughout the network and the computation technique is massively parallel. At the most, the output will degrade gracefully for larger defects in the network, compared to the catastrophic failure that is characteristic of the conventional digital computer, where computation is done in a serial manner.

Artificial Neural Network

12.3.3 Back Propagation Training

Back propagation (BP) is the most popular training method for a multi-layer feedforward network; therefore, the standard network (Figure 12.4) trained by this algorithm is often called the BP network. Rumelhart, Hinton, and Williams (1986) proposed the BP training method, although other workers made contributions to it independently. Basically, it is a generalization of the delta learning rule developed by Widrow and Hoft (1960) for Adaline training. Figure 12.9 shows the general flowchart for BP training of a network. In the beginning, the network topology with the number of layers, number of neurons in the hidden layer(s), and activation function is selected. The input/output example data patterns can be gathered from the experiment, or from the simulated system if a system model is available. The network is initialized with random positive and negative weights to avoid saturation before training starts. With one input pattern, the output is calculated (defined as forward pass) and compared with the desired output pattern. The weights are then changed until the error between the calculated pattern and the desired pattern is very small and acceptable. Similar training is done with all the patterns so that matching occurs for all the patterns. At this point, the network is said to have been trained satisfactorily to perform useful functions. If the error does not converge sufficiently, it may be necessary to alter the number of hidden layer neurons or add extra layers. Instead of selecting one pattern at a time in sequence, batch training can be used, where all the patterns are presented simultaneously and final weight updates are made after processing all the patterns.

The weight adjustment for reduction of error (usually a cost function given by the squared error) is done by the standard gradient descent technique, where the weights are iterated one at a time starting backward from the output layer. A backward trip for such calculations is known as a reverse pass. This is how the name "back propagation" originated. A round-trip of calculations, consisting of a pair of a forward pass and a reverse pass, is defined as an epoch. The algorithm is complex and usually a computer program implements it. The mathematical foundation of the algorithm is described below.

12.3.4 Back propagation Algorithm for Three-Layer Network

Consider the generalized three-layer network shown in Figure 12.10, where the first layer has n neurons (or nodes), the hidden layer has p neurons, and the output layer has q neurons. The bias network is not shown for simplicity. The sigmoidal-type activation function is assumed and is designated by block F for each neuron. There is no activation function for the first layer. The input and output vectors in the network are $X = [X_1 \ldots X_k \ldots X_n]$ and $Y = [Y_1 \ldots Y_m \ldots Y_q]$, respectively, where $Y_1 = F_{1.3}$, $Y_m = F_{m.3}$ and $Y_q = F_{q.3}$.

12.3.4.1 Weight Calculation for Output Layer Neurons

Consider first the problem of calculating the weights for the output layer. Let us take the neuron m and derive a mathematical expression for the iteration of its input weight $W_{lm.3}$, which

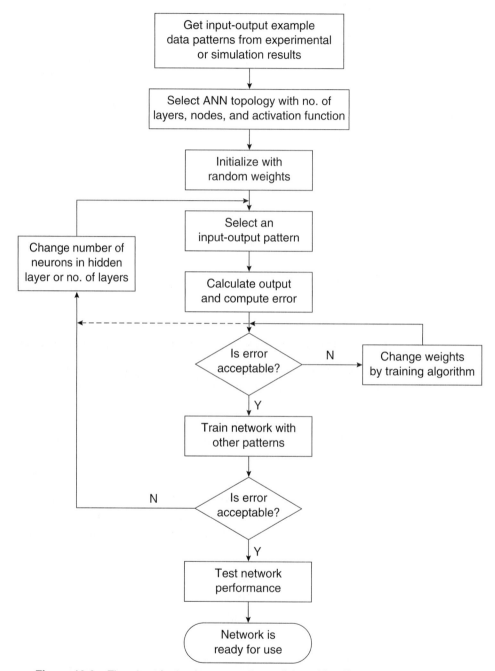

Figure 12.9 Flowchart for back propagation training of feedforward neural network

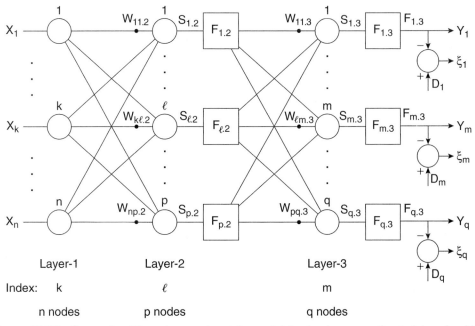

Figure 12.10 Generalized three-layer network for explaining back propagation training algorithm

receives signal $F_{l.2}$ from neuron l. The output of neuron m, $F_{m.3}$, is compared with the desired or target value D_m to calculate the error ξ_m as

$$\xi_m = D_m - Y_m \qquad (12.9)$$

where $Y_m = F_{m.3}$. The objective function to be minimized is defined as

$$\xi_m^2 = (D_m - F_{m.3})^2 \qquad (12.10)$$

The output $F_{m.3}$ and the corresponding ξ_m^2 will vary with the variation of weight $W_{lm.3}$, with all the input signals remaining constant. To minimize ξ_m^2 by the gradient descent method, the change in weight must be proportional to the rate of change of the square error with respect to that weight, that is,

$$\Delta W_{lm.3} = -\eta \frac{\partial \xi_m^2}{\partial W_{lm.3}} \qquad (12.11)$$

where η is the constant of proportionality defined as the learning rate. Therefore, the new weight is given as

$$W_{lm.3}(k+1) = W_{lm.3}(k) + \Delta W_{lm.3} \qquad (12.12)$$

$$= W_{lm.3}(k) - \eta \frac{\partial \xi_m^2}{\partial W_{lm.3}} \qquad (12.13)$$

where $W_{lm.3}(k)$ = old weight and k = number of iterations involved. The partial derivative in Equation (12.11) can be evaluated by the chain rule of differentiation as follows:

$$\frac{\partial \xi_m^2}{\partial W_{lm.3}} = \frac{\partial \xi_m^2}{\partial F_{m.3}} \cdot \frac{\partial F_{m.3}}{\partial S_{m.3}} \cdot \frac{\partial S_{m.3}}{\partial W_{lm.3}} \qquad (12.14)$$

where each term can be evaluated individually. The first term follows from Equation (12.10) as

$$\frac{\partial \xi_m^2}{\partial F_{m.3}} = -2(D_m - F_{m.3}) = -2\xi_m \qquad (12.15)$$

The second term follows the differentiation of the sigmoidal function (see Figure 12.3(d)), which can be derived in the form

$$\frac{\partial F_{m.3}}{\partial S_{m.3}} = \frac{\partial}{\partial S_{m.3}}\left(\frac{1}{1+e^{-\alpha S_{m.3}}}\right) = \alpha F_{m.3}(1 - F_{m.3}) \qquad (12.16)$$

Since the signal $S_{m.3}$ is contributed by all the input signals of the neuron m, we can write

$$S_{m.3} = \sum_{l=1}^{p} W_{lm.3} \cdot F_{l.2} \qquad (12.17)$$

where p = number of neurons in the middle layer. Taking its partial derivative with respect to $W_{lm.3}$ gives

$$\frac{\partial S_{m.3}}{\partial W_{lm.3}} = F_{l.2} \qquad (12.18)$$

which indicates that only $F_{l.2}$ contributes to the change in output when $W_{lm.3}$ is changed. Combining Equations (12.14)–(12.18) we get

$$\frac{\partial \xi_m^2}{\partial W_{lm.3}} = -2\alpha(D_m - F_{m.3})F_{m.3}(1 - F_{m.3})F_{l.2} \qquad (12.19)$$

$$= -\delta_{lm.2}F_{l.2}$$

or

$$\Delta W_{lm.3} = \eta \delta_{lm.3} F_{l.2} \qquad (12.20)$$

where

$$\delta_{lm.3} = 2\alpha(D_m - F_{m.3})F_{m.3}(1 - F_{m.3}) = 2\xi_m \frac{\partial F_{m.3}}{\partial S_{m.3}} \qquad (12.21)$$

The weight is adjusted in steps in Equation (12.12) with the help of Equations (12.20) and (12.21), where the $F_{l,2}$ value is known and the η and α values are assigned. A similar procedure is adopted to adjust all the weights of the output layer. Figure 12.11 illustrates the gradient descent minimization of ξ_m^2 by adjusting the weight $W_{lm.3}$. The initial operating point is 1, corresponding to weight W_0. The weight is increased in steps until the operating point goes to 0, where ξ_m^2 is at its minimum. The learning coefficient η determines the speed of convergence, or, how fast the optimum point is obtained. Normally, a large value of η is taken at the beginning, which is then reduced gradually (adaptive learning rate). One of the problems in gradient descent weight optimization is the local (or false) minimum shown in Figure 12.11 by the dotted U curve instead of the global minimum point 0. The operation may be locked at this point, giving a false ξ_m^2 minimization effect. To jump across this ditch, a momentun term is added on the right of Equation (12.13), which is given as follows:

$$W_{lm.3}(k+1) = W_{lm.3}(k) - \eta \frac{\partial \xi_m^2}{\partial W_{lm.3}} + \mu \left[W_{lm.3}(k) - W_{lm.3}(k-1) \right] \qquad (12.22)$$

where μ = momentum factor.

12.3.4.2 Weight Calculation for Hidden Layer Neurons

The calculation of weights for the hidden layer neurons is much more complex because there are no direct target values to calculate the errors. The error from the output has to be propagated backward, layer by layer, adjusting the weights of each layer in sequence. The procedure is essentially the same as before, but more complex. Consider the problem of calculating the weight

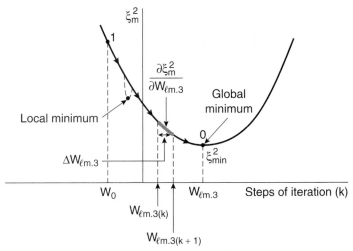

Figure 12.11 Minimization of square error ξ_m^2 by gradient descent method

$W_{kl.2}$ in the hidden layer of Figure 12.10. Note that when $W_{kl.2}$ is changed, the output of all the output layer neurons changes. Therefore, the objective function should consider the summation of all the square errors, that is,

$$\xi^2 = \sum_{m=1}^{q} \xi_m^2 = \sum_{m=1}^{q} \left[D_m - F_{m.3} \right]^2 \tag{12.23}$$

The weight update equation can be given as

$$W_{kl.2}(k+1) = W_{kl.2}(k) + \Delta W_{kl.2} \tag{12.24}$$

where

$$\Delta W_{kl.2} = -\eta \frac{\partial \xi^2}{\partial W_{kl.2}} \tag{12.25}$$

where η = learning rate. Again, by chain differentiation, we can write

$$\frac{\partial \xi^2}{\partial W_{kl.2}} = \sum_{m=1}^{q} \frac{\partial \xi_m^2}{\partial F_{m.3}} \cdot \frac{\partial F_{m.3}}{\partial S_{m.3}} \cdot \frac{\partial S_{m.3}}{\partial F_{l.2}} \cdot \frac{\partial F_{l.2}}{\partial S_{l.2}} \cdot \frac{\partial S_{l.2}}{\partial W_{kl.2}} \tag{12.26}$$

The first two terms of this equation are the same as Equations (12.15) and (12.16), respectively. Since $S_{m.3}$ is contributed by all the input signals, we can write

$$S_{m.3} = \sum_{l=1}^{p} W_{lm.3} \cdot F_{l.2} \tag{12.17}$$

Therefore,

$$\frac{\partial S_{m.3}}{\partial F_{l.2}} = W_{lm.3} \tag{12.27}$$

The last two differentials can be expressed, respectively, as

$$\frac{\partial F_{l.2}}{\partial S_{l.2}} = \alpha F_{l.2}(1 - F_{l.2}) \tag{12.28}$$

$$\frac{\partial S_{l.2}}{\partial W_{kl.2}} = X_k \tag{12.29}$$

Substituting Equations (12.15), (12.16), (12.27), (12.28), and (12.29) in (12.26), we get

$$\frac{\partial \xi^2}{\partial W_{kl.2}} = \sum_{m=1}^{q} (-2)\alpha(D_m - F_{m.3})F_{m.3}(1 - F_{m.3})W_{lm.3}\alpha F_{l.2}(1 - F_{l.2})X_k \tag{12.30}$$

Artificial Neural Network

The Equation (12.30) can be written in the form

$$\frac{\partial \xi^2}{\partial W_{kl.2}} = -\sum_{m=1}^{q} \delta_{lm.3} W_{lm.3} \frac{\partial F_{l.2}}{\partial S_{l.2}} \cdot X_k \qquad (12.31)$$

where

$$\delta_{lm.3} = -2\alpha \xi_m F_{m.3}(1 - F_{m.3}) \qquad (12.32)$$

Finally, from Equations (12.25) and (12.31), the expression for the change of $W_{kl.2}$ is given as

$$\Delta W_{kl.2} = -\eta X_k \sum_{m=1}^{q} \delta_{lm.3} W_{lm.3} \cdot \frac{\partial F_{l.2}}{\partial S_{l.2}} \qquad (12.33)$$

where the parameter values are known. Again, all the hidden layer weights are to be adjusted by a similar process. When training is completed for one input pattern, a similar procedure must be repeated all over again for other patterns in a back-and-forth manner. The total cumulative objective function for P patterns, which is to be minimized, is given as

$$\xi^2 = \sum_{p=1}^{P} \sum_{m=1}^{q} \xi_m^2 \qquad (12.34)$$

When ξ^2 has been minimized to the desired value, the training of the neural network is complete and the weights can be downloaded for practical implementation. Obviously, the training is a very laborious process and the help of a computer program is essential. Many other improved versions of the BP algorithm that tend to reduce the memory size and speed up the computation time have been proposed [5] in recent years. One of them is the Levenberg-Marquardt (L-M) algorithm, which is used in the Neural Network Toolbox in the MATLAB environment. The Toolbox is one of the development system tools available for neural network design and will be discussed later.

12.3.5 On-Line Training

The BP training method, as discussed above, is very time-consuming. Because of slow computation, training is usually done off-line by a computer program. This means that in the actual operation of the network, all the weights are fixed or non-adaptive. In many applications, such as power electronic systems, the network has to emulate nonlinear and time-varying functions where the functions might change depending on the plant operating condition and parameter variation. In such cases, the network requires continuous training on-line so that it correctly emulates the model. This type of ANN is called an adaptive network since the weights vary adaptively, depending on the plant condition and parameter variation. Besides, on-line training permits self-commissioning of a drive, where the controller ANN parameters can be varied adaptively. Drive self-commissioning was discussed in Chapter 8. Fast and improved versions of back propagation algorithm can be implemented on-line by a high-speed DSP to tune the ANN weights if the process speed is not very fast. It is true that DSP computation speed is improving and the cost of memory is falling. If the range of the plant parameter variation is known ahead of time, the ANN can be trained off-line with the nominal parameters and then the weights can be iterated on-line by a high-speed DSP. Training algorithm by EKF can be efficient for both off-line and on-line training, and it will be discussed later. Recently, a random weight change (RWC) algorithm has been proposed [6] to enhance the speed of on-line training.

12.4 OTHER NETWORKS

It was mentioned that the majority of applications use the BP type network. We will briefly discuss a few more topologies from the list given in the previous section that appear to have potential importance for power electronics applications. Particularly, the recurrent neural network will be discussed in some detail. Neuro-fuzzy systems will be introduced later.

12.4.1 Radial Basis Function Network

The radial basis function network (RBFN) is essentially identical to BP network, but it always has three layers. The structure of an RBFN with a 3-3-1 topology (bias circuit not shown) is shown in Figure 12.12. It is a fully connected feedforward network where the output layer neurons have linear characteristics, but the hidden layer neurons use a radial basis function as the activation function. This is usually a Gaussian type function, which is given by the expression

$$F = e^{-\frac{(X-\mu)^2}{\sigma^2}} \tag{12.35}$$

where X = input vector, μ = center of a region called the receptive field, and σ = width of the receptive field. The μ and σ values of all activation functions are usually staggered. If an input vector X lies in the center of a receptive field μ, then that particular hidden layer node will be activated. If the vector lies between two receptive field centers but inside their field widths σ, then both nodes will be partially excited. When the vector is located far away from all receptive fields, there will be no output of the network except the bias. In the practical design of an RBFN, the μ and σ parameters should be chosen such that there is overlap of the activation functions (as shown) and several radial basis neurons have some activation to each input, but all radial basis neurons are not highly active for a single input. Basically, an RBFN produces local mapping in contrast with the global mapping of a BP network. One advantage of an RBFN is that its training is much faster and easier than the BP network. In the first stage of training, the μ and σ values of all activation functions are set. In the second stage, the connection weights are trained by the BP method.

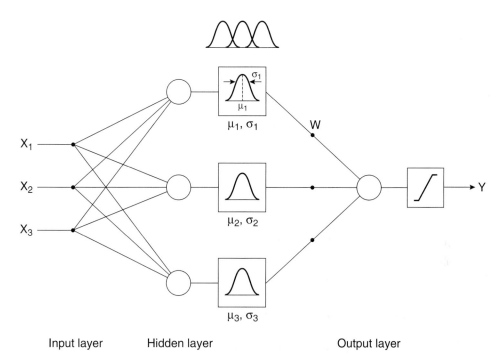

Figure 12.12 Radial basis function network with 3-3-1 structure showing overlap of receptive fields

12.4.2 Kohonen's Self-Organizing Feature Map Network

Kohonen's self-organizing feature map (SOFM) network was inspired by the concept that in the human brain, the different sensory inputs (visual, auditory, somatosensory, motor, etc.) are mapped into different areas of the cerebral cortex in an orderly fashion. Figure 12.13 shows a Kohonen network [4] with two-dimensional (4 × 4) output with 16 neurons. The input buffer layer has a vector of three inputs (x_1, x_2, and x_3). The single output layer of neurons in this network is called the Kohonen layer. The network is fully connected and has lateral feedback and bias input (not shown).

The SOFM network is trained on the basis of "competitive learning." This means that the neurons in the network compete among themselves to be activated or fired with the result that only one neuron or one per group is on or tuned at any one time. The neurons thus become selectively tuned to various input signal patterns in the course of the competitive learning process. The locations of the neurons tuned in this way (called winning neurons) tend to become ordered with respect to each other in such a way that a meaningful coordinate system for different input features is created on the layer of neurons. A SOFM is therefore characterized by the formation of a topographic map of the input patterns in which the spatial locations of the neurons correspond to the intrinsic features of the input patterns; thus the name "self-organizing feature map" is self-explanatory.

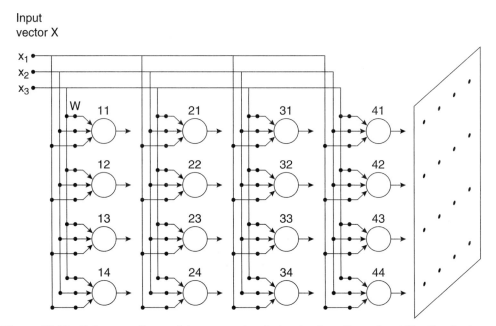

Figure 12.13 Kohonen self-organizing map network showing two-dimensional (4 × 4) output map

12.4.3 Recurrent Neural Network for Dynamic System

The feedforward networks discussed in the previous section simply give input/output static nonlinear mapping; in other words, the output at any instant depends on the input at that instant. In many applications, a neural network is required to be dynamic; that is, it should be able to emulate a dynamic system with temporal behavior, such as the identification of a machine model or estimation of flux. Such a network has a storage property like a capacitor or inductor. A dynamic neural network can be classified as follows:

- Recurrent network
- Time delayed network

The time-delayed network will be discussed later. The recurrent neural network (RNN) uses feedback from the output layer to an earlier layer, and is often defined as a feedback network. Figure 12.14 shows a general structure of a two-layer RNN for a dynamic system where the output signals are fed back to the input layer with a time delay. In this case, the output of a neuron not only depends on the current input signals, but also the prior inputs, thus giving temporal behavior of the network. This means that if, for example, a set of step function signals are impressed at the input, the response will reverberate in the time domain until a steady-state condition is reached. This is similar to the response of an R-L or R-C circuit. Often, it is defined as a real-time recurrent network The network shown in Figure 12.14, has a total of N output neurons

Other Networks

and M external input connections. The outputs are fed back as inputs with a unit time delay. Thus, there are all together $M + N$ inputs. $X(k)$ denotes the external input vector applied to the network at the kth discrete instant, and $Y(k + 1)$ denotes the corresponding vector neuron output produced one step later at discrete time $(k + 1)$. The input $X(k)$ and one step delayed output vector $Y(k)$ together constitute the total input vector given as

$$U(k) = [U_1(k)..........U_{N+M}(k)]^T$$
$$= [Y_1(k)......Y_N(k)\ X_1(k)....X_M(k)]^T \quad (12.36)$$

The output of the jth neuron at instant k is given as

$$Q_j(k) = \sum_{i=1}^{N+M} W_{ij}(k) U_i(k) \quad (12.37)$$

where W_{ij} = connecting weight between i^{th} input to j^{th} neuron and $U_i(k) = i^{th}$ input. In matrix form, the output is

$$\begin{bmatrix} Q_1 \\ . \\ . \\ . \\ Q_N \end{bmatrix} = \begin{bmatrix} W_{11} & . & . & . & W_{(N+M)1} \\ . & & & & . \\ . & & & & . \\ . & & & & . \\ W_{1N} & & & & W_{(N+M)N} \end{bmatrix} \begin{bmatrix} U_1 \\ . \\ . \\ . \\ U_{N+M} \end{bmatrix} \quad (12.38)$$

At the next step, the output of the neuron j is computed by passing the signal through activation function $F(.)$ as

$$Y_j(k+1) = F(Q_j(k)) \quad (12.39)$$

Equations (12.38) and (12.39) constitute the dynamic model of the network.

12.4.3.1 Training an RNN by the EKF Algorithm

To perform satisfactory input/output matching of dynamic functions, all the weights in the network must be trained with example data. Generally, the training of an RNN is more complex than that of a feedforward network. Recently, a powerful EKF algorithm has been proposed for training both feedforward and recurrent neural networks [7], and it will be discussed here briefly. The EKF was discussed in Chapter 8. It has been shown that the training time and training data can be substantially reduced by this algorithm compared to the BP algorithm, although both require the same partial derivative information. The EKF algorithm can be applied for off-line as well as on-line training.

The EKF algorithm is well-established for parameter identification and state estimation (observer) problems in the presence of noise. RNN training to determine the unknown weights can

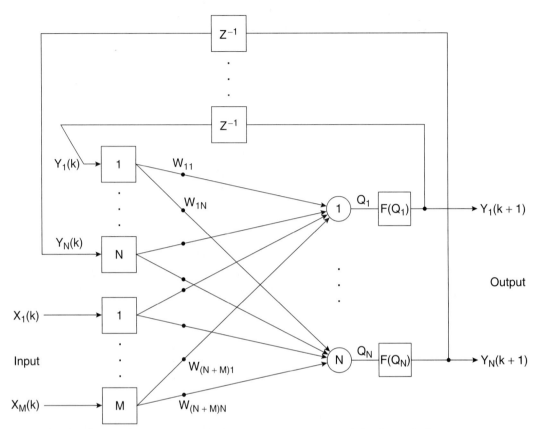

Figure 12.14 General structure of recurrent neural network for dynamical system

also be viewed as a parameter identification problem. Here, the problem involves the computation of partial derivatives of the network outputs with respect to the trainable weights, which are to be tuned. The training algorithm is formulated as a weighted least-square minimization problem, where the error vector is the difference between the functions of the network's output nodes and the desirable values of these functions. The desirable or target vector at time k is given by

$$d(k) = \begin{bmatrix} d_1(k) & d_2(k) & . & . & d_N(k) \end{bmatrix}^T \quad (12.40)$$

where the vector is of length N. Let the actual output $Y(k)$ of the network be represented by the vector $h(k)$ of the same length, that is,

$$h(k) = \begin{bmatrix} F(Q_1) & F(Q_2) & . & . & F(Q_N) \end{bmatrix}^T \quad (12.41)$$

Therefore, the error vector is given by

$$\xi(k) = d(k) - h(k) \quad (12.42)$$

The neural network training cost function $E(k)$ in this case is defined by

$$E(k) = \frac{1}{2}\xi(k)^T S(k)\xi(k) \tag{12.43}$$

where $S(k)$ = user-specified non-negative definite weighting matrix.

The network's trainable weights can be arranged into an M-dimensional vector $W(k)$. The EKF algorithm updates the network weight vector $W(k)$ at time step k. An error covariance matrix $P(k)$, which models the correlation between each pair of weights in the network, is stored and updated at every step. Matrix $P(k)$, at the start of the training procedure, is initialized suitably as a diagonal matrix $P(0)$.

The training procedure consists of the following steps:

1. At instant k, the input signals and recurrent node outputs are propagated through the network and the vector $h(k)$ in Equation (12.41) is calculated.
2. Error vector $\xi(k)$ in Equation (12.42) is calculated by comparing output $h(k)$ with $d(k)$.
3. The partial derivatives of each component of $F(Q)$ with respect to the weights of the network are formed and evaluated at the current weight estimates $W(k)$. These derivatives are arranged in the form of an $M \times N$ matrix $H(k)$ as follows:

$$H(k) = \begin{bmatrix} \dfrac{\partial F(Q_1)}{\partial W_{11}} & \cdots & \dfrac{\partial F(Q_N)}{\partial W_{1N}} \\ \cdot & \cdot & \cdot \\ \cdot & \cdot & \cdot \\ \dfrac{\partial F(Q_1)}{\partial W_{(N+M)1}} & \cdots & \dfrac{\partial F(Q_N)}{\partial W_{(N+M)N}} \end{bmatrix} \tag{12.44}$$

where $F(Q_1), F(Q_2) \ldots F(Q_N)$ are the respective neuron outputs.

4. Then, the estimated $\hat{W}(k)$ and $P(k)$ are updated by the following global EKF recursion equations [7]:

$$A(k) = \left[\eta(k)S(k)^{-1} + H(k)^T P(k)H(k)\right]^{-1} \tag{12.45}$$

$$K(k) = P(k)H(k)A(k) \tag{12.46}$$

$$\hat{W}(k+1) = \hat{W}(k) + K(k)\xi(k) \tag{12.47}$$

$$P(k+1) = P(k) - K(k)H(k)^T P(k) + Q(k) \tag{12.48}$$

where $\eta(k)$ = scalar learning parameter, $K(k)$ = Kalman gain, and $Q(k)$ = diagonal covariance matrix that provides a mechanism to attenuate the noise affecting the signals involved in the training process. This matrix helps to avoid numerical divergence of the algorithm and eliminates stopping at the local minima. The weights are updated by adding the product of $K(k)$ and $\xi(k)$ to the previous weight matrix. The gain $K(k)$ is also used to update the error covariance matrix $P(k)$.

12.5 NEURAL NETWORK IN IDENTIFICATION AND CONTROL

12.5.1 Time-Delayed Neural Network

It was mentioned previously that identification and control of a nonlinear dynamic system can be done by a dynamic neural network, which may have either a recurrent topology or a time-delayed topology. Figure 12.15 shows a time-delayed neural network (TDNN), which uses a single input signal $x(k)$, but parallel ANN input through a tapped delay line. The present and past time sequenced signals are propagated through the network and the output $y(k)$ is given by the expression

$$y(k) = F\left[\sum_{n=0}^{N} W_{nk} x(k-n)\right] \tag{12.49}$$

where $W_{nk} = [W_{0k}, W_{1k} \ldots W_{nk}]$ is the network weight vector and F denotes the activation function. The network can be trained by the static BP algorithm discussed before. The TDNN constitutes an important element in a dynamic system.

12.5.2 Dynamic System Models

Nonlinear dynamical systems can be modeled by nonlinear differential equations. These equations can be expressed in finite difference form for emulation by a neural network. It has been shown in the literature [8] that nonlinear SISO (single input/single output) systems can be modeled in one of the following forms:

- Model-1 – Time-delayed input function with time-delayed output:

$$y(k+1) = F[x(k), x(k-1), \ldots \ldots x(k-M)] + \sum_{i=0}^{N} \alpha_i y(k-i) \tag{12.50}$$

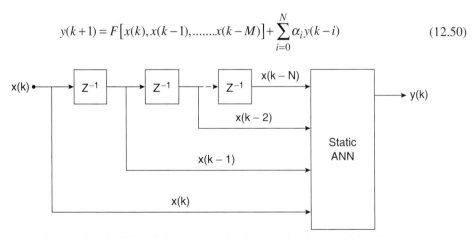

Figure 12.15 Time-delayed neural network using tapped delay line

- Model-2 – Time-delayed output function with time-delayed input:

$$y(k+1) = \sum_{i=0}^{M} \beta_i x(k-i) + F\left[y(k), y(k-1), \ldots y(k-N)\right] \quad (12.51)$$

- Model-3 – Time-delayed input function and time-delayed output function:

$$y(k+1) = F\left[x(k), x(k-1), \ldots x(k-M)\right] + G\left[y(k), y(k-1), \ldots y(k-N)\right] \quad (12.52)$$

- Model-4 – Function with time-delayed input and time-delayed feedback:

$$y(k+1) = F\left[x(k), x(k-1), \ldots x(k-M); y(k), y(k-1), \ldots y(k-N)\right] \quad (12.53)$$

where $[x(k)\ y(k)]$ = input/output data pair of the plant at time k, $F(.)$ and $G(.)$ = nonlinear functions, α_i = output gain factor, and β_i = input gain factor. The block diagrams of all four models are shown in Figures 12.16, 12.17, 12.18, and 12.19, respectively.

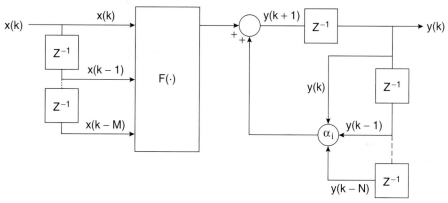

Figure 12.16 Model-1 of dynamic system

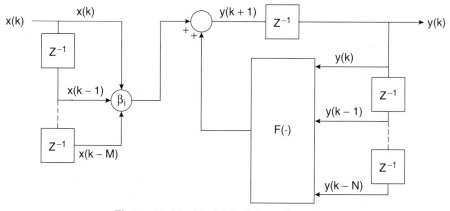

Figure 12.17 Model-2 of dynamic system

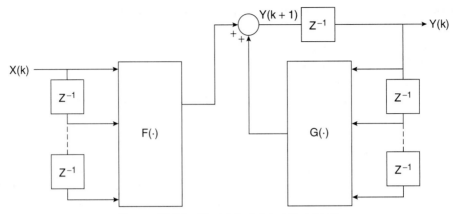

Figure 12.18 Model-3 of dynamic system

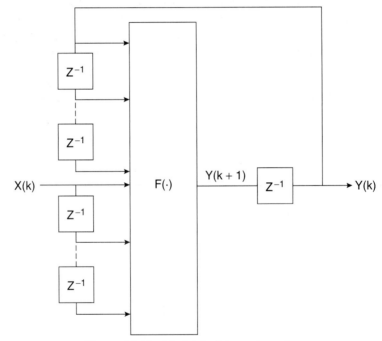

Figure 12.19 Model-4 of dynamic system

12.5.3 ANN Identification of Dynamic Models

How can neural networks be trained to emulate these dynamic models? Note that these block diagrams also represent neural networks incorporating the TDNN structure, as shown in Figure 12.15. For ANN training, the explicit mathematical model of the plant is not required, but the model structure with the order of the system must be known, and there should be adequate input/output data from an experimental system or system model simulation (if a mathematical model is available).

ANN training can be done either in parallel mode or series-parallel mode, as shown in Figures 12.20 and 12.21 (shown for Model-4 only), respectively. In both these methods, the input/output data are functions of time, satisfying the dynamic model of the plant, unlike the static data required to train a static network. Training in series-parallel mode can be done by using the static back propagation algorithm, but the parallel mode requires the dynamic BP algorithm [8] (modified form of static BP method) because of the feedback loop. In either case, the network is trained to match the temporal sequence of input/output patterns of the plant.

After satisfactory training, the ANN represents the plant dynamic system. In the parallel method, the target output from the plant is compared with the calculated ANN output at instant k, and the error $\xi(k)$ updates the network weights by the dynamic BP algorithm. In parallel mode, the error may not always converge and training of the network may fail. In the series-parallel model, the output of the plant (rather than the output of the identification model) is fed back into the identification model, as shown in Figure 12.21. The series-parallel training generally converges, giving satisfactory model emulation. However, if the model is not trained with sufficient accuracy, the stand-alone neural model may not be stable.

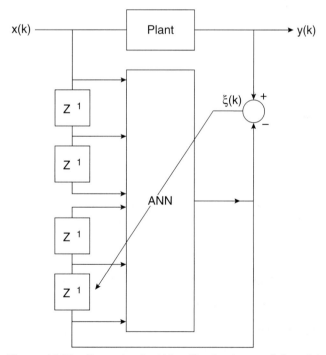

Figure 12.20 Dynamic plant identification by parallel model

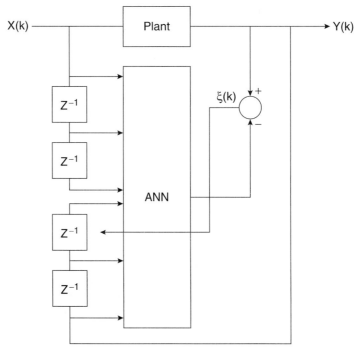

Figure 12.21 Dynamic plant identification by series-parallel model

The model identification discussed above is considered off-line. However, if plant parameters vary, the network should be trained on-line (adaptive model), which will track the varying plant model. In summary, the steps for an ANN identification of the plant with the help of time-delayed networks can be given as follows:

- Determine the structure and order of the dynamic plant model.
- Determine the topology of the ANN.
- Get temporal training data from experiment or simulation (if mathematical model is available).
- Using the back propagation algorithm (dynamic or static), train the network with the parallel or series-parallel method.
- Use the parameters and network weights to implement the actual system with a DSP or ASIC chip.

12.5.4 Inverse Dynamics Model

So far, we have discussed a forward model identification of the plant. It is also possible to identify an inverse model of the plant, as shown in Figure 12.22. In this case, the plant output data is impressed as the input of the ANN, and its calculated output is compared with the plant input (target) data. The resulting error trains the network. After satisfactory training, the model $F^{-1}(.)$ represents the inverse of the plant model $F(.)$.

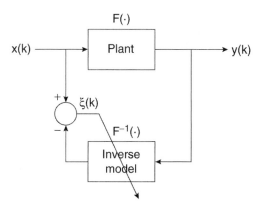

Figure 12.22 Inverse dynamic model training of plant

12.5.5 Neural Network-Based Control

Neural networks can be used for close loop control of a system. Figure 12.23 shows a neural controller, which can be trained to replace the controller shown above. If, for example, the controller is a fuzzy P-I type (with E and CE inputs) and the plant is a vector-controlled induction motor drive, the neural controller (which is a static type) can replace the fuzzy controller, thus simplifying the control implementation.

Figure 12.24 shows the adaptive control of a plant based on the inverse dynamic model of the plant. If an ANN with an accurate inverse dynamic model $F^{-1}(.)$ is available, it can be connected in the front end to cancel the plant dynamics $[F(.) \cdot F^{-1}(.) = 1]$. Then, ideally, the output signals will follow the input signals and no feedback control will be necessary. However, the output will deviate from the input if there is parameter variation of the plant or there is finite error in the reverse model. Figure 12.24 shows an inverse dynamic model-based control of a plant where the error e will constitute the model error and that due to plant parameter variation. The signal e adds with the model output $u(k)'$ to constitute the total input $u(k)$ as shown. The error e can also be used to tune the ANN on-line.

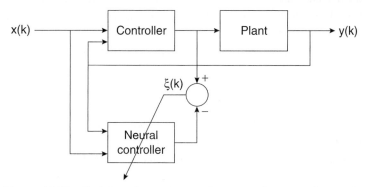

Figure 12.23 Training of neural controller to emulate actual controller

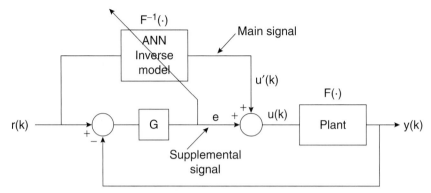

Figure 12.24 Inverse dynamic model based adaptive control of plant

Figure 12.25 shows a model referencing adaptive control using a neural controller. Here, the reference model output (target) is compared with the plant output, which has a parameter variation problem. The ANN controller is trained such that it, along with the plant model, always tracks the reference model. One problem in this direct MRAC control is that the plant lies between the controller and the error and there is no way to propagate the error backward through the controller. To overcome this problem, the indirect MRAC method shown in Figure 12.26 is used. In the beginning, the ANN identification model $F(.)$ is trained to emulate the forward model of the plant. This model is then placed in series with the ANN controller to track the reference model as shown. The tuning of the ANN controller is now convenient through the neural model $F(.)$ instead of through the plant. The ANN model can be updated periodically if there is plant parameter variation.

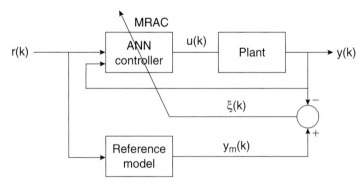

Figure 12.25 MRAC by neural network (direct method)

General Design Methodology

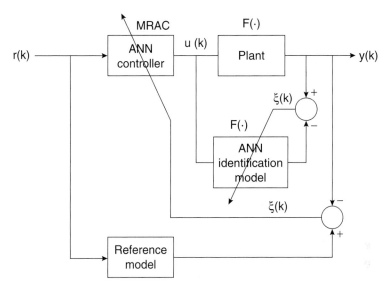

Figure 12.26 MRAC by neural network (indirect method)

12.6 GENERAL DESIGN METHODOLOGY

The methodology of designing a neural network is already evident from the above discussion and the flowchart given in Figure 12.9. The general steps can be summarized as follows:

1. Analyze the problem and find whether it has sufficient elements for a neural network solution. Consider alternative approaches. A simple DSP/ASIC-based direct solution may be satisfactory.
2. If the ANN is to represent a static function, then a three-layer feedforward network (Figure 12.4) should be sufficient. For a dynamic function, select either a recurrent network (Figure 12.14) or a time-delayed network (similar to Figures 12.15–12.19). Information about the structure and order of the dynamic system is required.
3. Select input nodes equal to the number of input signals and output nodes equal to the number of output signals and a bias source. For a feedforward network, select the initially hidden layer neurons typically mean of input and output nodes.
4. Create an input/output training data table. Capture the data from an experimental plant or simulation results, if possible.
5. Select input scale factor to normalize the input signals and the corresponding output scale factor for denormalization.
6. Select generally a sigmoidal transfer function for unipolar output and a hyperbolic tan function for bipolar output.
7. Select a development system, such as Neural Network Toolbox in MATLAB.
8. Select appropriate learning coefficients (η) and momentum factor (μ).

9. Select an acceptable training error ξ and a number of epochs. The training will stop whichever criterion was met earlier.
10. After the training is complete (see Figure 12.9) with all the patterns, test the network performance with some intermediate data points (interpolation).
11. Finally, download the weights and implement the network by hardware or software.

12.7 APPLICATIONS

Neural networks have been applied for various control, identification, and estimation applications in power electronics and drives. Some of these applications can be summarized as follows:

- Single or multi-dimensional look-up table functions
- Converter PWM
- Neural adaptive P-I drive controller
- Delayless filtering
- Vector rotation and inverse rotation in vector control
- Drive MRAC
- Drive feedback signal estimation
- On-line diagnostics
- Estimation for distorted waves
- FFT signature analysis of waves

In this section, we will briefly discuss some example applications from the literature.

12.7.1 PWM Controller

12.7.1.1 Selected Harmonic Elimination (SHE) PWM

The SHE PWM technique for a voltage-fed inverter was discussed in Chapter 5. It was shown that if there are five notch angles (α's) in a three-phase inverter, then the fundamental voltage can be controlled and four significant harmonics (such as the 5th, 7th, 11th and 13th) can be eliminated from the square wave, that is, the lowest harmonic present is the 17th. Figure 12.27 shows the precomputed notch angle curves with the modulation factor m. Normally, a look-up table is prepared in microcomputer memory and the angles, as functions of m, are retrieved for digital implementation. Instead of a look-up table, a feedforward neural network can be trained [10], as shown in Figure 12.28, with the data from Figure 12.27. After training, as the command modulation index m^* is impressed at the input, all the correct α angles are retrieved at the output. The advantage of an ANN in this case is that the α angles are easily interpolated by an ANN avoiding the need of a large precision look-up table memory. The network (1-8-5 structure) with a sigmoidal transfer function in each neuron was trained by the standard BP algorithm. The trained network was implemented in Intel's ETANN (Electrically Trainable Analog Neural Network) ASIC chip 80170NX for experimental verification with an

Applications

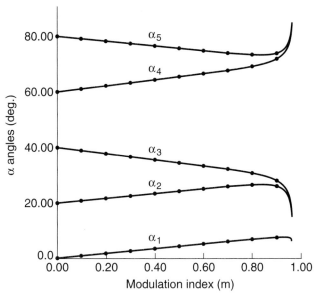

Figure 12.27 Notch angle chart with modulation factor for selected harmonic elimination of PWM inverter

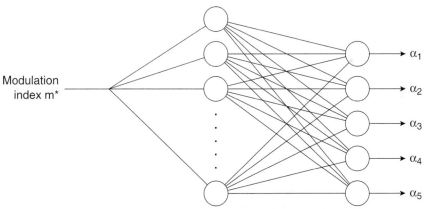

Figure 12.28 Feedforward neural network (1-8-5) for retrieving notch angles (bias circuit not shown)

inverter. The chip has 64 neurons and 10,240 EEPROM-based analog synaptic weights. (The chip was withdrawn from the market.)

12.7.1.2 Instantaneous Current Control PWM

Instantaneous phase current control of a three-phase, voltage-fed inverter by the hysteresis-band PWM method was discussed in Chapter 5. Instantaneous current control is also possible by a neural network-based PWM [11], as shown in Figure 12.29. The advantages of an ANN-based

PWM are constant switching frequency of the inverter, which unlike hysteresis-band PWM, does not vary with the dc supply voltage, machine parameters, and CEMF. In Figure 12.29, the network receives the phase current error signals (e_a, e_b, and e_c) through the respective gain or scale factor K. The output signals of the ANN are passed through hard limiters and sample & hold (S&H) circuits to impress the logic signals at the inverter input. The S&H elements are activated by signals SH_a, SH_b, and SH_c, respectively, as shown, with a sampling period T_c that corresponds to the switching frequency ($f_c = 1/T_c$) of the inverter. The binary values of S_a, S_b, and S_c constitute the eight switching states of the inverter. The neural network is trained such that if the error e exceeds the threshold value $\pm\delta$, the state change occurs in the respective output. If, for example, the threshold value of δ is 0.01 A, the output of a phase a channel S_a' will be 1 if $e_a > +0.01$ A and 0 if $e_a < -0.01$ A. For the range -0.01 A $< e_a < +0.01$ A, the output remains unchanged. The network is trained to satisfy the input/output vectors shown in Table 12.1, where $-\delta$ corresponds to state 0 and $+\delta$ corresponds to state 1. For balanced command currents and isolated neutral load, we can write

$$i_a^* + i_b^* + i_c^* = 0 \tag{12.54}$$

$$i_a + i_b + i_c = 0 \tag{12.55}$$

Subtracting Equation (12.55) from (12.54), we can write

$$e_a + e_b + e_c = 0 \tag{12.56}$$

To satisfy Equation (12.56), the current loop error signals cannot be of the same polarity. Therefore, states 1 and 8 in Table 12.1 are not valid for training. Table 12.1 indicates that if the S&H circuits of all the phases are activated at the same instant, the zero voltage vectors (111 or 000) cannot be established. To avoid this situation, sampling is made with a stagger of $T_c/3$ interval, as shown in the lower part of Figure 12.29. This modified sampling method permits all eight states of the inverter. The inverter switching through the zero voltage vector gives better PWM harmonic quality and reduces device voltage stress by avoiding abrupt voltage transition between positive and negative levels and instantaneous current reversal in the dc link.

12.7.1.3   Space Vector PWM

The space vector PWM technique for a three-phase isolated neutral load of a voltage-fed inverter was discussed in Chapter 5, and the discussion covered both the undermodulation and overmodulation regions. Instead of implementing SVM by DSP, it is possible to implement it by a feedforward neural network [12] because the SVM algorithm can be looked upon as a nonlinear input/output mapping. This means that the reference voltage vector V^* magnitude, and θ_e angle (see Figure 5.31) can be impressed at the input of the network and the corresponding pulse width pattern of the three phases can be generated at the output. To develop the network topology and generate training data, further theoretical discussion is necessary.

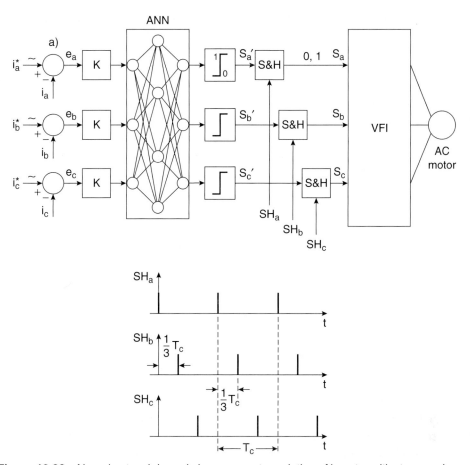

Figure 12.29 Neural network-based phase current regulation of inverter with staggered sampling

Undermodulation Region: In this region, we can write the following expressions (see Figure 5.31) from Equations (5.58), (5.59), (5.62), (5.63), and (5.64):

$$t_a = 2KV^* \sin(\frac{\pi}{3} - \alpha^*) \tag{12.57}$$

$$t_b = 2KV^* \sin\alpha^* \tag{12.58}$$

$$t_0 = T_c - (t_a + t_b) = \frac{T_s}{2} - (t_a + t_b) \tag{12.59}$$

where
  t_a = time interval of switching vector V_a that lags command vector V^*
  t_b = time interval of switching vector V_b that leads V^*
  t_0 = time interval of zero vector switching

Table 12.1 Neural Network Input/Output Training Data Vectors

State	Input			Output
	e_a	e_b	e_c	S_a', S_b', S_c'
1	δ	δ	δ	(1,1,1) (not valid)
2	$-\delta$	δ	δ	(0,1,1)
3	δ	$-\delta$	δ	(1,0,1)
4	δ	δ	$-\delta$	(1,1,0)
5	δ	$-\delta$	$-\delta$	(1,0,0)
6	$-\delta$	δ	$-\delta$	(0,1,0)
7	$-\delta$	$-\delta$	δ	(0,0,1)
8	$-\delta$	$-\delta$	$-\delta$	(0,0,0) (not valid)

$T_s = 2T_c = 1/f_s$ is the sampling time (see Figure 5.32)

$\alpha^* =$ angle of V^* in a $60°$ sector, and

$$K = \frac{\sqrt{3}T_s}{4V_d}$$

Therefore, time $T_{A\text{-}ON}$ in Figure 5.32 can be derived from Equations (12.57), (12.58), and (12.59) for sector 1 as

$$T_{A\text{-}ON} = \frac{t_0}{2} = \frac{T_s}{4} + K.V^*\left[-\sin(\frac{\pi}{3}-\alpha^*)-\sin\alpha^*\right] \quad (12.60)$$

Similar timing intervals can be derived for all six sectors, and correspondingly, the phase a turn-on time can be expressed as

$$T_{A\text{-}ON} = \begin{cases} = \frac{t_0}{2} = \frac{T_s}{4}+KV^*\left[-\sin(\frac{\pi}{3}-\alpha^*)-\sin(\alpha^*)\right] & S=1,6 \\ \frac{t_0}{2}+t_b = \frac{T_s}{4}+KV^*\left[-\sin(\frac{\pi}{3}-\alpha^*)+\sin(\alpha^*)\right] & S=2 \\ \frac{t_0}{2}+t_a+t_b = \frac{T_s}{4}+KV^*\left[\sin(\frac{\pi}{3}-\alpha^*)+\sin(\alpha^*)\right] & S=3,4 \\ \frac{t_0}{2}+t_a = \frac{T_s}{4}+KV^*\left[\sin(\frac{\pi}{3}-\alpha^*)-\sin(\alpha^*)\right] & S=5 \end{cases} \quad (12.61)$$

where the sector number S is indicated on the right. Because of symmetry, the corresponding turn-off time is given as

$$T_{A-OFF} = T_s - T_{A-ON} \tag{12.62}$$

Equation (12.61) can be written in the general form

$$T_{A-ON} = \frac{T_s}{4} + f(V^*) \cdot g(\alpha^*) \tag{12.63}$$

where $f(V^*)$ = voltage amplitude scale factor and

$$g_A(\alpha^*) = \begin{cases} K\left[-\sin(\frac{\pi}{3} - \alpha^*) - \sin(\alpha^*)\right] & S = 1,6 \\ K\left[-\sin(\frac{\pi}{3} - \alpha^*) + \sin(\alpha^*)\right] & S = 2 \\ K\left[\sin(\frac{\pi}{3} - \alpha^*) + \sin(\alpha^*)\right] & S = 3,4 \\ K\left[\sin(\frac{\pi}{3} - \alpha^*) - \sin(\alpha^*)\right] & S = 5 \end{cases} \tag{12.64}$$

which is defined as the pulse width function at unit amplitude $[f(V^*) = 1]$. In the undermodulation region, the scale factor is linear, that is, $f(V^*) = V^*$, which is shown in Figure 12.30 for dc voltage $V_d = 300$ V. Equation (12.63) for the undermodulation region has been plotted in Figure 12.31 as a function of the α angle in all six sectors. For the B and C phases, the corresponding curves are identical, but mutually phase-shifted by 120°.

At the upper limit of the undermodulation region, the curve will clamp at $T_s/2$ in the upper level and at 0 in the lower level, as indicated in the figure. If $V^* = 0$, T_{A-ON} always equals $T_s/4$.

Overmodulation Region: In overmodulation modes 1 and 2, as discussed in Chapter 5, rigorous calculations can be made for $f(V^*)$ and T_{A-ON} and plotted in Figures 12.30 and 12.31, respectively. However, it can be shown that the undermodulation curve in Figure 12.31 can be expanded to mode 1 and mode 2 by using a nonlinear scale factor $f(V^*)$, as shown in Figure 12.30, to get SVM with a linear transfer relation between the input and output voltages in the whole range. The idea is the same as that of sinusoidal PWM overmodulation, where the modulating voltage is magnified beyond the undermodulation range. At the upper limit of mode 2, $f(V^*)$ is ideally infinity and T_{A-ON} approaches a square wave.

Network Topology: The plots in Figures 12.30 and 12.31 give us sufficient information for the design and training of the network. Figure 12.32 shows the ANN topology, including the timer section at the output. There are essentially two subnets: the amplitude subnet and the angle subnet. The amplitude subnet receives the voltage vector magnitude as the input and solves for function $f(V^*)$, as shown in Figure 12.30. The angle subnet receives the angle θ_e^* at the input

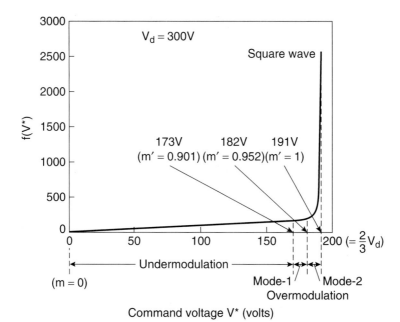

Figure 12.30 $f(V^*) - V^*$ relation in undermodulation and overmodulation regions of space vector PWM (V_d = 300 V)

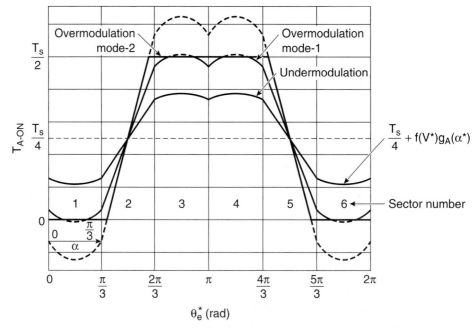

Figure 12.31 Calculated turn-on time $T_{A\text{-}ON}$ of phase a as function of θ_e angle in all six sectors

Applications

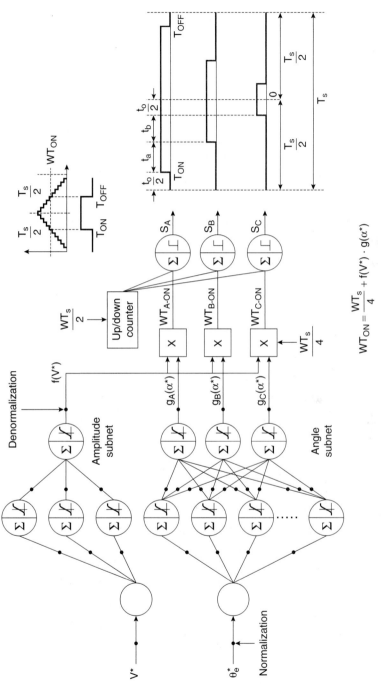

Figure 12.32 Neural network topology (2-20-4) for space vector PWM in undermodulation and overmodulation regions

and solves for digital equivalent of $g_A(\alpha^*)$, $g_B(\alpha^*)$, and $g_C(\alpha^*)$ at the output for the three phases. These are multiplied by $f(V^*)$ and added with $WT_s/4$ bias signal to generate the corresponding digital word for the three phases given by the general expression

$$WT_{ON} = \frac{WT_s}{4} + f(V^*)g(\alpha^*) \quad (12.65)$$

A single UP/DOWN-counter then helps to generate the symmetrical pulse widths for all three phases. Figure 12.33 shows the typical current waves for a volts/Hz-controlled (0–60 Hz) induction motor drive with an ANN-based SVM. The performances were found to correlate well with the corresponding DSP-based SVM.

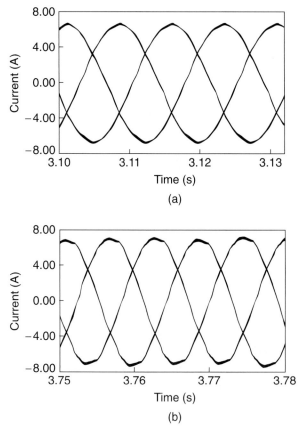

Figure 12.33 Current waves with neural network-based SVM (a) Undermodulation, (b) Overmodulation

12.7.2 Vector-Controlled Drive Feedback Signal Estimation

Figure 12.34 shows the block diagram of a direct vector-controlled induction motor drive, where a feedforward neural network-based estimator [13] estimates the rotor flux (ψ_r), unit vector ($\cos\theta_e$, $\sin\theta_e$), and torque (T_e) by solving the following equations, which were derived in Chapter 8:

$$\psi_{dm}^{\ s} = \psi_{ds}^{\ s} - i_{ds}^{\ s} L_{ls} \tag{8.33}$$

$$\psi_{qm}^{\ s} = \psi_{qs}^{\ s} - i_{qs}^{\ s} L_{ls} \tag{8.34}$$

$$\psi_{dr}^{\ s} = \frac{L_r}{L_m}\psi_{dm}^{\ s} - L_{lr} i_{ds}^{\ s} \tag{8.37}$$

$$\psi_{qr}^{\ s} = \frac{L_r}{L_m}\psi_{qm}^{\ s} - L_{lr} i_{qs}^{\ s} \tag{8.38}$$

$$\hat{\psi}_r = \sqrt{\psi_{dr}^{\ s^2} + \psi_{qr}^{\ s^2}} \tag{8.26}$$

$$\cos\theta_e = \frac{\psi_{dr}^{\ s}}{\hat{\psi}_r} \tag{8.24}$$

$$\sin\theta_e = \frac{\psi_{qr}^{\ s}}{\hat{\psi}_r} \tag{8.25}$$

$$T_e = \frac{3}{2}\left(\frac{P}{2}\right)(\psi_{dr}^{\ s} i_{qs}^{\ s} - \psi_{qr}^{\ s} i_{ds}^{\ s}) \tag{8.41}$$

A DSP-based estimator is also shown in the figure for comparison. Since the feedforward network cannot solve any dynamic system, the machine terminal voltages are integrated by a hardware low-pass filter (LPF) to generate the stator flux signals as shown. The variable-frequency, variable-magnitude sinusoidal signals are then used to calculate the output parameters by a feedforward network. Figure 12.35 shows the topology of the network where there are three layers, and the hidden layer contains 20 neurons.

The input layer neurons have linear activation characteristics (or else, no activation function), but the hidden and output layers have a hyperbolic tan-type activation function to produce bipolar outputs. Figure 12.36 shows the torque, flux, and unit vector signals of the estimator after successful training of the network with a large number of simulation data sets. The estimator outputs are compared with the corresponding outputs of the DSP-based estimator and show good accuracy. When tested with a low switching frequency of the inverter (2 kHz instead of 15

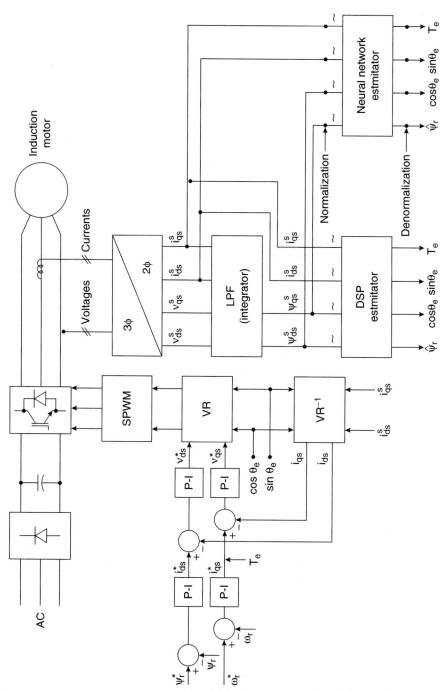

Figure 12.34 Vector-controlled induction motor drive with neural network-based estimator

Applications

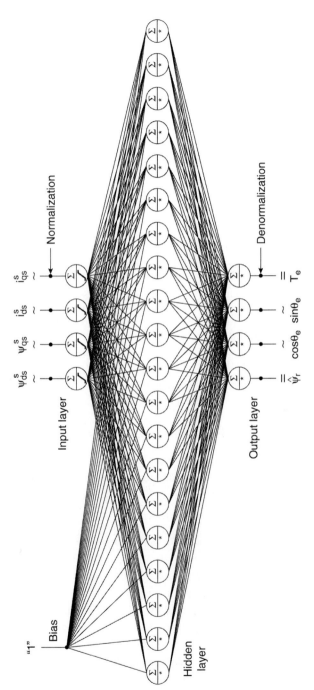

Figure 12.35 Neural network topology (4-20-4) for estimation

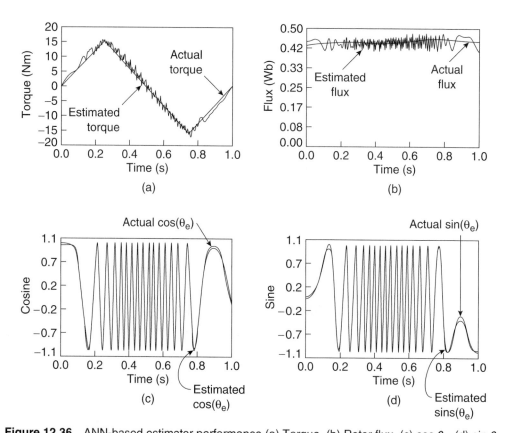

Figure 12.36 ANN-based estimator performance (a) Torque, (b) Rotor flux, (c) cos θ_e, (d) sin θ_e

kHz), the ANN-based estimator showed somewhat harmonic-immune performance. Inherent noise or harmonic filtering is one of the advantages of a neural network.

12.7.3 Estimation of Distorted Waves

The problem here is to estimate accurately the rms current I_s, fundamental rms current I_f, DPF, and PF for a single-phase, anti-parallel, thyristor-controlled R-L load with the help pf an ANN where the firing angle α, load impedance Z, and impedance angle $\varphi = \cos^{-1}R/Z$ are varying, but the supply voltage is constant. This can be viewed as a pattern recognition problem. We discussed similar estimation in Chapter 11 with the help of FL. The supply voltage is assumed to be a 220 V single-phase sine wave. Figure 12.37 shows the network structure for estimation [14], which has two hidden layers with 16 neurons in each hidden layer. The converter was simulated to get output waves for different values of α, φ, and Z_b/Z (where Z_b = base impedance), and the output parameters for training were calculated from the simulation waves with the help of a MATLAB program. Figure 12.38 shows the estimator's performance for the four output parameters, where each estimated curve was verified to be accurate with the corresponding calculated curve. In each case, the minimum α angle is restricted at the verge of continuous conduction. The $I_s(pu)$ and $I_f(pu)$ are converted to actual values by multiplying with the scale factor $|Z_b/Z|$ for the input condition $|Z_b/Z| = 1.0$. A numerical example is given below for calculation from Figure 12.38.

Input parameters: Firing angle $\alpha = 60°$
                        Impedance angle $\varphi = 30°$
                        Supply voltage $V_s = 220$ V
                        Base impedance $|Z_b| = 220\sqrt{2}$ Ω
                        Load impedance $|Z| = 50$ Ω

Estimated outputs from Figure 12.38:
    $I_s(pu) = 0.79$, i.e., actual $I_s = 0.79 \cdot 220 \sqrt{2}/50 = 4.915$ A
    $I_f(pu) = 0.78$, i.e., actual $I_f = 0.78 \cdot 220 \sqrt{2}/50 = 4.853$ A
    DPF = 0.84 (directly)
    PF = 0.82 (directly)

12.7.4 Model Identification and Adaptive Drive Control

This application describes an inverse dynamic model based indirect MRAC of a dc motor drive, where it is desirable that the motor speed follows a desired command speed trajectory. The

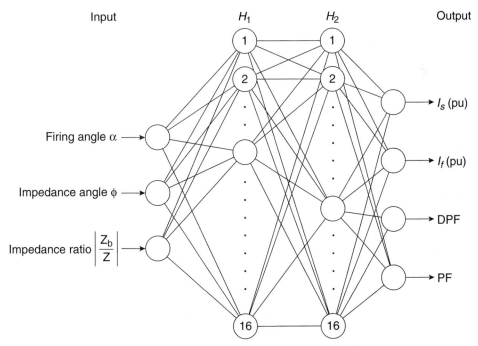

Figure 12.37 Neural network for estimation of triac controller line current

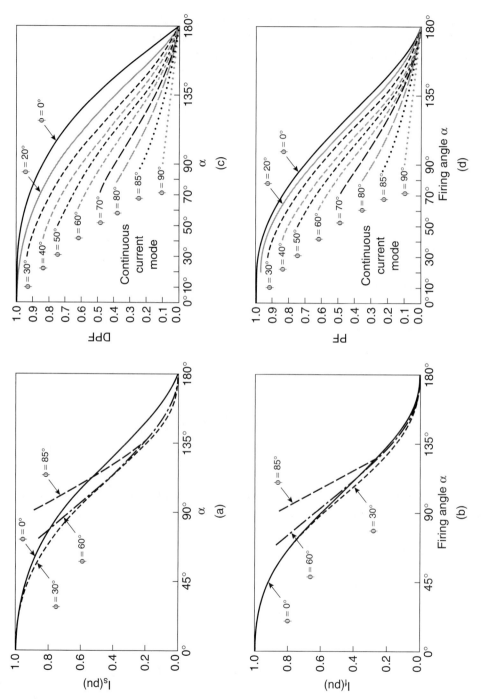

Figure 12.38 Estimator performance curves, (a) Rms current [$I_s(pu)$], (b) Fundamental rms current [$I_f(pu)$], (c) DPF, (d) PF

concept developed here can be easily extended to a vector-controlled ac drive. The motor model with the load is nonlinear and time-invariant, and the model parameters are unknown. The reference model that the motor is required to follow is given. The motor dynamics with the load are captured by an inverse motor model, which is represented by a feedforward neural network.

Figure 12.39 shows the block diagram of the complete control system [15]. The dc motor model represented by block F can be described by the following nonlinear dynamic equations:

$$v(t) = R_a i_a(t) + L_a \frac{di_a}{dt} + K\omega_r(t) \tag{12.66}$$

$$K i_a(t) = J \frac{d\omega_r}{dt} + B\omega_r(t) + T_L(t) \tag{12.67}$$

$$T_L(t) = \mu \omega_r^2(t)[Sign(\omega_r(t))] \tag{12.68}$$

where $v(t)$ = motor terminal voltage
$i_a(t)$ = armature current
$\omega_r(t)$ = electrical speed
$T_L(t)$ = load torque
R_a = armature resistance
L_a = armature inductance
K = CEMF and torque constant
J = moment of inertia
B = damping constant
μ = a constant

The fan load with square-law torque characteristics is assumed, and the load torque always opposes the direction of motion. The second-order discrete time model is derived by first combining Equations (12.66)–(12.68) and then replacing all continuous differentials with finite differences. The resulting equation is

$$\omega_r(k+1) = \alpha \omega_r(k) + \beta \omega_r(k+1) + \gamma [Sign(\omega_r(k))]\omega_r^2(k) \\ + \delta [Sign(\omega_r(k))]\omega_r^2(k-1) + \xi v(k) \tag{12.69}$$

or

$$v(k) = \frac{1}{\xi}\Big[\omega_r(k+1) - \alpha \omega_r(k) - \beta \omega_r(k+1) - \gamma [Sign(\omega_r(k))]\omega_r^2(k) \\ - \delta [Sign(\omega_r(k))]\omega_r^2(k-1)\Big] \tag{12.70}$$

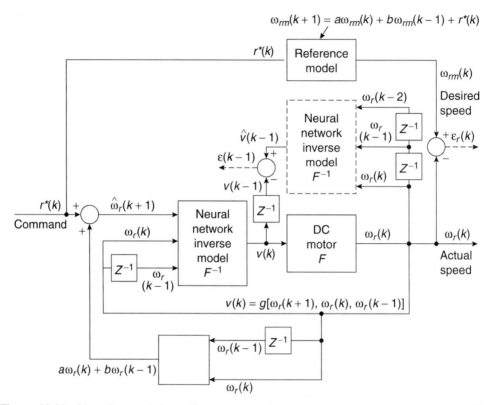

Figure 12.39 Neural network-based inverse dynamics adaptive speed control of dc motor drive

Equation (12.70) is nonlinear where $v(k)$ is a function of the variables $\omega_r(k+1)$, $\omega_r(k)$, and $\omega_r(k-1)$, i.e.,

$$v(k) = g\left[\omega_r(k+1), \omega_r(k), \omega_r(k-1)\right] \qquad (12.71)$$

or

$$v(k-1) = g\left[\omega_r(k), \omega_r(k-1), \omega_r(k-2)\right] \qquad (12.72)$$

which is a discrete model of the machine with unknown parameters. Figure 12.39 shows the block (dotted) of the neural network-based inverse model (F^{-1}) of the machine, which is first trained off-line by the input/output data generated by the machine. The network has no dynamics and it is represented by a three-layer, three-input, feedforward network with five hidden layer neurons, where the time-delayed speed signals are generated from the actual machine speed by a delay line as shown in the figure. The network output signal $\hat{v}(k-1)$ is compared with the target signal $v(k-1)$, which is generated with a one-step time delay from the machine input. After training, the inverse model is valid when the error signal $\varepsilon(k-1)$ becomes acceptably small. Once the inverse model F^{-1} is trained successfully, it is removed and placed in the forward path, as shown, so as to cancel the dynamics of the forward model F. The asymptotically stable second-order reference model shown in the figure is described by the equation

$$\omega_{rm}(k+1) = a\omega_{rm}(k) + b\omega_{rm}(k-1) + r^*(k) \quad (12.73)$$

where the model speed $\omega_{rm}(k)$ is to be tracked by the actual machine speed $\omega_r(k)$. The command signal $r^*(k)$ for the reference model can be calculated from the above equation for the desired speed trajectory. Considering that the tracking error $\varepsilon_r(k)$, as shown, tends to be zero, the speed at the $(k + 1)$ th time step can be predicted from the expression

$$\hat{\omega}_r(k+1) = a\omega_r(k) + b\omega_r(k-1) + r^*(k) \quad (12.74)$$

Therefore, the command signals for the inverse model corresponding to $\hat{\omega}_r(k+1)$, $\omega_r(k)$, and $\omega_r(k-1)$ can be synthesized and impressed on the inverse model controller to generate the estimated $v(k)$ signal for the motor, as shown in the figure. If there is a parameter variation in the machine, the inverse model can be updated by training it on-line, or the addition of a feedback loop, as shown in Figure 12.24, is required.

12.7.5 Speed Estimation by RNN

Speed estimation by the model referencing adaptive method was discussed in Chapter 8 (see Figure 8.47). The current model flux estimator (shown in the lower part) with an adaptive speed signal (ω_r) is a first-order dynamic system, and it can be implemented by the RNN shown in Figure 12.14. Figure 12.40 shows the RNN-based speed estimator, where the network replaces the adaptive current model in Figure 8.47. In this case, each output neuron uses the linear activation function. The solution of the voltage model generates the desired flux components. These signals are compared with the RNN output signals and the weights are trained on-line so that the error $\xi(k + 1)$ tends to zero. It is assumed that the training speed is fast enough so that the estimated speed and actual speed can track well. The current model equations can be discretized and written as

$$\begin{bmatrix} \psi_{dr}^s(k+1) \\ \psi_{qr}^s(k+1) \end{bmatrix} = \begin{bmatrix} 1-\dfrac{T_s}{T_r} & -\omega_r T_s \\ \omega_r T_s & 1-\dfrac{T_s}{T_r} \end{bmatrix} \begin{bmatrix} \psi_{dr}^s(k) \\ \psi_{qr}^s(k) \end{bmatrix} + \begin{bmatrix} \dfrac{L_m T_s}{T_r} & 0 \\ 0 & \dfrac{L_m T_s}{T_r} \end{bmatrix} \begin{bmatrix} i_{ds}^s(k) \\ i_{qs}^s(k) \end{bmatrix} \quad (12.75)$$

where T_s = sampling time, L_m = magnetizing inductance, and T_r = rotor time constant. The equation can also be expressed in the form

$$\begin{bmatrix} \psi_{dr}^s(k+1) \\ \psi_{qr}^s(k+1) \end{bmatrix} = \begin{bmatrix} W_{11} & W_{21} \\ W_{12} & W_{22} \end{bmatrix} \begin{bmatrix} \psi_{dr}^s(k) \\ \psi_{qr}^s(k) \end{bmatrix} + \begin{bmatrix} W_{31} & 0 \\ 0 & W_{32} \end{bmatrix} \begin{bmatrix} i_{ds}^s(k) \\ i_{qs}^s(k) \end{bmatrix} \quad (12.76)$$

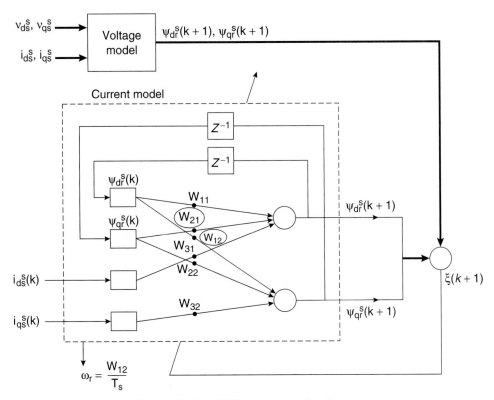

Figure 12.40 RNN-based speed estimator

where $W_{11} = 1 - T_s/T_r$, $W_{21} = -\omega_r T_s$, $W_{12} = \omega_r T_s$, $W_{22} = 1 - T_s/T_r$, $W_{31} = L_m T_s/T_r$, and $W_{32} = L_m T_s/T_r$. The RNN with a linear transfer function of unity gain satisfies Equation (12.76). Note that out of the six weights in the network, only W_{21} and W_{12} (circled in the figure) contain the speed term; therefore, for speed estimation, it is sufficient if these weights are considered trainable, keeping the other weights constant (assuming T_r and L_m are constants) for speed estimation. However, if all the weights are considered trainable, the speed as well as the rotor time constant can be tuned.

12.7.6 Adaptive Flux Estimation by RNN

In Chapter 8, we discussed programmable cascaded low-pass filters (PCLPFs) for stator flux estimation (see Figure 8.52). It is possible to implement this nonlinear dynamic system by an RNN [18]. For simplicity, let us consider only two stages of a PCLPF for flux estimation as shown in Figure 12.41. The functions to be implemented by the neural network are shown within the dotted enclosure. Each programmable low-pass filter stage (PLPF) is a first-order low-pass filter with time constant τ which is a function of supply frequency ω_e. The machine terminal voltages behind the stator resistance drop ($v_{ds}^{s''} \angle -\varphi_h$ and $v_{qs}^{s''} \angle -\varphi_h$) are being integrated to

Applications

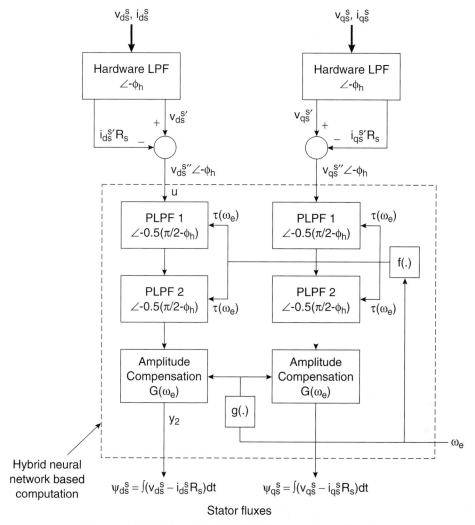

Figure 12.41 Two-stage PCLPF for stator flux estimation

estimate the respective stator fluxes ψ_{ds}^s and ψ_{qs}^s as shown. The series amplitude gain compensation G in each channel is also a function of frequency ω_e. Since the two integration channels are identical, let us consider only the ψ_{ds}^s estimation channel. Distributing the amplitude gain factor G equally between the two identical cascaded PLPF stages, the Laplace transfer function of each can be given as

$$\frac{Y_1(S)}{U(S)} = \frac{Y_2(S)}{Y_1(S)} = \frac{\sqrt{G}}{1+\tau S} \tag{12.77}$$

The corresponding discrete time equations are

$$y_1(k) = ay_1(k-1) + Ku(k-1) \tag{12.78}$$

$$y_2(k) = ay_2(k-1) + Ky_1(k-1) \tag{12.79}$$

where $K = \sqrt{G}T_s/\tau$, T_s = sampling time, $a = 1 - T_s/\tau$, $y_1(k)$ = output of the first stage (including $\sqrt{G}$), $y_2(k)$ = output of the second stage (including $\sqrt{G}$), and $u(k)$ = filter input. Equations (12.78) and (12.79) can be expressed in matrix form as

$$\begin{bmatrix} y_1(k+1) \\ y_2(k+1) \end{bmatrix} = \begin{bmatrix} a & 0 \\ K & a \end{bmatrix} \begin{bmatrix} y_1(k) \\ y_2(k) \end{bmatrix} + \begin{bmatrix} K \\ 0 \end{bmatrix} u(k) \tag{12.80}$$

The equivalent RNN will be represented by the following general equation:

$$\begin{bmatrix} y_1(k+1) \\ y_2(k+1) \end{bmatrix} = \begin{bmatrix} W_{11} & 0 \\ W_{21} & W_{22} \end{bmatrix} \begin{bmatrix} y_1(k) \\ y_2(k) \end{bmatrix} + \begin{bmatrix} W_{13} \\ 0 \end{bmatrix} u(k) \tag{12.81}$$

The upper part of Figure 12.42 shows the flux estimation of both channels using Equation (12.81), which is based on an RNN where the linear activation function is used. Weights W_{11}, W_{21}, W_{22}, and W_{13} vary with frequency and are trainable. For each sinusoidal input voltage wave at a certain frequency and with the corresponding calculated output flux wave as shown, the RNN is trained off-line by the EKF algorithm, as discussed before. The data for the training is generated by simulation in the frequency range of 0.01 Hz to 200 Hz with a step size of 1.0 Hz. If actual frequency falls within this step size, it is easily interpolated by the network. The trained weights as functions of frequency (slightly nonlinear) are shown on the right of the figure. An additional feed-forward network shown in the lower part generates these weight patterns as functions of frequency. The feedforward network uses a linear activation function at the output, but a sigmoidal function in the hidden layer. Note that ideally from Equation (12.80), $W_{11} = W_{22} = a$ and $W_{21} = W_{13} = K$. However, the RNN was trained in a general manner, without considering this symmetry.

12.8 NEURO-FUZZY SYSTEMS

So far, we have discussed the FL and neural network techniques independently. These techniques can be brought together into a hybrid neuro-fuzzy system to build a more powerful intelligent system with improved design and performance features. Research and development on the subject are topics of interest in the recent literature. In this section, we will discuss only one important topic, the adaptive network-based fuzzy inference system (ANFIS).

12.8.1 Adaptive Network-Based Fuzzy Inference System (ANFIS)

In Chapter 11, we stated that the design of MFs and the rule table of a fuzzy inference system was based on the experience of the operator or designer of the system. This means that there is no systematic method for the design of a fuzzy system. In a neural network, on the other hand, the

Neuro-Fuzzy Systems

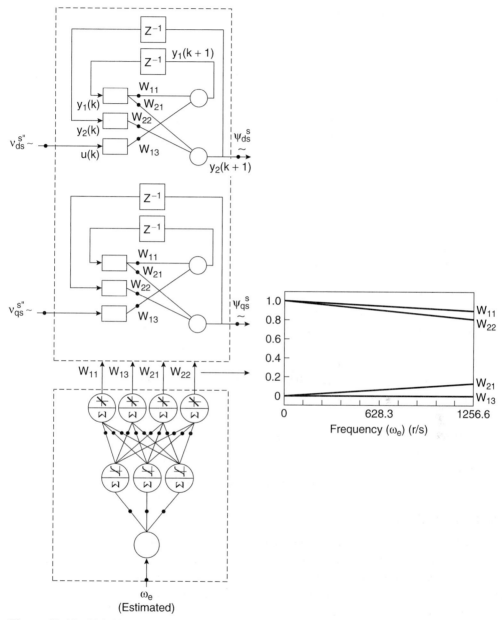

Figure 12.42 Hybrid recurrent neural network topology for stator flux estimation (showing the weights as function of frequency)

experimental or simulation input/output data can be used to design (or train) a network. The network then represents the model which satisfies the example data. In an ANFIS, as the name indicates, a fuzzy inference system is designed systematically using the neural network design method. This means that if the desired input/output data patterns are available for a fuzzy system, the MFs and rule table of the fuzzy model can be designed using the neural network training method.

Between the Mamdani and Sugeno methods of the fuzzy inference system, discussed in Chapter 11, usually the Sugeno method is used in an ANFIS. The literature on the ANFIS discusses both the zero-order and first-order Sugeno methods widely. In this section, we will review a simple zero-order Sugeno system only to understand the principle of the ANFIS.

Figure 12.43 (a) shows the zero-order Sugeno system (see also Figure 11.9), where X and Y are the input variables and F is the defuzzified output signal. $A_1, A_2, B_1,$ and B_2 are the triangular MFs for the antecedent part of the rules, and f_1 and f_2 are the output singleton MFs, as shown. The two rules which we are considering can be given as

Rule 1: IF X is A_1 AND Y is B_1 THEN $Z = f_1$
Rule 2: IF X is A_2 AND Y is B_2 THEN $Z = f_2$

The output F can be constructed as

$$F = \frac{W_1 f_1 + W_2 f_2}{W_1 + W_2}$$
$$= \frac{W_1}{W_1 + W_2} f_1 + \frac{W_2}{W_1 + W_2} f_2 \qquad (12.82)$$

where W_1 and W_2 are the DOFs of Rule 1 and Rule 2, respectively.

The corresponding ANFIS architecture is shown in Figure 12.43(b), where functions $A_1, A_2, B_1, B_2, f_1,$ and f_2 are being tuned. The feedforward neural network is shown with five layers where the output of the respective layers can be summarized as follows:

Layer 1 – Generate the membership grades:
$\mu_{A1}(X), \mu_{A2}(X), \mu_{B1}(Y)$ and $\mu_{B2}(Y)$

Layer 2 – Generate the DOF or firing strength by multiplication (π) or AND operation:
$W_1 = \mu_{A1}(X) \cdot \mu_{B1}(Y)$
and $W_2 = \mu_{A2}(X) \cdot \mu_{B2}(Y)$

Layer 3 – Normalize the firing strengths:
$\underline{W}_1 = W_1/(W_1 + W_2)$
$\underline{W}_2 = W_2/(W_1 + W_2)$

Neuro-Fuzzy Systems

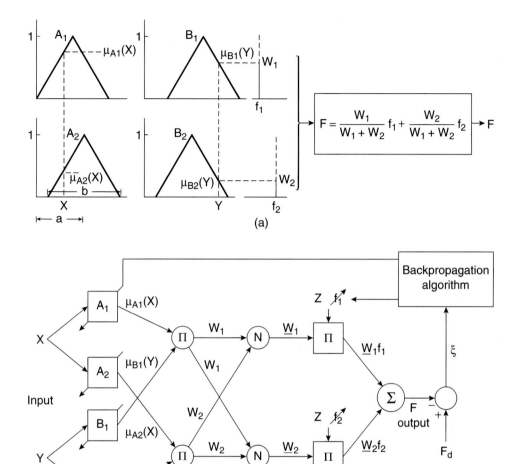

Figure 12.43 (a) Sugeno fuzzy inference system (zero-order), (b) Corresponding ANFIS structure

Layer 4 – Calculate rule outputs by multiplying with the consequent parameters:
$$\underline{W}_1 f_1 = \underline{W}_1 \cdot f_1$$
and $\quad \underline{W}_2 f_2 = \underline{W}_2 \cdot f_2$

Layer 5 – Sum all the component outputs:
$$F = \underline{W}_1 f_1 + \underline{W}_2 f_2$$

Each of the symmetric triangular MFs has again two unknown parameters, a (called peak) and b (called support), as shown in the figure. The equation of a function in terms of the a and b parameters is

$$\mu(X) = 1 - \frac{|X-a|}{0.5b}, \text{ if } |X-a| \leq 0.5b \qquad (12.83)$$
$$= 0 \quad \text{otherwise}$$

As indicated in the figure, the calculated output F of the network is compared with the desired value F_d, and the error signal is used to train the network parameters by the back propagation algorithm. The MATLAB-based Fuzzy Toolbox (Math Works) can be used to design an ANFIS. Usually, f_1 and f_2 are trained first. They are then followed by the design of the a and b parameters of the triangular MFs. Although triangular MFs have been considered, other types of MFs can also be used. Particularly, a Gaussian-type MF is more convenient because it is continuous, differentiable, and always has some finite value. Figure 12.43 contains only two MFs for each variable. Additional MFs can be added, as desired, for satisfactory training. Instead of a zero-order system, a first-order Sugeno system can also be trained by an ANFIS.

12.9 DEMO PROGRAM WITH NEURAL NETWORK TOOLBOX

12.9.1 Introduction to Neural Network Toolbox

The Neural Network Toolbox [5] in the MATLAB environment is one of the commonly used, powerful, commercially available software tools for the development and design of neural networks. The software is user-friendly, and being in the MATLAB environment, it permits flexibility and convenience in interfacing with other toolboxes in the same environment to develop a full application. The Toolbox can be used in an IBM-compatible PC, Macintosh, or a UNIX system. It supports a wide variety of feedforward and recurrent networks, including perceptrons, radial basis networks, BP networks, learning vector quantization (LVQ) networks, self-organizing networks, Hopfield and Elman NWs, etc. It supports the activation function types of bi-directional linear with hard limit (satlins) and without hard limit, threshold (hard limit), signum (symmetric hard limit), sigmoidal (log-sigmoid), and hyperbolic tan (tan-sigmoid), as previously mentioned. In addition, it supports unidirectional linear with hard limit (satlins) and without hard limit, radial basis and triangular basis, and competitive and soft max functions (not discussed in the text). A wide variety of training and learning algorithms are supported; the reader is referred to the user's guide for details.

The general design methodology of an ANN is given in Section 12.6, and the flowchart in Figure 12.9 should be followed for the development and design of the network. The training input/output data table is generated in MATLAB from experimental or simulation results, and training can be invoked in sequence or by a batch process. The number of training epochs and the objective function in the form of a sum-of-squared-error (SSE or ξ^2) are specified in the beginning. The training will stop whichever criterion is reached earliest. During training, the SSE vs. number of epochs may be continuously displayed. The training can be halted at any time to check the result and then it can be initiated again. After successful training, the performance is tested with input data (different from training data) at a faster rate to verify the learning performance. The performance curves can be generated and compared with the target functions using either MATLAB or SIMULINK.

12.9.2 Demo Program

A demo program was developed to clarify the understanding of the design of a simple feedforward neural network. An IBM-compatible PC with 700 MHz frequency was used for the training and testing.

As shown in Figure 12.44, the network received the signal X in radians (0 to 2π) at the input and implemented functions $\sin X$ and $\cos X$ at the output. A similar network was given for the illustration of $X - \sin X$ in Figure 12.5. The network has three layers, with the input node as a signal distributor only (sometimes called a two-layer network). The hidden layer has a hyperbolic tan-type activation function, whereas the output layer has a linear activation function because of nonlinear relation and bipolar outputs (the nonlinearity comes from the tan-sigmoid transfer function at the hidden layer). Network training was demonstrated with a variable number of neurons in the hidden layer until successful training was achieved. The Neural Network Toolbox in the MATLAB environment was considered for off-line training of the network according to the flowchart given in Figure 12.9. The detailed steps and commands can be found in the user's guide [5]. The Levenberg-Marquardt (L-M) algorithm, which is a faster back propagation technique,

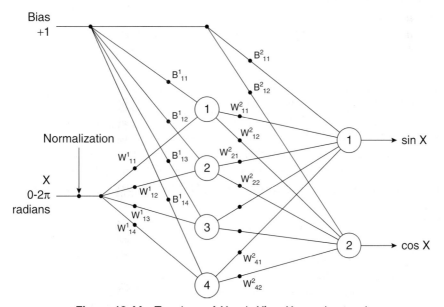

Figure 12.44 Topology of $X - \sin X / \cos X$ neural network

was adopted for the training. A MATLAB data table for $X - \sin X/\cos X$ was created for training the network with X at 0.2 radian intervals from 0 to 2π radians (i.e., 33 input/output data sets). The input data was normalized so that it varied in the corresponding range of 0 to 1.0. The training was done by the batch method, offering the whole table at the input simultaneously. A constant bias of 1.0 magnitude coupled all the neurons through a weight as shown. The initial learning coefficient η_0 was set to 1.0, which was decreased adaptively by the program as the training progressed. The momentum term (to avoid local convergence) was ignored for the simple program. In the beginning, only two neurons were considered in the hidden layer, and the specified values were SSE = 0.05 and number of epochs = 500. The SSE failed to converge, as shown in Figure 12.45(a). It practically locked the SSE at the value of 5.04 at the end of 75 epochs. To verify the training result, the network was simulated in SIMULINK, tested with input data at 0.01 radian steps in the 0–2π range repetitively, and compared with the calculated $\sin X/\cos X$ values. The test results are shown in Figure 12.45(b), indicating a poor correlation. Next, the hidden layer was added with an extra neuron (1-3-2) and training was continued with specified SSE = 1.0×10^{-3} and number of epochs = 99. Figure 12.46(a) shows the SSE, which locks to 3.47×10^{-2} at the end of 54 epochs, and (b) shows the corresponding test result. Evidently, the result improved significantly, but it is not yet satisfactory. The network was next trained with four hidden layer neurons and the corresponding results are shown in Figure 12.47, which indicate very good performance. The total training time for the simple network is around 10 seconds. The final training weights for the 1-4-2 network are given in Table 12.2.

12.10 GLOSSARY

Activation Function – A function that defines how the meuron's activation value is transferred to the output (also called Transfer Function).

ANN (artificial neural network) – A model made to simulate the biological nervous system of human brain.

Associative memory – A type of memory where an input pattern serves to retrieve related pattern.

Autoassociative memory – A type of memory where entering an incomplete pattern causes the output of a complete pattern, that is, the part that was entered plus the part that was missing.

Back propagation – A supervised learning method in which an output error signal is propagated through the network, altering the connection weights so as to minimize that error.

Back propagation network – A feedforward network that uses the back propagation training method.

Connection strength – Gain or weight in the connection (or link) between nodes through which data passes from one node to another.

Dendrite – An input channel of a biological neuron.

Distributed intelligence (or memory) – A feature in a neural network in which the intelligence (or memory) is not located at a single location, but is spread throughout the network.

Glossary

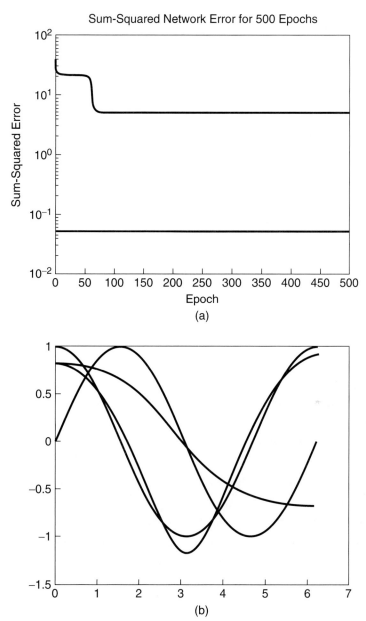

Figure 12.45 Training of 1-2-2 network (a) Training sum-squared-error (SSE) as function of number of epochs (final SSE = 5.04), (b) Test results of network

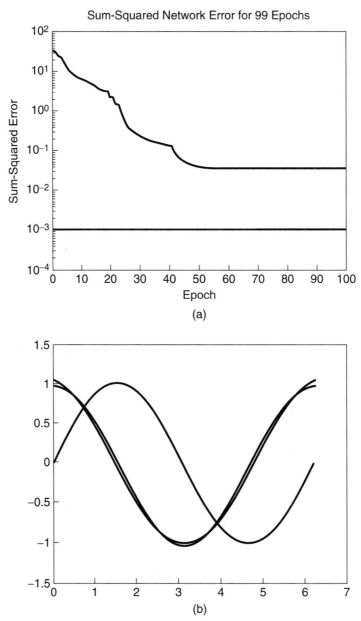

Figure 12.46 Training of 1-3-2 network (a) SSE as function of number of epochs (final SSE = 3.47×10^{-2}), (b) Test results of the network

Glossary

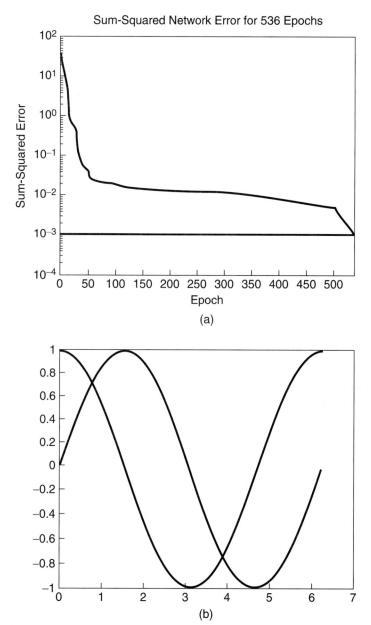

Figure 12.47 Training of 1-4-2 network (a) SSE as function of number of epochs (final SSE = 9.99X10^{-4}), (b) Test results of the network

Table 12.2 Table of Weights for 1-4-2 Network

$$W^1 = \begin{bmatrix} W_{11}^1 \\ W_{12}^1 \\ W_{13}^1 \\ W_{14}^1 \end{bmatrix} = \begin{bmatrix} -0.3413 \\ -0.3973 \\ 0.2569 \\ -0.2190 \end{bmatrix}$$

$$B^1 = \begin{bmatrix} B_{11}^1 \\ B_{12}^1 \\ B_{13}^1 \\ B_{14}^1 \end{bmatrix} = \begin{bmatrix} 1.1544 \\ 1.9764 \\ 0.2831 \\ 0.6153 \end{bmatrix}$$

$$W^2 = \begin{bmatrix} W_{11}^2 & W_{21}^2 & W_{31}^2 & W_{41}^2 \\ W_{12}^2 & W_{22}^2 & W_{32}^2 & W_{42}^2 \end{bmatrix} = \begin{bmatrix} 18.3102 & -3.9543 & 0.4590 & -19.2327 \\ -15.8852 & -9.5556 & 19.7069 & 43.8995 \end{bmatrix}$$

$$B^2 = \begin{bmatrix} B_{11}^2 \\ B_{12}^2 \end{bmatrix} = \begin{bmatrix} -0.7968 \\ -6.2782 \end{bmatrix}$$

Epoch – A computation cycle in back propagation training, which consists of a forward pass and a reverse pass.

Gradient descent method – A learning process that changes a neural network's weights to follow the steepest path towards the point of minimum error.

Heteroassociative memory – A network where one input pattern generates a different pattern at the output.

Learning – A process by which a network's weights are modified so as to give input/output pattern matching.

Learning rate – A factor (η) that determines the speed of convergence of the network.

Neuron – A nerve cell in a biological nervous system, or a processing element in an artificial neural network.

Pattern recognition – The ability to recognize an input pattern instantaneously without conscious thought (such as recognizing a face). This is the ability of a trained neural network (even in the presence of noise and distortion in the input pattern).

Perceptron – A neural network designed to resemble a biological sensory model.

Recurrent network – A neural network with internal feedback.

Sigmoid function – A nonlinear, differentiable activation function of a neuron that saturates at 0 and 1.

Squashing function – A transfer function where the output squashes or limits the output of a neuron between two asymptotes.

Supervised learning – A learning method where a teacher helps to correct the output.

Synapse – The junction in a dendrite which influences the signal flow magnitude.

Training – A process during which a neural network changes the weights in orderly manner to improve its performance.

Unsupervised learning – A method of self learning where there is no external influence to correct the network output.

Weight – An adjustable gain in the link between nodes in a neural network.

12.11 SUMMARY

This chapter gives a comprehensive review of the principles of neural networks and the network's application in power electronics and drives. The principle of the artificial neuron and its interconnection into a network were explained. A feedforward network with the BP algorithm, which is the most commonly used, was discussed in detail. Neuro-computing is fast compared to conventional computing because of massive parallel computation. Besides, it has the properties of fault tolerance and noise filtering. Neural network-based control (often called intelligent control) strictly does not need a mathematical model of a plant like a conventional control method does. The radial basis function network (RBFN) and Kohonen-type self-organizing feature map (SOFM) network were introduced briefly. A basic feedforward network permits static mapping between input and output patterns. A dynamic system can be implemented by a recurrent network or a feedforward network with external time-delay elements. An RNN and its training by the EKF method were discussed. A number of applications were described in power electronic systems using both feedforward and recurrent networks. The adaptive network-based fuzzy inference system (ANFIS), where a neural network is used in designing a fuzzy system, was briefly discussed. Finally, the MATLAB Neural Network Toolbox was introduced with a simple demo program, which was included for clarity of understanding. A glossary is given for the beginners in this area.

Neural technology, particularly neuro-fuzzy techniques, has advanced rapidly in recent years and its potential for application in power electronics systems seems enormous. The area is vast, and we gave only limited discussion on the subjects most relevant to the power electronics area. Neural networks can be implemented by high-speed DSP with serial computation, or by an ASIC chip with parallel computation. Unfortunately, ASIC chips currently available in the market are still very limited.

REFERENCES

1. B. K. Bose, "Expert systems, fuzzy logic, and neural network applications in power electronics and motion control", *Proc. Of the IEEE*, vol. 82, pp. 1303–1323, Aug. 1994.
2. L. H. Tsoukalas and R. E. Uhrig, *Fuzzy and Neural Approaches in Engineering*, John Wiley, NY, 1997.
3. J. W. Hines, *MATLAB Supplement to Fuzzy and Neural Approaches in Engineering, John Wiley*, NY, 1997.
4. S. Haykin, *Neural Networks, Macmillan*, NY, 1994.
5. Math Works Inc., *Neural Network Toolbox User's Guide*, 1998.
6. B. Burton, R. G. Harley, and T. G. Habetler, "The application and hardware implementation of continually online trained feedforward neural networks for fast adaptive control and prediction in power electronic systems", *Conf. Rec. of IEEE FEPPCON III*, pp. 215–224, South Africa, July 1998.
7. G. V. Puskorius and L. A. Feldkamp, "Neurocontrol of nonlinear dynamical systems with Kalman filter trained recurrent networks", *IEEE Trans. Neural Networks*, pp. 279–297, vol. 5, Mar. 1994.
8. K. S. Narendra and K. Parthasarathy, "Identification and control of dynamical systems using neural networks", *IEEE Trans. Neural Networks*, vol. 1, pp. 4–27, Mar. 1990.
9. K. J. Hunt et al., "Neural networks for control systems – survey", *Automatica*, vol. 28, pp. 1083-1112, 1992.
10. A. M. Trzynadlowski and S. Legowski, "Application of neural networks to the optimal control of three-phase voltage-controlled power inverters", *IEEE IECON Conf. Rec.*, pp. 524–529, 1992.
11. M. P. Kazmierkowski and D. Sobczuk, "Improved neural network current regulator for VC-PWM inverters", *IEEE IECON Conf. Rec.*, pp. 1237–1241, 1994.
12. J. O. P. Pinto, B. K. Bose, L. E. Borges, and M. P. Kazmierkowski, "A neural network based space vector PWM controller for voltage-fed inverter induction motor drive", *IEEE Trans. on Ind. Appl.* vol. 36, pp. 1628–1636, Nov./Dec. 2000.
13. M. G. Simoes and B. K. Bose, "Neural network based estimation of feedback signals for vector controlled induction motor drive", *IEEE Trans. Ind. Apl.*, vol. 31, pp. 620–629, May/June 1995.
14. M. H. Kim, M. G. Simoes and B. K. Bose, "Neural network based estimation of power electronic waveforms", *IEEE Trans. Power Electronics*, vol. 11, pp. 383–389, Mar. 1996.
15. S. Weersooriya and M. A. El-Sharkawi, "Identification and control of a dc motor using backpropagation neural networks", *IEEE Trans. Energy Conversion*, vol. 6, pp. 663–669, Dec. 1991
16. M. A. El-Sharkawi, "High performance drive of brushless motors using neural network", *IEEE-PES Summer Conf. Proc.*, July 1993.
17. B. Jayanand, *Simulation Studies on Speed Sensorless Operation of Vector Controlled Induction Motor Drives Using Neural Networks,* Ph.D. Thesis, IIT, Madras, 1997.
18. J. O. P. Pinto, B. K. Bose, and L. E. Borges, "A stator flux oriented vector controlled induction motor drive with space vector PWM and flux vector synthesis by neural networks", *IEEE IAS Annu. Meet. Conf. Rec.*, pp. 1605– 1612, 2000.

Index

A

Abbreviated rule language (ARL), 538
Absolute position encoder, 446-49
    analog resolver with decoder, 448-49
    optical encoder, 446-48
Absorption, 566
Ac equivalent circuit, induction motor slip-power recovery drives, 316-19
Ac machines, compared to dc machines, 29
Ac machines for drives, 29-97
    induction machines, 30-74
        axes transformation, 57-62
        constant volts/Hz operation, 44-45
        drive operating regions, 46-47
        dynamic d-q model, 56-57
        dynamic model state-space equations, 70-74
        equivalent circuit, 35
        equivalent circuit analysis, 35-39
        harmonics, 49-56
        NEMA classification of machines, 42
        rotating magnetic field, 30-33
        stationary frame—dynamic model, 67-70
        synchronous lay rotating reference frame—dynamic model, 63-67
        torque production, 33-34
        torque-speed curve, 39-41
        variable-frequency operation, 43-44
        variable stator current operation, 47-49
        variable-voltage constant-frequency operation, 42-43
    synchronous machines, 74-96
        developed torque, 79-80
        double reluctance machine, 94-96
        dynamic d^e-q^e machine model, 83-85
        equivalent circuit, 76-79
        Park model, 83-85
        permanent magnet (PM) machine, 86
        permanent magnet (PM) materials, 86-89
        salient pole machine characteristics, 80-83
        sinusoidal interior magnet machine (IPM), 89-93
        sinusoidal surface magnet machine (SPM), 89
        synchronous reluctance machine, 86
        trapezoidal surface magnet machine, 93-94
        variable reluctance machine (VRM), 94-96
        wound field machine, 74-76

ACSL, 265
Activation functions:
   defined, 684
   neurons, 628-29
Adaline, 629, 637
Adaptive control, 413-30
   classification of techniques, 414
   model referencing adaptive control (MRAC), 416-18
   self-tuning control, 414-16
      load torque disturbance compensation, 415-16
   sliding mode control, 419-30
      control principle, 419-24
         sliding trajectory control of a vector drive, 424-30
Adaptive flux estimation by RNN, 676-78
Adaptive network, 643
Adaptive resonance theory (ART) network, 630
Aggregation, defined, 619
AI, See Artificial intelligence
Air gap voltage, 83
Alphabet character recognition, by an artificial neural network (ANN), 634-36
Amplitude subnet, 663
Analog resolver with decoder, 448-49
Ancestors, subframes, 539
Angle control, 384
Angle subnet, 663
ANN, See Artificial neural network (ANN)
ANN-based fuzzy inference system (ANFIS), 678-82, 689
   Mamdani method, 680
   Sugeno method, 680
ANN identification of dynamic models, 652-54
Antecedent, 538
   defined, 619
Antecedent rule, defined, 555
Applications:
   current-fed converters, 272
   cycloconverters, 153
   diodes and phase-controlled converters, 99

expert system (ES), 546-55
   commercial ac drive product selection, 548-49
   configuration selection of a drive system, 549, 550
   control design, 554-55
   converter design, 552-54
   design of a drive system, 549, 550-55
   fault diagnostics, 547-48
   motor ratings design, 550-52
   P-I control tuning of a drive, 547-48
   simulation study, 554-55
induction motor drives, 333
neural networks, 658-78
   adaptive flux estimation by RNN, 676-78
   distorted wave estimation, 670-71
   model identification and adaptive drive control, 671-75
   PWM controller, 658-66
   speed estimation by RNN, 675-76
   vector-controlled drive feedback signal estimation, 667-70
slip-power recovery drives, 307
static Scherius drive, 326
voltage-fed converters, 191-92
Artificial intelligence (AI), 535-36
   defined, 555
Artificial neural network (ANN), 535, 625, 629-44
   adaptive network, 643-44
   alphabet character recognition, 634-36
   architecture of, 631-32
   back propagation algorithm for three-layer network, 637-43
   back propagation training, 637
      on-line training, 643-44
   defined, 684
   example application, 632
   feedforward neural network, training of, 632-36
   history of the technology, 626
   identification of dynamic models, 652-54
   learning methods, 634
   network models, 629-30

Index

Artificial neuron, 627-29
    structure of, 628-29
ASCI inverter-fed induction motor drives, 287
Associative memory, defined, 684
Associative property, 566
Auto-sequential current-fed inverter (ASCI), 285-87
Autoassociative network, 636
Auxiliary resonant commutated pole (ARCP) converter, 251-52
Axes transformation, induction machines, 57-62

B

Back propagation algorithm for three-layer network, 637-43
    weight calculation for hidden layer neurons, 641-43
    weight calculation for output layer neurons, 637-41
Back propagation, defined, 684
Back propagation network, 629
    defined, 684
Back propagation training, 637
Backward chaining, defined, 555
BASIC, 541
Bi-directional associative memory (BAM), 630
Bidirectional slip-power flow, with a cycloconverter, 328
Biological neuron:
    concept of, 626-27
    structure of, 627
Bipolar power or junction transistors (BPTs or BJTs), 2, 14-16
Bi-valued Boolean logic, 560
Black box, 560
Blaschke equations, 367, 406-8
    for stator flux vector estimation, 406
Boltzmann machine, 630
Boole, George, 536, 559
Boolean logic, defined, 555
Brain, 535-36
    memory/intelligence of, 625
Brain-state-in-a-box (BSB), 630
Brereton, D. S., 56

Bridge converters, concurrent and sequential control of, 140-41
Brushes, and dc machines, 29
Brushless dc motors (BLDMs), 17, 93, 445, 484, 486
Buffers, 630

C

CAD program compared to ES program, 537, 539
Cell, 627
Center of area (COA) defuzzification, 573-75
    defined, 619
Center-tapped inverters, 192-93
Cerebral cortex, 625
Circulating current mode:
    cycloconverters, 162-66, 166
        advantages of, 165
        disadvantages of, 166
Class A-E machines, 42-43
Classic AI, *See* Artificial intelligence:
Client interface, expert system (ES), 541-42
COA defuzzification, 573-75
    defined, 619
Cobalt-Samarium (CoSm) magnet, 88
Commutativity, 566
Commutatorless-brushless motor, 445
Commutatorless Kramer drive, 312-13, 322-23
    power flow diagram, 323-24
Commutators, and dc machines, 29
Competitive learning, SOFM network training and, 645
Complement, 565
Computer-integrated manufacturing (CIM), 547
Connection strength, 627
    defined, 684
Consequent, rules, 619
Constant volts/Hz operation, induction machines, 44-45
Controller, function of, 141-42

Converter control:
    cosine wave crossing control, 142-45
    phase-locked oscillator principle, 145-48

Converters:
    current-fed converters, 271-305
    cycloconverters, 153-90
    diodes and phase-controlled converter, 99-151
    force-commutated inverter, 285-87
    high-frequency non-resonant link converter, 252-53
    six-pulse center-tap converter, 136-37
    three-phase bridge converter, 128-32
    three-phase half-wave converter, 122-24
    thyristor converters, 112-41
    twelve-pulse converter, 137-40
    voltage-fed converter, 191-270

Coupling inductance, and EMI problems, 245

Current-fed converters, 271-305
    applications, 272
    control, 350-52
    force-commutated inverters, 285-87
        auto-sequential current-fed inverter (AS-CI), 285-87
    harmonic heating, 287-88
    inverters with self-commutated devices, 290-303
        double-sided PWM converter system, 299-302
        PWM inverters, 294-99
        PWM rectifier applications, 302-3
        six-step inverter, 290-94
    load-commutated inverters, 277-85
        single-phase resonant inverter, 277-81
        three-phase inverter, 281-85
    multi-stepped inverters, 289-90
    six-step thyristor inverter
        general operation of, 272-76
        inverter operation modes, 274-76
    torque pulsation, 287-88
    voltage-fed converters vs., 303-4

Current-fed inverter control, 350-52
    independent current and frequency control, 350
    speed and flux control in current-fed inverter drive, 351
    volts/Hz control of current-fed drive, 352

Current-fed inverter drive, vector control of, 384-85

Current vector control:
    sinusoidal IPM machine drives, 465-68
    synchronous reluctance machine drives (SyRM), 457-64
        constant d^e-axis current control, 458-59
        fast torque response control, 459-63
        maximum power factor control, 463-64
        maximum torque/ampere control, 463

Cycloconversion, basic principle of, 154-56

Cycloconverter drive:
    scalar control of, 507-10
    vector control of, 385-87, 510-13

Cycloconverters, 1, 153-90
    applications, 153
    circulating current mode, 162-66, 166
        advantages of, 165
        disadvantages of, 166
    circulating versus non-circulating current mode, 162-67
        blocking mode, 166-67
        circulating current mode, 162-66
    control of, 177-80
    cycloconverter circuits, 158-62
        three-phase bridge cycloconverter, 161-62
        three-phase half-wave cycloconverter, 158-61
    defined, 153
    DPF improvement methods, 180-85
        asymmetrical firing angle control, 180-83
        circulating current control, 183-85
        square-wave operation, 180
    high-frequency cycloconverters, 186-89
        high-frequency integral-pulse cycloconverter, 187-89
        high-frequency phase-controlled cycloconverter, 187

Index

Cycloconverters (*continued*)
    line displacement power factor (line DPF), 171-77
        theoretical derivation of, 173-77
    load and line harmonics, 167-71
        line current harmonics, 171
        load voltage harmonics, 167-70
    matrix converters, 185-86
    phase-controlled cycloconverters, 154-85
        cycloconverter circuits, 158-62
        operation principles, 154-56
        three-phase dual converter as cycloconverter, 156-58

D

d-q model, 56-57
Darlington transistor symbol, 16
Database, defined, 555
Dc current control, 372
Dc machines, compared to ac machines, 29
Dc motor speed control, PWM rectifier applications, 303
De Morgan's theorems, 566
Dead-time effect, 216-17
Declarative knowledge, 539
    defined, 555
Deep knowledge, 539
Defuzzification:
    defined, 619
Defuzzification methods:
    fuzzy logic (FL)
        center of area (COA) method, 573-75
        height method, 575
        mean of maxima (MOM) method, 575
        Sugeno method, 576
Degree of fulfillment (DOF), defined, 619
Degree of membership, defined, 619
Delta modulation, 211
Dendrite, defined, 684
Design methodology:
    expert system (ES), 546
    neural networks, 657-58
Developed torque, synchronous machines, 79-80

Device voltage and current ratings, three-phase bridge inverters, 203
Diode rectifier with boost chopper, 255-57
    single-phase, 255-57
    three-phase, 257
Diode rectifiers, 100-112
    displacement power factor (DPF), 108
    distortion factor (DF), 107-8
    power factor (PF), 107, 109
    single-phase bridge rectifier (*CR* load), 105-7
    single-phase bridge rectifier (*R*, *RL* load), 100-3
    single-phase bridge rectifier (*RL*, *CEMF* load), 104
    source inductance, 103-4
    three-phase full bridge rectifier (*CR* load), 112
    three-phase full bridge rectifier (*RL* load), 109-12
Diode symbol, and volt-ampere characteristics, 2
Diodes, 1, 2-4
    turn-off switching characteristics of, 3
Diodes and phase-controlled converters, 99-151
    applications, 99
    converter control, 141-50
        cosine wave crossing control, 142-45
        linear firing angle control, 142
        phase-locked oscillator principle, 145-48
    diode rectifiers, 100-112
    EMI problems, 148-49
    history of, 99
    line harmonic problems, 149-50
    thyristor converters, 112-41
        concurrent and sequential control of bridge converters, 140-41
        discontinuous conduction, 118-22, 132-36
        link leakage inductance analysis, 124-28
        single-phase bridge rectifier (*R*, *CEMF* load), 112-18
        six-pulse center-tap converter, 136-37
        three-phase bridge converter, 128-32
        three-phase bridge rectifier (*RL*, *CEMF* load), 122
        three-phase dual converter, 136
        three-phase half-wave converter, 122-24
        twelve-pulse converter, 137-40

Direct synthesis from state equations, 389-90
Direct torque and flux control (DTFC/DTC), 408-13
    block diagram, 410
    control strategy, 410-13
    torque expression with stator and rotor fluxes, 408-9
Direct vector control, 360-61, 360-63
    of line-side PWM rectifier, 378-80
Direct vector control with speed signal, 401-8
    drive machine start-up with current model equations, 404-8
    progammable cascaded low-pass filter (PCLPF) stator flux estimation, 401-4
Discontinuous conduction, thyristor converters, 118-22, 132-36
Displacement power factor (DPF), 108, 255
Distorted wave estimation, 606-9
    Mamdani method, 608
    neural networks, 670-71
    Sugeno method, 609
Distortion factor (DF), diode rectifiers, 107-8
Distributed intelligence, defined, 684
Distributive property, 566
Dither signal, 430
DOF (degree of fulfillment), defined, 619
Domain expert, defined, 556
Double Darlington transistors, 14
Double negation, 565
Double reluctance machine, 94-96
Double-sided PWM converter system, 299-302
Doubly-fed machine speed control by rotor rheostat, 308-9
Drive operating regions, induction machines, 46-47
Drive systems, uses of, 29
DSP-type TMS320C30 (Texas Instruments), 554
Dynamic braking, 253-54
Dynamic d-q model, 56-57
Dynamic d^e-q^e machine model, 83-85
Dynamic diagnostic system, 547
Dynamic model state-space equations, induction machines, 70-74
Dynamic network, 630
Dynamic neural networks, 646-50
Dynamic system models, neural networks, 650-52
DynaMind (NeuroDynamX), 658

E

Efficiency optimization control by flux program, 352-56
Eighteen-step inverter by phase-shift control, 209-10
Electronically commutated motor (ECM), 445, 484
Elman network, 682
EMI problems:
    diodes and phase-controlled converters, 148-49
    hard-switched inverters, 245
EMTP, 265
Epoch, 637
    defined, 688
Equivalent circuit:
    induction machines, 35
    synchronous machines, 76-79
    vector control, 358-59
Equivalent circuit analysis, induction machines, 35-39
ETANN (Electrically Trainable Analog Neural Network) ASIC chip 80170NX (Intel), 658-59
Executed rule, 538
Expert system (ES), 535-57
    applications, 546-55
        commercial ac drive product selection, 548-49
        configuration selection of a drive system, 549, 550
        control design, 554-55
        converter design, 552-54
        design of a drive system, 549, 550-55
        fault diagnostics, 547-48
        motor ratings design, 550-52
        P-I control tuning of a drive, 547-48
        simulation study, 554-55
    artificial intelligence (AI), 535-36
    basic elements of, 537
    defined, 536-37
    design methodology, 546

Index   697

ES program compared to CAD program, 537, 539
expert system shell, 543-45
    external interface, 543-44
    program development steps, 544-45
    shell features, 543
inference engine, 541
knowledge base, 537-41
    ES language, 540-41
    frame structure, 539-40
    meta-knowledge, 540
principles of, 536-42
user (client) interface, 541-42
user dialog, 542
Extended Kalman filter (EKF), 396-99
    method of parameter (state) estimation, 376

F

Fast-recovery diodes, 3-4, 4
Fault-tolerant neural network, 636
Feedback diode, 14
Feedback (FB) chopping mode, 486
Feedback vector control, 360, 360-63
Feedforward vector control, 360, 368-78
Field-oriented control, *See* Vector control:
Fired rule, 538
First breakdown effect, 14-16
Flux loop, 384
Flux programming efficiency improvement control, 585-91
    efficiency optimizer control block diagram, 589
    indirect vector-controlled drive, 588
    membership functions for fuzzy variables, 590
    on-line search method of, 587
    pulsating torque compensation, 589-91
    rule matrix for efficiency improvement, 590-91
Flux vector estimation, 363-68
    current model, 366-68
    vector control, 363-68
    voltage model, 363-66

Force-commutated inverters, 285-87
    auto-sequential current-fed inverter (ASCI), 285-87
FORTRAN, 541
Forward-bias safe operating areas (FBSOA), 16
Forward chaining, defined, 556
Forward pass, 637
Fractional horse power (FHP), 30
Frame, defined, 556
Freewheeling (FW) chopping mode, 486
    close loop current control in, 492
Full inverters, 193-97
    phase-shift voltage control, 195-97
Fully connected network, 631
Fuzzification, defined, 622
Fuzzy composition, defined, 622
Fuzzy control, defined, 622
Fuzzy expert system, 560
Fuzzy implication, defined, 622
Fuzzy logic (FL), 535, 539
    applications, 559-60, 582-609
        distorted wave estimation, 606-9
        flux programming efficiency improvement of induction motor drive, 585-91
        induction motor speed control, 582-85
        slip gain tuning of indirect vector control, 597-602
        stator resistance R_s estimation, 602-6
        wind generation system, 591-97
    defuzzification methods, 573-76
        center of area (COA) method, 573-75
        height method, 575
        mean of maxima (MOM) method, 575
        Sugeno method, 576
    fuzzy control, 576-81
        control implementation, 581
        control principle, 577-81
        historical perspective, 576-77
        purpose of, 576
    fuzzy (inference) system, 566-69
    fuzzy sets, 560-66
        membership functions, 561-63

Fuzzy logic (*continued*)
    general design methodology, 581-82
    implication methods, 569-73
        Lusing Larson type, 570-71
        Mamdani type, 569-70
        Sugeno (Takagi-Sugeno-Kang) method, 571-73
    principles of, 559-622
Fuzzy Logic Toolbox (MATLAB environment), 560, 581, 609-19
    demo program for synchronous current control, 613-19
    FIS Editor, 611
    graphical tools, 610-11
    Membership Function Editor (MFE), 611-12
    Rule Editor, 612
    Rule Viewer, 612-13
    Surface Viewer, 613
        user interface to control, 614
Fuzzy neural network (FNN), 630
Fuzzy rule, defined, 622
Fuzzy set theory, 559
    defined, 622
Fuzzy sets:
    absorption, 566
    associative property, 566
    commutativity, 566
    complement, 565
    De Morgan's theorems, 566
    distributive property, 566
    double negation, 565
    idempotency, 565
    intersection, 565
    multiplying by a crisp number, 565
    negation, 565
    operations on, 564-66
    power of a fuzzy set, 565
    product of two fuzzy sets, 565
    union, 564

Fuzzy speed control, 582-85
    algorithm for, 584-85
    design considerations, 584
    membership functions for, 583
    rule matrix for, 584
Fuzzy system, defined, 622
Fuzzy variable, defined, 622

G

Gate turn-off thyristor (GTOs), 2, 10-14, 210
    regenerative snubbers, 14-15
    switching characteristics, 11-14
    turn-off capability of, 10-11
    turn-on characteristics of, 11-13
General regression network, 630
Generator speed tracking control (FLC-1), 594-97
    block diagram of, 596
    membership functions for variables in, 597
    rule matrix for, 596
Genetic algorithm (GA), 535
Gradient descent method, defined, 688

H

H-bridge inverters, 193-97
    phase-shift voltage control, 195-97
Half-bridge inverters, 192-93
Hall sensors, 447, 490
Hard computing, 535
Hard-switched inverters, 245-47
    device stress, 245
    EMI problems, 245
    machine bearing current, 245-46
    machine insulation effects, 245
    machine terminal overvoltage, 246-47
    switching loss, 245
Harmonic heating, current-fed converters, 287-88

Index

Harmonics, 49-56
    harmonic heating, 49-53
    induction motor slip-power recovery drives, 321
    machine parameter variation, 53
    slot, 399
    torque pulsation, 53-56
Hebb's rule, 626
Height defuzzification, defined, 622
Height method, defuzzification, 575
HELP command, expert system (ES), 542
Heteroassociative memory, defined, 688
Heteroassociative network, 635-36
Hidden layer neurons, weight calculation for, 641-43
High-frequency cycloconverters, 186-89
    high-frequency integral-pulse cycloconverter, 187-89
        quasi-square-wave supply, 188-89
        sinusoidal supply, 187-88
    high-frequency phase-controlled cycloconverter, 187
High-frequency, non-resonant link converter, 252-53
High-level languages, 541
High-voltage integrated circuit (HVIC), 26
Holding current, 5
Hopfield network, 630, 682
HOW user command, expert system (ES), 542
Hysteresis band current control PWM, 210, 236-39

I

Idempotency, 565
IF...THEN production rules, 538
IGBTs (insulated gate bipolar transistors), 14
Implication methods:
    fuzzy logic (FL), 569-73
        Lusing Larson type, 570-71
        Mamdani type, 569-70
        Sugeno (Takagi-Sugeno-Kang) method, 571-73
Indirect vector control, 360, 368-78
Indirect vector control block diagram, with open loop flux control, 371

Indirect vector control slip gain tuning, 375-78
Induction machines, 30-74
    axes transformation, 57-62
    constant volts/Hz operation, 44-45
    drive operating regions, 46-47
    dynamic d-q model, 56-57
    dynamic model state-space equations, 70-74
    equivalent circuit, 35
    equivalent circuit analysis, 35-39
    harmonics, 49-56
        harmonic heating, 49-53
        machine parameter variation, 53
        torque pulsation, 53-56
    NEMA classification of machines, 42
    rotating magnetic field, 30-33
    stationary frame—dynamic model, 67-70
    synchronous lay rotating reference frame—dynamic model, 63-67
    torque production, 33-34
    torque-speed curve, 39-41
    variable-frequency operation, 43-44
    variable stator current operation, 47-49
    variable-voltage constant-frequency operation, 42-43
Induction motor drive control and estimation, 333-437
    adaptive control, 413-30
        classification of techniques, 414
        model referencing adaptive control (MRAC), 416-18
        self-tuning control, 414-16
        sliding mode control, 419-30
    direct torque and flux control (DTFC/DTC), 408-13
        control strategy, 410-13
        torque expression with stator and rotor fluxes, 408-9
    scalar control, 338-55
        current-fed inverters control, 350-52
        efficiency optimization control by flux program, 352-56
        voltage-fed inverter control, 339-49
    self-commissioning of a drive, 430-35
    with small signal model, 334-38

Induction motor drive control and estimation (cont.)
    vector control, 356-87
        of current-fed inverter drive, 384-85
        of cycloconverter drive, 385-87
        dc drive analogy, 356-58
        direct, 360-63
        equivalent circuit, 358-59
        feedback, 360-63
        feedforward, 368-78
        flux vector estimation, 363-68
        indirect, 368-78
        indirect vector control slip gain tuning, 375-78
        of line-side PWM rectifier, 378-80
        phasor diagram, 358-59
        physical principle of, 374-75
        principles of, 359-60
        sensorless, 388-408
        stator flux-oriented, 381-84
Induction motor drives, applications, 333
Induction motor slip-power recovery drives, 307-31
    ac equivalent circuit, 316-19
    doubly-fed machine speed control by rotor rheostat, 308-9
    harmonics, 321
    Kramer drive
        commutatorless drive system, 323-24
        performance characteristics of, 323
        speed control of, 322
    modified Scherbius drive for VSCF power generation, 328-31
    phasor diagram, 313-15
    power factor improvement, 322-24
    static Kramer drive, 309-13
    static Scherius drive, 324-31
        modes of operation, 326-28
    torque expression, 319-21
Inference, defined, 556
Inference engine, 541
    defined, 556
Injection of auxiliary signal on salient rotor, 399-401
Input/output mapping, 625, 633

Input ripple, three-phase bridge inverters, 202-3
Instantaneous current control PWM, 659-60
Insulated gate bipolar transistors (IGBTs), 14, 20-24
    defined, 20
    switching characteristics, 22-24
    thermal impedance, 22-24
Integrated gate-commutated thyristors (IGCTs), 25-26
Inter-group reactor (IGR), 136, 159, 164, 183
Interior permanent magnet (IPM), 439
Intersection, 565
Inverse dynamics model, neural networks, 654-55

K

Knowledge acquisition, defined, 556
Knowledge base, 537-41
    defined, 556
    ES language, 540-41
    frame structure, 539-40
    meta-knowledge, 540
    on-line alteration of, 539
Knowledge engineering, defined, 556
Knowledge representation, defined, 556
Kohonen layer, 645
Kohonen's self-organizing feature map (SOFM) network, 630, 645-46, 689
Kramer drive:
    commutatorless drive system, 323-24
    performance characteristics of, 323
    speed control of, 322
Kron Equation, 63-67
Kron, G., 56

L

Large band-gap materials for devices, 26
Latching current, 5
Learning, defined, 688
Learning rate, defined, 688
Learning vector quantization (LVQ) network, 630
Lenz's law, 33

Index

Levenberg-Marquardt (L-M) algorithm, 643, 683-84
Line-side PWM rectifier, vector control of, 378-80
Linear firing angle control, single-phase bridge converter, 142
Linguistic variable, defined, 622
Link leakage inductance analysis, thyristor converters, 124-28
LISP, 540-41
Load-commutated inverter (LCI) drive, 496-507
    control with constant turn-off angle, 498-501
    control with machine terminal voltage signals, 504-6
    delay angle control, 501-4
    phase-locked loop (PLL) angle control, 506-7
Load-commutated inverters (LCI), 276, 277-85
    single-phase resonant inverter, 277-81
        circuit analysis, 278-81
    three-phase inverter, 281-85
        lagging power factor load, 281
        over-excited synchronous machine load, 282-84
        synchronous motor starting, 284-85
Lock-out effect, 216-17
Luenberger observer, 392-96
Lusing Larson method, 570-71
Lyapunov's theorem, 394-95

M

Machine stator current magnitude, 384
Madaline, 629
MAGLEV, 440
Magnetic noise, 214
Mamdani method, 569-70
    ANFIS, 680
    distorted wave estimation, 608
Math Works (MATLAB), 265
MATLAB, 265, 541, 683
MATLAB/SIMULINK, 264-67
Matrix converters, 185-86
$MATRIX_X$, 265
MAX-MIN composition, 578

Mean of maxima (MOM) method, defuzzification, 575
Membership functions:
    defined, 622
    fuzzy sets, 561-63
        asymmetrical, 562
        closed, 562
        singleton, 562
        types of, 562-63
Meta-knowledge, 540
Meta-rule, defined, 556
Minimum ripple current PWM, 210
Model referencing adaptive control (MRAC), 376-78, 416-18
Model referencing adaptive system (MRAS), 390-92
    adaptation algorithm design, 391
Modified Scherbius drive for VSCF power generation, 328-31
Modular neural network (MNN), 630
MOS-controlled thyristors (MCTs), 24-25
Motoring and regenerative modes, three-phase bridge inverters, 201-2
Multi-layer Perceptron (MLP)-type network, 630
Multi-level inverter, 240-44
Multi-MW GTO current-fed drives with six-step waves, 293, 296
Multi-stepped current-fed inverters, 289-90
Multi-stepped inverters, 206-10, 288, 289-90
    eighteen-step inverter by phase-shift control, 209-10
    twelve-step inverter, 207-9
Multi-valued logic, 560
Multiplying by a crisp number, 565
MYCIN, 541

N

National Electrical Manufacturers Association (NEMA), 42-43
Negation, 565
NEMA classification of machines, induction machines, 42
Neodymium-iron-boron (Nd-Fe-B) magnet, 88-89

Neural Network Toolbox (MATLAB), 626, 643, 689
    demo program with, 683-84
    introduction to, 682-83
Neural networks, 625-82
    ANN identification of dynamic models, 652-54
    applications, 658-78
        adaptive flux estimation by RNN, 676-78
        distorted wave estimation, 670-71
        model identification and adaptive drive control, 671-75
        PWM controller, 658-66
        speed estimation by RNN, 675-76
        vector-controlled drive feedback signal estimation, 667-70
    design methodology, 657-58
    dynamic, 646-50
    dynamic system models, 650-52
    fault tolerance, 636
    in identification and control, 650-57
    input/output training data vectors, 662
    inverse dynamics model, 654-55
    Kohonen's self-organizing feature map network (SOFM) network, 645-46
    neural-network-based control, 655-56
    neuro-fuzzy systems, 678-84
        ANN-based fuzzy inference system (ANFIS), 678-82
    radial basis function network (RBFN), 644-45
    recurrent neural network (RNN), 646-50
        training by the EKF algorithm, 647-50
    time-delayed, 650
Neural technology, 689
Neuro-fuzzy systems, 560
NeuroDynamX's Dynamind, 658
Neuron:
    artificial neuron, 627-29
    biological neuron, concept of, 626-27
    defined, 688
    hidden layer, 630
    input layer, 630
    output layer, 630-31
    structure of, 626-29

Neurons, activation functions, 628-29
Neutral-point clamped (NPC) inverter, 240
Node, 627
Numeric computation-intensive, high-level languages, 541

O

Off-line diagnostics, 547
On-line diagnostics, 547
Optical encoder, 446-48

P

P-I-N structure, 2
Parallel-loaded resonance converter (PRC), 249
Parameter, defined, 556
Parasitic leakage, and EMI problems, 245
Park model, synchronous machines, 83-85
Park, R. H., 56
Park's transformation, 56
Partially connected network, 631
PASCAL, 541
Pattern classification property, 635
Pattern recognition, 625
    defined, 688
Perceptron, 626, 629, 630
    defined, 688
Permanent magnet (PM) machine, 86
Permanent magnet (PM) materials, 86-89
Permanent magnet synchronous machines (PMSMs), 439
Phase-controlled converter (PCC), 263
Phase-controlled cycloconverters, 154-85
    cycloconverter circuits, 158-62
    operation principles, 154-56
    three-phase dual converter as cycloconverter, 156-58
Phase-locked oscillator principle, 145-48
Phase-shift PWM, 206
Phase-shift voltage control, three-phase bridge inverters, 203-5
Phase shifter, 301

Index

Phasor diagram:
    induction motor slip-power recovery drives, 313-15
    vector control, 358-59
PMW-SW sequencing, sinusoidal IPM machine drives, 482-83
Polyphase converters, 122
Popov's hyperstability theorem, 418
Position encoders, absolute, 446-49
Power diodes, 2
    classification of, 4
Power diodes, classification of, 4
Power factor improvement, induction motor slip-power recovery drives, 322-24
Power factor (PF), diode rectifiers, 107, 109
Power integrated circuits (PICs), 26
Power MOSFETs, 17-19
Power of a fuzzy set, 565
Power semiconductor devices, 1-28
    bipolar power or junction transistors (BPTs or BJTs), 14-16
    classification of, 1-2
    defined, 1
    diodes, 2-4
    gate turn-off thyristor (GTOs), 10-14
    insulated gate bipolar transistors (IGBTs), 20-24
        defined, 20
        switching characteristics, 22-24
        thermal impedance, 22-24
    integrated gate-commutated thyristors (IGCTs), 25-26
    large band-gap materials for devices, 26
    MOS-controlled thyristors (MCTs), 24-25
    power integrated circuits (PICs), 26
    power MOSFETs, 17-19
    static induction transistors (SITs), 19-20
    thyristors, 4-8
        current rating, 8
        power loss and thermal impedance, 6-8
        switching characteristics, 6
        volt-ampere characteristics, 5
    triacs, 8-10

POWEREX fast-recovery diode type CS340602, 4
POWEREX SCR/diode module CM4208A2, 6-7
Precise computation, 535
Premise, 538
Probabilistic neural network (PNN), 630
Procedural knowledge, 539
Processing element (PE), 626, 627
Product of two fuzzy sets, 565
Programmable low-pass filter stage (PLPF), 676-77
PROLOG, 540-41
Proportional-integral (P-I) controller, 236
PSPICE, 265, 554
Pulsation torque, 53-56
Pulse amplitude modulation (PAM), 206
Pulse width modulation (PWM):
    classification, 210-40
    hysteresis-band current control PWM, 236-39
    principle, 210-40
    selected harmonic elimination PWM, 218-23
    sigma-delta modulation, 239-40
    sinusoidal PWM, 211-18
    sinusoidal PWM with instantaneous current control, 236
    techniques, 210-40
Pulse width modulation (PWM) rectifiers, 255-62
    diode rectifier with boost chopper, 255-57
        single-phase, 255-57
        three-phase, 257
PWM chopping mode, 486
PWM controller, 658-66
    instantaneous current control PWM, 659-60
    selected harmonic elimination (SHE) PWM, 658-59
    space vector PWM, 660-66
PWM converter as line-side rectifier, 258-62
    single-phase, 258-59
    three-phase, 259-62
        four-quadrant operation, 259-60
        line voltage sag compensation, 260
        programmable line power factor, 260

PWM inverters, 288, 294-99

    selected harmonic elimination PWM (SHE-PWM), 297-99

    trapezoidal PWM, 295-97

PWM rectifier applications, 302-3

    dc motor speed control, 303

    static VAR compensator/active filter, 302-3

    superconducting magnet energy storage (SMES), 303

R

Radial basis function network (RBFN), 629, 644-45, 689

Random PWM, 210

Random SPWM, 214

Random weight change (RWC) algorithm, 644

Recirculation network, 630

Rectifier Schottky diode type 6TQ045, 4

Recurrent network, defined, 688

Recurrent neural network (RNN), 646-50

    training by the EKF algorithm, 647-50

Reference contour, 426

Reference trajectory, 422

Regenerative braking, 254

Regenerative snubbers, 14-15

Resolver-to-digital converter (RDC), 448

Resonant inverters, 247-49

Resonant link dc converter (RLDC), 250-51

Reverse-bias safe operating areas (RBSOA), 16

Reverse pass, 637

Reverse slip-power flow, 324

Rheostatic control, of rotor resistance, 309

Ripple voltage, 202

Root frame, 539

Rotating magnetic field, induction machines, 30-33

Rotor resistance, electronic control of, 309

Rule base, defined, 556

Rule-based system, 538

Rule, defined, 556

S

SABER, 265

Safe operating areas (SOAs), 16, 17-19

Salient pole machine characteristics, synchronous machines, 80-83

Sample & hold (S&H) circuits, 660

Scalar control, 338-55

    coupling effect, 356

    current-fed inverter control, 350-52

    of cycloconverter drive, 507-10

    efficiency optimization control by flux program, 352-56

    voltage-fed inverter control, 339-49

Schottky diode, 4

Second breakdown, 16

Second breakdown effect, 14-16

Selected harmonic elimination (SHE) PWM, 210, 658-59

Self-commissioning of a drive, 430-35

    current model flux vector estimation, tuning, 434

    defined, 430

    flux control loop, tuning, 434

    friction coefficient, measuring, 434

    mechanical inertia, measuring, 434

    name plate machine parameters, feeding, 431

    rotor time constant, measuring, 433

    stator resistance, measuring, 431-32

    stator transient parameters, measuring, 432

    tuning current loops, 432

Self-commutated devices, inverters with, 290-303

    double-sided PWM converter system, 299-302

    PWM inverters, 294-99

        selected harmonic elimination PWM (SHE-PWM), 297-99

        trapezoidal PWM, 295-97

Index

PWM rectifier applications, 302-3
    dc motor speed control, 303
    static VAR compensator/active filter, 302-3
    superconducting magnet energy storage (SMES), 303
    six-step inverter, 290-94
    load harmonic resonance problem, 293
Self-organizing feature map (SOFM) network (Kohonen), 630, 645-46, 689
Self-tuning control, 414-16
    load torque disturbance compensation, 415-16
Sensorless control of PM synchronous motor drives, 515-29
    sinusoidal PM machine (PMSM), 522-29
        extended Kalman filter (EKF), 526-29
        inductance variation (saliency effect), 524-26
        terminal voltage and current sensing, 522-24
    trapezoidal SPM machine, 515-22
        stator third harmonic voltage detection, 519-22
        terminal voltage sensing, 515-19
Sensorless vector control, 388-408
    direct vector control with speed signal, 401-8
        drive machine start-up with current model equations, 404-8
        programmable cascaded low-pass filter (PCLPF) stator flux estimation, 401-4
    speed estimation methods, 388-401
        direct synthesis from state equations, 389-90
        extended Kalman filter (EKF), 396-99
        injection of auxiliary signal on salient rotor, 399-401
        model referencing adaptive system (MRAS), 390-92
        slip calculation, 388-89
        slot harmonics, 399
        speed adaptive flux observer (Luenberger observer), 392-96
Series-loaded resonance converter (SRC), 249
Shallow knowledge, 539
SHE-PWM, 297-99, 658
Shell, defined, 556
Shunt snubber, 11
Sigma-delta modulation, 211, 239-40
Sigmoid function, defined, 689
Silicon-controlled rectifiers (SCRs), *See* Thyristors:
SIMNON, 265
SIMULINK, 192, 265, 554, 615, 618, 683, 684
Single-phase bridge rectifier (*CR* load), 105-7
Single-phase bridge rectifier (*R, CEMF* load), 112-18
Single-phase bridge rectifier (*R, RL* load), 100-103
Single-phase bridge rectifier (*RL, CEMF* load), 104
Single-phase inverters, 192-97
    center-tapped inverters, 192-93
    full inverters, 193-97
    H-bridge inverters, 193-97
    half-bridge inverters, 192-93
Single-pole double-throw (SPDT) switches, 425
Singleton, defined, 622
Sinusoidal interior magnet machine (IPM), 89-93
Sinusoidal IPM machine drives, 465-68
    current vector control with maximum torque/ampere, 465-68
    field-weakening control, 468-71
    vector control with stator flux orientation, 468-83
        feedback signal processing, 477-79
        PMW-SW sequencing, 482-83
        square-wave (SW) mode field-weakening control, 479-82
Sinusoidal PM machine (PMSM), 522-29
    extended Kalman filter (EKF), 526-29
    inductance variation (saliency effect), 524-26
    terminal voltage and current sensing, 522-24
Sinusoidal PWM, 211-18
    dead-time effect and compensation, 216-18
    frequency relation, 214-16
    overmodulation region, 214
Sinusoidal PWM (SPWM), 210
Sinusoidal PWM with instantaneous current control, 210, 236

Sinusoidal SPM machine drives, 440-55
    absolute position encoder, 446-49
    analog resolver with decoder, 448-49
    open loop volts/hertz control, 440-44
    optical encoder, 446-48
    self-control model, 444-46
    vector control, 449-51
        field-weakening mode, 451-54
Sinusoidal surface magnet machine (SPM), 89
Six-pulse center-tap converter, 136-37
Six-step inverter, 290-94
    load harmonic resonance problem, 293
Six-step thyristor inverter:
    force-commutated inverter, 276
    force-commutated rectifier, 276
    general operation of, 272-76
    inverter operation modes, 274-76
    load-commutated inverter, 276
    load-commutated rectifier, 274-76
Sliding mode control, 419-30
    control principle, 419-24
    sliding trajectory control of a vector drive, 424-30
Sliding trajectory, 426-27
Sliding trajectory control of a vector drive, 424-30
    acceleration segment, 426
    constant speed segment, 426
    deceleration segment, 427-28
Slip calculation, 388-89
Slip gain tuning of indirect vector control, 597-602
Slip-power recovery drives, 307-31
    ac equivalent circuit, 316-19
    applications, 307
    doubly-fed machine speed control by rotor rheostat, 308-9
    harmonics, 321
    Kramer drive
        commutatorless drive system, 323-24
        performance characteristics of, 323
        speed control of, 322
    phasor diagram, 313-15
    power factor improvement, 322-24
    static Kramer drive, 309-13
    torque expression, 319-21
Slot harmonics, 399
Slow-recovery diodes, 3-4
SMES system, 303
Soft computing, 535, 633
Soft-switched inverters, 186, 249-53
    auxiliary resonant commutated pole (ARCP) converter, 251-52
    high-frequency non-resonant link converter, 252-53
    inverter circuits, 249-53
    principle, 249-53
    resonant link dc converter (RLDC), 250-51
Solid state dc relays, 17
Space-vector PWM, 224-36, 660-66
    converter switching states, 225-26
    implementation steps, 235-36
    linear region, 226-29
    network topology, 663-66
    overmodulation region, 229-35, 663
    undermodulation region, 226-29, 661-63
Space-vector PWM (SVM), 210, 224-36
Speed adaptive flux observer (Luenberger observer), 392-96
Speed droop correction in an open loop control, 340
Speed estimation by RNN, 675-76
Speed estimation methods:
    sensorless vector control
        direct synthesis from state equations, 389-90
        extended Kalman filter (EKF), 396-99
        injection of auxiliary signal on salient rotor, 399-401
        model referencing adaptive system (MRAS), 390-92
        slip calculation, 388-89
        slot harmonics, 399
        speed adaptive flux observer (Luenberger observer), 392-96
Speed loop, 384
Square-wave (six-step) operation, three-phase bridge inverters, 197-201

Square-wave (SW) mode field-weakening control, sinusoidal IPM machine drives, 479-82
Squashing function, defined, 689
Stanley, H. C., 56
Static induction transistors (SITs), 19-20
Static Kramer drive, 309-13
    one-quadrant speed control characteristics, 312-13
    regenerative braking, 312
    reversal of speed, 312-13
Static network, 630
Static Scherius drive, 324-31
    applications, 326
    compared to Kramer drive, 325-26
    modes of operation, 326-28
        subsynchronous motoring, 326
        subsynchronous regeneration, 326-28
        supersynchronous motoring, 328
        supersynchronous regeneration, 328
    modified Scherbius drive for VSCF power generation, 328-31
Static VAR compensator/active filter, PWM rectifier applications, 302-3
Static VAR compensators (SVCs), 11
Stationary frame—dynamic model, induction machines, 67-70
Stator flux-oriented vector control, 381-84
Stator resistance R_s estimation, 602-6
Stator third harmonic voltage detection, 519-22
Stepper motor drives, 17
Subframes, 539-40
Subharmonic method, 211
Suboscillation method, 211
Sugeno method:
    ANFIS, 680
    defuzzification, 576
    implication, 571-73
SUP-MIN composition, 578
    defined, 622
Superconducting magnet energy storage (SMES), 303
Supervised learning, 634
    defined, 689
Surface permanent magnet (SPM), 439

Switched reluctance machine (SRM), 439
Switched reluctance motor (SRM) drives, 529-32
Switching characteristics, thyristors, 6
Switching mode power conversion, 1
Switching mode power supplies (SMPS), 17
Symbolic logic, defined, 556
Symbolic processing languages, 540-41
Synapse, defined, 689
Synaptic weight, 627
Synchronous lay rotating reference frame—dynamic model, induction machines, 63-67
Synchronous machines, 74-96
    developed torque, 79-80
    double reluctance machine, 94-96
    dynamic $d^e\text{-}q^e$ machine model, 83-85
    equivalent circuit, 76-79
    Park model, 83-85
    permanent magnet (PM) machine, 86
    permanent magnet (PM) materials, 86-89
    salient pole machine characteristics, 80-83
    sinusoidal interior magnet machine (IPM), 89-93
    sinusoidal surface magnet machine (SPM), 89
    synchronous reluctance machine, 86
    trapezoidal surface magnet machine, 93-94
    variable reluctance machine (VRM), 94-96
    wound field machine, 74-76
Synchronous motor drive control and estimation, 439-534
Synchronous motor drives:
    applications of, 439-40
    control and estimation, 439-533
    sensorless control, 515-29
        sinusoidal PM machine (PMSM), 522-29
        trapezoidal SPM machine, 515-22
    sinusoidal IPM machine drives, 465-68
        current vector control with maximum torque/ampere, 465-68
        field-weakening control, 468-71
        vector control with stator flux orientation, 468-83

Synchronous motor drives (*continued*):
   sinusoidal SPM machine drives, 440-55
      absolute position encoder, 446-49
      open loop volts/hertz control, 440-44
      self-control model, 444-46
      vector control, 449-51
   switched reluctance motor (SRM) drives, 529-32
   synchronous reluctance machine drives (SyRM), 455-64
      current vector control, 457-64
   trapezoidal SPM machine drives, 483-95
      drive control, 490-92
      drive operation with inverter, 483-86
      extended speed operation, 494-95
      machine dynamic model, 489-90
      torque pulsation, 493-94
      torque-speed curve, 486-89
   wound-field synchronous machine drives, 495-515
      brush and brushless dc excitation, 495
      load-commutated inverter (LCI) drive, 496-507
      scalar control of cycloconverter drive, 507-10
      vector control of cycloconverter drive, 510-13
      vector control with voltage-fed inverter, 513-15
Synchronous reluctance machine, 86
Synchronous reluctance machine drives (SyRM), 455-64
   current vector control, 457-64
      constant d^e-axis current control, 458-59
      fast torque response control, 459-63
      maximum power factor control, 463-64
      maximum torque/ampere control, 463

T

Takagi-Sugeno-Kang method, 571-73
Three-level inverters, 240-45
   control of neutral point voltage, 243-45
Three-phase bridge converter, 128-32
Three-phase bridge inverters, 197-206
   device voltage and current ratings, 203
   input ripple, 202-3
   motoring and regenerative modes, 201-2
   phase-shift voltage control, 203-5
   square-wave (six-step) operation, 197-201
   voltage and frequency control, 205-6
Three-phase bridge rectifier (*RL, CEMF* load), 122
Three-phase dual converter, 136
Three-phase full bridge rectifier (*CR* load), 112
Three-phase full bridge rectifier (*RL* load), 109-12
Three-phase half-wave converter, 122-24
Thyristor converters, 112-41
   concurrent and sequential control of bridge converters, 140-41
   discontinuous conduction, 118-22, 132-36
   link leakage inductance analysis, 124-28
   single-phase bridge rectifier (*R, CEMF* load), 112-18
   six-pulse center-tap converter, 136-37
   three-phase bridge converter, 128-32
   three-phase bridge rectifier (*RL, CEMF* load), 122
   three-phase dual converter, 136
   three-phase half-wave converter, 122-24
   twelve-pulse converter, 137-40
Thyristors, 1, 4-8
   current rating, 8
   power loss and thermal impedance, 6-8
   switching chartacteristics, 6
   transient thermal impedance curve of, 9
   volt-ampere characteristics, 5
Time-delayed neural networks, 650
Torque expression, induction motor slip-power recovery drives, 319-21
Torque production, induction machines, 33-34
Torque pulsation:
   current-fed converters, 287-88
   harmonics, 53-56
Torque-speed curve, induction machines, 39-41
Training, defined, 689
Transfer function, 628, 684

Index

Trapezoidal SPM machine, 515-22
    stator third harmonic voltage detection, 519-22
    terminal voltage sensing, 515-19
Trapezoidal SPM machine drives, 483-95
    drive control, 490-92
        close loop current control in freewheeling mode, 492
        close loop speed control in feedback mode, 490-91
    drive operation with inverter, 483-86
        angle switch-on mode, 485-86
        PWM voltage and current control mode, 486
    extended speed operation, 494-95
    machine dynamic model, 489-90
    torque pulsation, 493-94
    torque-speed curve, 486-89
Trapezoidal surface magnet machine, 93-94
Triac symbol, and volt-ampere characteristics, 10
Triacs, 1, 8-10
Triangulation method, 211
Triple Darlington transistors, 14
Twelve-pulse converter, 137-40
Twelve-step inverter, 207-9
Two-layer network, 683

U

Union, 564
Universal power line conditioner (UPLC), 264
Universe of discourse, defined, 622
Unsupervised learning, 634
    defined, 689
User (client) interface, expert system (ES), 541-42

V

Variable-frequency operation, induction machines, 43-44
Variable reluctance machine (VRM), 94-96
Variable-speed, constant-frequency (VSCF) system, 177-79
    block diagram, 178
Variable stator current operation, induction machines, 47-49
Variable structure control system (VSS), 410
Variable-voltage constant-frequency operation, induction machines, 42-43
Vector control, 356-87
    of current-fed inverter drive, 384-85
    of cycloconverter drive, 385-87, 510-13
    dc drive analogy, 356-58
    direct, 360-63
    equivalent circuit, 358-59
    fast torque response control, 459-63
    feedback, 360-63
    feedforward, 368-78
    flux vector estimation, 363-68
        current model, 366-68
        voltage model, 363-66
    indirect, 368-78
    indirect vector control slip gain tuning, 375-78
    of line-side PWM rectifier, 378-80
    phasor diagram, 358-59
    physical principle of, 374-75
    principles of, 359-60
    sensorless, 388-408
        direct vector control with speed signal, 401-8
        speed estimation methods, 388-401
    sinusoidal SPM machine drives, 449-51
    stator flux-oriented, 381-84
    with voltage-fed inverter, 513-15
Vector control with stator flux orientation:
    sinusoidal IPM machine drives, 468-71
        feedback signal processing, 477-79
        PMW-SW sequencing, 482-83
        square-wave (SW) mode field-weakening control, 479-82
Vector-controlled drive feedback signal estimation, 670-71
Vector rotation (VR), 360
Volt-ampere characteristics, thyristors, 5
Voltage and frequency control, three-phase bridge inverters, 205-6
Voltage-controlled oscillator (VCO), 146, 148, 206

Voltage-fed converters, 191-270
    applications, 191-92
    current-fed converters vs., 303-4
    defined, 191
    dynamic braking, 253-54
    hard switching effects, 245-47
    MATLAB/SIMULINK, 264-67
    multi-stepped inverters, 206-10
        eighteen-step inverter by phase-shift control, 209-10
        twelve-step inverter, 207-9
    pulse width modulation (PWM)
        classification, 210-40
        hysteresis-band current control PWM, 236-39
        minimum ripple current PWM, 223-24
        principle, 210-40
        selected harmonic elimination PWM, 218-23
        sigma-delta modulation, 239-40
        sinusoidal PWM, 211-18
        sinusoidal PWM with instantaneous current control, 236
        space-vector PWM, 224-36
        techniques, 210-40
    pulse width modulation (PWM) rectifiers, 255-62
        diode rectifier with boost chopper, 255-57
    PWM converter as line-side rectifier, 258-62
        single-phase, 258-59
        three-phase, 259-62
    regenerative braking, 254
    resonant inverters, 247-49
    single-phase inverters, 192-97
        center-tapped inverters, 192-93
        full inverters, 193-97
        H-bridge inverters, 193-97
        half-bridge inverters, 192-93
    soft-switched inverters, 249-53
        auxiliary resonant commutated pole (ARCP) converter, 251-52
        high-frequency non-resonant link converter, 252-53
        inverter circuits, 249-53
        principle, 249-53
        resonant link dc converter (RLDC), 250-51
    static VAR compensators and active harmonic filters, 262-64
    three-level inverters, 240-45
        control of neutral point voltage, 243-45
    three-phase bridge inverters, 197-206
        device voltage and current ratings, 203
        input ripple, 202-3
        motoring and regenerative modes, 201-2
        phase-shift voltage control, 203-5
        square-wave (six-step) operation, 197-201
        voltage and frequency control, 205-6
Voltage-fed inverter control, 339-49
    current-controlled voltage-fed inverter drive, 346-48
    energy conservation effect by variable frequency drive, 342
    open loop volts/Hz control, 339-42
    speed control with slip regulation, 342-45
    speed control with torque and flux control, 345
    traction drives with parallel machines, 348-49
Voltage-fed inverter, vector control with, 513-15
Voltage model feedback signal estimation block diagram, 365
Von Neumann digital computer, 626
VSCF wind generation system, 387

W

Wave modulation, 206
Weight, defined, 689
WHY user command, expert system (ES), 542
Wind generation system, 591-97
    fuzzy control, 593-97
    generator speed tracking control (FLC-1), 594-97
        block diagram of, 596
        membership functions for variables in, 597
        rule matrix for, 596
    system description, 592-93
    wind turbine characteristics, 592

Wound field machine, 74-76
Wound-field synchronous machine drives, 439, 495-515
    brush and brushless dc excitation, 495
    control of with constant turn-off angle, 498-501
    load-commutated inverter (LCI) drive, 496-507
        control with constant turn-off angle, 498-501
        control with machine terminal voltage signals, 504-6
        delay angle control, 501-4
        phase-locked loop (PLL) angle control, 506-7
    scalar control of cycloconverter drive, 507-10
Wound rotor induction motor (WRIM), 495

Z

Zadeh, Lofti, 535, 559, 576-77
Zero-crossing method of terminal voltage sensing, 504-5
Zero current switching (ZCS), 249
Zero-order Sugeno system, 680-81
Zero voltage switching, 240-45, 249

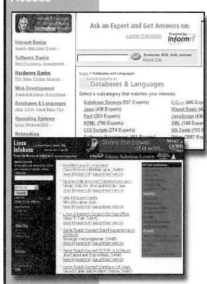

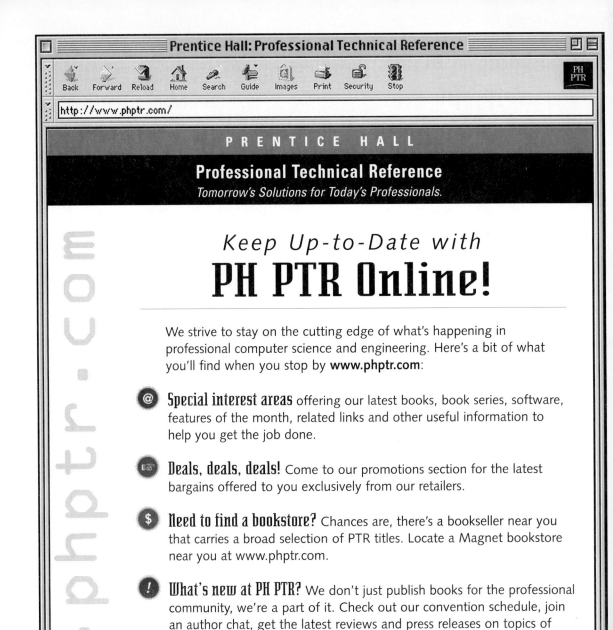